OTHER OUTSTANDING VOLUMES IN THE CHAPMAN AND HALL
ADVANCED INDUSTRIAL TECHNOLOGY SERIES

V. Daniel Hunt: SMART ROBOTS: A Handbook of Intelligent Robotic Systems

David F. Tver and Roger W. Bolz: ENCYCLOPEDIC DICTIONARY OF INDUSTRIAL TECHNOLOGY: Materials, Processes and Equipment

Roger W. Bolz: MANUFACTURING AUTOMATION MANAGEMENT: A Productivity Handbook

Igor Aleksander: ARTIFICIAL VISION FOR ROBOTS

D.J. Todd: WALKING MACHINES: An Introduction to Legged Robots

Igor Aleksander: COMPUTING TECHNIQUES FOR ROBOTS

Robert I. King and Robert S. Hahn: HANDBOOK OF MODERN GRINDING TECHNOLOGY

Robert I. King: HANDBOOK OF HIGH SPEED MACHINING TECHNOLOGY

V. Daniel Hunt: ARTIFICIAL INTELLIGENCE AND EXPERT SYSTEMS SOURCEBOOK

Standard Handbook of Industrial Automation

DOUGLAS M. CONSIDINE
Editor-in-Chief
*Registered Professional Engineer (California)
in Control Engineering*

GLENN D. CONSIDINE
Managing Editor

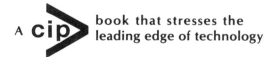
A book that stresses the leading edge of technology

CHAPMAN AND HALL
New York London Toronto Sydney

First published 1986
by Chapman and Hall
29 West 35 Street, New York, N.Y. 10001

Published in Great Britain by
Chapman and Hall Ltd
11 New Fetter Lane, London EC4P 4EE

© 1986 Chapman and Hall

Printed in the United States of America

All Rights Reserved. No part of this book may be
reprinted, or reproduced or utilized in any form
or by any electronic, mechanical or other means,
now known or hereafter invented, including
photocopying and recording, or in any information
storage or retrieval system, without permission in
writing from the publishers.

Library of Congress Cataloging-in-Publication Data

Standard handbook of industrial automation.

    (Chapman and Hall advanced industrial technology
series)
    Bibliography: p.
    Includes index.
    1. Factories—Automation—Handbooks, manuals, etc.
I. Considine, Douglas Maxwell. II. Considine, Glenn D.
III. Series.
T59.5.S68 1986   670.42'7   86-21602
ISBN 0-412-00831-9

About the Editors

Douglas M. Considine, a registered professional engineer (California), received a chemical engineering degree from Case-Western Reserve University. He is a Fellow of both the Instrument Society of America and the American Association for the Advancement of Science and a Senior Member of the American Institute of Chemical Engineers.

Among the technical management positions held by Mr. Considine were Process Design Engineer for The Lummus Company; Director, Process Engineering Applications, for Honeywell Inc.; Manager, Technical Planning, for P. R. Mallory (Dart & Kraft, Inc.); and Director, Advanced Control Engineering, for the Hughes Aircraft Company. As a consultant Mr. Considine has served numerous firms, including Beckman Instruments, Varian Associates, Ralph M. Parsons, General Signal, Milton Roy, and Dun & Bradstreet. He has also chaired numerous symposia on chemical process control and instrumentation.

Mr. Considine has written scores of technical articles and edited numerous handbooks and encyclopedias, including *Energy Technology Handbook, Process Instruments & Controls Handbook*, 3rd Edition, and *Chemical and Process Technology Encyclopedia* (McGraw-Hill); *Foods and Food Production Encyclopedia, Encyclopedia of Chemistry*, and *Van Nostrand's Scientific Encyclopedia*, 5th and 6th editions (Van Nostrand Reinhold); as well as major sections in chemical and plant engineering handbooks. He holds several patents for temperature and liquid level process controllers, is a coinventor of the laser fabric cutter, and was a pioneer in the development of the automated supermarket checkout stand.

Glenn D. Considine, a registered construction engineer (Georgia), is a member of both the American Society for Metals and the Institute of Food Technologists. He specializes in the management of information systems serving science, engineering, and technology. He has served as Managing Editor of numerous engineering and scientific handbooks and encyclopedias.

Contents

For the detailed contents of any subsection, consult the title page of that subsection. The alphabetical index follows Section 5.

Contributors	ix
Preface	xi
Introduction: Descriptions of Articles	xiii

1. BACKDROP TO AUTOMATION SYSTEMS — 1
- Automation—The Multifacted Technology — 3
- Annotated Glossary of General Automation Terms — 9

2. SENSORS AND MEASURING SYSTEMS — 15
- Motion, Position, Speed, Velocity—Geometric Variables I — 19
- Dimension, Displacement, Thickness—Geometric Variables II — 51
- Object Detection, Proximity, Presence, Nonpresence—Geometric Variables III — 73
- Machine Vision — 86
- Machine Vision—State-of-the-Art Systems — 110

3. CONTROL SYSTEMS — 133
- Control System Architecture — 136
- Programmable Controllers — 152
- Programmable Controllers—State-of-the-Art Systems — 169
- Programming the Programmable Controller — 189
- Sequence Controllers—Hardware/Software Trends — 201
- Expert System and Model-Based Self-Tuning Controllers — 216
- Numerical Control and Computer Numerical Control — 230
- Computer Numerical Control—State-of-the-Art Systems — 235

4. ACTUATORS AND MATERIALS—TRANSFER SYSTEMS — 257
- Robots in Perspective — 260
- Robot Technology Fundamentals — 262
- A Robot Dynamics Simulator — 321
- Control of Actuators in Multilegged Robots — 330
- Servomotor and Servosystem Design Trends — 336
- Stepper Motors and Controls — 346
- Linear and Planar Motors — 367
- Solid-State Variable Speed Drives — 376
- Materials Motion/Handling Systems — 386

5. INTERFACES AND COMMUNICATIONS — 423
- User Interfacing to Process Computer Controls — 424
- Local Area Networks (LANs) — 433
- Communication Standards for Automated Systems — 440

Index follows Section 5

Contributors*

ALLEN, D., NCR Corporation, Fort Collins, Colorado. (*Geometric Arithmetic Parallel Processor*)

ARNOLD, F., PMI Motion Technology, Division of Kollmorgen Corporation, Commack, New York. (*Stepper Motors and Controls*)

ARUM, H. R., Stock Drive Products, Division of Designatronics, Inc., New Hyde Park, New York (*Encoders*) (*Robot End-Effectors*)

AUVRAY, P., Sopelem, Levallois-Perret-Cedex, France. (*Optical Dimension Checking*)

BERNATH, M. S., Industrial Automation Systems, Gould Inc., Andover, Massachusetts. (*Cell Control*)

BLAESER, J. A., Industrial Automation Systems, Gould Inc., Andover, Massachusetts. (*Cell Control*)

BOYLE, J., Giddings & Lewis Electronics Company, Fond Du Lac, Wisconsin. (*CNC Gaging Systems*) (*Numerical Control and Computerized Numerical Control*)

BREEN, J. M., Adaptive Intelligence Corporation, Milpitas, California. (*Adaptive Assembly Robots*)

CARLSON, R. T., Cutler-Hammer Products, Eaton Corporation, Milwaukee, Wisconsin. (*Parts and Packaging Inspection Systems*)

CHAGGARIS, C. G., ORS Automation, Inc., Princeton, New Jersey. (*Robotic Acquisition of Jumbled Parts*)

CHAN, M., Video Logic Corporation, Sunnyvale, California. (*Motion System Analysis*)

CHESTER, G. L., Divelbiss Corporation, Fredericktown, Ohio. (*Programmable Controllers*)

COBB, J., Cognex Corporation, Needham, Massachusetts. (*Packaging, Parts, and Printing Quality Inspection Systems*)

COLONA, R. L., General Scanning Inc., Watertown, Massachusetts. (*Post-Objective Scanning*)

CONSIDINE, D. M., Editor-in-Chief. (*Automation—The Multifaceted Technology*) (*Annotated Glossary of General Automation Terms*) (*Communication Standards for Automated Systems*)

CONSTANTINO, P. J., Jervis B. Webb Company, Farmington Hills, Michigan. (*Power and Free Conveyors; Underhung Cranes and Monorail Systems; Inverted Power and Free Conveyors; Automatically Guided Vehicles*)

DAVIS, R., NCR Corporation, Fort Collins, Colorado. (*Geometric Arithmetic Parallel Processor*)

DOBSON, V. J., DynaPath Systems Inc., Detroit, Michigan. (*Computer Numerical Control*)

EBRAHIMI, N. D., Department of Mechanical Engineering, The University of New Mexico, Albuquerque, New Mexico. (*Control of Actuators in Multilegged Robots*)

ELLIOTT, R. A., Qualiplus USA, Inc., Stamford, Connecticut. (*Container Inspection Systems*)

FLACK, T., Numa-Logic Department, Industry Electronics Division, Westinghouse Electric Corporation, Madison Heights, Michigan. (*Programmable Controllers*)

FLINN, P. A., GMF Robotics Corporation, Troy, Michigan. (*Robots for Automotive and Other Industrial Uses*)

FORD, R. E., Bodine Electric Company, Chicago, Illinois. (*Stepper Motors and Controls*)

FOXX, D., Texas Instruments Incorporated, Johnson City, Tennessee. (*Programmable Controllers*)

GEBELEIN, R. E., Moore Products Co., Spring House, Pennsylvania. (*Automatic Gaging Systems*)

HANSEN, P. D., The Foxboro Company, Foxboro, Massachusetts. (*Expert System and Model-Based Self-Tuning Controllers*)

HARRIS, M. L., Octek Inc., A Foxboro Company, Burlington, Massachusetts. (*Package Inspection Systems*)

HENDRICKS, H. V., Programmable Control Division, Gould Inc., Andover, Massachusetts. (*Programmable Controllers*)

HICHBORN, Y., Video Logic Corporation, Sunnyvale, California. (*Motion System Analysis*)

HONCHELL, K., Cincinnati Milacron, Lebanon, Ohio. (*Industrial Robots*)

KENT, E. W., U. S. National Bureau of Standards, Washington, D. C. (*NBS Pipelined Image Processing Engine*)

KIM, L. K., Video Logic Corporation, Sunnyvale, California. (*Motion System Analysis*)

KIMBALL, K. E., Siemans Capital Corporation, Iselin, New Jersey. (*Programming the Programmable Controller, Sequence Controllers*)

KRASKA, P. A., Pattern Processing Technologies, Inc., Minneapolis, Minnesota. (*Machine Vision*)

KRAUS, T. W., Formerly, The Foxboro Company, Foxboro Massachusetts. (*Expert System and Model-Based Self-Tuning Controllers*)

LOYER, B. A., Semiconductor Products Sector, Motorola Inc., Phoenix, Arizona. (*Local Area Networks (LANs)*)

---

*Persons who authored complete articles or subsections of articles, or who otherwise cooperated in an outstanding manner in furnishing information and helpful counsel to the editorial staff.

MACALONEY, B., Industrial Automation Systems, Gould Inc., Andover, Massachusetts. (*Networking and Communications*)

MACKENZIE, P. C., Jervis B. Webb Company, Farmington Hills, Michigan. (*Power and Free Conveyors; Underhung Cranes and Monorail Systems; Inverted Power and Free Conveyors; Automatically Guided Vehicles*)

MACKIEWICZ, R. E., Sisco, Inc., Warren, Michigan. (*Programmable Controllers*)

MCCREADY, A. K., Chevron Research Company, Richmond, California. (*User Interfacing to Process Computer Systems*)

MAGER, K., Analog Devices Inc., Norwood, Massachusetts. (*Electronic Parts Assembly and Inspection Systems*)

MAHONEY, W. A., Intelledex Corp., Corvallis, Oregon. (*Robots*)

MANDRILLO, V., Eaton-Kenway, a subsidiary of Eaton Corporation, Salt Lake City, Utah. (*Automatic Storage/Retrieval—Warehouse Systems*)

MARCOULLIER, J., Mitsubishi Electric Sales America, Inc., Mount Prospect, Illinois. (*Programmable Controllers*)

MAZURKIEWICZ, J., Pacific Scientific, Rockford, Illinois. (*Resolvers/Synchros, Encoders, Tachometers*) (*Servomotor and Servosystem Design Trends—Microprocessors and Brushless Motors*)

MAZZA, T. A., Telemecanique, Arlington Heights, Illinois. (*Proximity Switches*)

MEHAN, G., Siemens Energy & Automation, Inc., Peabody, Massachusetts. (*Programming the Programmable Controller, Sequence Controllers*)

MELLISH, M. T., Industrial Automation Systems, Gould Inc., Andover, Massachusetts. (*Cell Control*)

MINNICH, C., MTS Systems Corporation, Minneapolis, Minnesota. (*Integrated Motion Systems*)

MINO, M. G., Ormec Systems Corp., Rochester, New York. (*Integrated Motion Systems*)

MOHR, E. D., Unimation Incorporated (A Westinghouse Company), Danbury, Connecticut. (*Robotic Technology*)

MOORE, M., Industrial Computer Group, Allen-Bradley Company, Highland Heights, Ohio. (*Programmable Controllers*)

MURRAY, J. J., The Superior Electric Company, Bristol, Connecticut. (*Stepper Motors and Controls*)

NEUMAN, C. P., Department of Electrical and Computer Engineering and the Robotics Institute, Carnegie-Mellon University, Pittsburgh, Pennsylvania. (*A Robot Dynamics Simulator*)

ORTEGA, D. G., Intelledex Incorporated, Corvallis, Oregon. (*Assembly Robots and Vision Systems*)

OSMAN, R. H., AC Drives, Robicon Corporation (A Barber-Colman Company), Pittsburgh, Pennsylvania. (*Solid-State Variable Speed Drives*)

RUSSELL, L., MTS Systems Corporation, Minneapolis, Minnesota. (*Electrohydraulic Stepping Cylinders and Motors*)

SABINASH, E. R., Industrial Control and Power Distribution Division, Cutler-Hammer Products, Eaton Corporation, Milwaukee, Wisconsin. (*Programmable Controllers*)

SCAFARO, A. E., Reliance Electric Company, Cleveland, Ohio. (*Programmable Controllers*)

SIEMENS ENGINEERING STAFF, Siemens Aktiengesellschaft, Erlangen, West Germany. (*Programming the Programmable Controller*)

SPEICHER, R. F., Micro Switch Division, Honeywell Inc., Freeport, Illinois. (*Photoelectric and Proximity Switches*)

STAFF, Anorad Corporation, Hauppage, New York. (*Positioners*)

STAFF, Daytronic Corporation, Dayton, Ohio. (*Displacement Transducers*)

STORMS, T., Hitachi America, Ltd., Tarrytown, New York. (*Robots and Robot Vision Systems*)

SULZER, E., Siemens Energy & Automation, Inc., Programmable Controls Division, Peabody, Massachusetts. (*Programming the Programmable Controller*)

THOMAS, D., NCR Corporation, Fort Collins, Colorado. (*Geometric Arithmetic Parallel Processor*)

THOMAS, J., SI Handling Systems, Inc., Easton, Pennsylvania. (*Automatically Guided Vehicles; Towline Conveyors; Track and Drivetube Conveyors; Automatic Storage/Retrieval Systems*)

VARNEY, T., Siemans Capital Corporation, Iselin, New Jersey. (*Programming the Programmable Controller, Sequence Controllers*)

VOLPE, G. T., University of Bridgeport, Connecticut. (*Linear and Planar Motors*)

WELLINGTON, J., U. S. National Bureau of Standards, Washington, D. C. (*NBS Pipelined Image Processing Engine*)

WERTH, L., Pattern Processing Technologies, Inc., Minneapolis, Minnesota. (*Machine Vision*)

WILDER, J., ORS Automation, Inc., Princeton, New Jersey. (*Robotic Acquisition of Jumbled Parts*)

WISMANN, D., PMI Motion Technology, Division of Kollmorgen Corporation, Commack, New York. (*Stepper Motors and Controls*)

ZIMMERMAN, C. K., Instrumentation and Control, Engineering Service Division, E. I. Du Pont de Nemours & Co. (Inc.), Wilmington, Delaware. (*Control System Architecture*)

Preface

The authors and editors of this *Handbook* have attempted to fill a serious gap in the professional literature on industrial automation.

Much past attention has been directed to the general concepts and philosophy of automation as a way to convince owners and managers of manufacturing facilities that automation is indeed one of the few avenues available to increase productivity and improve competitive position. Seventy-three contributors share their knowledge in this *Handbook*.

Less attention has been given to the "What" and "How" of automation. To the extent feasible and practical within the confines of the pages allowed, this *Handbook* concentrates on the **implementation of automation**. Once the "Go" signal has been given by management, concrete details—not broad definitions and philosophical discussions—are required. To be found in this distinctly different book in the field are detailed parameters for designing and specifying equipment, the options available with an evaluation of their relative advantages and limitations, and insights for engineers and production managers on the operation and capabilities of present-generation automation system components, subsystems, and total systems. In a number of instances, the logical extension of current technology into the future is given. A total of 445 diagrams and photos and 57 tables augments detailed discussions. In addition to its use as a ready reference for technical and management personnel, the book has wide potential for training and group discussions at the college and university level and for special education programs as may be provided by consultants or by "in-house" training personnel.

Concentrating on implementation, this *Handbook* will provide at least some of the answers to the question, "**What is the Bill of Material for an automated system?**"

DOUGLAS M. CONSIDINE
Editor in Chief

Introduction:

Descriptions of Articles

SECTION 1—BACKDROP TO AUTOMATION SYSTEMS

Automation—The Multifaceted Technology. The definition of automation, a concept that dates back several centuries, has changed over the years and is now associated mainly with modern manufacturing. The advantages claimed for automation today include improved productivity, enhanced product quality, and improvements in the status of workers. Limitations include high initial cost, vulnerability to down time, loss of flexibility, need for greater management attention, and, to some degree, the persistence of automation's image in the work force. Manufacturing automation has been built largely on the base provided by office automation. Manufacturing automation has developed along two principal paths, which reflect the rather distinct natures of two kinds of products—the fluid and bulk materials process industries on the one hand and the discrete-piece manufacturing industries on the other hand. Several scientific and technological developments have contributed to the feasibility of modern automation, including feedback, information and communication theory, sensor and measuring system developments, advancements in servopower, computer and memory power, digital technology, mechanization, systems and engineering analysis, information display technology, among others. In turn, automation has contributed to engineering and science, including the development of large and complex communication networks, vast improvements in the interface between operators and machines, the pioneering of new control concepts, such as the programmable controller, and the development of new outlooks in planning, designing, executing, and controlling manufacturing operations.

Annotated Glossary of General Automation Terms. Mainly during the past decade several broad concepts that apply directly or indirectly to automated systems have been generated. These concepts serve as useful guidelines and are easy to comprehend but not always easy to implement. Automation appears to be approaching maturity in terms of general concepts, but much refinement remains. The latter will arise from much needed specific experience gained from a wide variety of applications. Included in this glossary are such concepts as CAD (computer-aided design or drafting), CAE (computer-aided engineering), CAM (computer-aided manufacturing), CIM (computer-integrated manufacturing), FMS (flexible manufacturing system), Group Technology, Just in Time (delivery), MRP I (materials requirement planning), and MRP II (manufacturing resources planning), among others. Some of these concepts simply involve the application of difficult, time-consuming research on existing manufacturing processes and the application of common sense against the backdrop of newly found information. Manufacturers often find the application of planning that is based upon production information never previously tabulated or understood will produce excellent results—even in the absence of implementing extensive automation.

SECTION 2—SENSORS AND MEASURING SYSTEMS

Motion, Position, Speed, Velocity—Geometric Variables I. In the discrete-piece manufacturing industries, system geometry is nearly always of the utmost importance to maintaining product quality and throughput (productivity). Position is central among the geometric variables. Motion is comprised of a series of discrete positions. Position and motion systems may be classified in several ways: (1) By the excursion of the motion, as in point-to-point, or trajectory systems. There are also superimposed motions and physically fixed motion paths. (2) By the type of drive and control, that is, continuous or incremental motion systems, and newer approaches incorporating feedback enhancement. And (3) by the position sensors used. Traditional resolvers and synchros are described and diagrammed in detail, including operating equations and tabular summary of their physical parameters and specifications. Both absolute and incremental encoders are similarly described, accompanied by a tabular comparison of light sources for incremental encoders. A representative problem involving the application of an incremental encoder to a milling machine is explored. A summary of the state-of-the-art encoder hardware available is included. Resolvers/synchros and encoders are compared as an aid to procurement, emphasizing such factors as vibration and shock, contaminants, temperature range, accuracy available, mechanical packaging. The use of tachometers for providing feedback information in motion systems is described, along with the analysis of a representative tachometer application. Important characteristics of tachometers are included as an aid in specifying these devices—voltage constant, ripple, linearity, maximum operating speed, minimum load resistance, temperature stability, as well as the use of the self-contained brushless tachometer and speed ratio systems for certain applications. Linear/rotary position sensors are described and diagrammed in detail, these including inductive plate transducers, particularly as applied in computer numerical control, potentiometers, and pulse-type transducers. The basic types of electromechanical limit switches, including gravity-return and solid-state designs, are described and illustrated; principal specifying parameters are tabulated. Special attention is given to integrated motion systems that link motion control commands to computers and incorporate several microprocessors in their measuring circuits—factors that are proving the viability of computer integrated manufacturing (CIM) concepts. A diagram illustrates how the user can select from a variety of system architectures and hierarchies. A field of growing importance—motion system analysis—is described. Applications for these systems include diagnostics of paper machines, packaging equipment, vehicles and other machines that depend upon cams and gears that operate at high speeds, robotic systems, and weapons testing, among others.

Dimension, Displacement, Thickness—Geometric Variables II. Advancements in industrial metrology, accelerated by the requirements of automation, are reviewed. This includes description of the laser interferometer, which is capable of accuracies of about one part in ten million. Transducers, including linear variable differential transformers (LVDTs), linear variable reluctance transducers (LVRTs), linear transformers (LTs), impedance-type gaging transducers, linear potentiometers, strain gage load cells, and optical methodologies are described and diagrammed, including selected applications (manufacture of nonlinear springs, automobile spring assemblies, and excursive phenomena, such as wheel run-out and wobble). The modern version of statistical process control is described and illustrated in considerable detail, particularly as this discipline applies to dimension control. Basic statistical terms are described, including the mathematical equations that apply to such concepts as the inherent capability of a process, this capability in relation to specification limits, capability ratio, variation, kurtosis, skewness, normal distribution, among other concepts. Modern floor gages incorporating microprocessors are described, aided by photos and diagrams taken from CRT interfaces. Measuring machines incorporating inductive plate or eddy current transducers are described, including description of a six-axis coordinate automated part inspection machine. A noncontact, laser-controlled inspection system for on-line use is described. In addition to systems for measuring positions, hardware for achieving accurate position control, such as positioning tables and precise positioners, is described. Thickness measurement, of particular importance in the production of flat goods, includes discussions of nuclear radiation, ultrasonic, x-ray, and LVDT gages. The concept of waviness in the measurement of surface finishes is described.

Object Detection, Proximity, Presence, Nonpresence—Geometric Variables III. The many uses of object detectors are described. These include conveyor-associated applications, such as jam detection, empty-line detection, and automatic routing; safety and accident avoidance, such as light curtains; inspection of products, such as containers for proper fill levels and label locations; counting; sorting; hopper level control; and edge guidance, among other uses. Detectors are of two principal categories—contact and noncontact. The fundamentals of the photoelectric effect, the wide scope of photoelectric control applications, and scanning techniques (thru-scanning, reflective and retroflective) are described. Specular scanning (for shiny surfaces) and diffuse scanning for matte surfaces are delineated. Color differentiation for registration control is described. Factors affecting the selection of photosensors and scanning techniques are tabulated. Twelve representative applications of photoelectric controls are described and diagrammed. Factors important to the specification of light sources and sensors, including static sensitivity, dynamic sensitivity, speed of response, light history, effect of temperature, and color response are described. Also included are discussions of photodiodes, LED sources, and the use of fiber optics with photocells. Magnetic proximity switches, variable-reluctance sensors, magnetically actuated dry reed switches, inductive proximity sensors, Wiegand-effect switches, capacitive proximity sensors, magnetostrictive limit switches, and ultrasonic sensors are all described and several are diagrammed. The impact of machine vision systems on traditional object detection methods is described, and an on-line "inside can" application of image-processing techniques is described and illustrated.

Machine Vision. Part of the larger field of artificial visual perception, machine vision (MV)

Description of Articles xv

is defined and compared with human vision. Pattern recognition, the basis of MV, is described in detail, as are the fundamental applications of MV. The principles of extracting information from images are discussed. Computers versus adaptive hardware in MV systems are presented in a pro-and-con manner. Image data quantities that must be processed are described. Following a practical presentation of principles and theory, functional descriptions of machine vision systems are included. Factors that influence the specification of MV systems, such as optics/lighting, field of view, resolution, signal-to-noise ratio, time and temperature stability, and cost are delineated. Descriptions of line scanners, area type scanners, and solid-state cameras are included. Considerable attention is given to processing MV system information. Windowing and segmentation techniques are described. The elements of segmentation technology (algorithms, neighborhood processing, convolution, and string encoding) are described. The four major MV applications (inspection, location, recognition, and measurement) are described in considerable detail. The article includes over twenty diagrams to illustrate windowing techniques. Techniques of windowing and camera placement for printed circuit board inspection are explained in detail, using diagrams and magnified photos.

Machine Vision—State of the Art Systems. As a practical guide for specifying currently available machine vision systems, over fifteen successful MV systems are described and profusely illustrated with photos and diagrams. Gray-scale systems, for example, are meticulously diagrammed. Applications range from the use of MV systems in "smart" robots as well as for non-robotic uses, such as identification, classification, sorting and measuring, inspecting, verifying, and quality control. The legendary "bin-of-parts" problem is discussed. The recently perfected post-objective scanning technique is described, as are electronic innovations that have improved MV systems. These include the pipelined image processing engine, developed by the National Bureau of Standards, the geometric arithmetic parallel processor, which has markedly improved processing power and speed, and the concept of associative pattern processing, which increases the capability for processing tremendous amounts of data in less time than has been required traditionally.

SECTION 3—CONTROL SYSTEMS

Control System Architecture. In this article, the author traces the evolution of control system functionality and architecture—from stand-alone controllers to plantwide automatic control, which includes full integration of process control and plant business systems. The topic is approached in an exceptionally workmanlike manner for such a complex subject. This is aided by by a pyramid diagram, which shows process measurement and data input/output aspects of control systems as the base and plant and process management as the apex. Other functions of control systems are shown in layers rising from the base, including regulatory and other direct process control and process monitoring. Specific topics covered include data acquisition, remote information, and foreign device interfaces. Regulatory control, state-of-the-art and advanced systems, sequence control, and sequence-of-event recording are described. In the monitoring area, graphics, report generation, CRT operating stations, alarms, and trending are discussed. Other numerous functions in the automatic control hierarchy are presented, including process accounting, laboratory data collection, the engineer's interface, modeling and steady-state optimization, process data reconciliation, role of the master schedule, the data historian, product tracking, standard operating conditions/procedures, database management, and advisory control and alarming. Helpful and explicit tabular summations include: key factors that determine system architecture; evolution of system architecture, ranging from direct digital control (DDC) to supervisory control to distributed systems; key limitations in state-of-the-art distributed control systems; key limitations in state-of-the-art host computer systems; trends in distributed control systems. Future trends, including the greater use of personal computers and artificial intelligence, are explored.

Programmable Controllers. Developed essentially out of a need for more flexible and easier programmed controllers for the discrete-piece manufacturing industries, the author describes the characteristic functions of PCs. Basic design and operation of a programmable controller are discussed and diagrammed—including the processor section (power supply, memory, battery backup), the central processing unit (CPU), processor software, including user software. Programming and languages used receive detailed attention. Input/output (I/O) systems are carefully delineated. As a particular aid to specifying PCs, packaging characteristics are evaluated. Other factors in PC selection include how PCs tie into local and wide area networks, reliability, noise immunity, serviceability, fault tolerance, and graceful degradation.

Programmable Controllers—State-of-the-Art Systems. This article expands the coverage of PCs to presently available hardware and software. Specifications of several contemporary PCs are

xvi Description of Articles

given in much detail. This information will assist the engineer in determining what to look for in PCs for various applications and demands. Included are the most recent concepts, such as the cell controller, a factory-hardened unit that combines the features of a PC with those of a minicomputer.

Programming the Programmable Controller. In connection with manufacturing automation, software is the informal conversion of a technical problem into a sequence of instructions (program) which are "understood" by the PC. This article concentrates on software preparation, particularly by the user. A comparison of programming methods: (control system flow chart, CSF; ladder diagram, LD; and statement list, STL) is discussed. The author explains how structured programming enables software rationalization. The importance of documentation is stressed.

Sequence Controllers—Hardware/Software Trends. The author concentrates on the interaction of hardware and software in modern sequence controllers. Numerous diagrams pertaining to the programming of sequence controllers are given. Converting from traditional ladder programming to newer languages, such as GRAPH 5, is described in considerable detail. Some of the details delineated include linear sequence, simultaneous branching, synchronization, alternative branching, jump functions, wait and monitoring time, among others.

Expert System and Model-Based Self-Tuning Controllers. Time-varying process dynamics, variable operating conditions, nonlinear process dynamics, and lack of expertise during control loop commissioning have led to the current strong interest in self-tuning controllers. Economic incentives are the major thrust inasmuch as control loop performance directly affects product quality, energy consumption, product yield, production rate, pollution control, and plant safety. Self-tuning is carefully defined and explained in terms of a heat exchange problem. The two principal self-tuning approaches currently available (the expert system and the process model approach) are described. Each of these approaches is explained and diagrammed in an easy-to-understand manner. Specific topics include pattern recognition, knowledge-based rules, model-based adaptive control, open loop control, open loop control with feedback, minimum variance control, minimum variance with detuning, restricted complexity control, identification models (explicit, implicit), identifiers (least weighted squares, excursive least weighed squares, Kalman filter, restricted complexity delay identification)—these explanations are assisted by the inclusion of 63 mathematical equations.

Numerical Control and Computerized Numerical Control. The fundamentals and types of NC and CNC are described, including machine requirements, methods of position measurement, axis drives, programming, and part programming storage. Particular attention is given to the implementation of CNC and designing products for this important production technology.

Computer Numerical Control—State-of-the-Art Systems. This article concentrates on a review and evaluation of currently available major CNC systems, providing guidelines for the procurement of equipment. Four major application-dependent factors are explored—volume of work (rate of production), variety of work produced, complexity of workpiece, and immediate or future ties with computer integrated manufacturing (CIM) and flexible manufacturing systems (FMS). The manner in which CNC systems are marketed is reviewed. Contemporary system specifications are given in much detail, aided by tabular summaries and numerous diagrams and photos. The incorporation of microprocessors in currently available ultramodern systems is described. Programming features and diagnostics are included.

SECTION 4—ACTUATORS AND MATERIALS—TRANSFER SYSTEMS

Robots in Perspective. The very early history of robots is traced briefly—from the 1700s to the 1940s, when robots for industrial production were introduced on a very modest scale. The industrial trends from fixed or hard automation to flexible manufacturing systems (FMS) using robots and programmable controllers are described briefly. The objective of this very short article is one of introducing robots for the detailed technical coverage of the following article.

Robot Technology Fundamentals. Robots have developed from relatively simplistic systems to a current high degree of technical complexity and sophistication. Because of the rapidly evolving changes that have occurred, particularly during the past decade, a well-organized review of the fundamentals of robotics is in order for all but the most expert designers and users of these systems. Robots are classified from several points of view and features—axes motion; control systems, programming systems; load capacity and power required; dynamics properties, such as stability, resolution, repeatability, and compliance; and applications. Among detailed and diagrammed topics are degrees of freedom, revolute, cartesian, cylindrical and spherical (polar) coordinates. Attention is given to the concepts of world coordinates and tool coordinate systems, as well as to the robot's effective work envelope. Both non-servo and servo-controlled systems are described, with a tabular comparison of the two approaches. Definitions of point-to-point and continuous-path robots are given. The various levels of robot programming—from playback concepts to the use of more modern programming techniques are explored. Sample programs, including off-line, are given in considerable detail. Detail program specifications are presented. Robot drives (electric, hydraulic, pneumatic, and electromechanical) are described and compared. The principal dynamic properties of robots (stability—with solutions to oscillation problems; resolution—showing the differences between sliding and rotary joints; repeatability; and compliance) receive several pages of coverage. End-effectors (hands or grippers) are classified (mechanical, vacuum, magnetic, electromagnetic) are illustrated and described. Design guidelines for determining how mechanical grippers should grasp and the force required are an example of the type of detail included. Environmental concerns for the operation of robots in the workplace are described. The robot settings (work arranged around the robot, work brought to the robot, work that travels past the robot, and robots that travel to the work) are described and contrasted. The concept of work cells is delineated, including a listing of the criteria for establishing and utilizing a robotic work cell. The latter portion of the article is devoted to an overview (with highly detailed specifications) of some contemporary robots and their applications, such as assembly, inspection, machine loading, parts transferring, sealing, welding, and various material-handling operations. Advanced robotic systems and expanded uses, such as automatically adaptive (smart) robots, are covered.

A Robot Dynamics Simulator. The author of this article fully recognizes the complex (coupled and nonlinear) multivariable characteristics of all but the simplest robotic manipulators. A thorough description of VAST (Versatile Robot Arm Dynamic Simulation Tool) is described. This simulator has been designed to become the foundation of a robot-oriented computer-aided design (CAD) system. The system is described mathematically, with practical descriptions of its operation. Simulator inputs and outputs are summarized in tabular form. Simulation modes (open-loop waveform mode; closed-loop controller mode) are illustrated and compared. The implementation of robot control algorithms in VAST is described. In summary, VAST creates a user-friendly working environment in which to simulate robot dynamics and design and evaluate feedback controllers for robotic manipulators.

Control of Actuators in Multilegged Robots. Traditionally, the robot's area of operation (work envelope) is limited by the robot's reach. "Walking" robots, so to speak, may open up an entirely new area of applications for the robotic principle. Crisis management, where the presence of human operators may not be possible (nuclear reactor accidents, for example), is ready for the deployment of wide-ranging and controllable *legged robots*. For persons interested in expanding robotics, this comparatively short article covers kimenatic considerations, structural equilibrium, and other design factors. Explanations are supported by diagrams, equations, and examples.

Servomotor and Servosystem Design Trends. Following an updated perspective on AC and DC motors, the author explores, in particular, the operation, advantages, and control of brushless motors, including the use of Hall sensors. Torque, efficiency, package size, speed and inertia, versatility, life, and other performance characteristics of electric motors are delineated and also compared—one type to the next. Microprocessors in servo-control systems are delineated, including adjustable-speed brushless control, where it is shown how brushless servo-controls are software compensated, and thus require no pot adjustments. So-called intelligent controls that close the velocity loop and directly interface with the readily available programmable motion controllers to achieve position control are described and diagrammed. The author stresses that the intelligent digital microprocessor-based control system, coupled with brushless motors, constitutes an excellent solution for many automated applications.

Stepper Motors and Controls. Fundamentally high-precision devices, stepper motors are often the method of choice for many open-loop control applications. The article points out that the line of demarcation between servomotors (feedback-closed loops) and stepper motors for motion control is rapidly becoming less distinct, resulting from a change in design philosophy and improvements in stepper motors, which include higher incremental resolution, more stable torques at low speeds, and a reduction of previously speed-sensitive resonance. The principal categories of stepper motors are described in considerable detail, including variable-reluctance, permanent-magnet, and hybrid designs. Several tabular comparisons of these categories are given along with detailed specifications of each type, information that is important to the systems designer and specifier. Characteristics treated in detail include step angle, resolution, step accuracy, position accuracy, holding torque, pull-out and pull-in torque, slow speed range, minimum response time, single- and multiple-step response, ramping, damping, the latter including mechanical, electronic, capacitive, retrotorque, and delayed-last-step (DLSD) damping. Stepper motor controls are detailed. Transitions in stepper motor technology are described, where it is pointed out

xviii   Description of Articles

that actuation increments have shrunk in angular or linear dimension to a point where a sequentially pulsed stepper motor takes on the characteristics of a precision servomotor. In a subarticle, the theory, advantages, and applications of *microstepping* are described in considerable detail, with the use of equations and several diagrams to amplify the text. An ironless disc-rotor stepper motor is described. The article closes with a discussion of electrohydraulic stepping motors, including descriptions of operating principles and characteristics, and a tabular summary of the major mechanical and electrical parameters of electrohydraulic units.

Linear and Planar Motors. The three principal types of linear motors (AC, DC, and reluctance) are described, including principles of operation, backed up by a mathematical analysis and reference to applications, such as *X-Y* table positioning, where high acceleration, accuracy, and immunity to vibration are vital. The planar motor, which accomplishes *X-Y* motion in a plane without the need for gears or other mechanical couplings between motor and stage, is described. It is shown how a limited-motion fine positioner can be piggybacked onto a stage driven by the planar motor or any conventional drive mechanism using a rotary motor and lead screw. Actuators, such as the *piezo* (PZT) stack, terballoy magnetostrictive, or voice-coil actuators, are described. Motor performance test results, including stiffness, stalling characteristics, and frequency spectra, are delineated. The article ends with a mathematical derivation of the Sawyer motor force equation, as applied to planar motors. These motors are enjoying increased potential in very demanding situations, such as wafer/mask alignment in electronics manufacturing.

Solid-State Variable Speed Drives. These devices, of which there is a profusion of types suitable for nearly every type of electrical machine—from sub-fractional to multi-thousand horsepower ratings—have improved much during the past decade, as traced by this article. Particular attention is paid to the development of power-switching devices (gate-turn-off thyristors, transistor). DC and AC drives are each described from the most modern design viewpoint. The author stresses the many options available today to the system designer, with information that assists in making selection decisions.

Materials Motion/Handling Systems. Automatic handling of materials and workpieces has been a key element of automated systems for many years—long before the appearance of the much-discussed and dramatized robot. Progress in conveyors and other handling equipment has been steady for many years and, during the last few years, "smart" materials handling systems have become available. The overhead movement of materials by power-and-free conveyor systems and underhung cranes and monorail systems is described in detail, including generous photos and diagrams. Interesting interlocks for transferring carriers between crane runways or monorail spurs are described. Descriptions of floor-mounted conveyors include track-and-drive-tube conveyors, inverted power-and-free conveyors, as well as the less-traditional automatic guided vehicles (AGVs). Special attention is given to automated storage and retrieval systems (ASRS). Order-selection and order-filling systems are described in considerable detail. For warehousing, large, high-density, high-rise AS/RS configurations are described and diagrammed. The article demonstrates that materials handling is an intimate part of any automation concept today and that such equipment should not be treated in isolation.

SECTION 5—INTERFACES AND COMMUNICATIONS

User Interfacing to Process Computer Systems. Recent advances in computer technology and the adoption of hardware and software standards have provided the means to design highly functional interfaces for accessing a wealth of information maintained by the computer. The author stresses how easy-to-use interfaces can be constructed through the application of interactive graphic techniques in a distributed computing environment. The important backdrop of the computing evolution—past interfaces, greater regard for energy consumption, the introduction of microprocessors, changes in system architecture, and communications standards—is given in the first part of this article. Then the author addresses user interfacing concepts by developing such specifics as ergonomics, video displays, programmer's hierarchical interactive graphics systems (PHIGS), among others. This is followed by a discussion of user interactions—conversational, graphical, keyboard manipulation, touchscreens. Emphasis is placed on the use of event-driven software architecture, instead of brute force, time-interval scanning architecture.

Local Area Networks (LANs). Early communication needs of instrumentation and automation systems are tersely reviewed as a backdrop to the remainder of the article, which concentrates on the almost explosive introduction of local area networks (LANs) just a few years ago. The intercommunications needs of a generalized automated factory are described and effectively diagrammed. Particular attention is given to the effects of industrial environments on communications equipment. The basic topologies (point-to-point, star, and communications port with multiple drops are described briefly and diagrammed. The main thrust of the article concerns available, state-of-the art communications protocols, including IEEE-802.3 CSMA/CD (carrier sense multiple access with collision detection); IEEE-802.4 Token Bus; PROWAY; IEEE-802.5 Token Ring, all of which are described in considerable detail.

Communication Standards for Automated Systems. After a terse review of information needs, the quest for communications standards is explored. To put the current communication standards situation in perspective, early efforts dating back to the early 1970s are described. The work of the International Electrotechnical committee, the Instrument Society of America, the Institute of Electrical and Electronics Engineers, the International Purdue Workshop on Industrial Computer Systems, the International Standards Organization, the American National Standards Institute, the Computer and Business Equipment Manufacturers Association, the U. S. National Bureau of Standards, and international communication experts in developing present standards is described. A comprehensive tabular summary of MAP (Manufacturing Automation Protocol), first proposed by General Motors Corporation, explains the present status of MAP protocol on the basis of the seven layered OSI (Open System Interconnections) and relates the layers with uses and applications and the applicable protocols as of 1986. A rather exhaustive list of references is included for those *Handbook* users who wish to gain even further perspectives on the actions taken in the early-and-mid-1980s to achieve modern standards.

SECTION 1

Backdrop to Automation Systems

Douglas M. Considine
Editor-in-Chief

Section Contents

Automation—The Multifaceted Technology	3
Advantages/Limitations	3
Applications	4
Patterns of Industrial Production	5
Scientific Foundation for Automation	6
Contributions of Automation to Engineering and Science	7

Annotated Glossary of General Automation Terms	9
CAD (Computer-Aided Design)	9
CAE (Computer-Aided Engineering)	10
CAM (Computer-Aided Manufacturing)	10
JIT (Just-in-Time Concept)	11
Bridging CAD and CAM Systems	11
CIM (Computer-Integrated Manufacturing)	11
FMS (Flexible Manufacturing System)	12

Automation:
The Multifaceted Technology Overview

Automation, possibly more aptly termed *automation engineering*, is a design engineering philosophy that is directed toward enhancing the automatic nature (sometimes called *automaticity*) of a machine, process, or other type of work system[1]. The objective of automation is to cause the work system to be as automatic, that is, self-acting, self-regulating, and self-reliant, as may be possible—but against the practical backdrop of various economic, environmental, social, and other restraints. Because of these restraints, the work systems are only *partially automated*.

One definition of automation[2] was proposed in 1947 as *"the automatic handling of workpieces into, between, and out of machines."* As viewed in the last half of the 1980s, this is a limited definition, although still accurate as far as it goes. Some authorities claim that automation is a contraction of the more-difficult-to-say word *automization*. Still other scholars claim that automation was coined from *autom*a*tic* and *operation*. Even though the derivation of the word is not fully clear (and not important), it is well established that the practical application of automaticity to making and getting things done is centuries older than the words used.

As pointed out by Mumford[3], the curse of labor was described by the early Egyptians, who mentioned the daily hardships, the filth, the danger, the nightly weariness, of producing goods. Later the oppression of labor was recognized by the Greeks in the fifth century B.C. and by the Florentines in the twelfth century A.D. Prior to the last century or two, earlier people tended to look toward a force (leading to the modern concept of automation) that would abolish all work and, as described by Mumford, "the most desirable life possible would be one in which magical mechanisms or robots would perform all the necessary motions under their own power, without human participation of any kind. In short, the idea of the mechanical *automaton*, which would obey all orders and do all the work." Thus, the negative connotations of automation in terms of adverse effects on the economy of a human work force did not arise seriously until this present century.

Numerous scientific and engineering disciplines make up the technical foundation for automation. Very prominent are electronics; electrical, mechanical, chemical, metallurgical, and industrial engineering; measurement and control technology; computer, information, and communication sciences—all supported by the principles of physics and mathematics.

Advantages/Limitations of Automation

As is apparent from the numerous technical articles in this *Handbook*, advanced automated systems are available today, and further advances seem close at hand. Thus, a former question, "Is automation possible?" has been displaced by the query, "Is automation profitable?" As is essentially true of all business concerns, automation is welcomed most where it contributes to profit. Of the several dividends yielded by manufacturing processing automation, two are upper-most—improved productivity and better product quality.

1. *Improved productivity* of machines and people is a dividend that almost always translates into greater profitability and return on investment. Several factors enter into improved productivity, but two are most important.

   (a) *Increased production capacity*—more goods produced per manufacturing floor area, machines installed, and the human work force. In terms of machines, automation usually increases the duty cycle for

---
[1] Work, as used here, is the action or effort expended in production. Work refers to the application of machine energy, human energy (muscle and brainpower), and any other auxiliary energy used in the production of goods and/or services. Work may apply only to the manipulation of information, which occurs in data processing and office automation. Or it may apply to the manipulation of both information and materials to produce physical goods, that is, the products of industrial manufacturing and processing.
[2] D. S. Harder, who in 1947 was a vice president of Ford Motor Company.
[3] As found in "The Myth of the Machine: Technics and Human Development," by Lewis Mumford, Harcourt, Brace & World, Inc., New York, 1966.

machines, thus yielding more machine hours per day.

(b) *Better inventory control* (flow of materials and energy throughout the plant) of raw materials, goods in process, and finished goods. There is an axiom—"To automate well means to understand and plan exceedingly well." Some authorities have observed that just analyzing a plant's operations and procedures when considering further automation is very worthwhile even though only a limited amount of automation may be immediately installed. For the first time, such analysis may cause an in-depth understanding of the intricacies and interrelationships of a given production situation. A number of special techniques, most supported by excellent software for computerized analysis, have been developed in recent years. These include such concepts as *group technology*, *just-in-time* (JIT) systems, and other aspects of material resources and requirements planning (MRP). These concepts are described in a condensed fashion in the next article of this *Handbook* section.

2. *Enhanced product quality*, which improves competitive position and reduces waste and reworks. Improved competitive position naturally translates into higher volume and its usual attendant economic advantages.

It is interesting to note that some automation has entered the factory, not necessarily by choice, but rather by the force of improved manufacturing and processing operations that far exceed the limitations of human dexterity, awareness, cognition, speed, and strength, among other factors. Some manufacturing and processing variables, such as temperature, pressure, chemical composition, flow, weight, et al., are not directly measurable by people. Human inadequacies in these areas were among the first of the "external" forces that introduced a need for automation.

3. *Upward shift of workers' role*, that is, from numerous arduous, low-skilled duties to higher-skilled supervisory and maintenance responsibilities.

4. *Reduction of personal accidents* through the assumption of accident-prone duties by automated machines and processes.

Some of the limitations of automation include:

1. *High cost* of designing, building, and maintaining automated equipment. This cost is finding considerable relief because of the continuing lower cost of electronic components and equipment, although some of these savings are offset by continuing inflated costs of software. Successful efforts to date and that will continue into the future in terms of standardizing equipment, communication networks, and software will also relieve cost as a barrier to automation.

2. *Vulnerability to down time* because of increasing complexity of automated equipment. This vulnerability, however, is being reduced at an accelerated rate because of improved equipment self-diagnostics, fault-tolerance techniques, and more economic approaches to designing redundancy into automatic systems.

3. *Loss of flexibility*. This was a very important restraint on automation until relatively recently. Introduction and refinement of the concept of flexible manufacturing systems (FMS) has largely negated this restraint.

4. *More management attention*. Actions in highly automated systems occur sometimes at an almost unbelievably high rate and allow little or no time for human decisions. Currently, with state-of-the-art technology, a machine or process can be driven to make quite a lot of off-spec and scrap material before effective supervision can get into control of a runaway situation. Through the assistance of information networks, ranging from corporate to plant-wide to cells and individual machines and processes, managers can be apprised of factory floor situations on essentially a second-by-second basis. Thus, more and better management personnel is needed as a plant increases its content of automation. In the more distant future, a much greater portion of the almost instantaneously needed decision making will also be done automatically. However, assignment of this important responsibility to computers, needless to say, will require exceedingly careful attention and analysis by very sharp management personnel.

5. *Persistence of automation's negative image*. Surprising to many authorities has been the acceptance of automation technology by the labor force and the successful negotiation of new union contracts—even though the basic fact remains that jobs are eliminated by automation. Of course, automation also creates new and certainly higher-skilled jobs. As the public and the press and other media that serve it become better acquainted with the real nature of automation, earlier predictions of very adverse effects on the labor force will continue to be tempered. Fortunately, too, automation is frequently identified with the other aspects of so-called *high technology*, contributing to a reasonably good press for automation.

APPLICATIONS FOR AUTOMATION TECHNOLOGY

Nearly all human endeavors, including education, recreation, health care, national defense, communication, transportation, industrial manufacturing and processing, research and development, and business and commerce have been impacted by automation.

Office Automation

Sometimes simply referred to as computerization, office automation involves information as the input, the work in process, and as the final product. The information may be of many purposes and formats—payroll preparation, transportation reservations and scheduling, banking and security transactions, statistical and census compilations, inventory control, accounts receivable and payable, insurance risks and records, cost and price analysis, statistical quality control, electronic mail, and almost any activity that can be described as routine *paperwork*. Increased productivity per office worker is indeed a major advantage, but possibly more important is the rapidity with which information required to make business management decisions can be communicated over long distances and integrated with information from various institution and corporate entities.

Office automation has contributed in a very marked way to the furtherance of manufacturing and processing automation systems.

Manufacturing Automation

Manufacturing automation, in the long term, most likely will well exceed office automation in terms of investment. However, there will be so much blending, integration, and information exchange between the management of offices and factories that it will become increasingly difficult to determine any sharp demarcation between these two activities.

The tempo to automate production has hastened very much during the latter half of the 1980s, but what appears as intense activities now will pale in terms of investments in automation to be made during the remainder of this century. The somewhat lagging acceptance of automation on the part of the bulk of manufacturing industries is considered by many authorities as simply a "wait and see" attitude. Numerous segments of manufacturing are awaiting the experiences of the comparatively few leading users of the present time, notably the application of automation technology by the automotive and electronics industries, as examples of current leaders in the field. These industries have been under much pressure to improve both quality and productivity from forces that are national and international. Competitive pressures have warranted unusually high investments in manufacturing research and development. It is largely these industries, for example, that have funded advanced communication links and more effective robotization, including machine vision.

PATTERNS OF INDUSTRIAL PRODUCTION

Manufacturing automation has developed along two principal paths, which reflect the rather distinct natures of two kinds of products:

1. *Fluid and bulk materials process industries*, as typified by the chemical, petroleum, petrochemical, metals smelting and refining, and food-processing fields, among others, which largely react, separate, combine, and otherwise process materials in a liquid, slurry, gaseous, or vaporous state. During much of the manufacturing, raw materials, materials in process, and final products are in the form of fluids or bulk solids. Except at the molecular level, these materials are not in the form of discrete, identifiable pieces. Fluids and bulk materials are handled in enclosures, such as vats, bins, and other vessels, and are transported within pipes and atop bulk belt and other types of conveyors. A major exception in a number of these fluid/bulk industries is the final product, which may be a discrete can, box, tankcar, barrel, et al.

A rather high degree of automation has existed in the fluid/bulk industries for several decades, particularly since World War II, when many of the former batch processes became continuous in nature. Fluid/bulk industries traditionally have been capital rather than labor intensive. For many years and continuing into the present, the most commonly measured and controlled variables have been temperature, pressure, flow, and liquid level, and, as previously mentioned, these are quantities that essentially are impossible for humans to measure accurately, if at all, without the aid of instruments. The automation of measurement and control of these variables for many years was identified as *instrumentation and automatic control*—the term *automation* was rarely used in this regard.

2. *Discrete-piece manufacturing industries*, as typified by the manufacture of machines and parts, assemblies and subassemblies, et al., generally have been quite labor intensive because the production variables present, dimension, position, displacement, proximity, motion, and speed, have been at least partially within the grasp of measurement, and hence control by people. Technologically, too, it has been much more difficult to develop sensors to automatically measure and devices to automatically control, without human supervision, these manufacturing variables than, for example, the development of instrumentation for the fluid/bulk industries.

The progress of automation has been closely tied to the ease with which an operation may be automated. Thus, it is no surprise that automation of the fluid/bulk industries preceeded the discrete-piece industries by several decades.

This observation is further proved by a number of discrete-piece industries that currently remain well behind the leaders in automation.

6 Automation—The Multifaceted Technology

For example, still one of the most labor-intensive industries is the manufacture of garments and apparel. The skills of sewing have been very difficult to transfer to a machine control system. Unlike working with rigid materials, such as metals and plastics, textiles are soft, pliable, and, from the standpoint of machine design, much more difficult to manipulate. Further, the geometry of the parts of a garment and the dependence for appearance upon the nature of the seam for shape and drape are factors that do not enter in the assembly of something made from harder, more rigid materials. These kinds of difficult technical problems, coupled with an industry that is generally not accustomed to high capital expenditures, have substantially slowed the pace of automation in the garment and other like fields.

SCIENTIFIC FOUNDATION FOR AUTOMATION

Principal scientific and technological developments that contributed to the feasibility of automation have included:

1. *Feedback*, the fundamental principle and basic mechanism that underlies all self-regulating processes. Some experts have defined feedback as information about the output at one stage of a process or operation that is returned, that is *fed back*, to an earlier stage so as to influence its action and hence to change the output per se. Ingenious self-regulating devices and machines date back many years. The flyball governor, invented in 1788 to control Watt's steam engine, exemplifies the application of feedback long before a theory for feedback and closed-loop control was put forth. One of the earliest uses of closed-loop feedback was its application to the power steering of ships, adapted decades later to the power steering for automobiles.

2. *Information and communication theory* was not tackled formally until after World War II, when C. E. Shannon published "A Mathematical Theory of Communication" in 1948. In that same year N. Wiener published "Cybernetics or Control and Communication in the Animal and the Machine." The concepts put forth by Wiener stirred up excitement during that early period. Cybernetics essentially comprises three concepts: (1) animal or machine systems, (2) communication between systems, and (3) regulation or self-regulation of systems.

3. *Sensors and measurement systems* did not develop historically according to any particular master plan. Generally, sensors were developed so that more could be learned concerning the nature of physical and chemical phenomena—*not* as tools for achieving automation. Measurement of dimension and weight, for example, had its roots in antiquity, and its needs were largely the basis upon which early trade could be conducted. Although mechanically based sensors have and will continue to be used in automation systems, the measurement field progressed much more rapidly after the details of electromagnetics and electrical circuits were established earlier by such investigators as Ampère, Volta, and Ohm in the late 1700s and early 1800s—then to be followed in the first half of the 1800s by Faraday, Henry, Wheatstone, Kirchoff, and Maxwell. Even before the appearance of electronics, it usually was found much easier to measure and control a machine or process by electrical rather than mechanical, pneumatic, or hydraulic means. But in the absence of electronics, nonelectrical methodologies essentially by default became the approaches of choice. Even today wide use of mechanical, pneumatic, and hydraulic technologies persist. The comparatively new field of micromechanical sensors is successfully reestablishing some of the earlier nonelectronic approaches.

4. *Servopower*, electric, hydraulic, and pneumatic, made possible a host of actuators, ranging from valves, louvers, and dampers in the fluid/bulk industries and machine and workpiece positioners in discrete-piece manufacturing. Automation was assisted by the appearance of combined-technology devices, such as electromechanical, electrohydraulic, and electropneumatic relays and subsystems. The continuing progress in the design of electric motors, decreasing size and weight for a given horsepower rating along with increased energy efficiency are contributing to the furtherance of automation. During the past decade or two, outstanding progress has been made in DC and AC motor controls, in the refinement of stepping motors, and in the practical application of linear motors.

5. *Computer and memory power* have been of outstanding importance to automation even though these elements have not always been sophisticated. The Hollerith card that appeared in 1890 (frequently referred to for many years as the IBM card) most likely had its roots in the card-programmed Jacquard loom invented in 1801. In repeat-cycle automated machines, the memory required for operation in earlier machines was designed right into the mechanics of the machine—a practice that still can be found in printing and packaging machines, whose automaticity dates back a number of decades. As the degree of automaticity and complexity of a machine or process increases, there are continuing requirements for more information storage and retrieval at faster and faster rates. Prior to the entry of digital electronics, mechanical computing and memory systems (for example, desk calculators of just a few decades ago) were large, slow (in today's terms), and frequently quite difficult to alter (program). With the majority of

controllers of the last few decades being electronic, it is easy to forget that the earlier mechanical, pneumatic, and hydraulic controllers had to incorporate nonelectronic computers to calculate the error signal in a closed-loop feedback system. Actually, in the process control field prior to the appearance of electrical and electronic instruments, the words *memory* and *computing* were rarely used, even though all the elements were there under different designations.

6. *Digital technology*, which for practical purposes encompasses the advances of solid-state microelectronics, introduced vastly improved computing speeds for automated systems which, in combination with improved response speeds of detectors and sensors, greatly enhanced the performance of control systems. Modern computerization, of course, stems directly from digital technology. The two very marked trends of decreased size and cost for microelectronics have greatly influenced the availability of components in terms of application feasibility and economics. The question is sometimes asked, "Why is small size so important in regard to the electronic components widely used in automated systems?" First of all, size is directly related to the economics of component part production. Second, the example of having to mount detectors on robot arms (where the space available is limited) serves to answer the question from a practical applications standpoint. Obviously, many similar examples could be given.

7. *Mechanization*, presently simply taken for granted, was a major step toward automation. Mechanization was the logical next step toward automation after the emergence of metal hand tools (in contrast with the earlier stone and wood tools). Mechanization conferred the first degree of automaticity to a system.

8. *Systematization and engineering analysis* were and continue to be key elements for achieving successful automation plans and installed systems. As mentioned earlier, just good planning and thinking in depth about the prospects of automation for a manufacturing process can be extremely beneficial. Traditionally, production supervisory personnel have been the real storehouse of knowledge pertaining to all aspects of production—from incoming materials through warehousing and shipping.

Because advanced manufacturing automation minimizes (sometimes displaces) the subtleties of human judgment that can be applied directly on the factory floor in the form of minor machinery adjustments or procedural changes in the interest of maintaining smooth, uninterrupted production throughput, all of the vagaries of production have, prior to more extensive automation, been deeply implanted in the minds of production supervisors. This detailed, but very important information is not always easy to retrieve. As suggested by a major firm, one must "sweat out the details" if success is to be achieved via automation.

9. *Information display technology*, which has progressed beyond earlier expectations prior to the extensive use of the cathode ray tube, has contributed immeasurably to the expansion of automation technology—largely by automating the human/machine/process interface per se. Ingenuous ways of plotting and presenting information, now widely assisted by the use of color, have provided a way to interlock designing for manufacture with manufacturing itself in so-called CAD/CAM (computer assisted design/manufacturing).

Contributions of Automation to Engineering and Science

The prior recitation of the scientific and engineering developments upon which modern automation is based provides only part of the story. Within the past decade, with the firm establishment of automation in many major industries, the reverse transfer of technology has occurred at least to some degree. Pressures brought about by automation have impacted information communication—as represented by the possibility (once a dream) of integrating and interlocking manufacturing operations on a corporate and plantwide basis through the development of hierarchical two-way information transfer (communication) systems. This is exemplified by the great progress that has been made in the design of local area networks (LANs), which in turn are parts of wide area networks. Many examples can be given. One of the most recent and outstanding developments is MAP (Manufacturing Automation Protocol). See *Section 5* of this *Handbook*.

The concept of distributed control is another. Introduced in the mid-1970s, this control architecture combined three technologies—microprocessors, data communications, and CRT displays. Automation today is impacting on the design of future computers, on the development of more effective programming languages, the technology of expert systems, and although not exclusively, automation is a major source of pressure to develop the concept of artificial intelligence, which in past years has not exhibited the kind of practicality that is expected within the relatively near future. Progress is being made in the application of AI to machine vision in connection with the performance of robots.

Automation requirements of the automotive industry literally gave birth to the concept of the programmable controller as a replacement for electromagnetic relay systems. The acceptance of the programmable controller was almost immediate and over the past decade has expanded at a phenomenal rate.

Not the least of automation's contributions to technology has been its impact on the entire philosophy of manufacturing. For example, the concept of flexible manufacturing systems (FMS). This actually grew out of earlier dissatisfaction with attempts to automate various machines and processes. With the kinds of hardware available in the 1930s and 1940s, systems were essentially limited to *hard automation*, an approach that usually was advantageous only for high-volume, long-term production runs. In fact, the popular approach to automation in the 1940s and 1950s was to design a product *for* automation (there is still wisdom in this approach). It was found that products designed strictly with automation in mind often turned out unattractive aesthetically and minimized the options in design that the consumer expected. Although no universal automated system appears on the distant horizon, automation of the late 1980s is many times more flexible. The analytical planning required to create successful flexible manufacturing systems almost immediately led to the concept of computer-integrated manufacturing (CIM). This is the logical organization of individual engineering, production, and marketing and support functions into a computer-integrated system. Functional areas, such as design, inventory control, physical distribution, cost accounting, procurement, et al. are integrated with direct materials management and shop-floor data acquisition and control. Shop-floor machines serve as data-acquisition devices as well as production machines.

Annotated Glossary of General Automation Terms

Prior to the 1970s, the automation of industrial production was mainly an extension of mechanization, that is, the use of systems that did not incorporate feedback[1]. Attempts to automate were largely of an unplanned, scattered, piecemeal nature. Even by the late 1980s, just a few plants worldwide have been automated extensively across-the-board in a way that matches the rather distorted public image of automation on a grandiose scale. Plant-wide, all-at-once automation is found in but a comparative handful of plants that are either new grassroots facilities built from the ground up or plants that have been fully refurbished from the receiving to the shipping dock. In either case, such new or modernized facilities represent tremendous capital outlays that are well beyond the resources of most manufacturing firms.

Plant owners and managers have found that automation is best approached by stages in a carefully planned and tightly controlled manner. What has changed in the past few years is the attitude of top management, now characterized by greater motivation, courage, and confidence. Growing numbers of firms are pioneering automation on a vastly increased scale by targeting larger sections and departments of their plants—as contrasted with a former posture of experimentation and automation by trial in terms of a few machines or manufacturing islands.

The incentive to automate, of course, is fundamentally economic. Competitive pressures, frequently from the international marketplace, have been tremendous and largely unexpected. Thus, any endeavor that will trim costs in the long run, like automation, must be given the most serious of considerations. This factor accounts for the present *uneven* application of automation from one industry to the next. Those industries that have been hurt the most by competition will be found among the automation pioneers. Very large firms in these categories not only have invested heavily in the procurement of automation hardware and software but also have participated in a major way in automation R&D. This R&D has stemmed not only from newly established in-house groups but also from the corporate funding of outside research organizations, including academia. These efforts have been abetted by publicly funded programs, such as AMRF (Automated Manufacturing Research Facility) established at the National Bureau of Standards, and special programs sponsored by the Departments of Defense, Education, Energy, and Treasury. Similar programs have been set up by other governments, notably in Europe and Japan. Also, groups of firms have teamed together in cooperative automation R&D, as typified by CAM-I (Computer Aided Manufacturing-International, Inc., Arlington, Texas).

Achievements to date have contributed much to the confidence of management in automated production facilities. Not the least of past accomplishments is the establishment of a philosophical base for automation and the development of general concepts and guidelines. A degree of maturity in the generalized approaches to automation now exists. These are defined and discussed briefly in the remainder of this article. As pointed out in the *Preface* to this *Handbook*, the greatest need for automation information today has shifted from the general conceptual stage to one of specific implementation. With so much automation hardware and software now available, the field is open to exploitation by many firms that may not be severely threatened by competition or in the past have lacked the incentive and willingness to be automation pioneers. The field now appears to be ripe for a lot of "copycat" automation, which perhaps may be the next major phase in automation technology.

CAD (Computer-Aided Design)

This acronym can also be taken to mean computer-aided or computer-assisted drafting. Uncommonly, a combined acronym (CAD/D) may be used. This designates a system that assists only in the preparation and reproduction of drawings, but that also develops the information or intelligence associated with the drawing. Most CAD/D systems have six major components (four hardware; two software). See Ref. 1.

1. A central processing unit (CPU).
2. Storage—where drawings and graphics are stored electronically.

---

[1]As mentioned in the preceding article, automation in the bulk/fluid processing industries advanced much earlier and is in a much more mature state today as compared with the discrete-piece manufacturing industries.

3. Workstation—the interface between operator and computer.
4. Plotter station—where images stored in the computer memory are printed on drafting media.
5. Operating system (OS)—the master control program that coordinates the activities of all four of the aforementioned hardware components.
6. Application program—user software that creates a working environment for creating designs and preparing drawings.

Major Functions

CAD functions can be grouped into four principal categories:

1. *Design and Geometric Modeling.* In this function, the designer describes the shape of a structure with a geometric model constructed graphically on a cathode ray tube (CRT). The computer converts a picture into a mathematical model, which is stored in the computer database for later use.
2. *Engineering Analysis.* After creation of a geometric model, the engineer can calculate such factors as weight, volume, surface area, moment of inertia, center of gravity, among several other characteristics of a part. One of the most powerful methods for analyzing a structure is *finite element analysis*. Here, the structure is broken down into a network of simple elements and the computer uses them to determine stress, deflections, and other structural characteristics. The designer can see how a structure will behave before it is built and can modify it without building costly physical models and prototypes. The procedure can be expanded to a complete systems model, and operation of a product can be simulated. The topic is explored in much more detail in Ref. 2.
3. *Computer Kinetics.* The user can examine effects of moving parts on other parts of the structure or design and analyze more complex mechanisms.
4. *Drafting.* A CAD system can automatically draft drawings for use in manufacturing. Engineers can draw on geometric and numerically coded descriptions produced by CAD to create numerical control tapes, which permit direct computer control of shop machines, determine process plans and scheduling, instruct robots, computerize testing, and generally improve the management of plant operations.

CAE (Computer-Aided Engineering)

This acronym applies to the strictly engineering assistance obtained from CAD systems as just described.

CAM (Computer-Aided Manufacturing)

This acronym generally refers to the utilization of computer technology in the management, control, and operation of a manufacturing facility through the direct or indirect interface between a computer and the physical and human resources found in a manufacturing organization. Developments in CAM are found in four main areas:

1. *Machine Automation.* Originally confined to numerical control, machine automation has been expanded and now consists of a chain of increasingly sophisticated control techniques:
   (a) At the lower end of the scale is *fixed automation* with relays or cams or timing belts and timing chains. Relay logic has been extant in industrial production for decades. Essentially during the past two decades, many relay installations have been replaced by electronic means, notably in the form of programmable controllers.
   (b) Further up the scale of automaticity is plain numerical control (NC) whereby a machine is controlled from a prerecorded, numerically coded program for fabricating a part. In these systems machines were hard-wired and were not readily reprogrammable.
   (c) At a higher point in the scale of automaticity, the machine is directly controlled by a minicomputer, which stores machining instructions as software that is relatively easy to reprogram. Known as CNC (computer numerical control), this approach has the advantages of much higher storage capacity and greatly increased flexibility. Nearly all new numerical control systems today are CNC oriented. However, as recently as the late 1970s, CNC was considered a costly exception to the traditional approach. See separate articles in *Section 3* of this *Handbook*.
   (d) At the highest point in the scale of automaticity as presently viewed is the plant and even corporate-wide interconnection of machines on the floor with vast and complex information networks wherein decisions at the factory floor level are influenced by information flowing down from the corporate computer hierarchy—and, in the other direction, information from machines flows upward to enrich the database of the headquarters computer. This is further described under CIM.
2. *Robotics.* Robots are now used rather widely for performing materials-handling and manipulating functions in CAM systems. Robots can select and position tools and workpieces for CNC tools, operate such tools as drills and welders, or perform test and

# SECTION 2

Sensors and Measuring Systems

Section Contents

Position, Motion, Speed, Velocity 19	Linear Transformer (LT) 55
Classification of Position and	Impedance-type Gaging
Motion Systems 19	Transducer 55
By Excursion of the Motion 19	Linear Potentiometer 55
Point-to-point Systems 19	Thin-film Potentiometers 56
Path or Trajectory Systems 20	Strain Gage Load Cell 56
Superimposed Motions 20	Optical Methodologies 57
Physically fixed Motion Paths 20	Statistical Process Control 57
By Type of Drive and Drive	Background of Statistical Terms 57
Control	Capability and Control (Concept
Traditional Approach 20	of) 57
Trends 20	*Data Gage*™ 60
Feedback Enhancement 21	Automatic Gaging and Size-Control
Dual-mode Servos 21	Systems 64
By Sensors 21	Microprocessor-based Camshaft
By Generic Application Categories 21	Gage 64
Rotary (Angular) Position Sensors 22	Measuring Machines 65
Resolvers and Synchros 22	CNC Probe System 66
Resolver/Synchro-to-Digital	Eddy Current Inspection System 66
Converters 25	Noncontact, Laser-controlled
Encoders 28	Inspection System 68
Absolute Encoder 28	Positioning Tables and Precise
Incremental Encoder 29	Positioners 68
Comparison and Selection of	Thickness Measurement 69
Resolvers/Synchros and	Nuclear Radiation Thickness
Encoders 35	Gages 69
Tachometers 35	Ultrasonic Thickness Gages 70
Speed Ratio Systems 40	X-ray Thickness Gage 71
Linear/Rotary Position Sensors 40	Surface Finish Measurements 72
Inductive Plate Transducers 40	
Potentiometers 41	**Object Detection, Proximity, Presence,**
Pulse-type Transducers 41	**Nonpresence** 73
Electromechanical Limit Switches 43	Fundamental Detector
Solid-State Limit Switches 45	Methodologies 73
Integrated Motion Systems 46	Contact and Noncontact Detectors 73
Motion Systems Analysis 49	Photoelectric Devices 73
	Control System Configuration 74
Dimension, Displacement, Thickness 51	Operating Mode 74
Fundamental Sensor Methodologies 51	Scanning Techniques 74
Interferometer 51	Reflective and Retroreflective
Motion System Sensors 51	Scanning 75
Linear Variable Differential	Specular Scanning 76
Transformer (LVDT) 52	Diffuse Scanning 76
Linear Variable Reluctance	Color Differentiation (Registration
Transducer (LVRT) 53	Control) 77

Sensitivity Adjustment	77	Windowing the Scene	90
Light Sources and Sensors	77	Segmentation	90
Phototransistors and Photodiodes	78	The Recognition Process	90
Use of Fiber Optics	78	Application	90
Magnetic Proximity Switches	78	Computational Techniques	91
Variable-reluctance Sensors	78	Adaptive Hardware	91
Magnetically actuated Dry Reed Switches	81	Application Basics and Examples	92
Inductive Proximity Sensors	81	Specific Uses of Contrast/ Windowing Technique	92
Wiegand-effect Switches	81	Inspection	92
Capacitive Proximity Sensors	81	Surface Flaws	92
Magnetostrictive Limit Switch	83	Dimensional Tolerance Values	94
Ultrasonic Switches and Other Novel Approaches	83	Shape Verification	94
Impact of Machine Vision Systems	84	Location	94
		Orientation	96
		Measurement	99
		Recognition	99
Machine Vision	86	Implementation of Contrast/ Windowing Technique	104
Pattern Recognition in Machine Vision	86		
Basic Applications for Machine Vision Systems	87	**Machine Vision—State-of-the-Art Systems**	110
Artificial Vision System Basics	87	Gray-Scale Systems	110
Extracting Information from Images	87	Y-Axis Movement	111
Computers Versus Adaptive Hardware	88	Bin-of-Parts Problem	119
Data Differences	88	Post-Objective Scanning	121
Functional Description of Machine Vision System	89	Electronic Innovations for Machine Vision Systems	123
Line Scanners	89	Pipelined Image Processing Engine	123
Area Type Scanners	89	Geometric Arithmetic Parallel Processor	124
Solid State Cameras	89	Image Processing	129
Processing the Data	90		
Binary MV System	90		

Contributors*

ALLEN, D., NCR Corporation, Fort Collins, Colorado. (*Geometric Arithmetic Parallel Processor*)

ARUM, H. R., Stock Drive Products, Division of Designatronics, Inc., New Hyde Park, New York (*Encoders*)

AUVRAY, P., Sopelem, Levallois-Perret-Cedex, France. (*Optical Dimension Checking*)

BOYLE, J., Giddings & Lewis Electronics Company, Fond Du Lac, Wisconsin. (*CNC Gaging Systems*)

CARLSON, R. T., Cutler-Hammer Products, Eaton Corporation, Milwaukee, Wisconsin. (*Parts and Packaging Inspection Systems*)

CHAGGARIS, C. G., ORS Automation, Inc., Princeton, New Jersey. (*Robotic Acquisition of Jumbled Parts*)

CHAN, M., Video Logic Corporation, Sunnyvale, California. (*Motion System Analysis*)

COBB, J., Cognex Corporation, Needham, Massachusetts. (*Packaging, Parts, and Printing Quality Inspection Systems*)

COLONA, R. L., General Scanning Inc., Watertown, Massachusetts. (*Post-Objective Scanning*)

DAVIS, R., NCR Corporation, Fort Collins, Colorado. (*Geometric Arithmetic Parallel Processor*)

ELLIOTT, R. A., Qualiplus USA, Inc., Stamford, Connecticut. (*Container Inspection Systems*)

GEBELEIN, R. E., Moore Products Co., Spring House, Pennsylvania. (*Automatic Gaging Systems*)

HARRIS, M. L., Octek Inc., A Foxboro Company, Burlington, Massachusetts. (*Package Inspection Systems*)

HICHBORN, Y., Video Logic Corporation, Sunnyvale, California. (*Motion System Analysis*)

KENT, E. W., U. S. National Bureau of Standards, Washington, D.C. (*NBS Pipe-lined Image Processing Engine*)

KIM, L. K., Video Logic Corporation, Sunnyvale, California. (*Motion System Analysis*)

KRASKA, P. A., Pattern Processing Technologies,

---

*Persons who authored complete articles or subsections of articles or who otherwise cooperated in an outstanding manner in furnishing information and helpful counsel to the editorial staff.

Inc., Minneapolis, Minnesota. (*Machine Vision*)

MAGER, K., Analog Devices Inc., Norwood, Massachusetts. (*Electronic Parts Assembly and Inspection Systems*)

MAZURKIEWICZ, J., Pacific Scientific, Rockford, Illinois. (*Resolvers/Synchros, Encoders, Tachometers*)

MAZZA, T. A., Telemecanique, Arlington Heights, Illinois. (*Proximity Switches*)

MINNICH, C., MTS Systems Corporation, Minneapolis, Minnesota. (*Integrated Motion Systems*)

MINO, M. G., Ormec Systems Corp., Rochester, New York. (*Integrated Motion Systems*)

SPEICHER, R. F., Micro Switch Division, Honeywell Inc., Freeport, Illinois. (*Photoelectric and Proximity Switches*)

STAFF, Anorad Corporation, Hauppauge, New York (*Positioners*)

STAFF, Daytronic Corporation, Dayton, Ohio. (*Displacement Transducers*)

THOMAS, D., NCR Corporation, Fort Collins, Colorado. (*Geometric Arithmetic Parallel Processor*)

WELLINGTON, J., U. S. National Bureau of Standards, Washington, D.C. (*NBS Pipe-lined Image Processing Engine*)

WERTH, L., Pattern Processing Technologies, Inc., Minneapolis, Minnesota. (*Machine Vision*)

WILDER, J., ORS Automation, Inc., Princeton, New Jersey. (*Robotic Acquisition of Jumbled Parts*)

Geometric Variables I:

Position, Motion, Speed, Velocity

In the discrete-piece manufacturing industries as typified by metals, plastics, and other materials fabrication and assembly (machinery, automobiles, aircraft, appliances, electronic equipment, et al.), *system geometry* is *nearly always* of the utmost importance to maintaining product quality and throughput (productivity). In the flat goods and continuous-length product industries, geometric measurements, notably of thickness and caliper, are of equal importance in the control of film, foil, paper, wire, cable, and similar products. In the bulk/fluid industries (chemical, petroleum, food processing, among others), the measurement and control of liquid level (position of a fluid interface), the speed of bulk conveyors, and the volumetric feeding for fluid blending are but three examples of critical geometric variables that determine processing quality and profitability.

Among the geometric variables, described in this and the following two articles, position is central. Motion, speed, and velocity discussed in this article are derived from position—as are dimension, displacement, proximity, presence, and nonpresence of objects, which are described in the two following articles.

Motion may be described as the movement of an object from a starting position (origin) through an infinite series of points (positions) to a destination or target position. Such movement may be linear or curvilinear and may occur in one plane (two dimensions, x and y) or in space (three dimensions, x, y, and z, or $h2$). Motion is said to occur along a *path* or *trajectory*.

Speed describes the time required to move from one position to the next. Speed is a scalar quantity equal to the magnitude of velocity. *Velocity* is a vector quantity denoting both direction and the speed of a linear motion or denoting the direction of rotation and the angular speed in the case of rotation. Traditionally, industrial linear speeds are generally inferred from rotational measurements because of the manner in which most machines are designed—with rotating shafts, wheels, and gears to which speed transducers can be conveniently attached. *Acceleration*, the rate of change of velocity with respect to time, is a closely associated variable and is particularly important in connection with vibration measurements—important in devising methods to protect sensitive electronic equipment.

CLASSIFICATION OF POSITION AND MOTION SYSTEMS

1. *By excursion of the motion*
   (a) *Point-to-point Systems.* In numerical control (NC) systems, for example, the operations to be performed can be described in terms of discrete work or tool position coordinates without reference to the movement between these points. If there is movement in more than one axis at a time, coordination of the multiple movements is not necessary. Usually, point-to-point systems are used for moving the tool or work from point to point for actual machining operations, such as drilling, tapping, reaming, etc., to be done at the terminus after each axis movement has stopped.

   With a point-to-point robot, there are two main commands given: (1) The attitude of all limbs at the start of the move, and (2) the new attitude of those limbs when a particular move has been completed. While making the move as fast as possible and while moving all limbs simultaneously to fulfill a given command, there is no definition of the paths which the robot limbs will trace. In programming a robot of this type, however, the system designer must consider certain *intervening points* between start and destination (terminus). The robot may have to clear an object that may fall in its "direct line" path, or it may be desired for the robot arm to approach its destination at the best angle (in the instance of picking up a pallet, for example). Point-to-point robots can do any job performed by a limited-sequence robot. With sufficient memory capacity, these robots can handle jobs, such as palletizing, stacking, and spot welding. A limited-sequence robot (also called pick-and-place or "bang-bang") is nonservoed and capable of high speed and reliability. Point-to-point motions are encountered in fixed automation situations, such as those found in large transfer machines and warehousing storage and retrieval systems. Usually, fixed (stationary) position sensors can be used, including mechanical limit switches, proximity sensors, and photoelectric devices. However, in multiaxis systems involving rectilinear (x,y,z coordinates) in which movements are made in straight lines, feedback devices, such as resolvers, encoders, et al., may be used.

(b) *Path or Trajectory Systems.* Movement generally is curvilinear rather than linear (although the path may comprise an almost unlimited number of straight-line segments). In NC systems, for example, commonly referred to as *contouring* or *continuous-path* systems, the requirements ordinarily are to machine parts in irregular or curved shapes. The work or tool movements must be in a prescribed path continuously under control. This can be accomplished only through coordination of the movements of the various axes. Every point in the continuous path describing the part shape must be accounted for, whereas in point-to-point systems only the end points of a movement need be explicitly defined. In the case of continuous-path robots, not only the start and finish points are specified but also the path traced by the robot hand as it travels between these two extremes. Seam welding is an example of where a robot wields a welding gun and moves it along some relatively complex contour at the correct speed to produce a strong and neat weld. This requirement, of course, calls for a very substantial memory, as in the case of NC contouring. Contoured flame cutting is another application example. These preprogrammed contouring operations are not to be confused with contour-following situations where the part being operated on already has a contour in place in which a sensor simply follows (by optical, magnetic, et al. means) a scored or marked line (straight or curved) already on the part.

(c) *Superimposed Motions.* In special situations, it may be desirable to superimpose over the regular start-to-finish motion of a device an additional motion ("wobbling," for example) to a tool, such as a paint spray gun or polishing hand, to emulate the pseudogyratory motion of the human arm and hand.

(d) *Physically fixed Motion Paths.* In the motion excursions just described, the paths taken by tools and robot end-effectors are directed by instruments and controls, thus allowing great freedom for adjustment and change. The motion paths taken by conveyors, automatically guided vehicles, and rigid paths taken in fixed automation equipments, such as internal paths found in transfer lines and packaging and printing equipment, among many other examples, do not fall within the realm of what is generally understood today as motion control. Mention of these essentially unalterable paths of motion does serve to stress the importance of exquisite design of such elements initially because of the very high cost of changing the fundamental paths of motion on the factory floor.

2. *By type of drive and drive control*

A survey (mid-1980s) has shown that approximately half of the drives used in connection with motion systems for machines, robots, et al. are electric or electrohydraulic; about one-third are hydraulic; and about one-sixth are pneumatic. The choice of drive depends upon numerous factors, among the most important of which are the response required, the mass of the load, the environment of the installation, and cost. These and other factors are described in a series of articles on servomotor and servosystem design trends, stepper motors and controls, linear and planar motors, and solid-state variable-speed drives to be found in *Section 4* of this *Handbook*.

(a) *Traditional Approach.* Particularly with electric drives, the practice over many years has been that of furnishing a package that included the analog speed or motion controller, power amplifiers, as well as the motor drive. Generally, these packages were designed to accommodate a series of motor sizes, and they logically fell into two main categories—DC and AC motors. Also, in the past, there has been a marked distinction between the types of motion furnished by the drive, that is, *continuous motion* associated with the servomotor or *incremental motion* as characterized by stepper motors. Furthermore, as has been true of the measurement and control of other variables, the problems of position and motion systems tended to be solved in comparative isolation as contrasted with their being part of a much larger overall manufacturing control situation. Finally, past approaches tended to favor hardware rather than software solutions.

(b) *Trends.* With the advent of minicomputers and microprocessors (MPUs) in the comparatively recent past, the rather grandiose concepts of computer-integrated manufacturing (CIM) became viable and, of course, have impacted on position and motion control systems. For example, the prior marked distinction between *incremental* and *continuous* motion and of open- and closed-control loops has diminished. Continuous and incremental actuators can accept microcomputer-based control decisions expressed in digital commands.[1] The need to generate step-by-step load positioning has led not only to motors designed to be pulsed but also to the pulsing of motors originally designed to be run continuously. Thus, the distinction between stepper motors and continuous servomotors is fading. Stepper motor actuation increments have decreased in angular or linear dimension to the point where a sequentially pulsed stepper motor has the characteristics of a precision servomotor. Like the servomotor, the stepper moves smoothly in its target approach. It can maintain step count to ensure command positioning, while avoiding target overshoot. The final step can be as small as a 0.432 arc minute. This type of performance may diminish the need for feedback circuitry for many applications. However, this kind of sophistication is still in its early development and trial stages.

(c) *Feedback Enhancement*. The fundamentals of feedback control are shown in the accompanying diagram.

Microprocessors have made it possible to greatly enhance the feedback path of the closed loop. With the MPU, more decision-affecting information can be sent to the position controller, that is, more than simply the raw information from a sensor, such as a resolver. Stored software, some of which can be gained by experience with a given system during design and development (thus qualifying as "expert" information) can assist in computing error, monitoring loop gain, tailoring bandwidth, guarding stability, and supervising relations with other loops. This approach breaks with tradition where error signals have been fed as unconditioned inputs to the controller. Within the last few years, servo designers have found that by continuously adjusting an error signal, as the result of intelligence provided by experience, a number of errors can be removed, nullified, or moderated through the use of a feedback subloop. In essence, the position signal can be biased to remove a static friction error, or it may be integrated to produce, with the motor integrative effect, a zero velocity error servo[2].

(d) *Dual-mode Servos*. Serious consideration is being given to dual-mode servos; that is,

[1] Bailey (August 1985) reference listed.
[2] Bailey (February 1984) reference listed.

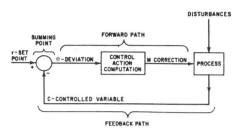

A feedback loop (closed loop) is a signal path which includes a forward path, a feedback path, and a summing point, and which forms a closed circuit. Because of the recirculation or feedback, the control action computation, which determines the corrective action required to maintain the measured variable at the setpoint, is critical to avoid cycling and instability. Where the feedback is positive, i.e., in phase with the deviation, the error is magnified and the process cannot be controlled. Only when negative feedback is present can the deviation be reduced. A reversal of phase must occur at the summing point in order that the effect of e (the control input) is opposite to that of c (the controlled variable). Feedback controllers produce a phase shift of 180° to give a negative feedback. In a comparatively new approach, advantage is being taken of conditioning the feedback path with information that goes well beyond simply producing a 'raw' error signal, as has been the traditional practice.

the load is driven to a prescribed error value in the feedback mode—then microstepped to zero error. A principal design criterion is to suspend internal commutation and feedback at a prescribed error value and then switch to phase-current ratio commutation. The error value at which microstepping starts can be algorized from load dynamics or loop stability, thus preventing overshoot as the last few microsteps are taken.

3. *By sensors*

A survey of the fundamental physical principles used for sensing position and its related variables could lead one to conclude that nearly all reasonably logical methodologies have been considered, are being studied, or are in use. This article and the subsequent two articles on geometric variables include electromechanical, electrooptical, electrohydraulic, electropneumatic, electromagnetic, purely electronic, purely optical, and purely mechanical approaches, among others. An exhaustive evaluation of what methodology may be best for a given application is well beyond the scope of this *Handbook*. Some comparisons of the more extensively used sensor approaches, such as resolvers/synchros versus encoders, are included. Impacting factors on sensor selection include simplicity, maintainability, resolution, cost, size, useful life and reliability, compatibility with digital information systems, commercial availability, and ruggedness where dictated by difficult installation environmental conditions, among other factors. The rather wide array of available sensors contributes to the difficulty of making procurement decisions.

ROTARY (ANGULAR) POSITION SENSORS

Traditionally, *resolvers* and *synchros* have been used in machine position control systems. These electric motorlike devices generally are rugged, reliable performers with sufficient resolution for most machine applications. Resolvers and synchros are rotary air transformers, the primary of which is a winding on the shaft (rotor) much like a motor. The secondaries (stators) are wound in the case, again much like a motor. Although there is a distinction, which will be described shortly, the terms *resolver* and *synchro* are sometimes loosely used interchangeably.

Rotary encoders also are widely used in connection with rotating motion systems. Through a coded rotating disk (metal with slits; glass with opaque lines) and an optical pickup system, encoders count pulses and thus translate rotary position (such as a shaft) into a digital address. Encoders are described in some detail shortly.

Resolvers and Synchros

These devices are used in factory applications that require small size, long-term reliability, and accuracy. Physically, typical units resemble small cylinders with diameters ranging from about $\frac{1}{2}$ to 4 inches (13 to 102 mm). The units look very much like small motors, with one end having terminal wires and the other end being a mounting flange with shaft extension. (See Fig. 1.)

Because resolvers and synchros produce an analog output, converters are required to provide a digital format for interfacing with digital numerical control and other systems. At one time these converters were comparatively large and expensive. Technological advances in electronics have solved these problems in recent years so that these devices, when equipped with synchro-to-digital (S/D) converters, are a practical approach to overall digital positioning systems and, for many applications, are preferred over encoders.

The difference between a resolver and a synchro is the number of stator windings. A resolver has two stator windings 90° apart, whereas a synchro has three windings 120° apart. In both devices, as the shaft turns, the relative positions of the rotor and stator windings change, and the root mean square (rms) voltage output of the stator winding varies as the sine of the angle between them. Only the ratio of the outputs is used. Although a synchro is a three-wire motorlike device, it is not a three-phase device. The AC outputs are either in phase or 180° out of phase. Phase shift does not change with angle except to reverse the phase at certain angles. (See Figs. 2 and 3.)

An AC reference signal is applied on one winding (primary). Note that either the rotor or the stator may be considered as the primary winding. This AC reference voltage, which can have a frequency of from 250 to 16,000 Hz, is then coupled to the secondary winding via

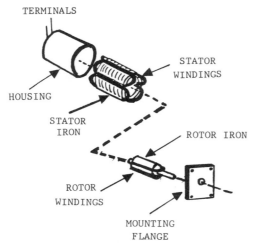

Fig. 1. Overall schematic diagram of a resolver or synchro.

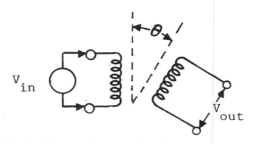

Fig. 2. The resolver/synchro is a rotating transformer. When one winding is excited with a signal, through transformer action, the second winding is excited. As the rotor is moved, the output signal changes and is directly proportional to the angle θ. Thus, resolver/synchro units are used to provide angular position feedback information.

Position, Motion, Speed, Velocity

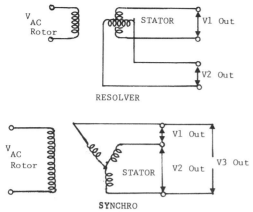

Fig. 3. The simplest resolver/synchro units contain a single winding on the rotor. More complex units employ two windings on the rotors (located 90° apart). The rotor is free to rotate inside the fixed stator assembly. The stator is constructed differently for a resolver than for a synchro. Resolvers contain a stator with two windings located 90° apart, whereas synchros contain three windings, electrically connected in a Y configuration.

transformer action. The amplitude of the signal picked up by the secondary winding depends on the angle between the primary (rotor) and the secondary (stator). A simplified unit would have an output signal, as shown in Fig. 4. Note that one electrical cycle of output voltage is produced for each 360° of mechanical rotation. This signal is then fed to additional instrumentation in the motion control system. Connection may be by brushes, slip rings, or another transformer-type winding. Brushes can introduce "bounce," which can be interpreted as an error signal by a digital controller. Brushes and slip rings can have contact problems at high speeds. In explosive atmospheres, contact wiping cannot be allowed. Consequently, a contactless approach is considered desirable in most situations.

Resolver Outputs. When a simple resolver—that is, one with only one primary winding and two secondary windings—is used and excited with an AC reference signal of E_i, then the outputs will be of the form:

$$E_{01} = KE_i \sin \theta \tag{1}$$
$$E_{02} = KE_i \cos \theta \tag{2}$$

where K = transformation ratio of the resolver (relationship between the primary and secondary windings).

θ = mechanical angle (position shaft is occupying).

Thus, a signal is provided that varies only as the angle θ varies. Between these voltage relationships, position and direction are derived via a resolver-to-digital converter, which transforms the signal into a digital representation of the angle θ. (See Fig. 5.)

A resolver-to-digital converter is shown in simplified form in Fig. 6. In operation, the outputs of the resolver windings E_{01} and E_{02} are fed into sine and cosine multipliers. Then the output is fed into a summing amplifier, which subtracts the two signals and provides a signal that is proportional to the angle θ. The phase detector demodulates this signal, using the resolver rotor voltage as a reference. This results in a DC signal that is proportional to the angle θ (an analog DC signal proportional to speed and direction). This DC signal drives a voltage-controlled oscillator which, in turn, drives an up/down counter to provide a digital output representing the angle θ. The signal from the up/down counter into the sine/cosine converters forces the output of the digital counter to be equal to the input angle θ (that is, if there are any differences between the input angle θ and the output digital code, this approach self-corrects).

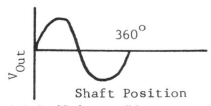

Fig. 4. A simplified rotor will have an output signal as shown here. Note that one electrical cycle of output voltage is produced for each 360° of mechanical rotation.

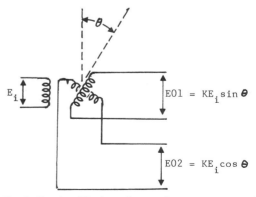

Fig. 5. If a simplified resolver (only one primary and two secondary windings) is excited with an AC reference signal, E_i, then the outputs will be as shown on the righthand side of this diagram.

24 Position, Motion, Speed, Velocity

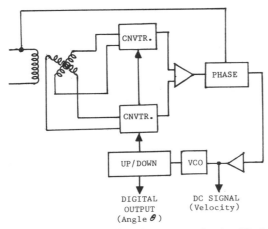

Fig. 6. A resolver-to-digital converter in simplified form. VCO = Voltage-controlled oscillator.

The advantages of utilizing this approach are: (1) With constant input, the output digital word continuously follows the input, with no phase lag between the digital word and actual shaft position (angle). (2) Voltage drops do not affect performance inasmuch as the operation depends only on the ratio between the two input signals. (3) The double integration makes this approach highly immune to noise. Any inductive noise spikes have equal plus-and-minus peaks that integration zeros out. (4) The analog DC voltage output is a direct indication of speed, or velocity, thus eliminating the need for a separate tachometer.

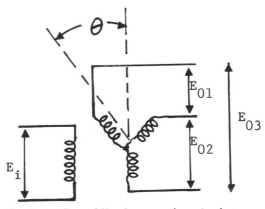

Fig. 7. Synchros differ from resolvers in that synchros have a three phase Y-connected stator. Exciting the synchro with an AC reference signal of E_i on the primary produces outputs of the forms given by equations 3, 4, and 5 in the text. A synchro-to-digital converter transforms the synchro output into a digital representation of the angle.

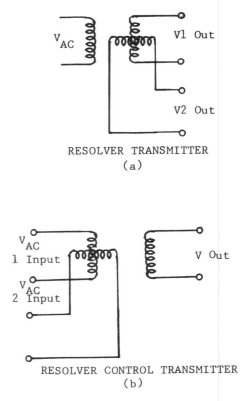

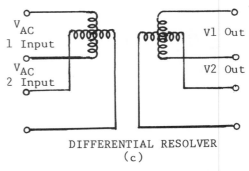

Fig. 8. Other types of resolvers: (a) Resolver transmitter, sometimes referred to by manufacturers as CX, RCX, or RX units. (b) resolver control transmitter, sometimes called CT, RC, or RCT units. (c) differential resolver, sometimes called RD or RC units.

Synchro Outputs. Synchros have a three-phase Y-connected stator. Exciting the synchro with an AC reference signal of E_1 on the primary will produce outputs of the form:

$$E_{01} = KE_i \sin \theta \quad (3)$$
$$E_{02} = KE_i \sin (\theta + 120°) \quad (4)$$
$$E_{03} = KE_i \sin (\theta + 240°) \quad (5)$$

where K = transformation ratio.
θ = mechanical angle (shaft position).

A synchro-to-digital converter will transform the synchro output into a digital representation of the angle. (See Fig. 7.)

Other Types of Resolvers/Synchros. There are other design formats for resolvers and synchros. The examples given in Fig. 8 feature resolvers, but it should be noted that the same principles will apply to synchros, keeping in mind that a synchro is a three-phase system.

The basic type of resolver described thus far also may be called a *resolver transmitter*—one-phase input and two-phase output (i.e., a single winding of the rotor is excited and the stator's two windings provide position information. (See Fig. 8[a].)

Another type of resolver is called a *resolver control transmitter*—two-phase inputs and one-phase output (i.e., the two stator windings are excited and the rotor winding provides position information). (See Fig. 8[b].)

A third type of resolver is sometimes called a *differential resolver*—two-phase inputs and two-phase outputs (i.e., two rotor windings are excited, and position information is derived from the two stator windings). (See Fig. 8[c].) It should be noted that some manufacturers call this simply a *resolver transmitter*.

Resolver/Synchro-to-Digital Converters. Usually referred to as S/D converters, there are two basic types: (1) *Tracking converters*, wherein the output tracks the input in real time. These are solid-state servo loops with feedback. Tracking converters have good noise rejection characteristics. Pickup on the synchro leads can be minimized by using twisted shielded cable. (2) *Sampling* converters that sample and hold the signals, generally on the carrier peaks, and then perform the actual conversion. The successive approximation method is usually faster and lends itself well to multiplexed systems. The latter often have a cost advantage. Sampling converters are more susceptible to noise than tracking units. Where long runs in noisy environments are required, careful shielding must be used.

Tracking synchro designs can track at rates typically up to 10% of the reference carrier frequency, or 50 rps (3000 rpm) in a 14-bit, 400 Hz system. Tracking converters are relatively insensitive to voltage and frequency variations and have excellent rejection of the effects of quadrature, harmonics, and noise. A major limitation is their relatively slow (hundreds of milliseconds) response to step changes. This makes them unsuitable for high-speed multiplexing.

Sampling converters perform the entire conversion process in very little time—typically 150 microseconds per channel. This high speed permits many conversions to occur in one carrier cycle, very important when converters are used in multiplexed systems. Sampling converters have at least two shortcomings: (1) *Staleness error*. This arises from the fact that data are sampled only once per carrier cycle. Where data are ran-

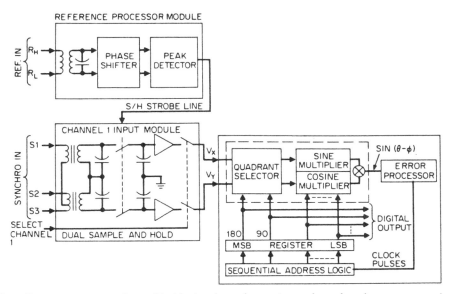

Fig. 9. Sampling converter samples and holds signals on the carrier peaks and performs a conversion before the next peak.

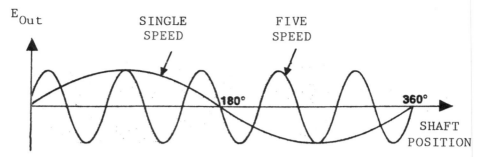

Fig. 10. Comparison of single-speed and five-speed multispeed brushless rotary transformers. Five-speed notation means that the resolver produces five complete electrical cycles of output signals for each 360° of mechanical shaft rotation. The multispeed feature enables system designers to eliminate or reduce various mechanical gears in the sensing system, resulting in improved system reliability and accuracy.

domly accessed and angle θ is changing, an error obviously will occur. This will depend on the time difference between when the data were sampled and when the data were accessed. (2) *Noise vulnerability* as previously mentioned. (See Fig. 9.)

Multispeed systems. In this type of system, two or more synchros are geared together (mechanically or electrically), usually with a gear ratio of a whole number. Common two-speed systems have ratios of 1:8, 1:10, 1:18, 1:32, and 1:36. Disregarding possible errors caused by backlash and gear nonuniformity, the accuracy of a two-speed synchro system will equal the synchro/resolver's accuracy divided by the speed ratio.

A two-speed tracking converter behaves like the single-speed version. Both are closed-loop servos, and both are free of velocity error. They present very recent "fresh" information. In systems where an even gear ratio is used, "false nulls" may occur. To enable even-ratio systems to function without the possibility of nulling at 180°, an angle offset and "stickoff voltage" may be introduced in the coarse channel. Two-speed systems also can be implemented by using two independent single-speed tracking converters and combining their digital outputs with either hard-wired discrete logic or by using a microprocessor. (See Fig. 10.) The schematic of a multispeed resolver transmitter is shown in Fig. 11.

Pancake Synchros and Resolvers. These components normally have large diameter and short length as compared with other servo components. The pancake design allows for greater diversity of mounting configurations and electrical parameters. Representative units are diagrammed in Fig. 12.

Overall Parameters. The overall physical parameters of a sampling of resolvers/synchros are summarized in Table 1.

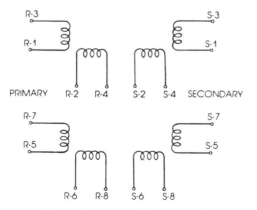

Fig. 11. Schematic of a multispeed resolver transmitter.

TABLE 1. General Range of Physical Parameters of Resolvers/Synchros

Speed	1X, 2X, 5X, 10X . . 32X
Frequency (kHz)	0.4 54
Input voltage (V)	10 26
Output voltage (V)	1.6 26
Input current (mA) (max)	1.8 53
Phase shift (degrees) $-$ = lag; $+$ = lead	-10.5 . . . . . . $+17.6$
Sensitivity (mV/degree)	17 . . . . 894 (multipole)
Friction (25°C; 77°F) (gm cm)	50
Friction (-25°C; -13°F) (gm cm)	100
Diameter (in/mm)	2/51 . . . . 3.5/89
Weight (oz/gm)	2.5/71 . . . . . 51/1443

Position, Motion, Speed, Velocity 27

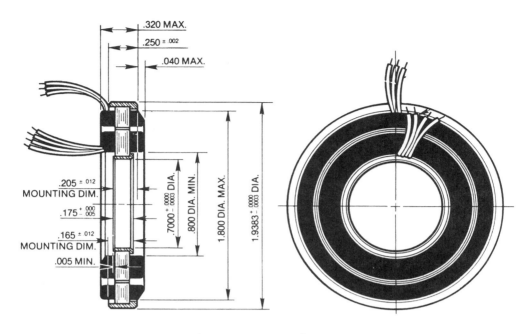

(Typical Size 20)

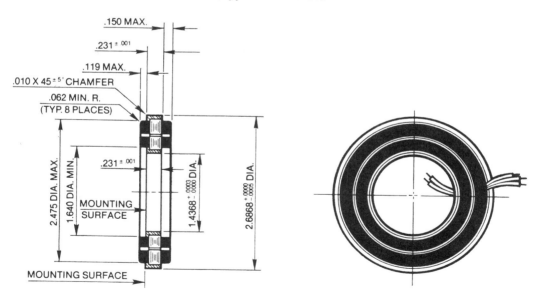

(Typical Size 27)

Fig. 12. Representative pancake type resolvers/synchros. (*Bowmar/Harowe, Harowe Servo Controls, Inc.*) (Dimensions and specifications subject to change.)

Encoders

An encoder is a mechanical-to-electrical conversion device that provides position, direction, speed, and displacement information. The rotary encoder, frequently used in automated systems, satisfies the IEEE definition of *encode*—"to produce a unique combination of a group of output signals in response to each group of input signals." Sometimes the comparison of an encoder with the familiar micrometer caliper (Fig. 13) is made. With the caliper, the micrometer screw is turned to accurately measure the dimension of a piece held between the jaws of the device. The micrometer barrel is divided so that measurements to an accuracy of one-thousandth of an inch (0.025 mm) or better can be made. Each revolution of the barrel advances the micrometer spindle 25 thousandths of an inch, thus requiring 40 complete revolutions of the micrometer screw to advance the micrometer spindle one lineal inch. Of course, the larger the barrel, the more divisions per turn to read, and consequently the greater the accuracy. Instead of manually reading and interpolating a scale, the encoder translates the simple analog rotation into discrete electrical signals that are directly related to shaft position and hence the distance traveled. Shaft encoders are of two basic types—*absolute* and *incremental*.

Absolute Encoder. An absolute encoder provides a unique output signal for each single or multiple revolution of shaft gearing. The device outputs a complete binary code (digital output) for each position. Absolute encoders are generally used in applications where position information rather than change in position is important. These devices have an individual digital address for each incremental move, and thus the position within a single revolution can be determined without a starting reference. By gearing two or more absolute encoders together, so that the second advances one increment for each complete revolution of the first (reminiscent of a mechanical counter), the range of absolute position can be extended.

The disk assembly is manufactured with a coded track pattern to provide a digital signal output (0-1 or on-off). The absolute shaft encoder uses either (a) contact (brush), or (b) non-

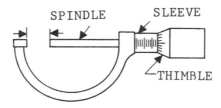

Fig. 13. An encoder is analogous in operation to a micrometer caliper.

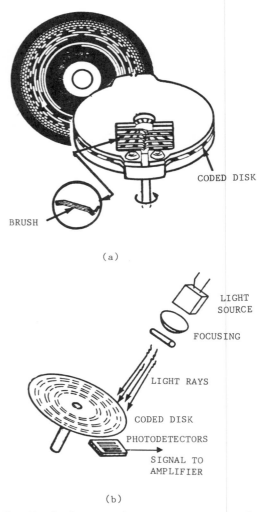

Fig. 14. Absolute encoder: (a) contact or brush-type, and (b) noncontact, photoelectric type.

contact schemes of sensing position. The contact type is shown in Fig. 14 (a). The device incorporates a brush assembly to make direct electrical contact with the electrically conductive paths of the coded disk for reading address information. The noncontact-type, shown in Fig. 14(b), uses optical means (commonly photoelectric) to sense position from the coded disk. In this case the disk consists of opaque and transparent (to light) segments. These segments are laid down in the same pattern as the electrically conductive paths used by the brush-type encoder.

The principle of operation is demonstrated by Fig. 15. The number of tracks may be increased as well as the segments around the disk until the

Position, Motion, Speed, Velocity 29

TABLE 2. Relative Advantages/Disadvantages of Light Sources for Incremental Encoders

Factor	LED	Lamp
Signal intensity	50% lower than lamp	High
Life	40,000 + hours	40,000 hours
Output variations with voltage	Moderate	High
Output variations with temperature	High	Nonexistent
Temperature cycle damage	Low	Moderate
Shock damage	Low	Moderate

number of graduations equals the desired resolution. Since position information is directly on the coded disk assembly, the disk has a built-in "memory system," and a system power failure will not cause this information to be lost. Thus, it is not necessary to return to a "home" or "start" position after reenergizing power.

Incremental Encoder. An incremental encoder produces a symmetrical pulse for each incremental change in position. Pulses from the incremental encoder are counted for each incremental movement from a calibrated starting point in an up/down counter to track position. The operating principle of an incremental encoder (also sometimes called *optical encoder* or *digital tachometer*) is shown and described by Fig. 16. The characteristics of the two principal light sources are summarized in Table 2. Another area that is application-dependent is the disk assembly. Depending upon the resolution and accuracy, the material used may limit the encoder for some applications. The disk can be made by using slits in metal or lines on glass. The metal disk is normally a low-resolution device. Glass provides higher resolution and accuracy but must be handled with care to avoid breakage.

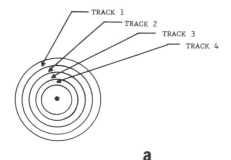

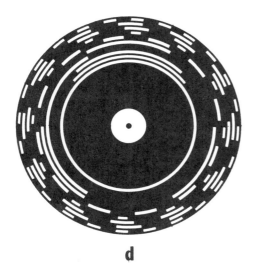

Fig. 15. Operating principle of absolute encoder: (a) Several concentric tracks (only four shown in oversimplified diagram) are present on the disk. (b) Portions of each track are either opaque or transparent to light emanating from a source. The pattern of opaque/transparent segments is designed so that for every degree of rotation (360°) of the disk, a unique coded address will be presented. The detail required is not shown in diagrams (b) and (c). When the disk is in position "A" as shown in (b) the segments on tracks 1 and 2 are transparent, thus each yielding an output of 0; the segment on track 3 is opaque, thus yielding an output of 1; the segment on track 4 is transparent, yielding an output of 0. Thus, the complete address is 0010. In (c), the disk has rotated so that tracks 1 and 2 are opaque and tracks 3 and 4 are transparent, providing the address of 1100. A reasonable facsimile of a full disk with ten tracks is shown in (d).

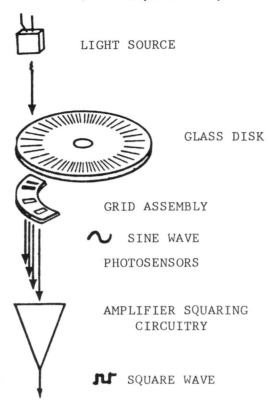

Fig. 16. Incremental encoder. This device provides an output pulse signal as the disk assembly rotates, thus total information is obtained by counting pulses. The disk is manufactured with opaque lines which are aligned with a grid assembly. A light source and photosensors complete the assembly. Light from a light emitting diode (LED) or tungsten filament lamp passes through the transparent segments of the disk and is sensed by photosensors. As the disk rotates, an alternating light/dark pattern is produced. The output of the photosensors is a sinusoidal wave which can be amplified in some situations. Electronic processing transforms this signal into a square-wave pulse for digital circuitry compatibility.

An incremental encoder can be either *unidirectional* or *bidirectional*. A unidirectional encoder yields information about speed or amount of displacement. A bidirectional encoder provides this same information as well as *direction* information, i.e., clockwise or counterclockwise rotation. Direction information is obtained by monitoring two signals electrically separated by 90°. As shown in Fig. 17, phase relationship between these two signals is utilized to determine rotation direction. Incremental encoders may have several tracks. As illustrated, a second track can be used for a "zero" index or "home" reference pulse.

Specifying parameters applicable to incremental encoders include: (1) *Line count*, which is the number of pulses per revolution. The number of lines is determined by the positional accuracy needed for a given application. Standard line counts commercially available range from 100 to 1000 pulses/revolution. In some specially designed "self-contained" encoders, higher line counts are obtainable. (2) *Output signal* can be either sine or square-wave. (3) *Number of channels*. Either one or two channel outputs can be provided. The two-channel version provides a signal relationship to obtain motion direction. In addition, a zero index pulse can be provided.

A typical servo application using digital feedback is shown by the block diagram of Fig. 18. The input command signal loads an up/down counter. The number of pulses in the counter represents the position the load must be moved to. As the motor accelerates (Fig. 19), the pulses emitted from the encoder continue at a faster rate until motor run speed is obtained. During the run, the pulses are emitted at a constant frequency directly related to motor speed. The

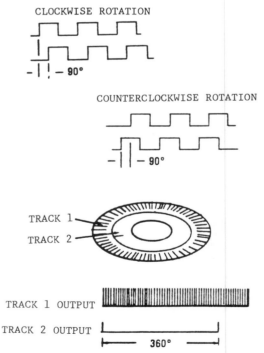

Fig. 17. Top of view shows phase relationship between two signals for determining rotation direction. Incremental encoders may have several tracks. As shown here, a second track can be used for a 'zero' index or 'home' reference pulse.

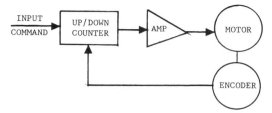

Fig. 18. Block diagram of servo system using digital feedback to an up/down counter.

counter counts down to "zero," and at a determined position, the motor is commanded to slow down. This is to prevent overshooting of the desired position. When the counter is within one or two pulses of the desired position, the motor is commanded to decelerate and stop. The load should now be in position.

Representative Problem. An incremental encoder is required on a milling machine to provide a digital readout display. The display must read directly in thousandths of an inch. The total travel of the milling machine bed is 36 inches. The travel is regulated by a precision lead screw, which moves the milling machine bed $\frac{1}{10}$th inch for every revolution (360°) of the lead screw. Since the display must read directly in $\frac{1}{1000}$ inch increments, the encoder must provide 100 pulses per revolution where each pulse represents .001 inch.

Solution. An encoder disk is connected to the shaft of the motor and the shaft is rotated. A pulse train is generated by photoelectric means, as previously described. These pulses are fed directly into an appropriate electric counter with digital display. Starting from a known reference position, the operator resets the counter to zero. The operator moves the milling machine bed from the zero position until the number 19.031 is shown on the counter. The operator is now exactly 19.031 inches from the zero position.

In some systems the number 19.031 is entered on the counter's preset function. When the counter counts 19,031 pulses, it stops the travel automatically. At this position, a hole is bored to a specific depth. An encoder on the "z" axis of the machine controls the drilling to a specified depth. Add to this an encoder for bed travel on the other axis, plus tape control for the preset functions and sequences, and automated numerical control is the result.

Encoder Interfacing. The square-wave output is derived, as previously mentioned, from electronic processing, or shaping circuitry within the encoder package. The output signal level is nominally 5V and zero (logic "1" is 2.4 V minimum and logic "0" is 0.4 V maximum). Signal distortion may be a result of cable capacitance (length)—the longer the cable, the more distortion. Beyond 30 feet (9 meters) in length, the signal must be reshaped if reliability is not to suffer. Good shielding must be used to keep noise to a minimum.

The sine wave output normally will be used where the designer performs the signal shaping somewhere else in the system, i.e., other than in the encoder package. Signal levels are typically 50 to 100 mV peak to peak into a 2K ohm load at 40 kHz. A disadvantage is susceptibility to electrical noise because the signal is at such a low level. Signal cables must be isolated from other AC lines and noise generators. Twisted, shielded wires should be used. Signal reshaping usually is not required for distortion, since the receiver is a signal shaper. Distortion is not significant where cable lengths do not exceed 30 feet (9 meters).

In addition to radiated noise, the encoder may be affected by transients caused by its power supply voltage. The regulation is typically specified at ±5% variation, without noise spikes. Spikes, of course, may damage the light source and encoder electronics.

Other Encoder Problem Areas. Of prime importance is the tolerance on the hub inner diameter onto which the glass disk is mounted, the motor shaft outer diameter, and the motor shaft's total indicated runout (TIR). If the hub is oversized and/or the shaft undersized, the unit will exihibit eccentricities when mounted and thus yield a moving center rather than a fixed center. Eccentric mounting causes a frequency-modulated signal to be superimposed on top of the encoder output signal. If the sensor reading the line count traces a path on the disk, as shown in Fig. 20, the resulting output signal will be frequency modulated. If the motor shaft has a

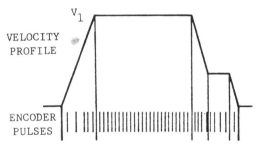

Fig. 19. As the motor accelerates, the pulses emitted from the incremental encoder (digital tachometer) continue at a faster rate until motor run speed is attained. During run, the pulses are emitted at a constant frequency directly related to motor speed. The counter counts down to 'zero' and, at a determined position, the motor is commanded to slow down—to prevent overshooting the desired position. When motor is within two pulses of desired position, the motor is commanded to decelerate and stop.

32 Position, Motion, Speed, Velocity

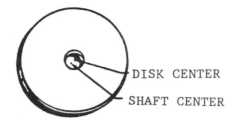

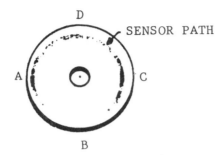

(Shaft Eccentricities)

(Frequency-modulated Output)

Fig. 20. Shaft eccentricities. If the hub is oversized and/or the shaft is undersized, the encoder will exhibit eccentricities when mounted, thus yielding a moving rather than a fixed center. Eccentric mounting causes a frequency-modulated signal to be superimposed on the encoder output signal. If the sensor reading the line count traces a path on the disk, as indicated here, the resulting output signal will be frequency modulated.

larger TIR, the disk assembly will "wobble," as shown in Fig. 21. This will produce an amplitude-modulated signal due to the varying illumination of the sensor. It can be corrected by (1) mounting the hub onto the motor shaft and machining, or (2) by utilizing dual optical pickoffs.

State-of-the-Art Encoder Hardware. Encoder kits or integrated optical encoders are available from numerous suppliers. A diagram and summary of parameters for a 28 mm diameter incremental optical encoder (kit form) is given in Table 3. The kit consists of the encoder body, a metal code wheel, and an emitter end plate. An

DISK WOBBLE

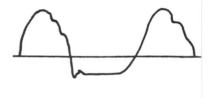

AMPLITUDE MODULATED OUTPUT

Fig. 21. Disk wobble. If the motor shaft has a large TIR (total indicated runout), the disk assembly will wobble. This will produce an amplitude-modulated signal due to varying illumination received by the sensors. This condition can be corrected either by first mounting the hub onto the motor shaft and machining, or by using dual optical pickoffs.

LED source and lens transmit collimated light from the emitter module through a precision metal code wheel and phase plate into a bifurcated detector lens. The light is focused onto pairs of closely spaced integrated detectors, which output two square-wave signals in quadrature and an optional index pulse.

New incremental optical encoders[3] take a modular form, which facilitates attaching to a precision servomotor shaft or embedding into an OEM motion device. A patented prealignment scheme eliminates an output phasing step during installation. Sensing of an LED source is done by a monolithic phototransistor array in radial configuration, a geometry that reduces effects of armature or shaft runout. The array chip lends uniformity from cell to cell, minimizing sensitivity to temporal and thermal drift. Frequency response is quoted at 100 kHz.

Low-cost modular optical encoders[4] are available in 1½- and 2-inch sizes and are designed for easy shaft installation. Typical OEM uses include capstan tape drives, plotter pens, and robot axis. A high-resolution, interpolation option enabling quadrature resolutions up to 12,700 cycles/turn, with pulse trains up to 50,000 counts, was recently introduced[5]. The manufacturer cites accuracy at

---
[3] PMI Motors, Syosset, New York.
[4] Honeywell Inc., Costa Mesa, California.
[5] BEI Electronics, Goleta, California.

TABLE 3. Modular Incremental Shaft Encoder/Motor Assemblies Available from Stock

DESIGN AND OPERATION

A shaft encoder is a component which translates the rotational movement of a shaft into an electrical waveform.

A shaft encoder used in a system such as a servo motor control enables the use of digital components in the loop, i.e., a microprocessor instead of servo amplifier, thus lowering the total system cost. A typical digital control loop is shown in Figure A.

The optical shaft encoder offers several advantages over other encoder types. It is noncontacting, thus it does not burden the system with added inertia and friction, and is inherently more reliable. The encoding speed is high and it offers high noise immunity.

The D32H85 . . . series is a family of modular incremental shaft encoders. Two similar channels whose outputs are in quadrature (90° phase difference) provide velocity and direction information. The output waveform is digital and is compatible with LSTTL logic.

The modular encoder kit is assembled from three parts:

1. The Encoder Body, which contains the phase plate, detectors and integrated circuits.
2. The Code Wheel, which is mounted on the system's shaft.
3. The Emitter End Plate, containing the light source (LED), which snaps onto the body to form a dust resistant unit.

These optical encoder kits are available in two sizes: 28 mm and 56 mm diameter.

The incremental shaft encoder operates by translating the rotation of a shaft into interruptions of a light beam that are then output as electrical pulses.

The light source is a Light Emitting Diode collimated by a molded lens into a parallel beam of light. The Emitter End Plate contains two or three similar light sources, one for each channel.

The standard Code Wheel is a metal disc that has N equally spaced apertures around its circumference. A matching pattern of apertures is positioned on the stationary phase plate. The light beam is transmitted only when the apertures in the code wheel and the apertures in the phase plate line up; therefore, during a complete shaft revolution, there will be N alternating light and dark periods. A molded lens beneath the phase plate aperture collects the modulated light into a silicon detector.

The Encoder Body contains the phase plate and the detection elements for two or three channels. Each channel consists of an integrated circuit with two photodiodes and amplifiers, a comparator, and output circuitry.

The apertures for the two photodiodes are positioned so that a light period on one detector corresponds to a dark period on the other ("push-pull"). The photodiode signals are amplified and fed to the comparator whose output changes state when the difference of the two photocurrents changes sign. The second channel has a similar configuration but the location of its aperture pair provides an output which is in quadrature to the first channel (phase difference of 90°). Direction of rotation is determined by observing which of the channels is the leading waveform. The outputs are TTL logic level signals.

The optional index channel is similar in optical and electrical configuration to the A and B channels previously described. An index pulse of typically 1 cycle width is generated for each rotation of the code wheel. Using the recommended logic interface, a unique logic state (P_0) can be identified if such accuracy is required.

The three-part kit is assembled by attaching the Encoder Body to the mounting surface using three screws. The Code Wheel is set to the correct gap and secured to the shaft. Snapping the cover (Emitter End Plate) on the body completes the assembly. The only adjustment necessary is the encoder centering relative to the shaft. This optimizes quadrature and the optional index pulse outputs.

Index Pulse Considerations

The motion sensing application and encoder interface circuitry will determine the necessary phase relationship of the index pulse to the main data tracks. A unique shaft position can be identified by using the index pulse output only or by logically relating the index pulse to the A and B data channels. On the three-channel encoder there is some adjustment of the index pulse position with respect to the main data channels. The position is easily adjusted during the assembly process as illustrated in the assembly procedures.

Offered by Stock Drive Products, Division of Designatronics, Inc., New Hyde Park, New York. Specifications subject to change.

TABLE 3. Continued

DEFINITIONS

Electrical degrees:
1 shaft rotation = 360 angular degrees
              = N electrical cycles
1 cycle   = 360 electrical degrees

Position Error:

The angular difference between the actual shaft position and its position as calculated by counting the encoder's cycles.

Cycle Error:

An indication of cycle uniformity. The difference between an observed shaft angle which gives rise to one electrical cycle, and the nominal angular increment of 1/N of a revolution.

Phase:

The angle between the center of Pulse A and the center of Pulse B.

Index Phase:

For counter clockwise rotation as illustrated above, the Index Phase is defined as:

$$\Phi_1 = \frac{(\phi_1 - \phi_2)}{2}$$

ϕ_1 is the angle, in electrical degrees between the falling edge of I and falling edge of B. ϕ_2 is the angle, in electrical degrees, between the rising edge of A and the rising edge of I.

Index Phase Error:

The Index Phase Error ($\Delta\Phi_1$) describes the change in the Index Pulse position after assembly with respect to the A and B channels over the recommended operating conditions.

Block Diagram

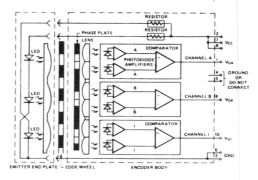

Motor Shaft Diameter	Cycles per Revolution
2 mm	100
3 mm	192
1/8 in	200
5/32 in	256
3/16 in	360
1/4 in	400
4 mm	500
5 mm	512

Output Waveforms

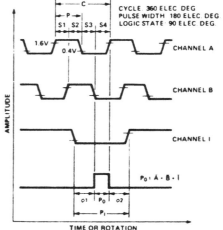

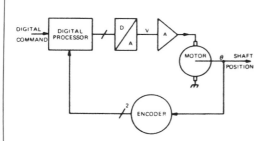

Fig. A. Digital Motor Control Block Diagram

±1 count/turn and repeatability with $\frac{1}{4}$ count. An index pulse (gated zero marker) of $\frac{1}{2}$ count may also be obtained. It is reported that the user may select from among over 127 interpolated resolutions.

A series of explosion-proof optical shaft encoders was recently announced[6]. These devices are Factory Mutual approved for Class 1, Div. 1, 2, Grps C, D; and Class 2, Div 1, Grps E, F. G—indoor/outdoor (NEMA for hazardous locations). This "EX" series is approved for petrochemical and paper manufacturing and offshore oil rigs.

Known as the "Pulsar" this is a low-cost optical encoder for measuring incremental shaft position or synchronizing shafts in variable-speed machinery[7]. Depending upon the particular model, the output may be from four to 12 square-wave pulses/revolution, powered by from 4.5 to 15V DC. The units have been designed to minimize effects of magnetic interference, speed variation, ferrous dust, and minor misalignment.

Improved isolation between output and internal circuitry, effected by a VMOS buffer stage that yields high noise immunity, has recently been offered[8]. Current sinking capacity of the VMOS output is increased by a factor of 3 from 16mA to over 48mA, thus allowing longer cables and heavier capacitive or resistive loads. A differential solar array that tracks evenly under variations of temperature, voltage, and light source ensures additional noise immunity.

An automotive gear plant has presented the contradiction of fixturing to 0.01 inch accuracy, yet machining to ±0.0001 inch with sigma six repeatability. This has been accomplished by modifying the machine tool controller[9] to reference workpiece surface rather than to absolute machine tool position. As pointed out by CE[10], when an adaptive depth option is elected, additional hardware is provided by the machine tool controller to monitor linear encoder input, which supplies a workpiece reference to enable surface detection by the tool. When detection occurs, encoder input to the controller is latched and a status bit is set. Software then scans the status bit during execution of all program steps that this option enabled. Indication by the status bit that workpiece surface has been detected during any program step causes a "GO TO" user-programmable step number command to be executed. The user program includes a remotely settable tool wear offset to be factored in.

In those infrequent situations where a drive motor may stall (because of rigorous environmental conditions as encountered, for example, in glass fiber manufacture), one systems maker[11] programs the indexer to recognize the absence of encoder pulses as an indication of a stalled motor. This action can be tied into a fault signal.

There are scores of suppliers of position control systems[12], many of whom furnish encoders. The brief descriptions given in the aforementioned several paragraphs have been selected to be representative but within the limited space of this *Handbook*; obviously this is just a sampling.

Comparison and Selection of Resolvers/Synchros and Encoders

In comparing various feedback devices, typically a resolver will be compared with an optical encoder. Basic concerns in the selection of these devices include:

1. *Vibration and Shock*. Resolvers perform reliably under loads up to 50g, whereas optical encoders are particularly sensitive due to tight tolerances between the disk and light sources.

2. *Contaminants*. Brushless resolvers are usually immune to contaminants and environment (humidity, oil, mist, frost, sand, dust, solvents, fungus, salt atmospheres, et al.) since there are no contacting electrical surfaces. With optical encoders, contamination will interrupt the light path and create unreliable data.

3. *Temperature*. Inasmuch as resolvers are transformers, operating temperature is limited by insulation and bearing lubrication. For an optical encoder, temperature causes expansion and contraction of the optical disk, aggravating mechanical alignments and electronics in the same location.

4. *Accuracy*. A resolver is an "infinite" resolution device with realizable resolution depending on resolver-to-digital converters. For higher accuracy, multispeed resolvers may be specified. Resolution of the encoder is determined by the number of lines on the disk. A high-resolution encoder can be quite large—with a diameter of 9 to 12 inches (230 to 305 mm).

5. *Mechanical Package*. The resolver's converter (electronics) is separate and requires a reference source. The encoder's electronics are contained in the same housing as the disk.

Usually when a position sensor must be mounted in an application subject to high vibration and/or contamination, a resolver may be the better choice.

Tachometers

In addition to the resolvers, synchros, and encoders just described, tachometers are used to provide feedback information in motion systems. Tachometers resemble a miniature motor, but the similarity stops there. In a motor a voltage is applied onto the terminals, and the interaction of

---

[6]Dresser Datametrics, Wilmington, Massachusetts.
[7]Servo-Tek Products, Hawthorne, New Jersey.
[8]Data Technology, Woburn, Massachusetts.
[9]Gould Motion Control Division, Racine, Wisconsin (C231 controller).
[10]*Control Engineering*.
[11]Compumotor, Petaluma, California.
[12]Reference to the "Control Products Specifier," published annually by *Control Engineering* features lists and addresses of such suppliers.

two magnetic fields results in shaft rotation. Large amounts of current must be passed through the windings in order to produce large amounts of torque. In a tachometer the gage of wire used is quite fine. Thus, the current-handling capability is small. In a tachometer the shaft is turned by some mechanical means, and a voltage is developed at the terminals—with the amplitude of the tachometer signal being directly proportional to speed.

In motion systems tachometers can function in a variety of ways: (1) They can be used as rate or speed indicators to provide data to a meter for visual speed readings, (2) they can be used to provide velocity feedback information, thus stabilizing a speed control system, or (3) they can be used as a feedback device to provide damping to a position servosystem, thus enhancing the overall stability of the system. Although it is possible to acquire rate feedback information in other ways, the analog tachometer provides the simplest, most direct method of doing this.

Representative Application. It is required to drive a load via a lead screw assembly at a constant rate of 12 inches (305 mm)/second in order to perform a laser trimming operation. To ensure proper cutting depth, accurate speed regulation must be maintained. With a pitch on the lead screw of 5 revolutions/inch, the motor must rotate at:

$$\frac{12 \text{ inches}}{\text{second}} \times \frac{5 \text{ revolutions}}{\text{inch}} \times \frac{60 \text{ seconds}}{\text{minute}} = 3600 \text{ RPM}$$

If a tachometer's output voltage gradient is 2.5 V/KRPM, the voltage read on the tachometer terminals should be:

$$3.600 \text{ KRPM} \times \frac{2.5 \text{ Volts}}{\text{KRPM}} = 9 \text{ V}$$

If the tachometer output voltage is indeed 9 V, then the tachometer and, therefore the motor, is rotating at the required 3600 RPM. With proper regulation and control from the servosystem, the voltage applied to the motor terminals will maintain the tachometer output voltage at 9 V, assuring the desired rotational speed. Although this example has been simplified, the basic concept of speed regulation via an analog device has been demonstrated.

Characteristics of Tachometers. The following terms are frequently used in describing specifications for tachometers.

Voltage Constant—this is also referred to as *voltage gradient*, or *sensitivity*. This represents the output voltage generated from a tachometer when operated at 1000 RPM, that is, V/KRPM. Sometimes converted, the parameter may be expressed in volts per radian per second, i.e., V/rad/sec.

Ripple—this is also called *voltage ripple* or *tachometer ripple*. Since tachometers are not an ideal device and inasmuch as manufacturing tolerances enter into the performance of the product, one should always be concerned with deviations from the norm. In an analog tachometer, when the shaft is rotated, a DC signal is produced, as well as an amount of an AC signal. This AC signal is superimposed on the DC level. It may be generated from winding imbalances, brush and/or commutator noise, among other factors. A low pass filter can be used to eliminate or reduce this noise component. In reviewing specific literature and specifications, the designer must be careful in interpreting the definition of ripple—because there are three methods of measuring and presenting relevant data: (1) *Peak-to-peak* ripple is expressed as a percent of the average DC level, (2) RMS (root mean square) is the ratio of the RMS of the AC component expressed as a percent of the average DC level, and (3) *peak-to-average ripple*—this is the ratio of maximum deviation from the average DC value, expressed as a percent of the average DC level. Along with this AC ripple, there is an associated ripple frequency. This frequency is determined by the construction of the tachometer, which includes the number of commutation bars, poles, and turns per slot.

Linearity—the ideal tachometer would have a perfect straight line for voltage versus speed. However, design and manufacturing tolerances enter in and alter the theoretically straight line. Linearity is a measure of how far away from perfect the actual situation is. Linearity is measured by driving the tachometer at various speeds, while simultaneously measuring both speed and output voltages. A plot is then made and compared with the theoretical curve. The maximum difference between the two curves is the linearity value expressed in percentage. At very high speeds (around 6000 rpm), aerodynamic lifting of brushes results in a drop of the output voltage versus speed characteristic. Along with linearity, there is an associated *reversibility error*, or direction error. This is an indication of what happens when the tachometer changes directions from clockwise to counterclockwise rotation. The reversibility error is the difference between CW and CCW voltage constants ($K_E(CW) = K_E(CCW)$), expressed as a percent of the nominal.

Maximum Operating Speed—this is the maximum recommended speed of the tachometer. Above this level, brush lifting may occur and will alter the output voltage. Maximum output voltage is also associated with this. If the maximum recommended operating speed is exceeded, excessive currents may flow from commutator bar to bar, resulting in unsatisfactory performance.

Minimum Load Resistance—this is the lowest recommended ohmic resistance that should be connected across the tachometer terminals. Lower values will result in higher currents flowing from the tachometer and consequent heating, excessive ripple, among other problems.

Position, Motion, Speed, Velocity 37

Fig. 22. Brush-type tachometers: (a) Above is shown a kit which includes rotors, magnets, and brush assembly inside a protective housing. The tachometer does not require any couplings or special adapters for mounting. Shown at the right is a fully self-contained unit. Shown in the middle is the tachometer mounted on the rear of a motor. Fig. 22 is continued on next page.

TABLE 4. Physical Parameters of Self-Contained and DC Tachometer Kit [illustrated in Fig. 22(a)]

Characteristics	Units	Self-contained DC tach configurations			
Voltage gradient (± 10%)	Volts/1000 RPM	7	20.8	50	100
Ripple (max.)	% Peak to Peak	2	2	2	2
Linearity	%	0.5	0.5	0.5	0.5
Temperature coefficient	%/°C	0.016	0.016	0.016	0.016
Max. armature temp.	°C	130	130	130	130
Armature inertia	oz-in-sec^2	.0020	.0025	0.0035	0.0057
	(gr-cm-sec^2)	(0.144)	(0.180)	(0.252)	(0.410)
Armature resistance	ohms	108	340	680	1360
Armature inductance	Millihenry	26	190	350	720
Max. operating speed	RPM	5000	5000	5000	2500

Characteristics	Units	Kit configurations	
Voltage gradient (± 10%)	Volts/1000 RPM	7	20.8
Ripple (max.)	% Peak to Peak	2	2
Linearity	%	0.5	0.5
Temperature coefficient	%/°C	0.016	0.016
Max. armature temp.	°C	130	130
Armature inertia	oz-in-sec^2	0.0015	0.0018
	(gr-cm-sec^2)	(0.108)	(0.133)
Armature resistance	ohms	108	340
Armature inductance	Millihenry	26	190
Max. operating speed	RPM	5000	5000

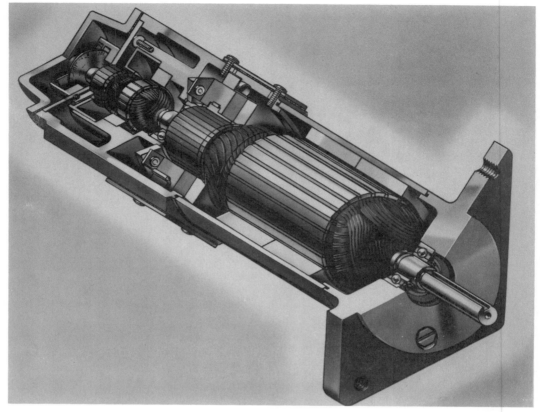

Fig. 22. (b) Cutaway view of a servomotor showing analog tachometer mounted in a protective cover. A digital encoder is mounted on the rear of the unit. This configuration provides velocity information and position data. (*Pacific Scientific*.)

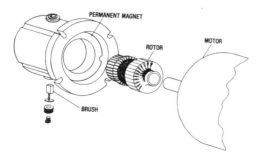

Fig. 23. A typical motor end-bell mounting surface used with the DC tachometer kit shown in Fig. 22(a).

Temperature Stability—as temperature of the tachometer changes, so will the output voltage. In a permanent-magnet tachometer, the major cause is the change of magnetic flux with temperature. Temperature stability is expressed as a change in output voltage per degree Celsius.

Brush-type tachometers are illustrated in Fig. 22. The DC tachometer kit shown in Fig. 22(a) consists of the rotor, magnets and brush assemblies inside a protective housing, and a compression ring. The tachometer does not require any couplings or special adapters for mounting. The compression ring is first inserted in the rotor's hollow shaft, and then the rotor is press fitted on the motor rear shaft extension. This procedure ensures that the rotor fits securely on the motor shaft. A typical motor end-bell mounting surface is shown in Fig. 23. The self-contained DC tachometer, also shown in Fig. 22(a), has its own shaft and bearings contained in a

Position, Motion, Speed, Velocity 39

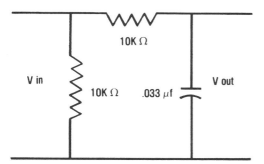

Fig. 24. Typical filter used to measure ripple voltage [in connection with DC tachometers shown in Fig. 22(a)].

housing that has been sealed with O rings between the shell and endcaps. It incorporates a machined mounting surface, built-in terminal board, and conduit connector, design factors that contribute to mounting flexibility. The high voltage gradients (available on some models) are well suited for lower-speed applications. The typical filter used to measure ripple voltage, used for both the self-contained unit and the kit, is shown in Fig. 24. Specifications for the self-contained and DC tachometer kit, shown in Fig. 22(a), are given in Table 4.

Self-Contained Brushless Tachometer. The DC tachometer shown in Fig. 25 is brushless. Internal electronics produce a DC voltage proportional to shaft angular velocity. Low ripple and absence of brushes make the unit attractive for velocity feedback in velocity control systems and for rate damping in position control systems.

Fig. 25. Self-contained brushless tachometer. Internal electronics produce a DC voltage proportional to shaft angular velocity. Low ripple and absence of brushes makes the tachometer well suited for velocity feedback in velocity control systems and for rate damping in position control systems. Modular design offers mounting flexibility in connection with end-cap of standard AC or DC motors. Unit has less than ±4% ripple (peak-to-peak with 1.2kHz), and 3db small signal bandwidth. Wide speed range—12,000 to as low as 1 RPM. Bidirectional operation—low reversibility error of less than ±.05%. Rare earth magnetics permit a low temperature coefficient (.02%/°C). Input voltage: ±12VDC (both + and − voltages required). Maximum output voltage: ±9VDC for 12-volt supply. Weight: 4 pounds (1.8 kg). Note optional digital encoder mounted on rear of unit. (*Pacific Scientific*)

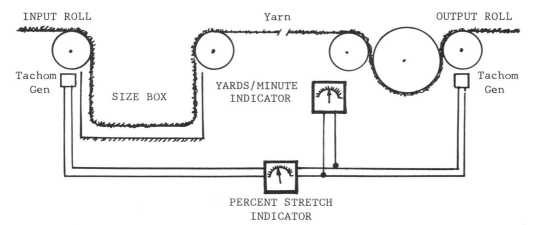

Fig. 26. Speed ratio system. Two tachometer generators used to measure speed differential (consequently *stretch*) between two processing rolls in textile industry. Numerous similar examples exist in the paper, steel, and other continuous-length industries. A system of this type obviously can easily be fully digitized.

Speed Ratio Systems. As shown by Fig. 26, by using two DC tachometer generators connected to a ratio meter mechanism, measurements that depend on differential processing speed, such as "percent stretch" and "ratio of draw," can be measured and controlled through additional elements in the system. The system of Fig. 26 is a textile application, but similar systems are found in the paper and steel industries. Usually the minimum generator speed must be 400 RPM or more because of voltage requirements of the indicator. Full-scale range limits are from 10% to 100% shrink. Percent stretch = output − input × (100% input). If the input generator = 100 units/minute and the output generator reads 125 units/minute, the percent stretch = 125 − 100/100 = 25%. Through suitable switching arrangements, the outputs of several pairs of generators may be selectively fed through the ratio and production rate instruments to provide readings from various sections of multi-stage machines.

LINEAR/ROTARY POSITION SENSORS

A number of techniques for measuring position and position-associated variables can be used in connection with linear and/or rotary motions.

Inductive Plate Transducers

The rotary form of the inductive plate position transducer is essentially a two-phase synchro or resolver, whose windings have been projected onto a linear motion. As shown in Fig. 27, an inductive plate includes an etched stator winding that has been projected on a dimensionally stable nonconducting surface. The rotor associated with this transducer is constructed in a like manner. Variations in inductor displacement are averaged over a large number of inductors by summing the voltages from a like number of coils located on the rotor plate. Thus, the reproduced rotor and stator inductors need not be printed with a positional accuracy equivalent to that of the final transducer. Either a trigonometrically related transformer amplitude analog system or a clock-pulse-derived phase analog system can be used to supply sine and cosine voltages to the two windings contained in the stator plate. This is the same arrangement used with rotating resolvers and synchros previously described. Essentially, this transducer is a two-phase synchro or resolver whose windings have been projected onto a linear medium. The advantage of the device is elimination of gear backlash between the positioned machine member and the movable element of the transducer. A related type of device uses bifilar windings on a rod and sensing sleeve.

Widely used is the linear version of the inductive plate, as shown in Fig. 28. In this configuration a slider moves across the scale, but with an air gap between the two. Since there is no physical contact, there is no apparent wear on the feedback device. The slider is attached to a movable push rod, which can transverse up to 500 inches (1270 cm)/minute and attain a total of 2.5 mil travel cycles without replacing the seal. Each scale is laser-checked to ensure an accuracy within ±0.001 inch (±0.025 mm). The scales are manufactured in 10-inch and 250 mm lengths and

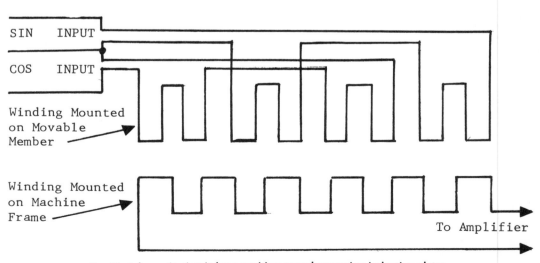

Fig. 27. Schematic circuit for a position transducer using inductive plates.

Potentiometers

Multiple-turn potentiometers are used in machine control systems that do not require an accuracy greater than the positional equivalent of ±0.001 inch (0.025 mm). Circuitry for a rotary potentiometric position transducer is shown in Figs. 30 and 31. Potentiometric devices have the advantage of low cost but suffer from problems of durability and accuracy. Wear can quickly become a significant factor in device accuracy. *Resistofilm*™, developed in the early 1980s[14], has been used in rugged environmental applications, such as in the instrumentation for automobiles. This is a conductive plastic material that combines carbon powder with a plastic binder. The film is bonded to a flexible high-temperature substrate to form a flat track for a wiping contact. A signal applied to terminations on the wiping contact and the conductive track will vary with relative motion. Accuracy from 1 to 3% is claimed for this approach.

A line of linear potentiometers[15] incorporating conductive plastic (*MystR*™) produce a DC output signal that requires no amplification. Claimed incremental sensitivity is 0.00005 inch (0.001+ mm). Minimum operating life is set at 10 million cycles at a speed of 2 inches (50.8 mm)/second. Travel ranges over eight models from 10 to 50 inches (254 to 1270 mm).

Linear potentiometers for the measurement of position and displacement take numerous forms, depending on intended application. These devices are discussed in the next article in this Handbook devoted to dimensional metrology, displacement, and thickness measurement.

Pulse-type Transducers

The two main categories of pulse generators are (1) magnetic, and (2) optical.

Magnetic Pickup. This device is essentially a coil wound around a permanent magnet probe. When ferrous objects, such as gear teeth or turbine blades, are passed through the probe's magnetic field, the flux density is modulated, inducing AC voltages in the coil. One complete cycle of voltage is generated for each interruption or object passed. If the objects, such as gear teeth, are evenly spaced on a rotating shaft, the total number of cycles will be a measure of the total rotation, and the frequency of the AC voltage will be directly proportional to the rotational speed of the shaft. A sectional view of a magnetic pickup is shown in Fig. 32. If a gear with 60 teeth is selected to measure the revolutions per minute of a rotating shaft, the output frequency (Hz) will be numerically equivalent to the revolutions

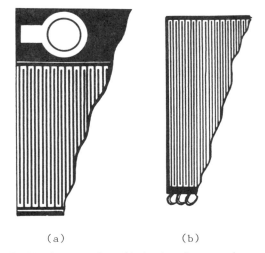

Fig. 28. Linear version of inductive plate transducer: (a) Scale. Standard sections of the scale are 10 inches (250 mm) long. They are 2.3 inches (58.4 mm) wide and 0.375 inch (9.5 mm) thick. They are made up of a copper pattern bonded to heat-treated steel. (b) The slider is 4 inches (101.6 mm) long, 2.875 inches (73 mm) wide, and 0.375 inch (9.5 mm) thick. (*Linear Inductosyn*, licensed under patents of Farrand Industries.)

can be placed adjacent to one another for long travels—up to 200 feet (60 meters). Transducers of this type are used on jig borers, horizontal boring machines, contouring machines, boring and turning machines, drilling machines, milling machines, positioning tables, vertical turret lathes, grinders, and horizontal turret lathes, among others. The induction plate transducer is also available in the form of a tape for ready mounting to machines. The tape[13] yields a bidirectional linear accuracy of ±0.002 inch (0.05 mm) when installed on slides, arms, or machine members. The tape can be mounted on a nonmachined surface and aligned with self-contained leveling screws. The self-contained unit incorporates the tape, movable slider assembly, and built-in scale amplifier. The tape comprises a copper scale pattern etched on a spring steel tape.

The use of a programmable controller with an inductive plate transducer is shown in Fig. 29.

Reported accuracies for the *Inductosyns* are from ±3 to ±1 arc second for disks from 3 to 12 inches (76.2 to 304.8 mm) in diameter. Three-inch disks have 360 poles (180X), while a 12-inch disk may have 1,024 poles (512X). Linear *Inductosyns*, as previously described, are available in bar or tape formats. The accuracy for a 10-inch (250.4 mm) bar scale is ±0.000040 inch (0.001 mm). For a 100-foot (304.8 meters), the accuracy is 0.0001 inch/foot (0.008 mm/meter).

---

[13] *Numerislide Spar*, Giddings & Lewis Electronics Company, Fond Du Lac, Wisconsin.

[14] New England Instrument Co., Natick, Massachusetts.
[15] Waters Manufacturing Inc., Wayland, Massachusetts.

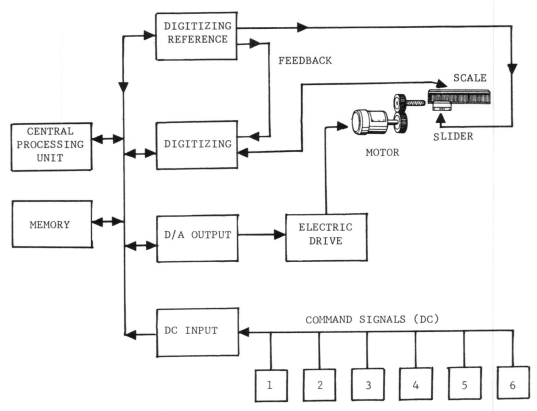

Fig. 29. Programmable controller for machines and processes, using inductive plates. (*Giddings & Lewis Electronics Company.*)

per minute—a situation that may allow frequency meters to be used without calibration. For high rotational speeds, the use of a gear with a smaller number of teeth may be specified. The magnetic pickup circuit shown in Fig. 33 contains its own signal conditioning circuitry for generating a clean square-wave output pulse (+5 V) for each ferrous discontinuity passing the head of the pickup. The output is either on or off, depending on the presence or absence of ferrous material. The unit senses motion down to "zero velocity" and always produces a pulse train of constant amplitude, irrespective of the rotational speed of the gear. Magnetic pickups like this are sometimes used in connection with turbine-type flowmeters.

A magnetic noncontacting device for speed and motion measurement is shown and described in Fig. 34.

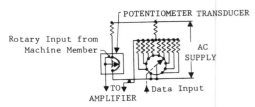

Fig. 30. Basic circuit of a multiturn potentiometer as used in a position transducer.

Fig. 31. Decade switch circuit for a multiturn potentiometer.

Position, Motion, Speed, Velocity 43

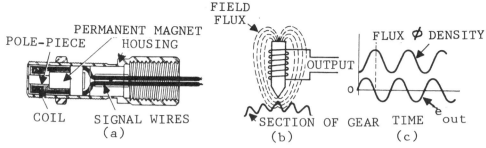

Fig. 32. Magnetic pickup: (a) Sectional view of the pickup; (b) placement of the pickup with a small air gap between the pickup and gear teeth; (c) an output wave form which is a function not only of rotational speed but also of gear teeth dimensions and spacing, pole piece diameter, and the air gap between the pickup and the gear tooth surface. The pole piece diameter should be less than or equal to both the gear width and the dimension of the tooth's top (flat) surface. The space between adjacent teeth should be approximately 3 times the diameter. Ideally, the air gap should be as small as possible, typically 0.005 inch (0.13 mm). (*Daytronic.*)

Electromechanical Limit Switches

These devices have been used for decades. For many years mechanical limit switches were key elements (and continue to be) for automating machines, conveyor systems, and other motion-dependent manufacturing operations. Even prior to the appearance of solid-state circuitry, many advancements were made in the design of these switches. Stress on design for miniaturization commenced in the 1940s. Because there are so many repetitive operations in automated systems, limit switches must be capable of reliable performance over millions of actuations, often in relatively dirty and severe industrial environments. For positive safety interlocks, a lever-actuated limit switch is still more reassuring to many control engineers than an unseen software command, particularly when the programming of a microprocessor controlled system is subject to revisions each time a design change is made. It must be realized that an inadvertently dropped program step, which may go undetected during normal operation, may have life- or equipment-threatening side effects.

The principal categories of limit switch actuators are (1) rotary operating heads, (2) plunger

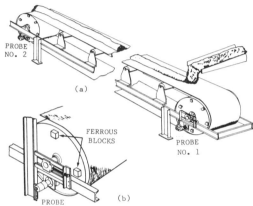

Fig. 34. Magnetic noncontacting device for speed and motion measurement. These units are particularly useful for *slowdown* indication of conveyors and other rotating manufacturing and processing machinery. The system is rugged and primarily intended for use with heavy equipment in bad environments. (a) Conveyor loss-of-motion detection requires use of two probes, one on the head pulley, another on the tail pulley. By computing the plus or minus speed relation of the tail to the head pulley and comparing this value with percentage slip points, slip protection is ensured. A minimum percent speed feature of the system checks for mechanical failure between the motor and the tail pulley. Each of these features has its own time delay after startup and one common delay to ignore nuisance alarms and/or shutdowns. The four alarms are fed to a first-out annunciator and latch in the output circuit. Reset can be manual or automatic. (b) Probe and ferrous block mounting detail. A noncontacting tachometer can be calibrated in the field—from 0 to 8ppm through 0 to 720ppm equal to 4 to 20 mA output. The parts-per-million input is equal to the number of ferrous objects per revolution times the revolutions per minute of the machine. Note carefully attached ferrous blocks which are detected by the probe. (*Milltronics.*)

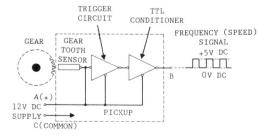

Fig. 33. Circuit of 'zero velocity' magnetic pickup. (*Daytronic.*)

44 Position, Motion, Speed, Velocity

TABLE 5. Maximum Contact Ratings of Some Heavy-duty, Plug-in Limit Switches

AC volts	Current, A			Voltamperes		DC volts	DC current, A
	Make	Break	Cont.	Make	Break		
	One normally open, one normally closed, contact on same polarity						
	NEMA A600 rating						
120	60	6	10	7200	720	120	0.2
240	30	3	10	7200	720	240	0.1
480	15	1.5	10	7200	720	240	0.1
600	12	1.2	10	7200	720	240	0.1
	Two normally open, two normally closed, each pole on same polarity						
	NEMA B600 rating						
120	30	3	10	3600	360	120	0.1
240	15	1.5	10	3600	360	240	0.05
480	7.5	0.75	10	3600	360	240	0.05
600	6	0.60	10	3600	360	240	0.05

operating heads, and (3) wobble lever operating heads. These configurations permit switch actuation by moving parts and moving machine members of various sizes and shapes and for approaching the switch from different angles and directions. Side and top rotary switches are shown schematically in Fig. 35. Various kinds of levers used with rotary switches are shown in Fig. 36. The adjustability of rotary switches is shown in Fig. 37. Plunger switches are shown in Fig. 38 and wobble actuators in Fig. 39.

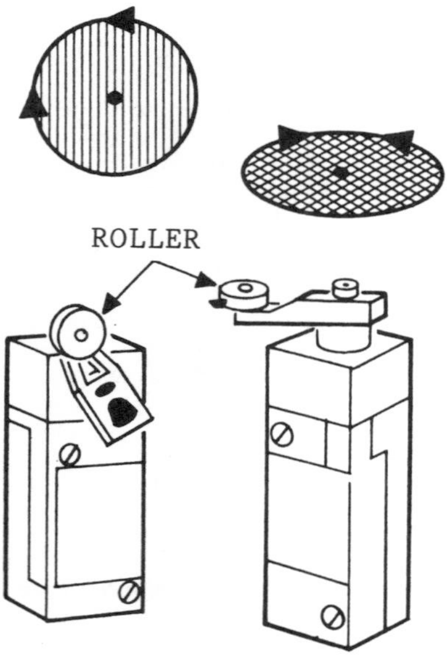

Depending on the particular configuration, limit switches vary in dimension: Height, 2 to 6 inches (51 to 152 mm); width, 0.6 to 3.2 inches (15 to 81 mm); length, 1.6 to 3.4 inches (41 to 86 mm). The upper operating temperature range, depending upon particular design, is 160°F (71°C) to 250°F (121°C), and the lower operating temperature range is −25°F (−32°C) to 35°F (2°C). Housing materials are frequently zinc or aluminum. Electrical characteristics of some heavy-duty plug-in, watertight and oiltight limit switches are shown in Table 5. Additional characteristics are shown in Table 6.

Gravity Return Switches. Unlike standard switches with spring return mechanisms, a switch is available with *gravity return*. The weight of the actuating lever must provide the force to restore

Fig. 35. Rotary operating heads on limit switches may be (a) side rotary, or (b) top rotary. (*Micro Switch*.)

TABLE 6. Some Physical Parameters of Heavy-Duty, Plug-in, Watertight and Oiltight Limit Switches

STANDARD. Overtravel 60° minimum, pretravel 15° maximum, differential travel 3° (single-pole) and 7° (double-pole) maximum.

LOW-DIFFERENTIAL TRAVEL DESIGN. Overtravel 68° minimum, pre-travel 7° maximum, differential travel 3° (single-pole) and 4° (double-pole) maximum.

LOW-OPERATING TORQUE DESIGN. Overtravel 60° minimum, pretravel 15° maximum, operating torque 1.7 inch-pounds (0.19 Newton-meter) maximum.

LOW-TORQUE, LOW-DIFFERENTIAL TRAVEL DESIGN. Overtravel 68° minimum, operating torque 1.7 inch-pounds (0.19 Newton-Meter) maximum. Differential travel 3° (single-pole) and 4° (double-pole).

SEQUENCE ACTION DESIGN. Delayed action between operation of two poles, in each direction. Overtravel 48° minimum.

CENTER NEUTRAL DESIGN. One set of contacts operates on clockwise rotation, and another set on counterclockwise rotation. Overtravel 53° minimum.

MAINTAINED CONTACT DESIGN. Operation maintained on counterclockwise rotation; reset on clockwise rotation, and vice versa. Overtravel 20° minimum.

Position, Motion, Speed, Velocity 45

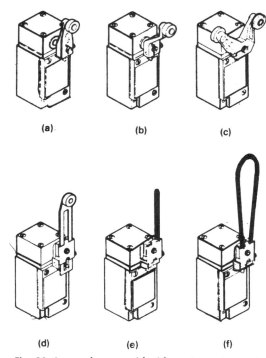

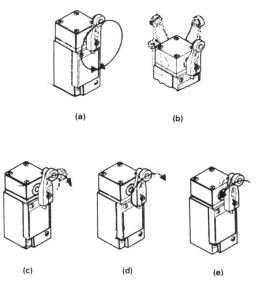

Fig. 36. Levers for use with side or top rotary actuated switches are available in a variety of sizes and materials. Rollers may be on either side of the lever to most effectively match the external actuating mechanism. They permit a wide range of cam tracking possibilities: (a) A standard roller lever with a fixed 1.5 inch (38.1 mm) radius. (b) An offset roller lever with a fixed 1.5-inch (38.1 mm) radius. (c) A yoke roller lever used with side rotary maintained switches where a reciprocating actuator operates the switch in one direction and reverses it when moving in the other direction. (d) An adjustable-length roller lever with an adjustable radius from 1.5 inches (38.1 mm) to 3.5 inches (88.9 mm). (e) A rod lever which may be formed by the user. The hub permits lever length to be adjusted. A flexible spring or rod is available. (f) An adjustable loop lever for accommodating certain types of external actuating mechanisms. (*Micro Switch.*)

it to the free position. The very small 5 inch-ounce (0.035 Newton-meter) operating torque is useful in some conveyor applications because it permits operation with small or lightweight objects.

Solid-State Limit Switches. In addition to the traditional types of electromechanical switches just described, units employing Hall-effect position and vane sensors are available. See Fig. 40. The external design of these switches parallels that of the switches just described. The computer-compatible digital output of these switches

Fig. 37. The actuation of limit switches with rotary levers is adjustable for operation clockwise, counterclockwise, or in both directions. (a) The lever locks in any position, 360° around the shaft. (b) The head may be positioned and locked in any of four 90° positions. (c) Clockwise rotation. (d) Clockwise and counterclockwise rotation. (e) Counterclockwise rotation. (*Micro Switch.*)

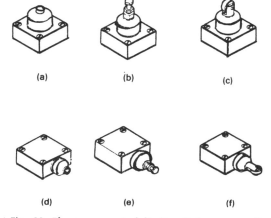

Fig. 38. Plunger-operated limit switches are available in several configurations to accommodate various external actuating mechanisms. (a) Top plunger, (b) adjustable top plunger, (c) top roller plunger, (d) side plunger, (e) adjustable side plunger, and (f) side roller plunger. (*Micro Switch.*)

46 Position, Motion, Speed, Velocity

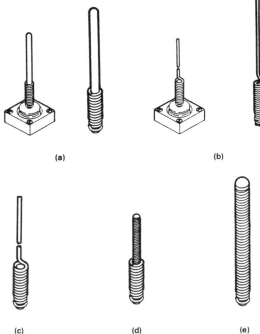

Fig. 39. Wobble actuators. (a) A wobble stick mounted on a switch head with detail of stick at right. (b) A cat whisker mounted on a switch head with detail of whisker at right. (c) Spring wire, (d) cable, and (e) coil spring. (*Micro Switch*.)

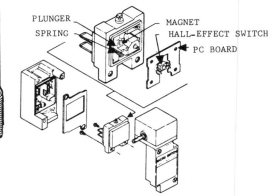

Fig. 40. Solid-state limit switch that incorporates a Hall-effect switch. (*Micro Switch*.)

makes it possible to interface directly with most electronic circuits, discrete transistor circuits, microprocessors, and integrated logic circuits (register-transistor logic, diode-transistor logic, transistor-transistor logic, and Hall-effect transistor logic), and silicon controlled rectifiers (SCRs). Switches incorporating the Hall-effect principle[16] cannot be used in areas where high magnetic fields are present. Also, wiring for these devices should not be run in the same conduit with high (AC or DC) power lines. Use of the Hall-effect device eliminates any problems of contact bounce sometimes encountered with mechanical designs. The supply voltage is 5V DC or higher.

INTEGRATED MOTION SYSTEMS

As positioning and motion control systems have become more sophisticated with emphasis on tying motion control commands to computers, the incorporation of more and more MPUs in the measuring and control circuits—factors that are proving the viability of computer-integrated manufacturing (CIM) concepts—changes have occurred within the firms that manufacture and market both hardware and software. For example, some manufacturers of programmable controllers have added specialty departments or divisions to concentrate on motion systems; a number of independent firms have been formed to tie together the elements obtained from a number of suppliers, and large users have established special staffs to concentrate on the design, evaluation, and operation of motion systems. These are trends that probably will continue into the 1990s and beyond.

There is also a trend among the specialty houses to concentrate on select applications for motion systems. For example, the motion engineering of machine tools is one area of specialization[17]. (See Fig. 41.)

A programmable motion controller (PMC) has been designed[18] to operate as an intelligent slave for controlling velocity and position in response to high-level ASCII commands. In a typical application, the PMC interfaces with a servodrive, a servomotor, a DC tachometer, and an incremental position encoder with quadrature outputs to create a closed-loop digital positioning servo that controls the position and velocity of a mechanical load. Position stepping rates are available up to 285kHz, and encoder resolution is in excess of 100,000 counters per revolution. The system can perform variable indexes at rates in excess of 3000 indexes/minute. See Fig. 42.

The heart of the PMC system is an *Intel 8085* microprocessor, giving each axis of motion con-

---

[16] When a steady current flows in a steady magnetic field, emfs are developed that are at right angles both to the magnetic force and to the current and are proportional to the product of the intensity of the current, the magnetic force, and the sine of the angle between the directions of these quantities. This is known as the *Hall effect*.

[17] MTS Systems Corporation, Machine Controls Division, Minneapolis, Minnesota.

[18] ORMEC Systems Corporation, Rochester, New York.

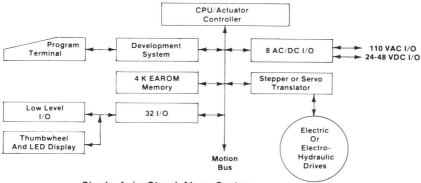

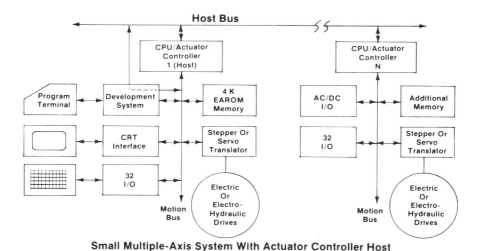

Fig. 41. Microcomputer hardware engineered as a family of building blocks which can be combined to configure either stand-alone pre-engineered controllers or larger more customized multi-axis systems. The blocks consist of numerous drive interface, control and computing, memory, I/O, and display modules. Each system can incorporate either user or factory-generated software packages. All system modules requiring communication with the system CPU are designed with a remote host bus hardware interface for independent bidirectional communication. This allows the user to select from a variety of system architectures and hierarchies. (INCOL/470 Hardware System, MTS Systems Corporation, Machine Controls Division, Minneapolis, Minnesota.)

48   Position, Motion, Speed, Velocity

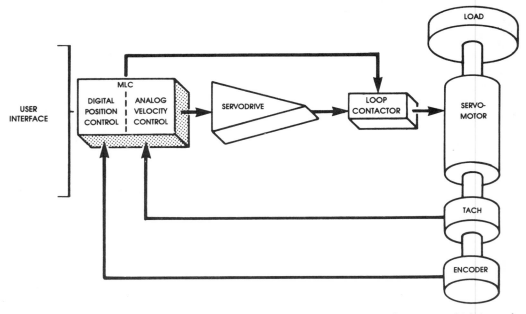

Fig. 42. The motor loop controller (MLC) is a fundamental building block of the system which is used to create a closed-loop digital-position servo. When interfaced with a servodrive and servomotor, a DC tachometer (optional) and an incremental position encoder with quadrature outputs, the system controls the position of a mechanical load. The MLC is a printed circuit module which allows design engineers to conveniently develop high-performance positioning systems. By using an MLC to close the velocity and position loops of a servomotor, a designer can apply a closed-loop positioning system instead of a stepping motor system. The MLC combines analog and digital circuitry to use feedback signals from a DC tachometer and digital position transducer to control velocity and position of a servomotor. (*Programmable Motion Control System—PMC—Ormec Systems Corp.*)

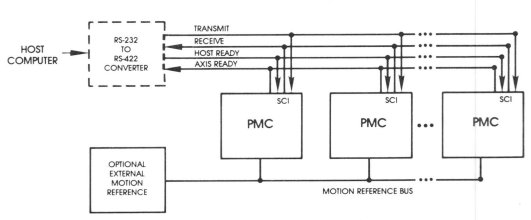

Fig. 43. Multi-axis implementation (up to 32 axis) of the programmable motion controller (PMC). (*Ormec Systems Corporation.*)

trol the processing power of a personal computer. The CPU processes instructions stored in its firmware (EPROM), allowing the PMC to interpret the high level set of user-oriented motion control commands of the special language of the PMC. Motion control programs are stored in up to 2k bytes of nonvolatile memory (EEPROM). As shown in Fig. 43, to communicate with the PMC, the host sends a byte (character) to it and waits for that character to be processed. When the host completes a command by sending a command terminator, the PMC executes that command and returns a prompt character to the host, signifying its completion. Simply by sending ASCII characters to the PMC, the host can execute PMC positioning commands directly. Only high-level commands and status requests are required from the host computer, since the PMC isolates it from the real-time servicing requirements of the servo system. This allows a single microcomputer host to manage several high-performance servomotor systems. The PMC's RS-422/449 serial communications bus allows PMCs to be separated by distances of more than 1000 feet (305 meters), which is very useful for distributed multi-axis systems. Up to 32 devices can be connected to the bus with twisted pair or mass termination cable. Thus, any host computer implementing a serial communications channel with RS-422 interface can connect directly to the bus, or a host with an RS-232 serial channel can be connected to the bus with a serial communications bus interface card.

Motion Systems Analysis

Prior to improving or designing a new automated system that involves complex, often extremely fast linear and rotary motions, it is sometimes mandatory to study the motions in the utmost of details. Effective, computerized simulations are not always a satisfactory substitute for studying the motions of *real* machines and processes. The human eye has an exposure time of approximately 0.1 second. If an engineer looks at a point on a line moving at any speed in excess of 25 feet (7.62 meters) per minute, the eye captures a blurred image. For diagnosis of machine motions obviously the unaided human eye is insufficient. The use of video tape systems has gained wide acceptance. Another, more traditional approach is the use of high-speed film. With the exception of some applications, high-speed video tape has a number of advantages over 16 mm film systems. Video tape has a longer recording time, which can be continuous and interrupted only by rewind time on some systems. This feature provides the opportunity to view infrequent random events that can be very irksome problems for the motion diagnostician. With video tape, elapsed time between frames, which are counted and appear on the monitor, provide a measure of the frequency of the events that might otherwise appear random. In contrast, if 20 pictures of the same event within a very

Fig. 44. High-speed video-tape instant playback motion analysis system. (*Instar*™, *Video Logic Corporation*.)

short period are needed, high-speed film may be the appropriate answer.

A motion analysis system may be visualized as a closed-circuit television (CCT) system which records motion for subsequent viewing. Critical aspects of motion analysis are: Frame rate (pictures per second), picture quality, and slow motion and stop action ability. One system[19] can record 120 *full screen* pictures per second or 240 *split-screen* pictures per second. The system features a variable-speed slow-motion playback with high motion resolution (3% to 15% infinitely variable). The system is designed for adverse environments and is quite portable. Television pictures can be in black and white or color. As shown in Fig. 44, the system consists of a high-resolution camera, a video tape recorder, a CRT monitor, operating controls, and a high-intensity stroboscope[20].

Some applications for motion analysis include diagnostics of paper machines, packaging equipment, vehicles and other machines that depend upon cams and gears that operate at high speeds, robotic systems, weapons testing, among numerous others.

[19] Video Logic Corporation, Sunnyvale, California.
[20] Stroboscopy is described in some detail in Sec. 9, pp. 9.85–9.88. in the "Process Instruments and Controls Handbook," 3rd Ed. (D.M. Considine, Editor), McGraw-Hill, New York, 1985.

ADDITIONAL READING

Bailey, S. J.: "Multistepping on the Digital Bus: The New Look in Incremental Motion," *Cont. Eng.*, **29** (8) 68–72 (1982).

Bailey, S. J.: "Programmable Motion Control Paces Factory Automation Efforts," *Cont. Eng.*, **30** (4) 107–110 (1983).

Bailey, S. J.: "Step Motion—Open-Loop Path to Productive Precision," *Cont. Eng.*, **30** (7) 73–77 (1983).

Bailey, S. J.: "Servo Design Today: Software-Driven Adaptations in the Feedback Loop," *Cont. Eng.*, **31** (2) 67–72 (1984).

Bailey, S. J.: "Step Motion Control 1985: Direct Digital Incrementing with ServoLike Performance," *Con. Eng.*, **31** (8) 49–52 (1985).

Bailey, S. J.: "Position Transducers: Strategic Components in Motion Control Loops," *Cont. Eng.*, **32** (14) 41–45 (1985).

Berris, R., and D. Hazony: "Discrete Pulses Put Induction Motors into the Stepping Mode," *Cont. Eng.*, **29** (1) 85–86 (1982).

Brenza, R. M.: "Digital Tracking Controls Motor Rotor Positions," *Con. Eng*l, **29** (1) 144–146 (1982).

Chan, M.: "New Views on High-Speed Motion Analysis," *Production Engineering*, (reprint pp. 1–4) (1982), Video Logic Corp., Sunnyvale, California.

Corrado, J. K.: "Low-Cost, Color, Video, Motion-Analysis System," *Design News* (March 12, 1984).

Gauen, K.: "Designing a DC Servo Position Control Using a Microcomputer," *Cont. Eng.*, **30** (7) 80–83 (1983).

Glass, M.: "Synchro and Microprocessor Combine for Versatile Multi-Turn Position Sensing," *Cont. Eng.*, **30** (5) 84–87 (1983).

Keehbauch, T. J.: "Programmable Position Control Uses Standard Induction Motor as Servo," *Cont. Eng.*, **31** (1) 108–110 (1984).

Kompass, E. J.: "Separate Controller Aids Tailoring of Motor Drives to Position Control Applications," *Cont. Eng.*, **32** (8) 72–73 (1983).

Miller, T. J.: "Digital Positioning Trades Hardware for Software," *Cont. Eng.*, **29** (9) 59–61 (1982).

Miller, T. J.: "Step Motor Controller Options Fit Package to Application," *Cont. Eng.*, **31** (8) 81–82 (1984).

Morris, H. M.: "Robotic Servo Control Systems Need Accurate Positional Feedback Inputs," *Cont. Eng.*, **31** (1) 90–92 (1984).

Tome, D. M.: "Adaptive Control Sensors for Manufacturing Systems: An Overview," *Cont. Eng.*, **31** (8) 78–79 (1984).

Geometric Variables II:

Dimension, Displacement, Thickness

Metrological standards and practices grow in importance as more and more automation is applied to manufacturing and processing facilities. Automation enhances the chances of generating scrap and rejects at an exceedingly rapid rate. Higher throughputs and productivity have developed a critical need for very efficient on-line inspection of parts and materials with no degradation of measurement accuracy and resolution. It is interesting to note that one of the forerunners of modern automation was the mass production line introduced several decades ago. These lines could not have succeeded without the concept of interchangeable parts. In turn, this concept would not have been viable had not concentrated attention been paid to establishing in-plant metrological laboratories and the development of more precise measuring techniques—improved interferometers, gage blocks, optical gratings, optical comparators, and the like[1].

Although increased automation of production processes is occurring at an accelerated rate, much progress remains to be made in automating metrological inspection operations on the plant floor. To be sure, there are many impressive automatic parts gaging systems, many of which tend to be fixed automation, as well as on-line thickness-measuring systems for the sheeted goods, paper, and metal foils industries, but much progress remains to achieve more on-line stations. Most of the hardware is on hand, including systems using laser optics and advanced electronic sensors. In the relatively near future, as attention to quality control becomes even greater and as cost effectiveness can be improved, many of the new techniques will be integrated in on-line inspection stations.

FUNDAMENTAL SENSOR METHODOLOGIES

Interferometer

An interferometer is a precision instrument that uses the interference of light waves as the basis for measuring distance (length and associated dimensional variables). The optics of the interferometer are designed so that the variance of known wavelengths and path lengths within the instrument permits accurate measurement of distances.

Length is expressed in terms of the *meter*. The meter is defined as 1 650 763.73 wavelengths in vacuum of the orange-red line of the spectrum of krypton 86. A new definition of the meter based on a very accurate measurement of the speed of light was adopted by the General Conference on Weights and Measures in 1983. The meter has been defined as the length of path traveled by light in vacuum during a time interval of 1/299 792 458 of a second. Practical realization of this definition will be through time-of-flight measurements, by frequency comparison with laser radiations of known wavelengths, or through interferometric comparisons with radiations of stated wavelengths, such as that from krypton 86.

Closely associated geometric variables include: *area*, expressed in terms of the *square meter* (m^2). *Volume*, expressed in terms of the *cubic meter* (m^3). The liter is a special name for the cubic decimeter (0.001 m^3).

With the introduction of the laser interferometer some years ago, the interferometer came closer to the factory floor. See descriptions later in this article under "Positioning Tables and Precise Positioners." The principle of the laser interferometer is given in Fig. 1.

The fundamental position transducers found in motions systems are described in the preceding article in this *Handbook*. These include resolvers and synchros, encoders, inductive plate transducers, rotary potentiometers, limit switches, and others. There are other transducers for the measurement of dimension and displacement that are not always necessarily associated with motion systems. Some of these are described in the next several pages.

---

[1] These instruments are described in some detail in the "Process Instruments & Controls Handbook," (D. M. Considine, Editor-in-Chief), 3rd Ed., pp. 9.53-9.62, McGraw-Hill New York, 1985.

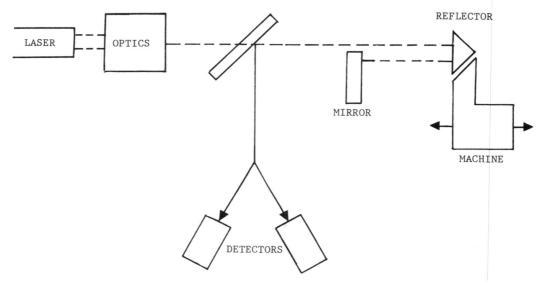

Fig. 1. Principle of the laser interferometer capable of accuracies of about one part in ten million. Earlier devices required that the laser beam be reflected off a moving mirror that traveled along a track at least one meter long. Later devices require that the mirror travel only a few centimeters.

Linear Variable Differential Transformer (LVDT)

These transducers depend upon inductance effects. A sectional view of an LVDT is shown in Fig. 2. A magnetic nickel-iron core, supported by a nonmagnetic push rod, moves axially within the cylinder in exact accordance with mechanical displacement of the probe tip. As shown by Fig. 3, with AC excitation of the primary coil, induced voltages appear in the secondary coils. Because of the symmetry of the magnetic coupling to the primary, these secondary induced voltages are equal when the core is in the central (*null* or *electric zero*) position. When the secondary coils are connected in series opposition, the secondary voltages cancel and (ideally) there is no net output voltage. If, however, the core is displaced from the null position, one secondary voltage will increase while the other decreases. Since the two voltages no longer cancel, a net output voltage is the result. If the transducer is properly designed, this output will be exactly proportional to the magnitude of the *displacement*, with a phase polarity corresponding to the direction of displacement. (See Fig. 4.) A miniaturized displacement transducer designed for measurement applications in the range up to ±5 mm, where small size is a prime requirement, is shown in Fig. 5. Transducers with ranges down to ±2.5 mm (and less) are available. For longer-stroke applications, ranges up to ±1 inch (25.4 mm) and higher are available. These devices have a useful temperature range of −40 to +100°C (−40 to +212°F). Depending on the particular design, the excitation frequency ranges from 0.4 to 10 kHz. Linearity ranges from 0.1 to 5% of range.

Unlike strain gages, LVDTs are not furnished with meaningful calibration data. System sensitivity is a function of excitation frequency, cable loading, and amplifier phase characteristics, among other factors. It is general practice to calibrate each LVDT-cable instrument system after installation. LVDTs have a wide variety of applications, including industrial gaging (thick-

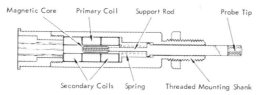

Fig. 2. Sectional view of a LVDT linear displacement transducers. (*Daytronic*.)

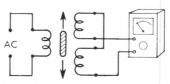

Fig. 3. Schematic of operating principle of a LVDT linear displacement transducer. (*Daytronic*.)

Dimension, Displacement, Thickness 53

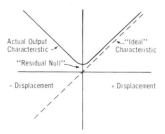

Fig. 4. Diagram showing the *null* or *electric zero* position of a LVDT linear displacement transducer. (*Daytronic.*)

ness, taper, et al.), through structural testing. The units are available with several tip configurations so that numerous applications can be handled.

Although the LVDT is used to electronically measure linear displacement, such as actuator or mechanism positioning, the device is also used to measure displacements in mechanical type instruments, such as pressure sensors. The LVDT can also be used as an acceleration-measuring device where displacement of a seismic mass produces a linear displacement of a magnetic core within the windings. Three representative applications of displacement measurements by the LVDT are shown in Figs. 6, 7, and 8. See also Fig. 33 later in this article which demonstrates the use of a pair of LVDTs for thickness measurement.

Linear Variable Reluctance Transducer (LVRT)

The LVRT is useful for measurement of stress, thickness, vibration, shock, and contour and is used to provide electrical feedback for valves and

Fig. 5. Miniaturized LVDT linear displacement transducer. Probe (unextended) is 28.15 mm (1.10 inch) long; extended, 96 mm (3.74 inches) long. The diameter of the tip is 4 mm (0.16 inch). Transducer parts are made of stainless steel. (*Daytronic.*)

actuators in servocontrol systems used in manufacturing machinery. As shown in Fig. 9, the variable reluctance unit forms half of a four-arm bridge. For many servo applications, a center-tapped transformer is used to form the reference half of the bridge. The output signal then is

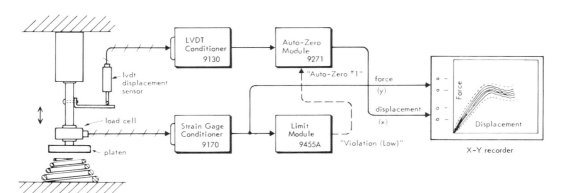

Fig. 6. In monitoring the production of critical nonlinear springs, quality control procedures require that the force-displacement characteristics of successive lot samples be plotted on a single sheet of an *X-Y* recorder. Ideally, all plots should fall within the narrow band that defines the desired nonlinear characteristic. This method allows quick spotting of any aberrant units. Such determinations normally are difficult (and rather subjective), because any variation in spring height will cause the displacement plot to start at some point other than the origin of the *X-Y* graph. The system shown here solves this problem by automatically rezeroing the displacement channel for each measurement. (*Daytronic.*)

54 Dimension, Displacement, Thickness

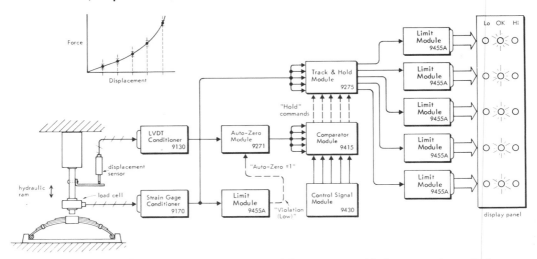

Fig. 7. Use of a LVDT and strain gage to test an automobile spring assembly for proper force-displacement characteristics requires but two seconds. When the descending ram makes initial contact with the part, the displacement channel is instantly autozeroed. As the displacement signal subsequently reaches each of five test points preset on the signal module, the comparator module will issue a logic 'Hold' command. In this way, the system captures the five individual force values corresponding to the five displacement test points. These held values are then examined by five limit modules for conformance to appropriate high and low limits at each point. Logic outputs are transmitted to a display panel, where acceptance or rejection of parts is indicated. (*Daytronic.*)

produced between the transformer center tap and the transducer centerpin. The signal phase changes 180° at midrange position. A variable resistive ratio divider or a variable inductive ratio divider may be used to complete the bridge. The null position is adjustable to any point within the transducer range.

The winding is continuous over the length of a spool-type bobbin. There are no intermediate partitions that divide the bobbin segments, as in the case of LVDTs. The advantages of the LVRT include excellent linearity over long strokes—up to 24 inches (610 mm) and good temperature stability. A disadvantage is that the overall length must be double the total stroke. The device can withstand rapid temperature changes from -160 to $+450°$ F (-107 to $+232°$ C). Encapsulation of the coils in an inert epoxy resin hermetically seals the unit and protects it against vibration and shock. The working portion of the probe can be made of magnetic stainless steel (nonmagnetic stainless steel can be used for the extension). Standard probe diametric clearance is 0.001 to 0.020 inch (0.025 to 0.5 mm). Carrier frequencies from 1000 to 20,000 Hz may be used. Optimum results are obtained with a carrier frequency of 3000 to 5000 Hz. Operation with other frequencies results in slight changes in the maximum excitation voltage, phase shift, and linearity. The size of the devices (less probe extension, but includ-

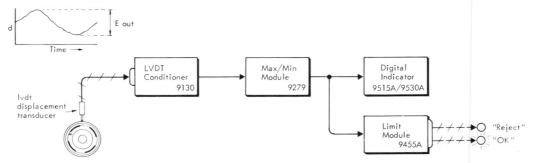

Fig. 8. Use of a LVDT for measuring excursive phenomena, such as run-out, wobble, and looseness of a rotating part. Here, the TIR (total indicated runout) of an automobile wheel is checked. (*Daytronic.*)

Dimension, Displacement, Thickness

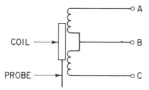

Fig. 9. Circuit of a linear variable reluctance transducer (LVRT). (*Gulton.*)

ing the connector) ranges from 0.25 × 2.2 inches (6.4 × 56 mm) to 0.5 × 10.2 inches (12.7 to 259 mm) in standard configurations. The linearity is ±1% or better.

Linear Transformer (LT)

This is a special synchro consisting of a salient-pole rotor and a single-phase stator, distributively wound. The winding on the stator is designed to produce an output voltage that varies linearly with rotor position. This linear function is valid only within a restricted band about the zero position—generally ±50° or ±85°—which is known as the *excursion range*. Past the excursion range, the output voltage no longer is linear and tends to become sinusoidal. The LT is used in some instances as a replacement for a potentiometer. It is an infinite resolution device and also has the advantage of being constructed in the same manner as other servo system components. Thus, service behavior of the LT is the same as that of the other elements in the system.

Impedance-type Gaging Transducer

The gage head contains two coils with a sintered iron core centered between them. This core is attached to the gage spindle and moves axially, as shown in Fig. 10. Its position relative to the coils affects their impedance. The coils and a symmetrical transformer in the oscillator form a bridge. When the core rests equally between the two coils, the bridge is balanced and the output signal from the gage head is zero. When the core is displaced, the impedance of the coils is changed. The impedance of one is reduced, while that of the other is increased, generating a signal in proportion to the amount of displacement of the core. The signal is amplified and rectified for display (calibrated in units of length). For portability, battery power can be used. Differential amplifiers are available to determine the difference between, or the sum of, two measurements, using two gage heads simultaneously. Typical applications include checking roundness, concentricity, parallelism, thickness, cam contours, tapers, flatness, and squareness—without requiring precision fixturing. The difference technique is well suited for the selection of parts with a specific clearance where the actual size itself is not critical; for example, in mating the inside diameter of a cylinder with the outside diameter of a piston.

Linear Potentiometer

The linear potentiometer is among the oldest of electrical techniques for measuring position and displacement. The simplest, least costly form is a single length of wire along which a slider or other form of moving device contacts the wire. The position of the slider determines the effective length of the conductor. Hence, a change in electrical resistance or voltage drop is related to the position or displacement of the slider. Simple wire conductors have been replaced over the years by wire-wound, thin-film, or printed circuits because these techniques make possible a considerably greater length of resistor within the same linear length. Thus, the voltage drop per unit length is larger, enabling stronger, more useful signals.

An example of a linear displacement transducer of the wire-wound potentiometer type used for rugged environmental applications is shown in Fig. 11. The basic design is cylindrical because often the device is used in connection with hydraulic actuators. The device comprises: (1) two resistance elements *j*, which are mounted (molded) along with slide bars *d* in the element block *s*, which in turn is contained within the outer case *n*. Wiper assemblies *h* are attached to wiper carrier *e* and are aligned to coincide with the resistance elements and contact bars immediately opposite each other. (See sectional view.) The wiper carrier is secured to the actuat-

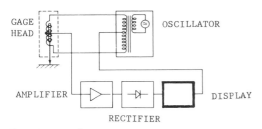

Fig. 10. Impedance-type gaging system. (*Brown & Sharpe.*)

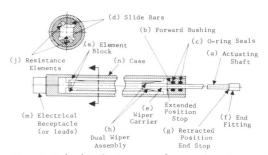

Fig. 11. Cylindrical wire-wound potentiometer-type linear-displacement transducer. (*Gulton.*)

56 Dimension, Displacement, Thickness

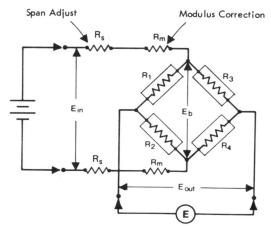

Fig. 12. Representative strain gage transducer (*Daytronic*). The resistance strain gage is an electrical sensing device that varies its resistance as a linear function of the deformation (strain) of the structural surface to which it is bonded. This effect may be used to measure weight, pressure, torque, and displacement by sensing the deformation of calibrated beams, diaphragms, or other flexures to which mechanical force is applied. Through proper flexure design and gage placement, a linear relationship can be achieved between the applied force and the sensed strain. If gages are connected in a balanced Wheatstone bridge circuit as shown here and excited by an AC or DC voltage, the transducer will produce an electrical output which is a direct linear function of the excitation voltage and the magnitude of the applied mechanical input:

$$E_{out}(mV) = E_{in}(V) \times K \times F/100$$

where K = Calibration Factor (mV/V, full scale); and F = Input variable (% of full scale).

ing shaft *a* in such a manner that it eliminates backlash and yet allows the shaft to rotate freely 360° when installed. This is necessary in many cases because the end fitting *f* provides a thread for attachment to a moving member and free rotation of the shaft is necessary to avoid deflection of the wiper system or dislocation of the wipers from the element and contact bars. The forward end of the transducer provides a bush *b* into which are inserted a pair of O-ring seals, providing adequate sealing against intrusion of salt spray, sand, dust, and moisture. These transducers are 0.5 and .75 inch (12.7 and 19.1 mm) in diameter, depending on application. Larger sizes are generally used where motion is in excess of 6 inches (152 mm). In addition to industrial machines, these devices find applications in aircraft, missiles, and valve indicators.

Thin-film Potentiometers. These are generally of three types: (1) *Cermet*—a ceramic-metal mix that provides very long life and is capable of withstanding high temperatures (250°C; 482°F). The material provides continuous or stepless output and is used primarily in trimming potentiometers where linearity is not a major problem. (2) *Conductive plastic*—generally a carbon particle and plastic mix that provides continuous, stepless output and comparatively long life (in excess of 3 million cycles). (3) *Metal-deposited Thin Film*—a homogeneous metal generally vacuum-deposited onto a glass substrate. The deposit may be extremely thin—on the order of 5000 Å, giving 0.001 to 0.002 inch (0.025 to 0.05 mm) resistances of 20,000 to 1 million ohms, or thicker films giving resistances of 100 to 1000 ohms. These types generally are not associated with long life, but they do provide a continuous output.

Strain Gage Load Cell

Although strain gages are not immediately associated with motion, position, and displacement systems, they will often use the displacement or distortion of a structural material as a measure

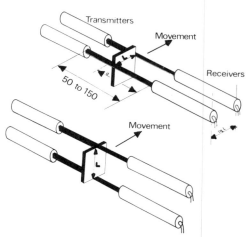

Fig. 13. Optoelectronic system for contact-free dimensional control of parts in movement on production lines. The system is particularly suitable for control applications with medium and large production batches. Optical beams are sent by transmitters and detected by receivers. These beams are so arranged that all are partially intercepted by the part at a certain moment during its displacement. At this instant, the various receivers automatically trigger measurement of residual luminous flux values. Each measurement or pair of measurements is then processed electronically for comparison with tolerance limits adjusted by calibration on potentiometers. The system can be tied to control circuits for generating alarm signals or machine stop. The system also is used to establish statistical data, sliding average values, standard dispersions, et al. (*Optal*™, *Sopelem*.)

of weight, torque, tension, and other mechanical forces[2]. A typical strain gage transducer is diagrammed in Fig. 12. Strain gages frequently are used in connection with LVDTs, previously mentioned. See Figs. 6 and 7 given earlier.

Optical Methodologies

In addition to the laser interferometer previously discussed, there are numerous optical means for checking and measuring dimension, displacement, and thickness. Many of these systems are refinements of the methodologies described in the next article of this *Handbook* dealing with proximity, presence, and nonpresence detection systems. A recently introduced French system[3] that combines optics and electronics for contact-free dimensional control of parts in movement on the production line is shown in Fig. 13.

STATISTICAL PROCESS CONTROL

The concept of statistical quality control dates back to the early 1900s, if not further. It is a discipline used by manufacturing management that involves the collection, organization, analysis, and interpretation of masses of data pertaining to those physical and other parameters of a part, subassembly, or assembly that, in the long run, determine the quality of a product—actually spreading from the receipt of raw materials, through various stages of production, to the shipping dock. To maintain high end-product quality, obviously many inspections (measurements) must be made directly on the factory floor, or representative samples must be examined by satellite laboratories. Where economically feasible, the ideal arrangement, of course, is a closed-loop system that immediately feeds back quality information to control a machine or process. The next best arrangement, if complete automation cannot be justified, is the automation of data collection and the display of information to production managers in what might be termed a "predigested" and convenient format for analysis and decision making.

Background of Statistical Terms[4]

Following is an abridged glossary of terms used in statistical process control. Symbols used in the equations presented are defined as follows:

N = Number of data points
R = Range
$\bar{R}$ = Average range
S = Standard deviation
X_i = Value of a specific data point
$\bar{X}$ = Average value
USL = Upper specification limit
LSL = Lower specification limit

Capability and Control (Concept of): A process is in control if the only sources of variation are common causes. (See also definition of *Variation*.) The mean and spread of such a process will appear stable and predictable over time. (See Fig. 14.)

The fact that a process is in control does not imply that it will yield only good parts, i.e., parts within the specification limits. Control only denotes a stable process. The size variation due to the common causes may be so large that some parts are outside the specification limits. Under these conditions, the process is said to be *not capable*.

Capability is the ability of the process to produce parts that conform to blueprint specifications. (See Fig. 15.)

Cp—Inherent Capability of the Process: Cp is the ratio of the tolerance to 6 sigma. The formula is:

$$Cp = \frac{USL - LSL}{6S}$$

The Cp ratio is used to indicate whether a process is capable.

1.33 or greater	Process is capable.
1.0 to 1.3	Process is marginally capable—should be monitored.
1.0 or less	Process is *not* capable.

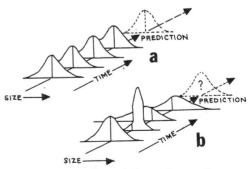

Fig. 14. Concept of capability and control: (a) Process is in control—stable and predictable over time. (b) Process exhibits lack of control—it is neither stable nor predictable over time. (*Moore Products Co.*)

---

[2]Considerable detail on the design, operation, and application of strain gages can be found in the "Process Instruments & Controls Handbook," 3rd edition (D. M. Considine, Editor-in-Chief), pp. 8.1–8.18, McGraw-Hill, New York, 1985
[3]Optal™, Sopelem, Levallois-Perret-Cedex, France.
[4]Information furnished by the engineering staff of Moore Products Co., Spring House, Pennsylvania.

58 Dimension, Displacement, Thickness

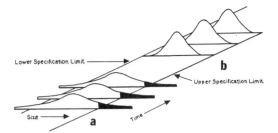

Fig. 15. Demonstration of process variations: (a) Process is in control, but not capable, indicating variation from common causes is excessive. (b) Process in control, and capable, with variation from common causes is reduced. (*Moore Products Co.*)

It should be noted that Cp does not relate the mean to the midpoint of the tolerances. If the mean is not at the midpoint, out-of-tolerance parts may still be probable even if the process is capable.

Cpk—Capability in Relation to the Specification Limits: Cpk relates the capability of a process to the specification limits. The formula is CpK = lesser of

$$\frac{USL - Mean}{3S} \quad or \quad \frac{Mean - LSL}{3S}$$

The Cpk value is useful in determining whether a process is capable and is producing good parts based on the specification limits. Values for the Cpk index have the following meaning:

Greater than 1.0	Both of the 6 sigma limits fall within the specification limits. The process is both capable and producing good parts (99.73% or greater).
1.0	At least one of the 6 sigma limits falls directly on the specification limit.
0.0 to 1.0	At least one of the 6 sigma limits falls outside the specification limit. The process is either not capable or the mean is not at the midpoint of the specification limits.
Less than 0.0	The mean is outside of the specification limit.

Capability Ratio: This is the inverse of Cp. The formula is:

$$\frac{6S}{USL - LSL}$$

The value for this ratio can be thought of as the portion of the part tolerance consumed by six sigma. A common interpretation for the various values of the capability ratio is:

.50 or less	Desirable
.51 to .70	Acceptable
.71 to .90	Marginal
.91 and greater	Unacceptable

Common Cause: A source of random variation that affects all the individual measurements in a process. The distribution is stable and predictable. (See also *Special Cause and Variation* later in this glossary.)

Control Limits—LCL, UCL: The upper and lower control limits are values used to determine whether or not a process is in control. The values are inherent to the process and should not be confused with specification limits.

Histogram: A chart that plots individual values versus the frequency of occurrence and is used for statistical data analysis.

Individual: A single measurement of a particular process characteristic.

Kurtosis: This is an indication of whether the data in a histogram have a normal distribution. Specifically, it is a measure of the flatness or "peakedness" of a curve. The formula for kurtosis is:

$$\sum_{i=1}^{n} \frac{(X_i - \overline{X})^4}{4S^4}$$

Values for kurtosis have the following meaning:

3	Normal distribution.
Less than 3	*Leptokurtic curve*, i.e., the curve has high peak (data are concentrated close to the mean).

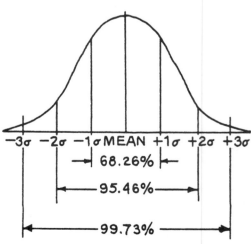

Fig. 16. Normal distribution curve. (*Moore Products Co.*)

Greater than 3   *Platykurtic curve*, i.e., the curve has a low peak (data are disbursed from the mean).

Note: In connection with a later description of the *Data Gage*™[5], the following form of the equation is used:

$$\frac{\eta^3 \sum_{i=1}^{n} X_i^4 - 4\eta^2 \sum_{i=1}^{n} X_i \sum_{i=1}^{n} X_i^3 + 6\eta \left(\sum_{i=1}^{n} X_i\right)^2 \sum_{i=1}^{n} X_i^2 + 3 \left(\sum_{i=1}^{n} X_i\right)^4}{\left[\eta \sum_{i=1}^{n} X_i^2 - \left(\sum_{i=1}^{n} X_i\right)^2\right]^2}$$

Mean—X: This is the mean or arithmetic average value of the data. The process mean is the average of all the process data. The subgroup mean averages just those values in the subgroup. The formula is:

$$\frac{1}{\eta} \sum_{i=1}^{\eta} X_i$$

Median—X: This is the value of the middle individual when the data are arranged in order from lowest to highest. If the data have an even number of individuals, the median is the average of the two middle values.

Mode: This is the most frequently occurring value, i.e., the highest point on a histogram.

Normal Distribution: Data are often summarized graphically as a means of better understanding and analyzing the variation. A plot of the frequence of occurrence of a particular variable is one such commonly used tool for analysis. If a definite pattern emerges from the data, this plot is referred to as a distribution.

Many distributions have been identified and named inasmuch as their pattern of distribution is repeatable and certain mathematical characteristics can be defined for each distribution. One of the most commonly occurring patterns is the normal distribution, as shown in Fig. 16. The normal distribution describes many natural and human-caused phenomena.

The mean and standard deviation define a specific normal distribution. Knowing these values, one can predict the total spread of expected outcomes. It is important to note that 99.73% of the population lies between -3 standard variations (often called -3 sigma) from the mean and $+3$ standard deviations ($+3$ sigma) from the mean. This is the basis for many statistically calculated indications of the status of a process, including capability and control limits.

It is equally important to note that if the distribution of a process is not normal, a number other than 99.73% of the parts will fall within six standard deviations. Because of this, many of the statistical control indicators calculated from a *nonnormal* distribution will not accurately describe that distribution. However, often a distribution is close enough to normal that the errors are insignificant.

Pareto Chart: A chart that ranks events according to a specified parameter, such as frequency or cost. *Note:* In connection with the later description of the *Data Gage*™, the instrument ranks rejected individuals by frequence of occurrence.

Process Quality Coefficient: Also referred to as *process capability coefficient*, this denotes the area under the normal curve that falls within the specification limits. It is expressed as a percentage. The calculation is a multistep procedure:

1. Calculate the distance of the mean from the specification limits in terms of standard deviations:

$$*L_u = \frac{USL - \bar{X}}{S} \qquad L_L = \frac{\bar{X} - LSL}{S}$$

2. Determine the area outside of the specification limits expressed as a fraction of the normal curve**. See equation below†.

---

*For this equation to be valid, the values for L_u and L_L must be greater than or equal to 0. If either is negative, the corresponding area is calculated, using $A = 1 - A'$, where A' is calculated using $L' = L$.

**When a tolerance limit is unused (has a U prefix), the area under the normal curve that is outside that limit does not enter into the process quality calculation.

---

[5]Moore Products Co., Spring House, Pennsylvania.

$$\dagger A \text{ UPPER} = \frac{1}{\sqrt{2}} \cdot e^{-\left(\frac{L_u^2}{2}\right)} \cdot \left[B_1 Y_u + B_2 Y_u^2 + B_3 Y_u^3 + B_4 Y_u^4 + B_5 Y_u^5\right]$$

$$A \text{ LOWER} = \frac{1}{\sqrt{2}} \cdot e^{-\left(\frac{LL^2}{2}\right)} \cdot \left[B_1 Y_L + B_2 Y_L^2 + B_3 Y_L^3 + B_4 Y_L^4 + B_5 Y_L^5\right]$$

where: $Y_u = \dfrac{1}{1 + R \cdot L_u}$ $\qquad Y_L = \dfrac{1}{1 + R \cdot LL}$

$R = 0.2316419$
$B_1 = 0.31938153$
$B_2 = -0.356563782$
$B_3 = 1.781477937$
$B_4 = -1.821255978$
$B_5 = 1.330274429$

Dimension, Displacement, Thickness

3. Process Quality = $[1 - (A_{Upper} + A_{Lower})]100\%$

Range—R: This is the difference between the highest and lowest value in the group. The average of subgroup ranges is denoted as $\bar{R}$.

Run Chart (xR Chart): A chart that displays the most recent individual measurement (x) and the absolute difference from the previous measurement (R) on a consecutive basis.

Sample: A sample is a synonym of a subgroup in process control applications. Sometimes, however, "sample" is used to denote an individual reading within a subgroup. Because this can be confusing, "subgroup" is the preferred description.

3 Sigma: Plus or minus ($\pm$) 3 sigma are the two specific points on a normal distribution centered about the mean, and 99.73% of the population will fall between these values. Since this is essentially the entire population, $+3$ sigma and -3 sigma represent the probable range of variation.

$$+3 \text{ sigma} = \bar{X} + 3S$$
$$-3 \text{ sigma} = \bar{X} - 3S$$

Skewness: This is an indication of whether the data in the histogram have a normal distribution. Specifically, it is a measure of symmetry. Values for skewness have the following meaning:

0	Symmetrical distribution.
Greater than 0	Positive skewness, i.e., the distribution has a longer "tail" to the positive side (the median is greater than the mode).
Less than 0	Negative skewness, i.e., the distribution has a longer "tail" to the negative side (the median is less than the mode).

The formula for skewness is:

$$\sum_{i=1}^{n} \frac{(X_i - \bar{X})^3}{3S^3}$$

Note: For convenience in connection with the *Data Gage*™, the following form of the equation is used:

$$\frac{\eta^2 \sum_{i=1}^{n} X_i^3 - 3\eta \sum_{i=1}^{n} X_i \sum_{i=1}^{n} X_i^2 + 2\left(\sum_{i=1}^{n} X_i\right)^3}{\sqrt{\eta \sum_{i=1}^{n} X_i^2 - \left(\sum_{i=1}^{n} X_i\right)^2}}$$

Special Cause: A source of nonrandom or intermittent variation in a process. The distribution is unstable and unpredictable. It is also referred to as an *assignable cause*. (See also terms "Common Cause" and "Variation" given in this list.)

Specification Limits—LSL, USL: The upper and lower specification limits (blueprint tolerances) are values that determine whether or not an individual measurement is acceptable. They are engineering tolerances established external from the process and should not be confused with control limits.

Standard Deviation: This is a measure of the variation among the elements in a group. The formula for standard deviation is:

$$S = \sqrt{\sum_{i=1}^{n} \frac{(X_i - \bar{X})}{\eta - 1}}$$

Note: For convenience with the *Data Gage*™, the following form of the equation is used:

$$\sqrt{\frac{\sum_{i=1}^{n} X_i^2 - \left[\left(\sum_{i=1}^{n} X_i\right)^2 / \eta\right]}{\eta - 1}}$$

Subgroup: This is a group of individual measurements, typically from two to ten, used to analyze the performance of a process.

Variation (Concept of): Variation occurs in manufacturing because the process conditions are not exactly identical for any two parts. If variation did not exist, obviously quality would not be a problem. Every piece would be identical, so all the components would be assembled and test correctly, *without exception*. Since differences do exist, the goal of SPC (Statistical Process Control) is to reduce the variation as much as possible, thereby improving product consistency and quality.

For the purposes of SPC, the causes of variation are divided into two classifications: (1) Common causes and (2) special causes. Common causes of variation influence the size of every part. These sources of variation are inherent to the process and are predictable. When a process is subject only to common causes of variation, it is said to be *in control*. Special causes of variation arise on an irregular basis and generally influence only some of the product. A broken tool, a worn bearing, a faulty part blank, or a misadjustment by the operator are examples of special causes. When a process is subject to special causes of variation it is said to be *out of control*.

XR Chart: A chart that displays the subgroup mean and range on a consecutive basis. Upper and lower control limits are plotted to assist in the analysis of the process.

Data Gage™

The SPC data gage shown in Fig. 17 can be used in three different modes: (1) *Static gaging*—the part status is determined from a single set of readings usually taken when the part is at rest in the fixture. The status of each measured parameter can be based on a single circuit or derived from several circuits. Among parameters commonly measured are those shown in Fig. 18. (2) *Dynamic gaging*—each circuit takes readings continuously throughout the gaging cycle. A

Dimension, Displacement, Thickness 61

Fig. 17. Example of SPC (Statistical Process Control). (*Data Gage*™, *Moore Products Co.*)

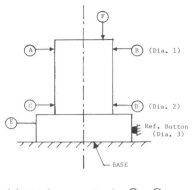

CHECKS:
1. Dia. 1 (maximum)   Dia. 1 = Ⓐ + Ⓑ
2. Dia. 1 (roundness)   Roundness = TIR (Dia. 1)
3. Dia. 2 (maximum)   Dia. 2 = Ⓒ + Ⓓ
4. Dia. 2 (average)   Dia. 2 (average)
5. Taper Dia. 2 - Dia. 1   Taper = Dia. 2 - Dia. 1 (maximum)
6. Dia. 3 (average)   Dia. 3 = Ⓔ average
7. Squareness - Small   Squareness (See zig zag below)
   Dia. to Base   [(Ⓐ + Ⓓ) - (Ⓑ + Ⓒ)]
8. Height (maximum)   Height Ⓕ
9. Runout Dia. 1 to Dia. 3   Runout = Ⓑ TIR
10. Runout Dia. 2 to Dia. 3 Runout = Ⓓ TIR
                             TIR = Total Indicated Runout

Fig. 18. Representative dimensional parameters measured by (*Data Gage*™, *Moore Products Co.*)

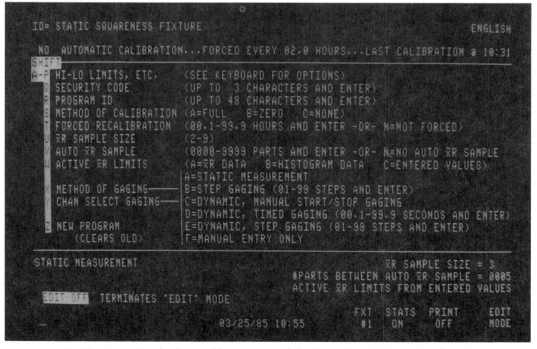

Fig. 19. Prompting display used by operator of the *Data Gage*™. (*Moore Products Co.*)

62 Dimension, Displacement, Thickness

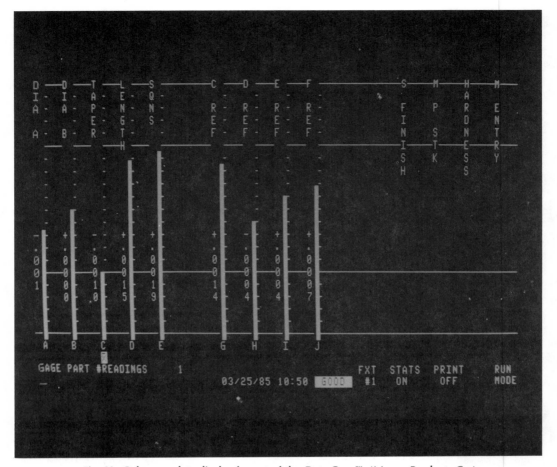

Fig. 20. Columnar data display format of the *Data Gage*™. (*Moore Products Co.*)

built-in timer or manual keystroke defines the cycle. Part characteristics, such as maximum or minimum dimensions and runouts, can be monitored in this mode. (3) *Step gaging*—several readings are used to determine the status of each parameter. Readings are taken each time the "GAGE PART" key is pressed so that subjective operator influences can be removed or interrupted surfaces can be checked. Once the required number of readings has been taken, the part status is determined.

Sixteen analog inputs can be monitored simultaneously. Inputs may be any combination of LVDTs, half-bridge transducers, transduced air signals, or any other .5 to 10V analog signal. Simple software is used for easy programming. The operator need understand only basic gaging principles and answer questions presented on setup displays. Special worksheets are included to assist in organizing the part, master, and fixture data into a logical format. The instrument's setup screens duplicate the worksheet format. Automatic calibration routines utilize either a zero master or a pair of minimum and maximum masters. This procedure compares the master readings to the known master values. The gain and zero of each circuit are automatically adjusted to correct for probe wear or circuit drift. The calibration of all (16) circuits can be handled simultaneously without operator adjustment of pots or knobs.

Examples of displays furnished by the Data Gage™ are shown in Figs. 19 through 23. The semiautomated instrumental gaging system, particularly from the standpoint of information handling and presentation and ease of manual operation, is representative of numerous tools that, even though not fully automated, make automation of floor operations viable and the entire process faster and more efficient.

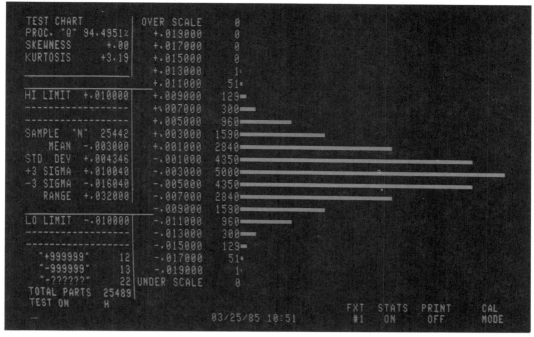

Fig. 21. Histogram produced by *Data Gage*™. (*Moore Products Co.*)

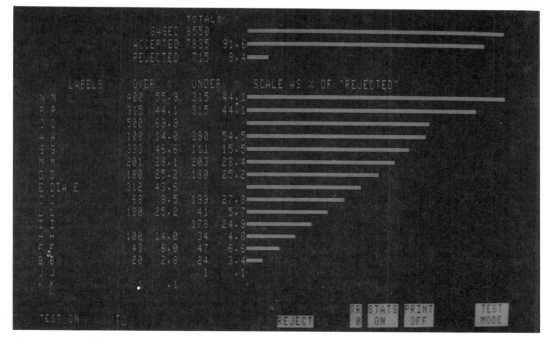

Fig. 22. Pareto diagram display format produced by *Data Gage*™. Listing includes total parts gaged, parts accepted, and parts rejected, with the parameters for which parts rejected in descending order of frequency. The most important corrections become more directly obvious. (*Moore Products Co.*)

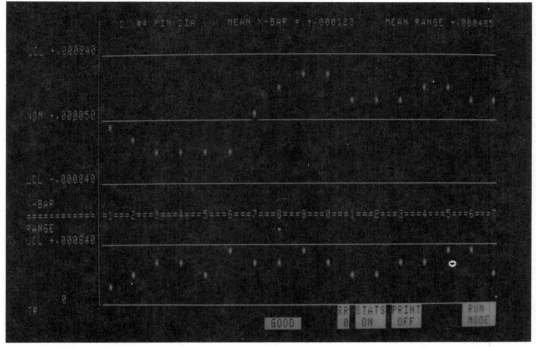

Fig. 23. Example of X̄R chart produced by *Data Gage*™. (*Moore Products Co.*)

AUTOMATIC GAGING AND SIZE-CONTROL SYSTEMS

Prior to the mid 1970s, a large percentage of automatic gaging systems were pneumatic. Although no longer the system of choice for a majority of users, some systems are in use. (See Fig. 24.)

Popular sensors and transducers used in state-of-the-art electric and electronic gaging systems include LVDTs, half-bridge transducers, and inductive-bridges, described earlier in this article. Machine vision is also impacting the field. See two articles on machine vision later in this *Handbook* section.

The use of impedance-type dimension gages is illustrated in Fig. 25.

Automatic gaging systems on the factory floor are used to provide signals and controls for a variety of applications, including (1) adjusting movement of cutting or forming tools, (2) signaling for replacement of worn tools, (3) stopping machines, (4) warning when part-size trend approaches minimum or maximum specification limits, (5) segregating parts at various stages of production, (6) connecting or straightening parts automatically, (7) matching parts for selective assembly, (8) inspecting finished parts, (9) classifying and matching parts, and (10) selectively packaging parts.

Microprocessor-based Camshaft Gage

Although short of being fully automated, the gaging system for inspecting camshafts right on the factory floor, shown in Fig. 26, is exemplary of how gage manufacturers have taken full advantage of the latest electronic technology, including microprocessors in the circuit and advanced color display formats[6]. All of the critical dimensions of the cam shaft, as shown in Fig. 27, are inspected *simultaneously* on the production floor. With a system of this type, it is no longer necessary to limit difficult characteristics, such as lobe profile, to sampling checks in a laboratory environment. Tooling mounts the camshaft on centers. All surfaces are tracked by customized followers, which closely represent the actual usage in the engine. The inspection cycle is programmed and is fully automatic. Although in most installations the operator presents and removes

[6]Moore Products Co., Spring House, Pennsylvania.

Dimension, Displacement, Thickness

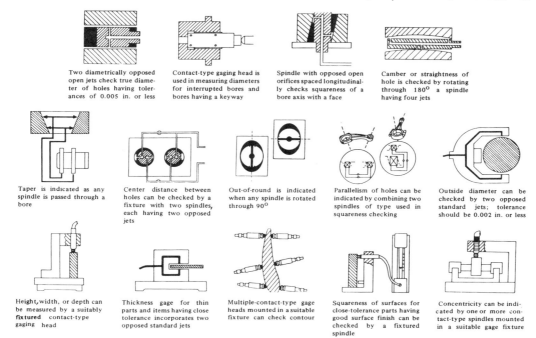

Fig. 24. Basic applications of pneumatic gaging. For many years, pneumatic gaging predominated the automatic gaging field. Pneumatic means are still used, but now are in a minority position. A pneumatic gaging system consists essentially of components that provide a constant pressure air supply, an indicating means, and a metering orifice. The principle of operation is based on the effects of varying the flow of air from the metering orifice—by the obstruction of the part being measured. An obstruction causes a build-up of pressure to the regulated value and flow through the system drops. Over a significant range of values in such a system, there exists a linear relationship between flow or pressure and the size of the escape orifice. In some gaging applications, the linear relationship can be equated to the clearance that separates a sized metering orifice and an obstruction. In automatic systems pneumatic-to-electric transducers are frequently used.

the part piece, mechanical feed can be added for complete automation of the inspection and sorting functions. Gaging and readout is digital, the display is 8-color graphic with printed outputs. Statistical analysis charts are provided in a number of formats.

Measuring Machines

The measuring machine with digital readout has been available for many years. The workpiece is mounted on the table in the conventional manner, and a probe is moved in the x, y, and z axes until it is positioned in a reference hole or against a reference surface. The readout then is cleared to zero by pushing a reset button. In earlier versions of the machine, the probe is manually moved to the first check location, and the amount of movement in each axis is displayed. The output signals are interfaced directly with electronic counters and computers, thus adding the use of storage banks, logic, and capability to perform mathematical computations. The clear-to-zero feature allows measurement to begin at any point. Readout is automatic to any point within the measurement range. All readouts are progressive from zero and are preceded by a + or − sign to indicate position.

The coordinate measurement system may be a steel grating with the reading head mounted so that the short glass grating in the reading head can be superimposed over the steel grating at a slight angle. As shown in Fig. 28, a beam of light is passed through the glass grating, reflecting back from the steel grating onto four photocells. The relative movement of the two gratings produces optical patterns or fringes. As these fringes move, they interface with the beam of light being reflected from the steel grating. The fringe pattern travels across the grating at right angles to the physical motion. Photocells in the reading head detect the interference fringes (dark and light bands) and convert the changes in light intensity into electrical signals, thus counting the fringes and indicating the precise amount of movement by digital display.

The coordinate measurement system may use other physical principles. For example *Induc-*

66 Dimension, Displacement, Thickness

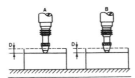

By placing two gage heads A and B parallel and on the same side of a workpiece and master with gage settings properly actuated, the indicator will show the difference between A and B. If both master and workpiece are affected by the same source of error D (such as temperature), the difference will remain the same. Such a setup can be used for measuring roundness, parallelism, and flatness.

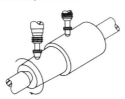

With the setup shown above, only the difference in concentricity is shown on the indicator. If both parts are out of round in same amount, the reading is not affected.

By placing two gage heads parallel on a tapered part, it is possible to check the degree of taper as compared with a master part, without regard to diameter.

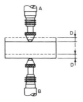

By placing two gage heads A and B perpendicular and opposite to the workpiece with gage settings properly attached, the indicator will show the sum of A and B. If the workpiece is displaced either upward or downward, the indicator reading will not change. In this example, gage head A will move a distance +D while gage head B moves a distance −D so that the reading of the thickness of the part does not change because of such displacement of the part. Such a setup can be used to measure thickness or diameters without the need for precision fixtures.

With this setup, the diameter of a workpiece is compared with that of a master without the necessity of placing the part in a precise fixture, since upward or downward displacement does not affect the accuracy of the reading.

Fig. 25. Use of gaging heads (impedance-type) for measuring fundamental dimensional parameters. Principles shown hold for manual, semi-automated, or automated systems. (*Brown & Sharpe*.)

tosyn™ plates may be used. This inductive-plate method is described in some detail under "Inductive Plate Transducers" in the preceding article of this *Handbook*.

The *NumeriProbe* system[7] is used in conjunction with computer numerical control of machine tools. This gaging system allows the CNC programmer to include commands in the part program to probe workpiece features, thus reducing operator intervention, providing finished part inspection information, and increasing the probability of machining the part to the specified tolerance. On machines equipped with automatic tool changers, the probe can be loaded into the tool storage matrix and automatically selected and loaded into the machine by the tool changer. On machines with turrets, the probe is permanently assigned to a specific turret station.

The sensor unit is an omnidirectional switch that is activated when the probe stylus contacts the object being measured. The instant the stylus is deflected in any direction, a signal is transmitted to the CNC control. The CNC control, upon receiving the transmitted signal, automatically stores the coordinate positions of the moving axes of the machine slides at that instant. By comparing the actual touch location with the programmed location in the part program, the measuring routine determines appropriate compensation, which is then used to augment the CNC program or determine the error, if any, in position, skew, or size.

The in-process measurement can be used for: (1) verification of fixture position, (2) part alignment and setup, (3) inspection of part geometry—position or diameter of hole, squareness, parallelism, and straightness, (4) part identification, (5) automatic tool offset—length and diameter, and (6) workpiece stock check.

Up to three axes can be probed simultaneously. A printout of all probe data can be obtained.

Eddy Current Inspection System. A six-axis coordinate measuring machine and eddy current inspection system has been developed[8]. The system inspects the dimensional features of large jet

---

[7]Giddings & Lewis Electronics Company, Fond du Lac, Wisconsin.
[8]Anorad Corporation, Hauppauge, New York.

Dimension, Displacement, Thickness 67

Fig. 26. Microprocessor-based semi-automatic cam shaft gage which operates on the factory floor. The gage presents analytical charts in different formats on an 8-color screen. Difficult characteristics, such as lobe profile, are measured, eliminating need for sampling checks in a separate metrology laboratory. (*Moore Products Co.*)

engine parts and detects minute superficial cracks on external surfaces and in the numerous internal air cooling hole bores. The system is capable of inspecting a 500-pound (227 kg) workpiece, measuring 36 × 36 inches (915 × 915 mm) in diameter and height or less. The programmable measuring area (x, y, and z axes) is 48 × 48 × 36 inches (1220 × 1220 × 915 mm), respectively. Air bearings ride on a flat granite surface guided on precision straight edges. Ball screw drives reach a speed to 8 inches (203 mm) per second with linear encoder position resolution of .0001 inch (.0025 mm). The x-axis carriage, which travels under a y-axis cross bridge, includes a 25-inch (635 mm) diameter worm gear driven rotary table, offering positioning resolution to .002 degrees (7.2 arc seconds) at speeds up to 20 RPM. Mounted to the x carriage is a tool-holding nest that can store up to four inspection instruments (eddy current devices or Renishaw type touch probes) for automatic tool changing.

To align the probe in a position normal to the compound angles on the contour of the part, the spindle of the z-axis combines two rotary positioning tables, mounted 90° to each other on a stiff angle bracket. These tables are set for 360° and 120° of rotation. A universal tool-gripping device is mounted to the spindle of the z axis and

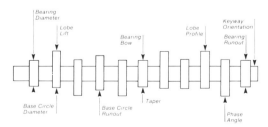

Fig. 27. Critical dimensions checked by cam shaft gage. (*Moore Products Co.*)

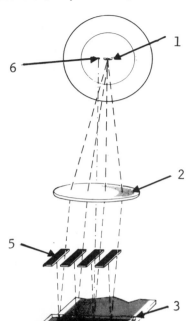

Fig. 28. Optical system used in some measuring machines: (1) Line filament, (2) collimating lens, (3) index grating, (4) scale grating, (5) photocell strips, (6) principal focus of lens.

automatically picks up the selected tool, locks it in position, and removes it from the holding nest. Preprogrammed subroutines are initiated through a microprocessor-based controller, configured to interface with the user's host computer.

Because of the large envelope size and the use of dimensionally stable granite as the machine base, the system measures 13.5 feet (4115 mm) high, 8 feet (2440 mm) long across the bridge, and 10.5 feet (3200 mm) in base length. The granite structure weighs 25,000 pounds (11340 kg). Total gross weight is in excess of 14 tons. The six-axis system has numerous standard features, such as home reference on all axes, electrical and mechanical limits, linear and circular interpolation, accordion type bellows on linear axes, and vibration isolation mounts.

Noncontact, Laser-controlled Inspection System. A system has been developed[9] for automated inspection of two- and three-dimensional objects. The system features multiple lenses, a laser interferometer positioning control system, autofocusing, air bearings, and an integral vision computer that permits operator use without any prior programming knowledge. The vision computer features a resolution of 256 × 256 pixels to 512 × 512 pixels, depending upon camera selection. Standard linear measurement accuracy is ±0.5 micrometer utilizing a 0.1 micrometer resolution laser interferometer system. Standard travel is 24 × 24 inches (610 × 610 mm) up to 36 × 50 inches (915 × 1270 mm). Lens magnification of 1.3X, 2.5X, 10X, and 40X can be selected automatically.

A machine of this type is used for the inspection of precision artwork, measurement of specific features and items on surface mount boards, three-dimensional inspection of high-precision parts. The system has the capability of performing inspection simulation or previewing the inspection (dry run). Graphic or numeric options allow the operator to monitor the inspection process. For extremely accurate measurements, compensation is included for the thermal expansion of the part being inspected.

POSITIONING TABLES AND PRECISE POSITIONERS

Although there always has been a requirement for positioning systems with a resolution of ±.0001 inch/inch (mm/mm), it was not until relatively recently that such requirements became more widespread—particularly by the electronic components manufacturing industry, the use of laser cutting for numerous applications, and other examples from new technology. In the electronics manufacturing and allied microfabrication and micromanufacturing industries, there also has been an increased need for very precise, small-size position-control setups.

CNC Inspection Workstation. Traditionally, computerized numerical control, (CNC) has been applied to large, massive machine tools with distances of several feet (one meter plus) involved in the movement of the x, y, and z axes. Much smaller units are now used both for production and inspection tasks. One workstation[10] has been designed for performing noncontact optical inspection of a variety of parts and tools. In one unit two standard linear positioning tables (x, y) are combined with a rotary table (ϕ axis), a manual workpiece locating fixture, and a computerized axis control system. The latter provides motion control of the workstation from a down line host computer system. An x-mounted optical detector is mounted at a right angle to the adjustable workpiece centering fixture. The x-axis has

---
[9]Anorad Corporation, Hauppauge, New York.
[10]Anorad Corporation, Hauppauge, New York.

12 inches (305 mm) of travel and is driven by a linear DC motor. The z-axis has 24 inches (610 mm) of travel. This combination of positioning tables is mounted on a precision lapped vertical granite base for stability and longterm accuracy.

Linear Positioning Tables. Outstanding refinements have been made in linear motors during the past decade. See also article on "Linear and Planar Motors" in *Section 4* of this *Handbook*. For example, a very high-speed positioning table with a linear DC motor has been developed for use in automatic assembly, robotic applications, and electronic, electromechanical, and semiconductor manufacturing. The positioning table[11] achieves a constant velocity of 60 inches (1524 mm) per second over a distance of 3 inches (76 mm), with acceleration and deceleration rates of 1 g. Advantages claimed for the linear motor include reduced maintenance through elimination of drive lead screws, nuts, pillow blocks, and bearings. The linear motor's permanent magnet is affixed to the slide of the table, and the coil assembly is fastened to the table base.

A laser cutting machining center has been designed for metal part fabrication and is capable of accurately positioning and moving metal sheets, permitting the laser to cut shapes, notches, holes, and intricate contouring at selected speeds. Laser technology is favored for some piece part manufacture because it eliminates the need for more costly traditional tooling, such as blanking dies. The system[11] consists of two standard 14 × 36 positioning tables capable of locating a 36-inch (915 mm) square metal sheet anywhere beneath the laser cutting head. The system has two axis movement from a zero datum point and features a resolution of ±.0001 inch/inch (mm/mm). Table speeds are programmable, with regulated cutting speeds up to 400 inches (10,160 mm) per minute. The machine is located on a granite bridge and base for stability. The controller features linear and circular interpolation for laser cutting circles, as small as .050 inch (1.27 mm) and special shapes.

THICKNESS MEASUREMENT

Thickness may be defined as a dimension from one surface to an opposite surface, as distinguished from length and width. Automated thickness measurement is widely used in the continuous production of sheets and webbed materials (sometimes called flat goods or continuous-length products), both metallic and nonmetallic, such as paper and plastics, coated materials, aluminum foils, et al. Many types of transducers, including some of those described in the earlier portions of this article, are used. Frequently, however, thickness is measured by some form of radiation (nuclear, x-ray, and ultrasonic), as described in the following paragraphs.

Nuclear Radiation Thickness Gages

Several types of noncontact gaging techniques are used: (1) single-measuring assemblies for sheet thickness control, (2) double-measuring assemblies for control of coatings, (3) a single-gage system for controlling coatings on paper and plastics, and (4) a single-gage system for coating control on sheet steel.

A basic single-head gage is shown in Fig. 29. This holds a beta radioisotope source beneath the sheet to be measured and a detector cell located above. The gage can measure a single point or scan the sheet automatically. Transmitted radiation is converted into an electric current and amplified for readout or automatic control. Measurement is basis weight (weight per unit area), such as ounces per square foot. The precision is ±1% of range.

[11] Anorad Corporation, Hauppauge, New York.

A double-gage system is shown in Fig. 30. This system is used to measure and control coatings on paper, textiles, and plastics. Measurement is made of the substrate before the coating, and a second measurement is made after coating. The difference between the two weights determines the net coated weight. The system normally is used when the coating is 20% or more of the substrate.

Gages are matched on dwell cycles so that the same areas are read before and after coating. Signals are transmitted to a difference computer for readout. These can be recorded or used as the basis of automatic coating control. Uncontrolled short-term variations are not properly subtracted from the second signal unless a memory-delay unit is installed.

One system designed to overcome the disadvantages of a two-gage system measures and controls coating on paper and plastics with a

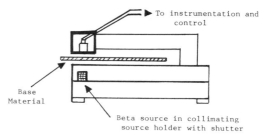

Fig. 29. Basic single-head beta gaging system for sheet thickness control. (*Ohmart.*)

70 Dimension, Displacement, Thickness

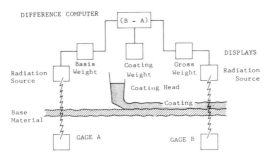

Fig. 30. Double-gage beta-radiation system for control of coatings on paper, textiles, and plastics. (*Ohmart*.)

single gage. A photoelectric effect involving photoabsorption of energy is used. A low-energy gamma-emitting source is utilized. The system can be used when the atomic number of the coating material is at least twice the atomic number of the base material. The preferential absorption principle detects only radiation absorbed by the ingredient with the highest atomic number—which must be the coating. Low energy (less than 0.5mV) from the radioisotope source restricts the system to limited total mass—base material and coating combined. The total weight per unit area and mass coefficient of the materials combination must be calculated for acceptable absorption units.

Another system uses a phenomenon known as *nuclear fluorescence* to measure and control coatings on sheet steel or aluminum. This system measures only coating weight regardless of base material hardness or thickness. Control and readout are in absolute (not relative) terms. The system may be applied to single- or double-sided coating.

Nuclear fluorescence is produced when gamma radiation excites electrons in the metal of the coated strip. As shown in Fig. 31, excitation is continuous, but each occurrence is only temporary. Electrons returning to the original unexcited state produce a characteristic low-energy radiation called fluorescence. The radiation spectrum produced is returned to the detection system. Each fluorescence has its own energy level, which is differentiated by two filters and two ion detection chambers at the measuring point. Detectors are of opposite polarity. One filter permits all back-scattered radiation to reach an ion chamber; the other filter permits only radiation from the metal strip to reach the other chamber. Hence, the base-metal-strip signals cancel one another, leaving only the coating signal as an output from the detectors. Precision is ±0.02 oz/square foot (6.1 grams/square meter). The area measured at any one point is 3×5 inches (76×127 mm), but can be reduced to $\frac{3}{4} \times 5$ inches (19×127 mm) in some instances. These systems bear a strong resemblance to other nuclear systems used for weighing and bulk-solids level detection.

Ultrasonic Thickness Gages

These gages find extensive use in the measurement of wall thickness and are valuable in measuring most structural materials that are good transmitters of sound. Almost all metals, plastics, and ceramics and various composite materials can be measured easily. Measurements are instantaneous and can be made from one side of the part. Suitable transducer configurations also permit readings at elevated temperature for the determination of corrosion losses in chemical equipment while on-stream.

These gages utilize ultrasonic vibrations, most commonly in the portion of the frequency spectrum between 1 and 15 MHz. Sound waves in these frequencies have certain characteristics that make them suitable for making measurements and finding defects. The sound beam can readily pass through most structural materials. It reflects from acoustic boundaries, either internal flaws or geometric boundaries, such as the back surface of the part. Sound travels at a characteristic velocity in various materials; this velocity is dependent on the elastic properties and the density of the material, and this permits timing of the wave propagation through the material.

The basic elements of an ultrasonic gaging system consists of a *generator* to produce high-frequency electronic vibrations, which in turn activate the crystal transducer. The transducer converts the electric signals to mechanical vibrations or ultrasound, which is introduced into the part. Echoes from various boundaries in the part are picked up by the same transducer and are reversibly converted from mechanical vibrations to an electric signal. This is further amplified and processed, and the information is displayed in one of several types of data presentations. Certain repetitive tests can be automated with the application of advanced electronic technology.

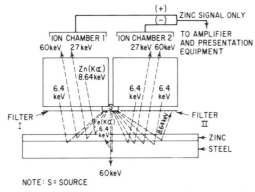

Fig. 31. Nuclear fluorescence gage for controlling coatings, such as zinc, on sheet steel. (*Ohmart*.)

Ultrasonic gages are of two basic types: (1) resonance and (2) pulse-echo. Each has distinct electronic characteristics and readout methods.

Resonance gages produce a frequency-modulated continuous wave signal. This provides a corresponding swept frequency of sound waves, which are introduced into the part. When the thickness of the part equals one-half wavelength, or multiples thereof, standing wave conditions or mechanical resonances occur. The frequency of the fundamental resonance, or the difference frequency between two harmonic resonances, is instrumentally determined. The thickness is determined by the formula:

$$Th = \frac{Vel}{2F}$$

where Th = thickness of part under transducer
Vel = speed of sound in the material
F = frequency, Hz

The most common readout is a large CRT. Readings are instantaneous, and thickness variations can be monitored as the transducer is scanned over the part. Accuracy is usually ±1% of nominal thickness.

Pulse-echo gages operate somewhat like a refined sonar system. Very short electric pulses, usually at discrete frequencies, are generated. These produce short acoustic pulses from the transducer, which in turn pass through the material and reflect from the boundaries, much as in the resonance technique. The *transit time* of the pulses through the material is measured. Thickness is determined by:

$$Th = \frac{Vel \times T}{2}$$

where Th = actual thickness of part under transducer or within sound beams
Vel = velocity of sound in the material
T = transit time of sound pulse through one round trip in the material, seconds

The system lends itself well to battery-operated portable equipment. Accuracies between ±0.5 and 1% of full scale are usually obtainable.

A limitation of ultrasonic gaging is the need for continuous coupling of the sound beam between the transducer and the part. Sound cannot pass across an air-solid or air-liquid boundary. Liquid coupling, either a continuous thin film or some other type, is required. In some instances, complete immersion of the transducer and the material is required. In others, commonly used on production lines, a bubbler or partially contained water column provides the continuous coupling path.

X-ray Thickness Gage

Noncontact x-ray thickness gages measure the thickness or density of hot or cold materials while in motion or when stationary. Steel, aluminum, brass, copper, glass, paper, and rubber—in continuous strip or sheets—as well as plastic films, foils, and material coatings are typical of applications for x-ray thickness gaging. (See Fig. 32.) The gage is set to the desired thickness standard. Deviation in material thickness from nominal is indicated in percent or thousandths or millionths of an inch. A sample piece of the material to be measured or a reference standard is used to calibrate the gage. The gaging signals can be used to operate a variety of accessory instrumentation, including automatic controllers, recorders, totalizers, classifiers, markers, and sorters.

An x-ray gaging system comprises three basic units: (1) a scanning unit that contains the x-ray generator and a detecting unit, (2) an operator control station, (3) a power unit. The scanning unit generally is a C- or O-frame mounted in a stationary position or on a traversing track—as on a steel rolling mill, process, or inspection line.

X-ray gages are accurate to ±0.25% of the thickness being measured or better. Repeatability of readings is 0.01%. Speed of response is very important in keeping fast-moving strip and sheet stock within tolerance. For example, in a mill running 6000 feet/minute (1829 meters/minute), a gage with a 50-millisecond response time reacts to a change in material thickness within 5 feet of the material passing. In contrast, a gage with a 150 millisecond response will permit 25 feet (7.6 meters) of material to pass before indicating a thickness change. Continuous process x-ray gaging is an integral part of a computer-operated rolling mill. Such systems will include a digital display of gage setting, compensation for varying temperatures and alloys of steel, fully automatic calibration to maintain accuracy, and electric-motor-operated positioning systems for locating the scanning unit over the steel. A system may also record the difference in thickness between the edge and the center (crown) of the steel. Anticipatory gaging systems foresee an out-of-tolerance trend or change in material thickness

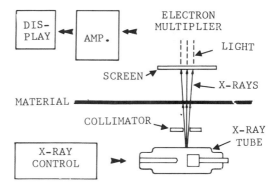

Fig. 32. Principal elements of an x-ray thickness gage.

72 Dimension, Displacement, Thickness

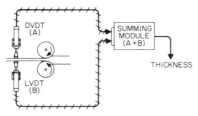

Fig. 33. Two linear variable differential transformers (LVDTs) and a scanning module for measuring thickness. (*Daytronic*.)

and use the signals to actuate screw-down, speed, and tension controls to keep the material on gage. Automatic control permits rolling to close tolerances, maximum on-gage length per ton, and less scrap at both ends of a coil or run.

LVDT Gages

The principle of the linear variable differential transformer, described earlier in this article, also can be used for determining thickness. (See Fig. 33.)

Surface Finish Measurements

Any surface produced by maching departs from the perfect form for many reasons, such as inaccuracies in the machine tool, deformation of the work under the cutting force, and irregularities caused by vibration. In considering the typical surface, two features are of importance: (1) height of the irregularities (essentially variations in thickness), and (2) amount of separation between such irregularities, causing roughness and waviness. Examples are given in Fig. 34. With increasing automation of machining, attention to surface finish takes on increased importance, even though it remains essentially a sophisticated laboratory operation. Without close laboratory supervision, particularly in the production of fine machinery parts by automated machine tools, wasted materials can be created at a rapid rate.

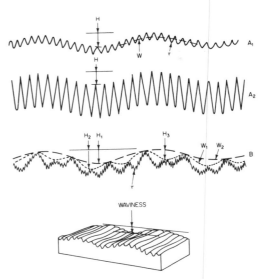

Fig. 34. Waviness. When measuring across the roughness lay, several types of waviness can be found. The profile A_1 has a wavy undulation W on which is superimposed roughness r of smaller amplitude. The profile A_2 is of the same kind, but the roughness is larger in amplitude than the waviness. In these cases, the mean line and crest line waviness are substantially the same. The crest line is irregular and rarely identical with its mean line. Roughly equivalent representations of waviness often can be obtained by filtering. On ground surfaces, the crest spacing of closely spaced waviness surfaces often is in the region from 0.02 to 0.1 inch (0.5 to 0.25 mm), and the height (essentially the thickness) is generally less than half the overall height of the grinding texture. Surfaces also may be encountered where the spacing and height are the same, yet the surfaces are quite different. Complex parameters such as these will benefit from semi-automated instrumental systems (particularly from standpoint of data handling and analysis), but still remain essentially outside the scope of what may be considered advanced automation.

ADDITIONAL READING

Bailey, S. J.: "Position Transducers: Strategic Components in Motion Control Loops," *Cont. Eng.*, **32** (14) 41–45 (1985).

Corser, T., and A. A. Seireg: "Optimizing: A Design for Production, Inspection, and Operation," *CIME*, **4** (2) 18–27 (September 1985).

Jutila, J. M.: "New Dimensions in Optical Gaging," *InTech*, **27** (6) 9 (1980).

Meyers, M.: "Optical Centering and Auto-Focus for Automatic Laser PC Board Drilling and Cutting," *Anonews*, **3** (2) 1 (June 1984), Anorad Corporation, Hauppauge, New York.

"3 Axis CNC Inspection Workstation," *Anonews*, **3** (3) 3 (September 1984).

"Five-Axis System for Laser/Ultrasonic Machining, Assembly, and Inspection Operation," *Anonews*, **3** (4), 2 (December 1984).

"Six Axis Coordinate Measuring Machine and Eddy Current Inspection System," *Anonews*, **4** (1), 1 (March 1985).

Non-Contact Inspection System Integrates Vision with Laser Controlled Positioning Tables," *Anonews*, **4** (3) 1 (August 1985).

Staff: "Surface Roughness Averages for Common Production Methods," *Met. Prog.*, **118** (2), 51 (1980).

Staff: "New Device Measures Laser Wavelengths More Accurately," *Cont. Eng.*, **27** (1) 44 (1980).

Staff: "1986 Control Products Specifier," *Cont. Eng.*, **32** (13) 51–52 (Displacement and Position Instrumentation), 168 (Optical Measurement Equipment) (November 1985).

Tome, D. M.: "Adaptive Control Sensors for Manufacturing Systems: An Overview," *Cont. Eng.*, **31** (8) 78–79 (1984).

Geometric Variables III:

Object Detection, Proximity, Presence, Nonpresence

As pointed out in an earlier article on "Position, Motion, Speed, Velocity," *position* is central among the geometric variables. Generally[1], in position control and motion systems, control action is taken to establish a position or series of positions that lie along a trajectory or path. In these cases the exact coordinates of position are usually paramount. In *object detection*, exactness of position is not usually the case (but there are exceptions)—the primary purpose being that of detecting the presence or absence of an object.

The needs for object detectors are several: (1) conveyor-associated applications, such as jam detection or protection, empty line detection, automatic routing, (2) safety and accident avoidance—to detect human hands and fingers where machines are manually loaded or unloaded and during maintenance procedures, (3) inspection of products—containers filled to the proper levels? labels properly in place or missing? incorrect closures? open-flap detection and folding and wrapping imperfections, detection of web breaks, and plating and coating imperfections, (4) counting—to detect missing parts and to measure throughput, (5) sorting—by size, color, and other parameters, (6) as hopper level detectors and as feed cut-off controllers, and (7) a host of miscellaneous applications, such as edge guidance, remote door openers, and overhanging roof detectors. In industry, as well as in the lay world, these same kinds of devices play a major role as security monitors, warning of the unauthorized presence of persons and actions in designated secure areas.

It is interesting to note that object detection technology, although continuously refined from an engineering standpoint, dates back in principle for several decades and was one of the forerunners of modern industrial automation. Whereas today robots seem to be in the forefront in terms of public recognition, it was not too many years ago that automated conveyor lines, packaging equipment, and the like were the operations proudly shown to visitors by plant managers.

FUNDAMENTAL DETECTOR METHODOLOGIES

Very few of the reasonably viable physical phenomena for detecting objects have been overlooked—a wide potpourri of detection methods is commercially available. A 1985 survey indicated that inductive and photoelectric detectors predominate, but other physical realms include electrical, electromagnetic, optoelectronic, radiant energy, air flow, and sonic approaches. Also very popular among industrial users are capacitive, Hall-effect, Wiegand-effect, and, to a lesser extent, magnetostrictive approaches.

Contact and Noncontact Detectors

A rather clean categorization of object detectors is: (1) contacting types, where the sensor makes actual physical contact with the object, and (2) noncontacting types, where the object only need be in the vicinity of the sensor. Depending on the type of sensing technology used, the vicinity can range from a millimeter or so up to several hundred feet (meters), although these are extreme cases.

Electromechanical (limit) contact switches are described in the preceding article.

Photoelectric Devices

Discovered by Hertz in 1887, the *photoelectric effect* was not fully explained until 1921, when Einstein formulated the photoelectric law. The photoelectric effect is manifested by changes in electrical characteristics of certain substances when they are subjected to radiation, generally in

---

[1]Because of the complexities of industrial instrumentation and automation, it is difficult to safely state sweeping generalizations. Areas of applications and families of technologies are becoming increasingly interdisciplinarian. All of the geometric variables, as described in this and the preceding two articles, blend together almost too well to treat them in neat editorial niches.

the form of light or the near visible. The effect, first understood in terms of metals, causes the irradiated material to lose bound electrons, which are given off with a maximum velocity proportional to the frequency of the radiation, i.e., to the entire energy of the photon. The Einstein law, first verified by Millikan, states

$$E_k = h\nu - \omega$$

where E_k = maximum kinetic energy of emitted electron
h = Planck constant
ν = frequency of radiation (frequency associated with absorbed photon)
ω = energy necessary to remove electron from system, i.e., the photoelectric work function for the surface of the emitting substance

An inverse photoelectric effect results from the transfer of energy from electrons to radiation.

The principal aspects of the photoelectric effect of interest industrially are as follows: (1) *photoconductivity*, evidenced by an increase in electrical conductivity of a material upon the absorption of light (or other electromagnetic radiation), and (2) the *photovoltage* or *photovoltaic effect*, wherein the energy of photons is converted to electrical voltage by the substance receiving the radiation. Photoconductivity is the basis of the operation of photocells and phototransistors utilized in industrial photoelectric switches. Principal applications for photovoltaic cells have been in aerospace applications, satellites for power, and the solar energy field.

Scope of Usage. Photoelectric controls respond to the *presence* or *absence* of either opaque or translucent materials at distances from a fraction of an inch (a few mm) up to 100 or even 700 feet (30 to 210 m). Photoelectric controls need no physical contact with the object to be triggered—important in some cases, such as those involving delicate objects and freshly painted surfaces. Some of the more common applications include thread break detection, edge guidance, web break detection, registration control, parts ejection monitoring, batch counting, sequential counting, security surveillance, elevator and conveyor control, bin level control, feed and/or fill control, mail package handling, and labeling, among many others.

Photoelectric Control System Configuration. A self-contained control includes a light source, a photoreceiver, and the control base function, which amplifies and imposes logic on the signal to transform it into a usable electrical output. A *modular control* uses a light source-photoreceiver combination or reflective scanner separate from the control base. Self-contained retroreflective controls require less wiring and are less susceptible to alignment problems, while modular controls are more flexible in permitting remote positioning of the control base from the input components and hence are more easily customized.

Photoelectric controls are further classified as nonmodulated or modulated. *Nonmodulated* devices respond to the intensity of visible light. Thus, for reliability, such devices should not be used where the photosensor is subject to bright ambient light, such as sunlight. *Modulating controls* employing light-emitting diodes (LEDs) respond only to a narrow frequency band in the infrared. Consequently, they do not recognize bright, visible ambient light.

Controls typically respond to a change in light intensity above or below a certain value of threshold response. However, certain plug-in amplifier-logic circuits cause controls to respond to the rate of light change (transition response) rather than to the intensity. Thus, the control responds only if the change in intensity or brightness occurs very quickly (not gradually).

Operating Mode. Both modulated and nonmodulated controls energize an output in response to:
1. A light signal at the photosensor when the beam is not blocked (light-operated, LO).
2. A dark signal at the photosensor when the beam is blocked (dark-operated, DO).

Although some controls have built-in circuitry that determines a fixed operating mode, most controls accept a plug-in logic card or module with a mode selector switch that permits either light or dark operation.

In addition to a light source, light sensor, amplifier (in the case of modulated LED devices), and power supply, a complete system includes an electrical output device (in direct interface with logic level circuitry—the output transistor of a DC-powered modulated LED device or of an amplifier-logic card).

Scanning Techniques. There are several ways to set up the light source and photoreceiver to detect objects. The best technique is that one which yields the highest signal ratio for the particular object to be detected, subject to scanning distance and mounting restrictions. Scanning techniques fall into two broad categories: (1) thru (through) scan, and (2) reflective scan.

In *thru (direct) scanning*, the light source and photoreceiver are positioned opposite each other, so light from the source shines directly at the sensor. The object to be detected passes between the two. If the object is opaque, direct scanning will usually yield the highest signal ratio and should be the first choice. (See Fig. 1.)

As long as an object blocks enough light as it interrupts the light beam, it may be skewed or tipped in any manner. As a rule, the object size should be at least 50% of the diameter of the receiver lens. To block enough light when one is detecting small objects, special converging lenses

Object Detection, Proximity, Presence, Nonpresence 75

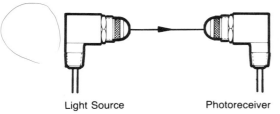

Light Source   Photoreceiver

Fig. 1. In direct (or *thru*) scan configuration, the light source is aimed directly at the photoreceiver. Sometimes the configuration is referred to as the *transmitted beam system*. (*Micro Switch*.)

for the light source and photoreceiver can be used to focus the light at a small, bright spot (where the object should be made to pass), thereby eliminating the need for the object to be half the diameter of the lens. An alternative is to place an aperture over the photoreceiver lens in order to reduce its diameter. Detecting small objects typically requires direct scan.

Because direct scanning does not rely on the reflectiveness of the object to be detected (or a permanent reflector) for light to reach the photosensor, no light is lost at a reflecting surface. Therefore, the direct scan technique permits scanning at farther distances than reflective scanning. Direct scanning, however, is not without limitations. Alignment is critical and difficult to maintain where vibration is a factor. Also, with a separate light source and photoreceiver, there is additional wiring, which may be inconvenient if the application is difficult to reach.

In *reflective scanning*, the light source and photoreceiver are placed on the same side of the object to be detected. Limited space or mounting restrictions may prevent aiming the light source directly at the photoreceiver, so the light beam is reflected either from a permanent reflective target or surface, or from the object to be detected, back to the photoreceiver. There are three types of reflective scanning: (1) *retroflective scanning*, (2) *specular scanning*, and (3) *diffuse scanning*.

Retroflective Scanning. With retroflective scanning, the light source and photosensor occupy a common housing. The light beam is directed at a retroreflective target (acrylic disk, tape, or chalk)—one that returns the light along the same path over which it was sent. (see Fig. 2.) Perhaps the most commonly used retrotarget is the familiar bicycle-type reflector. A large reflector returns more light to the photosensor and thus allows scanning at a further distance. With retro targets, alignment is not critical. The light source-photosensor can be as much as 15° to either side of the perpendicular to the target. Also, inasmuch as alignment need not be exact, retroreflective scanning is well suited to situations where vibration would otherwise be a problem.

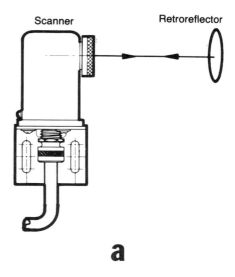

a

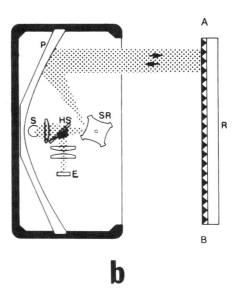

b

Fig. 2. (a) Reflected beam (retroreflective scan) system in which the light source and photoreceiver are contained in a single enclosure. This simplifies wiring and avoids critical alignment of the source and sensor. (*Micro Switch*.) (b) By adding a rotating-mirror wheel (SR), a parabolic reflector (P), and a semitransparent mirror (HS), a parallel-scanning beam can be obtained. This beam moves at high speed from A to B, thus forming a "light curtain," any interruption of which is detected and signaled by a relay, S. E is the photoreceiver. (*Sick Optik Electronik*.)

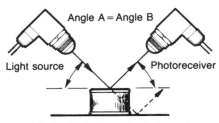

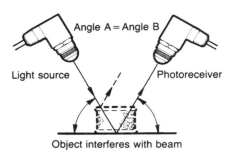

Fig. 3. Specular scan technique uses a very shiny surface, such as rolled or polished metal, shiny plastic, or a mirror to reflect light to the photosensor. (*Micro Switch*.)

Retroreflection from a stationary target normally provides a high signal ratio so long as the object passing between the scanner and the target is not highly reflective and passes very near the scanner. Retroreflective scanning is preferred for the detection of *translucent objects* and ensures a higher signal ratio than is obtainable with direct scanning. With direct scanning, the "dark" signal may not register very dark at the photosensor, because some light will pass through the object. With retroreflective scanning, however, any light that passes through the translucent object on the way to the reflector is diminished again as it returns from the reflector. The system is also useful where retroreflective tape or chalk coding can be placed on cartons for sorting. Retroreflective scanning is normally used at distances up to 30 feet (9 m) in clear air conditions. As the distance to the target increases, the retro target should be made larger so that it will intercept and return as much light as possible. Single-unit wiring and maintenance are secondary advantages of retroreflective scanning.

Specular Scanning. The specular scan technique uses a very shiny surface, such as rolled or polished metal, shiny plastic, or a mirror to reflect light to the photosensor. (See Fig. 3.) With a shiny surface, the angle at which light strikes the reflecting surface equals the angle at which it is reflected from the surface. Positioning of the light source and photoreceiver must be precise. Mounting brackets, which firmly fix the light source–photoreceiver relationship, must be used. Also, the distance of the reflecting surface from the light source and photoreceiver must be consistently controlled. The size of the angle between the light source and photoreceiver determines the depth of the scanning field. With a narrower angle, there is more depth of field. With a wider angle, there is less depth of field. For a full-level detection application, for example, this means that a wider angle between the light source and photoreceiver allows detection of the fill level more precisely.

Specular scanning can provide a good signal ratio when required to distinguish between shiny and nonshiny (matte) surfaces, or when using depth of field to reflect selectively off shiny surfaces of a certain height. When monitoring a *nonflat*, shiny surface with high and/or low points that fall outside the depth of field, these points appear as dark signals to the photosensor.

Diffuse Scanning. Nonshiny (matte) surfaces, such as kraft paper, rubber, and cork, absorb most of the incident light and reflect only a small amount. Light is reflected or scattered nearly equally in all directions. In diffuse scanning, the light source is positioned perpendicularly to a dull surface. Emitted light is reflected back from the target to operate the photoreceiver. (See Fig. 4.) Because the light is scattered, only a small percentage returns. Therefore, the scanning distance is limited (except with some high-intensity modulated LED controls), even with very bright light sources. It is often difficult to obtain a sufficient signal ratio with diffuse scanning when

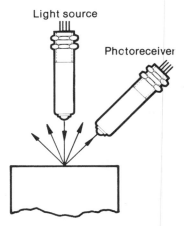

Fig. 4. Diffuse scan is used in registration control and to detect material (corrugated metal, for example) with a slight vertical flutter—which may prevent a consistent signal with specular scan. Alignment is not critical in picking up diffuse reflection. (*Micro Switch*.)

TABLE 1. Factors Affecting Selection of Photosensor and Scan Technique in Registration Control Applications

Background	Mark	Photosensor	Scan technique
Clear film	Black, blue, red	Any	Direct scan
White (kraft paper, metal foil)	Black, blue	Phototransistor or CdSe photocell	Diffuse scan
	Red	CdS photocell with blue-green filter	Diffuse scan
Black, blue, or other dark colors	Red	Any	Diffuse scan
Red	Black, blue	Any	Diffuse scan

the surface to be detected is almost the same distance from the sensor as another surface (for instance, a nearly flat or low-profile cork liner moving along a conveyor belt). Contrasting colors can help in such situations.

Diffuse scanning is used in registration control and to detect material (corrugated metal, for example) with a slight vertical flutter—which might prevent a consistent signal with specular scanning. Alignment is not critical in picking up diffuse reflection.

Color Differentiation (Registration Control). In distinguishing color, as in registration mark detection, contrast is the key. High contrast (dark color on light, or vice versa) provides the best signal ratio and control reliability. Therefore, if possible, bright, well-defined, contrasting colors should be considered in the interest of the registration control system. Diffuse scanning is normally used to detect color change. (See Table 1.)

When the background is clear (transparent), the best method is to detect any color mark with direct scanning. When the background is a second color, contrasts such as black against white usually ensure a sufficient signal ratio (difference between dark and light signals) to be handled routinely with diffuse scanning. Red, or a color that contains considerable red pigment (yellow, orange, brown) on a white or light background is a special case. In such instances, a photoreceiver with a cadmium sulfide cell for detecting red marks is preferred because it makes red appear dark on a light background.

A retroreflective scanner with a short-focal-length lens (but without a retrotarget) can be used to detect registration marks. It is placed near the mark and is actually used in the diffuse scan technique. If a retroreflective scanner is used to detect marks on a shiny surface, the scanner should be cocked somewhat off the perpendicular to make certain that only diffuse reflection will be picked up. Otherwise, the shiny surface of the mark could mirror-reflect so brightly that it would overcome the dark signal that a CdS cell normally receives from red. This would mean a light signal from both the background and the mark. In detecting colors, a rule of thumb is to use diffuse (weakened), rather than specular (mirror) reflection.

Sensitivity Adjustment. Most photoelectric controls have a sensitivity adjustment that determines the light level at which the control will respond. Conditions which may require an adjustment of sensitivity include: (1) detecting translucent objects, (2) a high speed of response, (3) a high cyclic rate, (4) line voltage variation, and (5) a high electrical noise atmosphere.

Light Sources and Sensors. Early photoelectric control systems used incandescent light sources and traditional photocells—a combination still used. A photocell changes its electrical resistance with the amount of light that falls on it. A number of photocells have been used over the years for different applications (photoelectric controls, copying machines, television pickup tubes, etc.). Widely used for photoelectric controls are cadmium sulfide and cadmium selenide cells. During recent years, phototransistors and photodiodes have become available as sensors—and LEDs have been used as light (infrared) sources. There are several advantages in using the more recent hardware.

Photocells. There are at least four parameters that are important in the operation of a photocell: sensitivity, speed of response, light history effect, and effect of temperature.

Static sensitivity—expresses the resistance of the cell at a given light intensity; the lower the resistance, the more sensitive the photocell. So long as the resistance falls within the range of the control unit with which the photocell is used, static sensitivity is usually not important.

Dynamic sensitivity—is an expression of the ratio of photocell resistance at one light level to its resistance at a different level; the greater the ratio, the higher the sensitivity. This is a much more useful expression of photocell sensitivity.

Speed of response—is the time it requires to produce a change in resistance in response to a given change in light intensity. Although all photocells are fast, they require a finite amount of time to respond. Speed of response depends on the amount of light falling on the cell; the greater the light intensity, the faster the response.

Light history—is a characteristic of a photocell that has been kept dark or light for extended periods. Such a cell overresponds to a change in light before returning to its normal response (somewhat analogous to the response of the human eye to sudden changes in light). Usually this is not a significant consideration in industrial applications.

Effect of temperature—increases the resistance of a photocell with increase of temperature for a given level of illumination. The temperature effect is smaller when the level of illumination is high than when it is low.

Color response—affects the performance of a photocell. Photocells generally used in industry have a far greater response to red and in the infrared range than in the blue-violet range.

Phototransistors. A phototransistor produces a collector current that is a function of both base current and light. Since the base lead of a phototransistor is usually left unconnected, only variations in light intensity produce variations in current output. There are several differences between the phototransistor and the photocell: (1) Current output of a phototransistor is largely independent of the voltage across it, whereas that of a photocell is not. As a result, controls designed to work with photocells will not necessarily work well with phototransistors, and vice versa. (2) The response of phototransistors is affected by changes in temperature, but in a way opposite to that of photocells; the higher the temperature, the higher the current output. (3) Phototransistors have a polarity that must be observed; photocells do not. (4) Phototransistors respond to light much faster than photocells, but typically have a lower sensitivity.

Photodiode response is narrower than that of the phototransistor, making the diode more effective in blocking stray light from incandescent, sun, or other sources.

LED Sources. The useful life of an LED is estimated at 100,000 hours, which is at least ten times that of an incandescent lamp. However, incandescent lamps are still frequently used because they have a spectrum from the ultraviolet to the visible to the infrared, allowing a wide range of colored targets to be detected. LEDs have the advantage that they can be modulated directly, whereas incandescent lamps require a mechanical chopper. Silicon phototransistors and photodiodes are excellent matches for infrared LEDs because their greatest sensitivity peaks almost match precisely at the transmitter's (LED) wavelength.

Use of Fiber Optics with Photocells. Fiber-optic bundles can be added to existing photoelectric switches to provide object sensors, and these can be combined to implement logic functions. Such systems are useful for applications that require several sensing inputs and one or more outputs to interface with microcomputers. Program selection permits use of the LO or DO mode and allows operation of any channel for a predetermined time, thus avoiding sequential channel operation in fixed time frames. These systems frequently find application where a programmable controller is not warranted because of cost or complexity. Input can be from a relay or switch contacts, transducers, memory devices, CMOS, or TTL. The output section provides a channel signature for each emitter and detector pair, resulting in the capacity of actuating one or more output devices.

Applications of Photoelectric Controls. As illustrated by Fig. 5, the applications for photoelectric controls in automated systems seem to be limited only by the ingenuity of the control and system engineer. Most of these applications can be served by other types of proximity sensors, but there are exceptions.

Magnetic Proximity Switches

There are four principal types of magnetic proximity switches: (1) variable-reluctance-type sensors, the operation of which depends on the interruption of a fixed magnetic field (circuit) by a ferrous actuator, (2) magnetically actuated dry reed or mercury switches, (3) Hall-effect sensors[2], and (4) Weigand-effect sensors.

Variable-reluctance Sensors. The principle of operation of variable-reluctance position (presence) sensors is shown schematically in Fig. 6. These transducers convert motion (rotating, sliding, oscillating) into electrical control signals. As shown in Fig. 6(a), with no actuating object in the vicinity of the sensor (pole piece plus coil plus magnet), the path of magnetic flux is undisturbed. As an object approaches and passes near the pole piece, the flux path is distorted. This system is often used in connection with rotating equipment for speed measurement (tachometry) where discontinuities, such as gear teeth, shaft keyways, drilled holes in steel plates, etc., alter the magnetic flux in proportion to rpm. Sensors are available in active and passive forms. Passive sensors require no external electric power. The output signal is an alternating current, the waveform of which is a function of the actuator, usually sinusoidal. The amplitude and frequency of the output signal are both proportional to the surface speed of the actuator as it passes the sensor's pole piece. The active configuration requires a DC power supply. The output signal is a pulse train whose amplitude is constant over the operating range for a fixed supply voltage level. Active magnetic sensors provide usable output signals at very low actuator speeds and at relatively large air gaps between the sensor pole piece and the actuator. They produce a logic-level output signal directly compatible with digital instrumentation.

---

[2] Hall-effect switches are discussed under "Solid-State Switches" in a prior article in this *Handbook* section on "Position, Motion, Speed, Velocity."

Object Detection, Proximity, Presence, Nonpresence 79

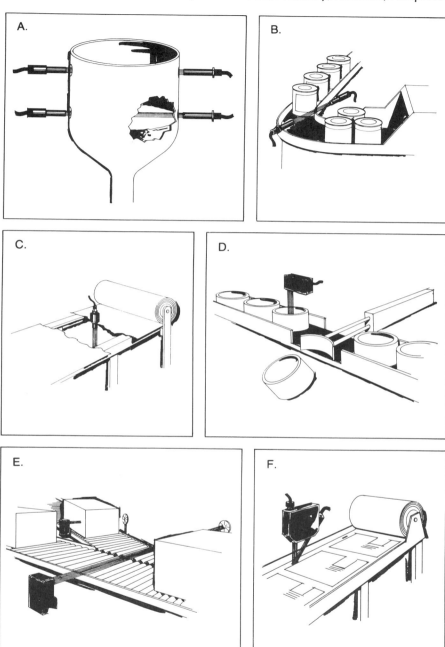

Fig. 5. Representative applications of photocell detectors in automated systems: (A) Two light source–photoreceiver pairs are used to keep hopper fill level between high and low limits. (B) Counting products is a common application of photoelectric controls. Counting batches or groups of cans or other items prior to packaging or group processing is also common. (C) A photoelectric control operating on reflected light is a simple way to detect a web break. An alternative is to put a light source above the web, and a photoreceiver below. (D) Dark caps are checked for white liners by a photoelectric scanner. The scanner activates a mechanism that rejects caps without the liners. (E) To prevent collisions where two conveyors merge, each

(*Continued on next page*)

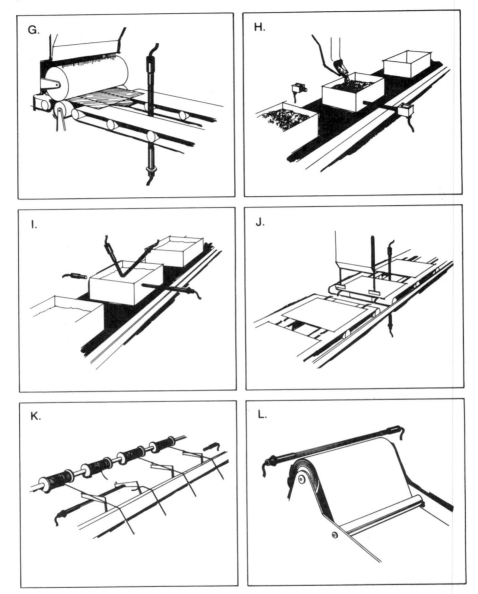

(Continued from previous page)
conveyor is monitored by a control that powers the other conveyor when its own conveyor is cleared. (F) A tubular light source and photoreceiver in a specially designed bracket detect registration marks to initiate any related operation, such as printing, cutoff, or folding. (G) Gluing, buffing, or flattening can be done efficiently by controlling the pressure rollers or buffer with a photoelectric light source and photoreceiver that detect the product to be processed. (H) Using logic for one-shot pulse output, a photoelectric control slows a conveyor and fills the carton which has interrupted the light beam. (I) Two light source-photoreceiver pairs work together to check fill level. The box-detecting pair turns on, or enables, the fill inspection pair—thereby preventing the inspection pair from mistaking the space between boxes as an 'improper fill.' (J) Light source and photoreceiver placed near a guillotine are used to detect products and operate the blade for cutting the link between products. (K) Thread break detection made possible when the photoelectric beam is interrupted by a lightweight flag riding on the taut thread. (L) The size of a paper or fabric roll can be controlled by positioning a light source and a photoreceiver so the roll diameter blocks the beam. *(Micro Switch.)*

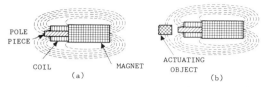

Fig. 6. Schematic representation of the action of a variable-reluctance object sensor: (a) A sensor with no actuating object in the magnetic field, (b) actuating object in field alters the voltage generated at coil terminals. The voltage is proportional to the rate of change of magnetic flux.

Magnetically-actuated Dry Reed Switches. Generally consisting of a thin reed (wire) contained in a hermetically sealed container (encapsulated), this type of switch is both inexpensive and rugged. Whenever an activating magnet approaches the critical range of the switch, a contact closure is made. Life expectancy usually is in excess of 20 million operations at contact ratings of about 15VA. These switches generally can operate loads directly. Since the actuating magnet (powerful alloy magnets now usually used) can be installed on a rotating or reciprocating object, the switch can be used in a wide variety of applications for counting, positioning, and synchronizing. Contact closure speeds can be up to 100 per second. At one time, mercury switches with flexible electrodes that can be attracted by the proximity of a magnet were more widely used than at present. Methods of actuation of encapsulated reed switches are shown in Fig. 7.

Inductive Proximity Sensors

In an inductive proximity sensor, an electromagnetic field (radio frequency, rf) is generated by an oscillator circuit. When a metal object enters the effective field generated by the sensor, a countercurrent (eddy current) is set up in the metal object. This causes a voltage drop in the oscillator. This drop is sensed by the detector, which triggers the output. The output can be used for many industrial control purposes. (See schematic diagram of a typical sensor given in Fig. 8.) The sensors are available in three basic configurations: (a) cylindrical—a two-piece mounting clamp with socket head screws allows the installed sensor to be moved to the desired position, (b) threaded—the installed sensor can be rotated to the desired position and held in place with two flat nuts, and (c) rectangular—the sensor has slots in mounting base, which allow adjustment to desired position after installation. Nominal detection ranges from about 2 mm to about 20 mm, depending on specific design. Some units are adjustable from 10 to 50 mm. A standard target is considered to be mild steel, with a factor of 1.00. This varies with other materials—stainless steel, Type 340, 1.03; stainless steel, Type 302, 0.85; nickel, 0.85; brass, 0.50 to 0.54; aluminum, 0.47 to 0.50; and copper, 0.40 to 0.46.

The point at which a target will be detected is influenced by the type of metal, its size, and its surface area. Targets may approach axially (head on) or laterally from left to right and detected at the point where they first touch the envelope of the sensing curve. Curves for standards and extended ranges are given in Figs. 9 and 10. Effect of temperature on the operating points is shown in Fig. 11.

Wiegand-effect Switches

A Wiegand wire is a small-diameter wire that has been selectively work-hardened so that the surface and the core of the wire differ in magnetic permeability. When subjected to a magnetic field, the wire emits a well-defined pulse that requires little signal conditioning. This pulse induces a voltage in the surrounding sensing coil. The wire is insensitive to polarity and emits a pulse whether the magnetic field is flowing from north to south, or vice versa. A Weigand proximity sensor senses the presence or absence of ferromagnetic material.

Capacitive Proximity Sensors

The basic element of a capacitive proximity sensor is a high-frequency oscillator containing a capacitor, one of the plates of which is built into the end of the sensor. When oscillating, a field is created around this free capacitor plate. When an object is placed in the field, the amplitude of the oscillator output changes. These oscillations are rectified and smoothed by an integrating coupling stage. The resulting DC signal is fed to a trigger circuit, which switches an output transistor. The sensing zone or envelope of the sensor is influenced by the physical properties of the object being sensed in the following ways: (1) Nonconducting materials, such as glass and plastics, are sensed by a change in dielectric characteristics. Since this change is small, the sensing ranges are necessarily limited. (2) Conducting materials are sensed by a change in dielectric characteristics as well as by an additional disturbance of the noise field caused by terminal conductivity. (3) Materials containing both conducting and nonconducting properties, especially if grounded, are sensed by a combination of the foregoing characteristics, as well as by absorption. These conditions produce the greatest switching distance for a given sensor. Because capacitive sensors are so markedly affected by material characteristics, an estimate of the switching range requires knowledge of the medium to be measured. In one popular model[3], the disk sensor is $1\frac{3}{16}$ inch (30 mm) in diameter. Disk size and shape can be altered to provide improved sensitivity, as determined by

---

[3] *Proximitrol*™, Automatic Timing & Controls.

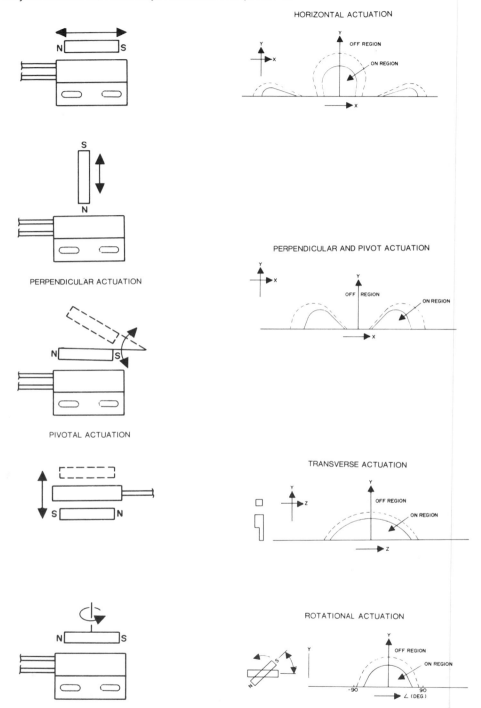

Fig. 7. Methods of actuation of encapsulated reed switches. These switches are used to detect the closeness of two mechanical parts by having the sensor mounted on one part and the actuator/permanent magnet or shield plate mounted on the other part. (*Hamlin, Inc.*)

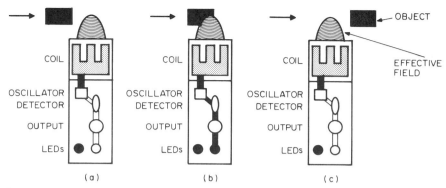

Fig. 8. Inductive proximity sensor: (a) Object approaching sensor, (b) object in field of sensor, (c) object beyond sensor. (*Cutler-Hammer*.)

the shape of the object being detected, its dielectric constant, and the operating distance. The operating range is from 0.2 to 5 inches (5 to 127 mm), speed of response is up to 200 operations per second, capacitance range is 0.1 to 2 pF, and a change in probe capacitance of 0.01 pF can be detected. The temperature range is 32 to 131°F (0 to 55°C). Minimum object size can be as small as 0.1 cubic inch (1.5 cubic centimeter).

Magnetostrictive Limit Switch

Of limited application to date, the principle of magnetostriction was discovered in 1858 and represents a change in dimension of a material as the magnitude or direction of magnetization in a crystal is changed. The principle was introduced in the limit switch field in the mid-1970s. The active element is a helical spring that is subjected to axial extension or compression caused by mechanical displacement. The magnetostrictive transducer is considered to have good potential where high-resolution information is required over a substantial range. (See Garshelis [1977] reference in list at end of article.)

Ultrasonic Switches and Other Novel Approaches

The presence of objects can be detected by return echoes from reflecting materials in a form of sonar. Similarly, rf reflection (radar) can be applied. A number of the large, well-established suppliers serving the proximity switch field have been joined by scores of newer, smaller firms that are busily engaged as of the mid-1980s in applying "smart" electronics to new and old concepts. In the Bailey (September 1985) and Morris (1986) references listed, a number of these newer approaches are described.

One firm[4] has developed a lacquer detection system (inside coating of cans). The system is based on the principle that ultraviolet light is reflected by uncoated metal surfaces, such as aluminum or steel, but is absorbed by lacquer. The system measures the amount of light reflected from a can and, through a series of statistical routines performed by a microprocessor, the sys-

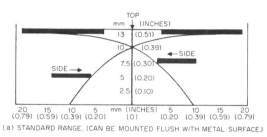
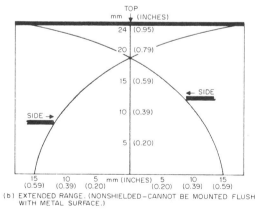

Fig. 9. Sensing distance—typical response curve for a target of mild steel (45 × 45 × 1 mm) (1.77 × 1.77 × 0.04 inch). Black rectangular areas represent objects approaching the sensor from the top or sides. (*Cutler-Hammer*.)

[4,5]*Qualiplus*™ *USA, Inc.*, an affiliate of Continental Packaging Co., Stamford, Connecticut.

84 Object Detection, Proximity, Presence, Nonpresence

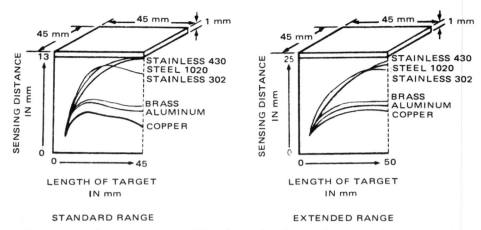

STANDARD RANGE   EXTENDED RANGE

Fig. 10. Effect of target size and target material on the sensing distance for some inductive proximity sensors. (*Cutler-Hammer*.)

tem is able to distinguish areas of metal exposure as small as 0.5 square inch (3.2 square centimeters), depending on location. The detector-ejector is built as a unit and located above the conveyor systems. Speeds up to 1500 cans/minute are achieved. Also performed by the microprocessor are: (1) electronic compensation for UV light source aging, (2) eject control to adjust ejector timing to the actual conveyor speed, (3) counting of tested, ejected, and knocked-over cans, and (4) printer control and data supply to a host computer.

Impact of Machine Vision Systems. Normally, one equates machine vision systems with considerable complexity and relatively high cost. But there are opportunities for simpler, lower-cost vision systems that accomplish some of the tasks traditionally assigned to photoelectric and proximity switch systems. The cost differential, however, still remains quite large. One supplier[5] introduced (in the mid-1980s) an on-line can inspection system. The system is based on image-processing techniques and can recognize 256 gray levels (as compared with 64 shades for the human eye). Incorporated in the system is a CCD (charge-coupled device) matrix array camera and an image-processing computer. (See Fig. 12.) Similar systems have been developed for detecting and removing foreign objects from cut tobacco and for inspecting empty bottles for chipped glass flaws, inside contamination, the presence of residual liquid, and remnants of labels (not removed by cleaning). The bottle inspection system uses a patented linear conveyor system that operates up to speeds of 100,000 units/hour. Total changeover time for different container sizes is about five minutes.

See also the next two articles in this Handbook section that deal with machine vision systems.

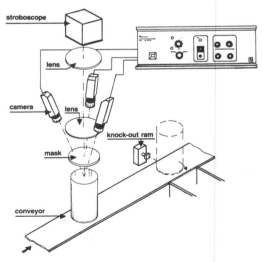

Fig. 12. On-line "inside can" inspection system based on image-processing techniques. (*Qualiplus*™ *USA, Inc.*, an affiliate of *Continental Packaging Company*.)

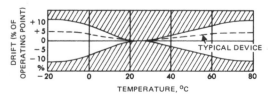

Fig. 11. Effect of temperature on the operating points of some inductive proximity sensors. (*Cutler-Hammer*.)

ADDITIONAL READING

Andriev, N.: "On-Line Optical Gaging," *Cont. Eng.*, **27** (11) 78 (1980).

ATC: "Timers and Counters, Photoelectric Controls, Proximity Switches," *Pubn. TS-1*, Automatic Timing and Controls Co., King of Prussia, Pennsylvania. (Revised periodically)

Bailey, S. J.: "Optical Sensors Critical to Future Productivity," *Cont. Eng.*, **29** (1) 72 (1982).

Bailey, S. J.: "High Speed Production Lines Benefit from Latest Object Motion Detectors," *Cont. Eng.*, **32** (9) 137–141 (1985).

EC HDBK: "Magnetic Sensors—Technical Applications Handbook," Electro Corporation, Sarasota, Florida (1981).

Garshelis, I. J.: "Magnetostriction Put to Work in Industrial Limit Switching," *Cont. Eng.*, **24** (10) 60 (1977).

Krueger, A. H.: "Applying Fiber Optics to Photoelectric Switches," *Cont. Eng.*, **27** (8) 61 (1980).

Micro Switch: "Photoelectric Controls," *Pubn. 60, Micro Switch* (A Honeywell Division), Freeport, Illinois. (Revised periodically)

Morris, H. M.: "You Can't Automate If You Don't Know Where Things Are," *Con. Eng.*, **33** (2) 67–69 (1986).

Tome, D. M.: "Adaptive Control Sensors for Manufacturing Systems," *Cont. Eng.*, **31** (8) 78–79 (1984).

Warner: "Photoelectrics Solve Motion Control Problems 16 Ways," *Pubn. P-721*, Warner Electric Brake & Clutch Company, Beloit, Wisconsin. (Revised periodically)

Machine Vision

By Larry Werth[1]

Machine vision (MV) is part of the larger field of artificial visual perception, which may be defined as seeing, analyzing, and interpreting patterns, scenery, juxtapositions, dimensional magnitude, color, and many other characteristics of the visual environment. This is done with the partial or total substitution of the human visual system by instruments (frequently optical) and computers and/or electronic subsystems.

Human visual communication (information transfer) with the outside environment depends upon the interactions (predominantly absorption, reflection, and refraction) of light (visual) radiation with physical objects (in point, two or three dimensions), enabling the individual person to cope with the surrounding environment in a safe and efficient manner.

The vision activity itself, whether human or machine, is complemented by some processor (the human brain or electronic device) to determine the identity of the image. This determination is accomplished by interpreting the visual pattern or pattern recognition.

Basically pattern recognition is the identification of sensed information. MV is only one implementation of pattern recognition. The information in any pattern recognition process is a set of sensed data that exist in the environment. The device that performs pattern recognition is called a *pattern processor*. The most extraordinary pattern processor in existence is the human mind. Its function is to recognize patterns of data in our physical environment. The musical tones of a familiar melody are such a pattern data set; so are the words in a sentence that we have been trained to understand or the view our eye encounters when we see someone that we recognize. The human brain, in processing pattern data, does an amazing job of sorting out extraneous information in the input and quickly identifying what is present.

The study of pattern recognition goes back well over 60 years. The more one studies it, the more one finds out how fundamental the process really is to the human brain function. The amount of information that human systems take in through the senses is amazing, and yet the brain seems to accommodate it all in parallel without any apparent problems.

Any pattern recognition system will contain the same three basic elements that sense, process, and apply input data. First, the sensing element records the stimuli and presents it to a processing element. After interpreting the input and identifying or matching it to some *prelearned* pattern, the processing element represents the identification data to an application element which, in turn, reacts to the situation. (See Fig. 1.)

An example of a human pattern recognition occurs during facial recognition. The eye senses visual pattern data, the mind performs recognition, and the vocal cords greet the approaching person.

The primary objective of pattern recognition is to classify a given unknown pattern as belonging to one of several classes of patterns. Pattern recognition developed before a number of the most important of its current uses. The applications of pattern recognition are many and varied. In some problems, such as character recognition, the patterns are easily generated and recognized by humans, and the basic goal has been to improve human-machine communication. In other situations the patterns are difficult for humans to recognize, as, for example, the interpretation of electrocardiograms. The wide range of applications, as well as the relationship to diverse disciplines including communications, control, computers, artificial intelligence, linguistics, and biology, has caused the field of pattern recognition to be very broad.

Pattern Recognition in Machine Vision

The problems in MV are very similar to the rest of the pattern recognition field. The principal steps of a MV system are: (1) sensing (camera), (2) processing (computer or other electronic device, and (3) applying (presenting processed information for a specific need).

MV started with the development of robotics over 20 years ago. At that time, *Unimation, Inc.* developed and installed the world's first industrial robot. Researchers soon realized that many major technical advances in the form of flexible manufacturing systems (FMS) would be limited without an effective capability for the interaction between machines and their external environment. In the

---
[1] President, Pattern Processing Technologies, Inc., Minneapolis, Minnesota.

Analysis → Process Data → Control Output

Fig. 1. Three basic operations of a pattern recognition system.

late 1960s and 1970s experiments in image processing and machine vision showed that the concepts were possible and financially plausible.

It was not until 1978, when the Symposium on Computer Vision and Sensor Based Robots was held at the General Motors Research Center in Warren, Michigan, that one could honestly say that the MV industry had any credibility. As the industry entered the 1980s it rapidly expanded with new companies, technological advances, and a user community ready to try the concept. Now, as the technology enters the second half of the decade, many users have become experienced in machine vision applications and the real benefits of the technology are becoming more commonplace.

Machine vision systems today can operate effectively off-line or in real time in connection with target objectives that may be stationary or moving. Furthermore, such systems are not necessarily confined to light radiation in the visual range, but they may use infrared, ultraviolet, x-radiation, and other parts of the electromagnetic spectrum. Techniques developed for systems in the visual range such as pattern recognition can also be applied to systems involving tactile rather than radiation-sensitive sensors.

Basic Applications for Machine Vision Systems

Principal areas of use for machine vision systems include industrial manufacturing and processing, military and space programs, and human needs.

Research and development in these areas essentially has proceeded unilaterally, although there is considerable interdependence upon the findings and progress in each area. Included in the military field are tracking systems such as the heat-sensitive systems that lock onto jet engine exhausts; infrared spectroscopes (circa 1940s); automatic navigation systems based upon celestial star patterns; target acquisition and range finding; detection of battlefield movement; and various reconnaissance objectives. In the industrial area, particularly during the 1980s, much effort has been directed toward improving machine vision. Ironically, artificial vision systems for the blind and partially blind are still in a rudimentary state and essentially will depend upon progress made in the more profitable vision systems for industry. Systems partially developed include the means for reading documents and speaking out what is read and automatic "guide dog" navigation systems.

Artificial Vision System Basics

Regardless of artificial vision systems' design complexity, cost, and practicality, they all begin with the use of images.

What the human eye or vision system "sees" may be called an *image*. Images are important because of: (1) their impressionistic values in an aesthetic sense (not of concern here), and (2) the information which they contain. An image may be recorded as recollections in the human mind or optically, chemically, and electronically through the use of various image-capturing cameras and camera-like techniques. An image may be exceptionally large, such as the earth or portion of the earth as viewed from a satellite or monstrous galaxy as seen through a telescope, or something of microscopic size as seen by optical or electronic means.

The information content of an image falls into three basic categories:

1. *Geometry*, which, in turn, portrays shape, position, dimension, and a number of properties, such as density and texture, which can be inferred from the known geometry in some cases.
2. *Color*, which is very helpful when present. There are, of course color-blind persons and instruments that preclude its use.
3. *Movement*, which is present in two or more images. It represents a situation within the scene.

The schemes used for extracting the foregoing information from an image are described shortly.

Extracting Information from Images

As previously described, the human eye or the artificial vision system commences with "seeing" an image. As with other forms of industrial instrumentation, the vision system senses (reads may be a better term) the image, but before the data obtained from the image can become meaningful, the sensed information must be compared with some form of standard or prelearned pattern. These combined actions of sensing and comparing constitute measurement. For example, in a thermocouple thermometer, the measured signal is an emf, which is the electrical analog of the temperature applied to the thermocouple. Extracting useful information from an image is grossly more complex because most images of concern are quite complex. In fact, the human vision system will require much expertise and quite a lot of time to compare a valid dollar bill with an excellent counterfeit bill. Artificial vision systems can make this comparison in a fraction of a second. And, obviously, for dynamic industrial applications, the response time of a machine vision system must be extremely brief.

From years of experience, through pattern recognition, people learn to recognize thousands of objects that may be part of an image: trees, houses, autos, and so on, including George Washington on the U. S. dollar bill. The comparison of what is seen and what is stored in memory in the brain is fast and essentially automatic in a biological sense. For an artificial vision system, image data as gathered by an electronic camera must be compared electronically with information in some form of electronic memory. Numerous techniques have been developed to at least partially duplicate the human image comparison system, and that is why artificial vision is considered by some experts to be a form of artificial intelligence.

Computers Versus Adaptive Hardware

Generically the machine vision systems operate similar to any pattern recognition system. First, during an analysis step, an image appearing on some video camera is turned into a set of electronic signals compatible with the system's processor.

In the second step, where the electronic data are compared with some known or "learned" pattern, recognition occurs. Most systems currently on the market are computationally based and accomplish this step by running a computer program on some general-purpose computer. This means taking the electronic image data, translating it into some mathematical features, and then comparing these features with some reference set.

The programs used to accomplish this recognition step are often written by the users, and the computers used to run them must be fast, large, and expensive in order to provide realistic performance.

As machine technology advances in the areas of camera input and external control, the major difference in various systems will be increasingly in the *processing techniques*.

There are two development approaches being pursued by the industry in the area of processors:

1. Higher-performance computational systems.
2. Adaptive hardware approaches.

To understand these development approaches in more detail, one needs to know the basic reason behind the dual approaches. It is due to a basic difference in data structure between pattern data and computational data.

Data Differences

When the first vision systems came on the market, they were controlled by general-purpose computers running software that performed pattern recognition.

The application of the general-purpose computer to MV was the only choice the original pioneers in the industry had. It was the only data-processing device available. The problem is that the computer was designed to process computational data, *not* patterns of data. Computers do a very good job of processing data such as financial balance sheets or a scientific problem like linear regression analysis. But that is not the kind of data the computer encounters when it tries to process visual scenes from a solid state camera.

The data from a video camera in an MV system is pattern data, and it has very different characteristics from computational data. The first characteristic is the tremendous amount of data involved as compared with a typical computational data problem.

The process of converting an image on camera into pixel data creates a huge data volume, which in turn causes a bottleneck in the computational process. The following illustrates how much data are generated by a solid state black and white camera:

IMAGE DATA QUANTITY

484 × 320 pixel density
154,880 pixels per frame
6 bits of grey level data per pixel

929,820 pixel values per frame
30 times a second refresh rate

27,878,400 bits per second

Even this number is conservative because it addresses only the data rate and not the combinations of possibilities that can occur in patterns of data. And the future only holds more of the same in terms of increasing data. MV systems using multiple cameras with concurrent processing of the data will see increasing application in the future. Color is another data capacity problem for the future. Data volumes from color cameras can be three times greater than black and white cameras. And, eventually the addition of higher pixel density cameras, like the 1024 × 1024 unit, will further complicate the problem. The bottom line is that future automation vision tasks will produce even more data and more problems for the computationally based vision systems.

Equally serious as the amount of data is the nature of the data itself. The pattern data from the camera is inherently different from the computational data for which the computer was designed.

First, computational data typically consist of data arranged in a predefined pattern or format while the information in a visual image typically consists of arbitrary patterns. For example, if a programmer was handed a disk with data on it and told to determine what information it contained, he would first need to know the format or pattern of data on the disk (i.e., ASCII, binary, integers, floating point, etc.). Without this information the programmer may conclude the task is nearly impossible. However, this is the type of task required by computer programs when applied to pattern recognition. The point is that computers typically operate on data in a predetermined pattern while in machine vision it is the pattern itself which must be determined.

Second, the computational process operates in logical/mathematical statements and when applied to pattern recognition, one must express the pattern in the same descriptive language available to the program (i.e., mathematical or logical relationships between data). The problem is that mathematical models have a very limited ability to describe visual patterns of data except for simple visual patterns created by holes (i.e., equation of a circle) or edges (equations for a line). Typical visual patterns do not have simple mathematical descriptions. As an extreme example, consider the task of using a mathematical model to describe a person's facial features.

A third point to consider is that visual data are highly redundant while the computer is optimally designed to process data that are highly *non-redundant* (i.e., every bit means something). Taking the computer perspective that every bit is significant means that the huge amount of data described earlier must be processed as significant data. The point is that computers are designed to optimize the information capacity of input data while the data from a visual image are extremely redundant.

A fourth point to consider is that an emphasis of MV is to automate human vision in the factory. It is clear that human visual processing architecture is nothing like computer architecture. More likely, the human visual processing architecture embodies some form of adaptive statistical process. The general applicability of computer-based vision may be limited except in very simple, low speed, or specifically controlled applications with dedicated software.

The basic difference in pattern data and computational data has made the struggle to make computationally based pattern recognition techniques work in the machine vision environments a challenging one. However, there are successes in using this approach, and future developments continue to promise better performance. The adaptive hardware approach, or development of unique architecture systems designed to process pattern data, is still probably a better alternative. What this takes, however, are real "inventions," that is, unique ways of applying electronics to recognition tasks. Unfortunately, real inventions are not common[2]. Before discussing these technologies, a better understanding of MV theory will be helpful.

FUNCTIONAL DESCRIPTION OF MACHINE VISION SYSTEM

The objective of an MV system's sensing component (camera) is to convert an image into a set of compatible electronic signals. In most MV systems, this is done with a video camera. The sensors for a vision system must be carefully matched to the requirements of the vision problem addressed. Factors which usually need to be considered include: (1) optics/lighting, (2) field of view, (3) resolution, (4) signal-to-noise ratio, (5) time and temperature stability, and (6) cost. Although there are special-purpose photosensors in use, this article is confined to scanning sensors.

Line scanners include solid state arrays, flying spot scanners, and prism, mirror, or holographically deflected laser scanners. These scanners are fast, high-resolution devices that are relatively free of geometric distortion in one dimension. In order to capture a complete two-dimensional scene, the second dimension must be obtained either by motion of the object past the scanner or by mirror or prism deflectors. Mechanical motion tends to slow down data acquisition and, in some cases, produces geometric distortion.

Area type scanners used in early systems utilized closed-circuit television cameras with vidicon image sensors. These sensors are inexpensive and provide high resolution (300 to 500 television lines). However, they suffer from geometric distortion, time and temperature instability, lag (requiring several television frames for complete erasure), and sensitivity to nearby magnetic fields.

Solid state cameras represented a major advance in sensing technology. They digitize a scene onto an array of photosensitive cells. Charge coupled devices (CCDs) or charge injection devices (CIDs) are used in these cameras. The arrays form a pixel grid containing the data currently appearing on camera. A pixel is a picture element, a small region of a scene within which variations of brightness are ignored. The solid state camera comes in a variety of pixel densities, including 128×128, 256×256, and 512×512. A 1024×1024 pixel camera is in development.

The primary advantages of solid state cameras are: (1) The image has a fixed geometry in respect to the sensing elements, and (2) sensing elements or pixels have a fixed geometry between themselves. These factors make precise measurements possible. In the earlier vidicon tube systems, the image could stretch, bend, and be distorted. A CCD or CID image is effectively more stable.

As indicated earlier, the limiting factor of cameras for image systems is not necessarily the resolution or speed. Presently, the capability of camera technology to furnish data is beyond the ability of most computationally based processors to handle the information, especially when multi-

---

[2] However, one such product is the Associative Pattern Processor (APP) invented by Pattern Processing Technologies, a machine vision supplier in Minneapolis, Minnesota.

ple cameras are used. Sensors can detect information in real time a lot faster and in larger quantities than the present processors can analyze it in real time.

Processing the data, the second part of the MV process, can take place once the image has been converted into a set of electronic signals. This processing really occurs in two parts: (1) The image must be converted into the language of the computer, which is number or mathematics oriented; and (2) the data must be matched to some prelearned pattern. The image analysis is probably the greatest challenge facing computationally based MV systems. The technique used in image analysis is a science in itself. Again, the challenge is to reduce the amount of data involved.

Since representing all the combinations possible in every pixel of every scene is not realistic, practical, or always necessary, short cuts are often taken to reduce the data volume. One data reduction technique used by a large number of systems is the conversion of each pixel point into a binary value. As the name implies, binary systems evaluate each pixel as black or white. A threshold-adjusting capability often allows users to select what intensity of signal is to be the black/white border.

A **binary MV system** reduces the pixel data, but it also reduces the accuracy of the analysis. **Grey level systems,** on the other hand, can interpret each pixel's value as a specific grey tone. These systems vary in precision with each pixel point being evaluated as 16, 64, or even 256 different values.

Windowing the scene is another method of reducing the total data required. In other words, only a portion of the pixel data need be analyzed to determine its formula. Since analysis of an entire scene is often unnecessary for proper recognition, the use of a window can do a great deal to eliminate unneeded image data and will be discussed a bit later.

Segmentation is the most common means of data reduction. This technique divides the image data into areas of interest and then interpolates surrounding pixels into those areas. There are several ways to segment an image, but they all make the analysis less accurate. Segmentation procedures are all accommodations to the basic problem, namely, too much data to be processed and too little computational power. Segmentation methodology includes at least four procedural options:

1. **Algorithms** are a form of segmentation that works with an already data-reduced binary image. Algorithms, in general, are a set of mathematical models that can be used to describe an image. A very popular set of algorithms is the *SRI Algorithms,* developed at Stanford Research Institute. They consist of about 50 different features that are extracted from a binary image such as the size of blank areas (holes) or objects (blobs), their centroids or perimeters. Many MV systems on the market today use SRI algorithms working with binary images.

2. **Neighborhood processing** is an averaging method. By treating each pixel value as if it were part of a group or neighborhood, this grey level data reduction technique can change the otherwise complicated image into areas with well-defined lines. Each pixel's point value is calculated by considering the value of neighboring pixels. The process effectively averages the scene into regions or areas.

3. **Convolution**. If an image is represented by the rate of light change per pixel instead of light density, the image will look like a line drawing because the greatest intensity of a pixel change is at the boundaries.

4. **String Encoding**. In run length, string, or connectivity encoding, the values of the first and last pixel positions of each scan line are compared to see if they are equal and therefore belong to the same region. The tables generated by this process also consider the vertical changes in pixel state and, as could be expected, are rather short because most simple binary images contain relatively few transition borders.

The Recognition Process. Once the user figures out how to describe the image mathematically, the second half of the processing step can occur. This is the implementation of the mathematical description of the visual data into the recognition process. The programming of a general-purpose computer to interpret vision algorithms takes programming expertise that is costly. Using this computational approach can also mean that performance levels can be too slow for real time operation unless the application involves only very basic procedures. Some MV systems employ parallel processing and pipeline processing to increase computing power. These systems show promise, but they may provide only temporary relief because gains in processing power are quickly taken up by application complexity and increasing data demands.

Application. The final element in the MV system is the application. This is the part of the system that interfaces with the external world. It may guide robots, provide pass/fail signals in inspection stations, or measure the size of milled components. Regardless of the application, however, the way the MV system communicates to the real world of the factory environment is fairly standardized. The common interfaces include RS232 and IEEE 488.

All tasks that MV performs can be categorized into four groups: (1) inspect, (2) locate, (3) measure, and (4) recognize. If any application is analyzed into the basic components, it can be considered one of a series of these basic steps.

First and foremost, machine vision can **inspect** to find out if parts are acceptable. Inspection tasks, in turn, can be categorized into one of four functions: (1) surface flaw detection, (2) presence/absence feature detection, (3) dimensional tolerance verification, and (4) shape verification.

The second general MV task is related to finding *out where a part is* in terms of its position or orientation on the system's monitor. This function is called *locating* and is predominantly used in robotic applications.

Measurement data is the third group and tells what *size* an object is, and, finally, the recognize function *identifies* the viewed object.

COMPUTATIONAL TECHNIQUES

Since the first MV systems on the market were computationally based, this discipline has had the most time to develop. However, it was apparent early that the running of software to perform vision analysis was too time-consuming for many applications. To accelerate the computational steps, supplemental processing engines, such as array or systolic processors, are used.

The basic mathematics needed to define the image on camera is currently going through evolution and change. The algorithms have also gone through changes over the years, and the original SRI versions, previously mentioned, are often too simple and basic for complex applications. Many machine vision suppliers are developing their own proprietary algorithms for the purpose of creating a competitive difference and advantage.

In early MV products it was an accepted fact that the customers were required to do the vision analysis programming or pay substantially for the MV supplier to do it. The heavy reliance on programming in early systems meant that they were also very niche-oriented and not general-purpose. This is why many of the MV suppliers are doing specialized products oriented to specific industries. As the industry matures, there are more "canned" software routines available to do this part of the task. The advent of this generic software that is capable of identifying specific categories of shapes, objects, etc., can be said to add a rudimentary artificial intelligence to the MV process. This development in computational MV analysis schemes is very promising because it not only reduces the amount of basic programming needed to make MV systems more flexible, but also points the way to a developing technology in the future of machine vision.

As these canned programs become perfected and the computers needed to run them continue to become more powerful and smaller, a natural evolution will be to combine the two in super-sophisticated parallel processor schemes where individual microprocessors and their programs can be applied to "regions" of an image. The end result of this development will be in VLSI (very large scale integration) designs that accommodate this technique.

ADAPTIVE HARDWARE

The most exciting advances in MV technologies will be in the inventions of new technologies to bypass the computational bottlenecks, i.e., processing technologies designed specifically for processing patterns of data versus computational data[3].

A basic premise in considering specialized hardware is that it will always be faster to execute electronic hardware performing MV functions than it will ever be to use a computer executing instruction steps in software no matter what advances are made in the computational arena.

The hardware example used here is the Associative Pattern Processor (APP) being marketed by Pattern Processing Technologies (PPT). The APP is a statistical encoding system that compares the contents of two memories to find a match or recognition state. The first memory is an image buffer that contains the pixel data for every point of some current image and the second memory (response memory) contains condensed mathematical representation of all previously trained images.

The condensation of the image data is done by proprietary circuitry within the image analyzer and incorporates a statistical phenomenon of repeating values that occurs when large amounts of data are sampled selectively. This is the key to the APP's performance level while it seemingly stores such a small percentage of the total data available.

The APP processes pattern data in some ways similar to how the human mind does. The human mind learns a pattern by storing a condensed version of it in biological neural networks and then recognizes it in subsequent encounters by matching the condensed version of the pattern.

The APP system works similarly. Video signals for camera images are stored in the image buffer (the equivalent of the retina in the human

---

[3]One of these inventions (developed by Pattern Processing Technologies, Inc.) will be discussed here. This particular approach may not be the only one to develop, but it can serve as a current example here of how the MV industry may eventually cope with the challenges associated with real-time performance levels.

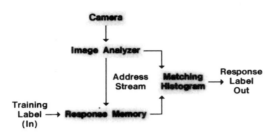

Fig. 2. Essential elements of the *Associative Pattern Process* (APP).

system). Any image that can appear on a video camera represents a unique pattern of grey levels, which is refreshed at a 30-millisecond rate in the image buffer. (See Fig. 2.)

The image analyzer constantly samples the contents of the image buffer and condenses the data into a loop of repeating sample addresses. This sampling loop contains a minute fraction of the total addresses available, but identifies the image very precisely.

The technique used for creating the APP loop is via proprietary hardware. The origin of the looping theory is based on a mathematical theory of statistical neurodynamics, which concludes in part that if a randomly interconnected mass of neurons were activated at some state and allowed to progress to subsequent states, the states would settle to a repeating sequence or loop of states.

The repeating pattern loop for any given image is a tremendous condensation of the data for the whole image, but is still unique to an image and accurately identifies it. The analysis of the image and creation of its loop continue constantly at the refresh rate of the image buffer and speed of the camera. The loop automatically changes with changing images.

Pattern recognition of this real-time device is accomplished by constantly comparing the live-image fingerprint with those stored previously in the response memory. When the loops match, recognition occurs.

APPLICATION BASICS AND EXAMPLES

The discussion here will revolve around the four basic application categories previously mentioned—inspection, location, recognition, and measurement. The approach here is to present a method of accomplishing each of these tasks by using a method called a "contrast sensing and windowing technique." This discussion strips off the technology issue and addresses the problem by application because any technology that can sense contrast and window an image can use this technique.

As an example, the APP can sense contrast and window images efficiently. It can break up an image into 1024 windows of user-defined rectangular dimensions and then check for contrast in each at speeds as fast as the access time of its memory.

When studying this applications material, special note should be made that the mathematics and programming so often feared is not mentioned. The reason is that since these applications start from a point where a system already can window and sense contrasts with simple operator commands, it occurs at a stage in the use of MV that is subsequent to whatever mathematics may be required.

First, the concepts of contrast and windowing need definition. Contrast is the most fundamental element of visual information. A scene with uniformity of brightness (one solid color) provides no information. Once an object is placed in the scene and creates contrast, there is information present that an MV system should be able to extract.

The second essential concept is *windowing*. Since a solid state camera can produce volumes of information too substantial to evaluate as a whole, the image is broken up into small segments and only the contents of those segments are evaluated. Another important aspect of windowing is to use the concept to find a part in a scene. This is done by finding a contrast within a prescribed window. This ability to locate a part in a field of view is of obvious importance in real-life factory environments where parts are not always precisely situated.

Specific Uses of Contrast/Windowing Technique

In applying the technique, generally one takes a part that represents the "standard" (that is, the way the part should appear and identify the critical features that make it "correct") by placing small inspection windows over the contrast produced by these features. This establishes a *test mask* for matching subsequent parts. If a test part does not match contrast levels in the test mask, then a defect is identified. This general procedure can be applied to each of the major categories of machine vision tasks previously mentioned.

Inspection: First, inspection windows are placed over an image of a good part, creating a mask that identifies the characteristics necessary for a part to successfully pass inspection. Then, during subsequent testing, manufactured parts are compared with the test template to detect surface flaws or presence/absence of parts, check dimensional tolerance, and verify shape.

Surface flaws can be detected because they

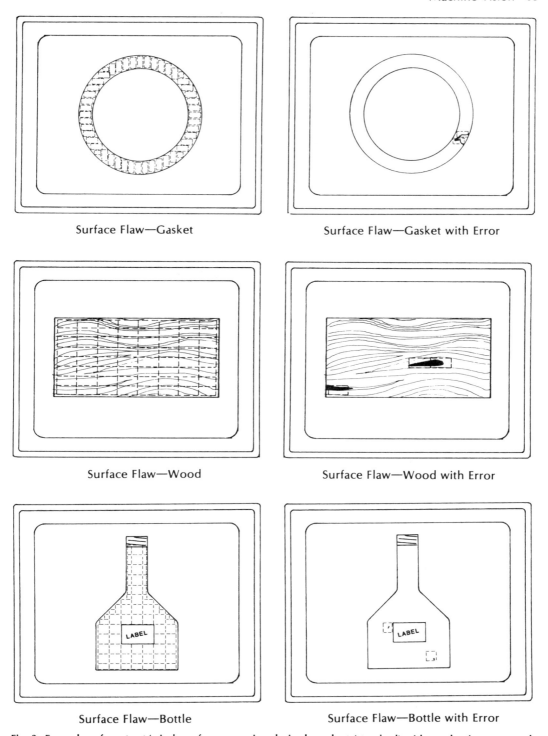

Fig. 3. Examples of contrast/windows for comparing desired product (standard) with production-run products for the purpose of detecting flaws.

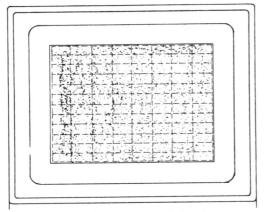
Surface Flaw—Textured Coatings

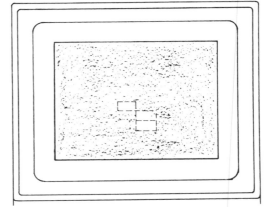
Surface Flaw—Textured Coating with Imperfection

Fig. 3. Continued from previous page.

exhibit a sharper contrast change than a flawless surface. Examples include:

Foreign materials, distortions, or cracks in glass.

Scratches, dents, or cracks in surfaces.

Voids in application of coatings, such as paints or adhesives.

The contrast/windowing system detects surface flaws by placing a grid of *contrast windows* over the inspection surface. Each window independently detects the presence or absence of contrast within a small region of the total surface, which provides two visual properties that are also characteristic of human vision:

1. Immunity to general variations in brightness of a surface.
2. Immunity to nonuniform brightness.

Several illustrations follow that demonstrate techniques for the distribution of windows over different surfaces to accomplish inspection by **surface flaw detection**. Examples of surface flaw detection of a gasket, wood, bottle, and textured coatings are given in Fig. 3. Detection of the presence or absence of parts from an otherwise acceptable product, as is frequently encountered in the electronics industry, is shown in Fig. 4.

Checking of **dimensional tolerance values** using the contrast/windowing technique verifies distances between the edges of the object. This is possible because the edges of objects will produce sharp contrasts in a visual image. To set up a test template, one places windows at the dimensional tolerance limits of each edge on the image of a good part on the visual monitor. Subsequent testing will indicate where parts do not match the test template. See Fig. 5.

To compensate for **variation in part position** during on-line operation, a locate window placed over a reference feature of the part will first find the exact position of the referenced feature and the other windows will shift automatically.

Shape verification or the **physical contour** of an object will create various contrasts on its visual image. Contrast/windowing uses windows to sense these contrasts and therefore sense the contour of the object. Windows are placed to surround contours of interest on an image of a good part on the visual monitor. In some applications the object may be nonrigid and the size of the window regulates the degree of acceptable variability. The examples shown in Fig. 6 illustrate the use of the technique for shape verification.

Location. The ability to find where a part or assembly is located within the camera's field of view is the second major task for which MV systems can be used efficiently. *Locate functions* are the tasks most often associated with machine vision applications because this function is essential to the guidance and control of industrial robots and the compensation of part position in semifixtured part presentation. The locate function is also one of the most difficult for MV products to accommodate because robots require virtually real-time operation. Traditional computationally based MV systems have difficulty in performing at the speeds required in the real-time environment.

The locate command compensates for part position variations by automatically positioning other test window masks. This capability allows the application of other MV functions on ran-

Machine Vision 95

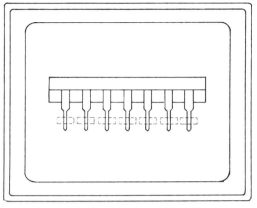

Presence/Absence—Electrical Pins

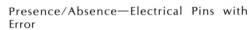

Presence/Absence—Electrical Pins with Error

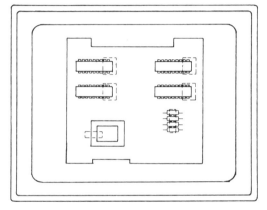

Presence Absence—Circuit Board

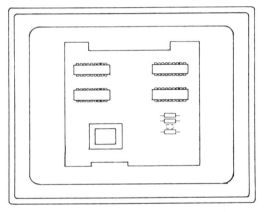

Presence/Absence—Circuit Board

Fig. 4. Contrast/windowing inspection can sense the presence or absence of features to identify defective parts and assemblies.

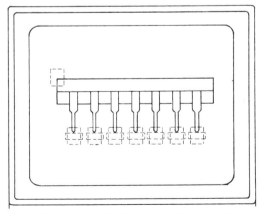

Dimensional Tolerance—Pin Length

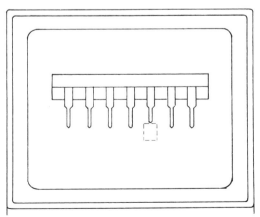

Dimensional Tolerance—Pin Length with Error

Fig. 5. Checking of dimensional tolerance values using the contrast/windowing technique verifies distance between the object's edges.

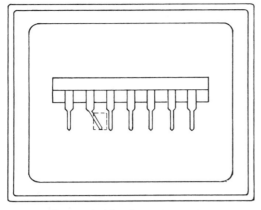
Shape Verification—Bent Pins

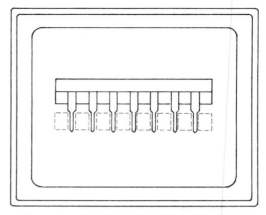
Shape Verification—Bent Pins with Error

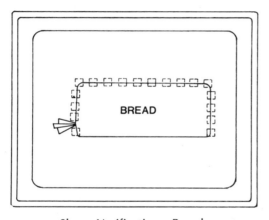
Shape Verification—Bread

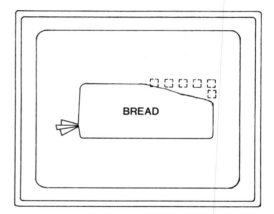
Shape Verification—Bread with Error

Fig. 6. Examples of shape veritification using the contrast/windowing technique.

domly placed parts. This capability adds manufacturing flexibility because parts need not be fixtured or constantly oriented for MV operations to work properly.

Information about the position of a feature within the field of view is important to many vision tasks. The windows identify the XY coordinate position and size of a feature within the range of the locate window boundaries. The locate window defines the range over which a search for the feature will occur. It then provides the position and size of a feature within it. The features are typically edges, corners, and objects.

Positional information provided by the locate window is valuable generally for two purposes:

1. To automatically offset other windows to compensate for part position variation during inspection, as in the dimensional tolerance example previously described.

2. To make XY positional information and object size information available to robots, programmable controllers, and computers.

The examples shown in Fig. 7 illustrate the use of locate windows to find an object, a corner, and the orientation of an object in the camera's field of view.

The **orientation** of a feature within the field of view typically refers to its rotated position. Many times multiple locates will provide sufficient information about the rotated position of a part. In some cases, finding the orientation of a part or feature occurs in two steps: (1) A locate window provides the XY coordinates and size of the features, and (2) other locate or contrast windows are placed relative to the feature to provide additional orientation information. The second step of window placement is an external controller function (e.g., programmable con-

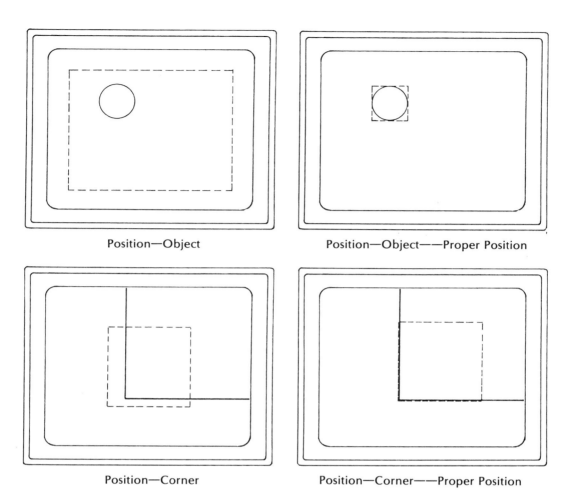

Position—Object

Position—Object——Proper Position

Position—Corner

Position—Corner——Proper Position

Fig. 7. Locate windows are used to find an object and a corner.

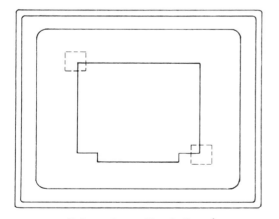

Orientation—Circuit Board

This view shows the position of locate windows on an image properly positioned. The locate windows surround the area where the edge is expected.

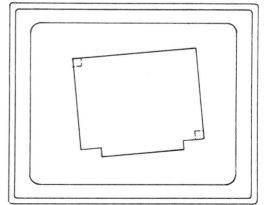

Orientation—Circuit Board

Upon request by a controller, the specified window provides the XY coordinates of the corner. The controller uses the position of data to determine orientation.

Fig. 8. Example of windowing technique for determining the orientation of an object.

98   Machine Vision

Measurement—Diameter

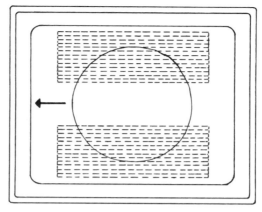

This is a view of the window grid set up to measure round objects. As each passes the grid, its diameter will trigger measurement windows.

Measurement—Width

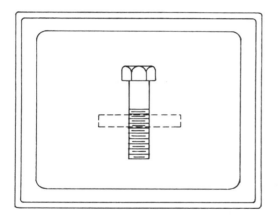

This is a view of the window placement necessary to measure the dimension of a bolt.

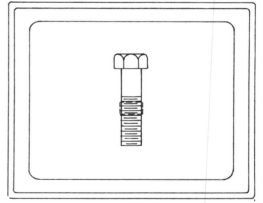

During subsequent execution of the measurement function, the specified locate window provides the X distance in pixels of a rectangle which converges to the sides of the object.

Fig. 9. Examples of windowing techniques for measuring objects.

Object Recognition

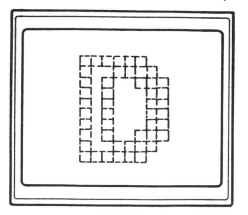
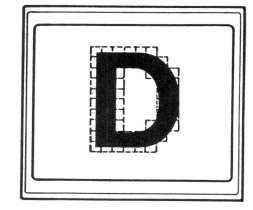

Operation

—Controller initiates request for recognition.
—APP compares window mask to test object for shape identification.

Set-Up

—View objects to be recognized.
—Create window mask over each image pattern.
—Assign a test number.
—Repeat for each object.

Fig. 10. Example of use of windowing technique for a recognition task.

troller, computer, robot controller). This technique is illustrated in Fig. 8.

Measurement. The gathering of measurement data is the third major task of machine vision. Accomplished to the accuracy of individual pixel values, the actual physical dimensions measured are dependent upon the camera's position relative to the test part and the optical characteristics of the lens. The use of MV measuring tasks can provide better accuracy, precision, and speed than can other methods.

By using locate or contrast windows, a contrast/windowing technique can provide measured distances in *numbers of pixels*. Each locate window responds with the number of pixels spanning the X and Y dimensions of a rectangle that exactly surrounds an object within the locate window. The measured size of an object can be determined directly from the number of pixels scanned. The locate windows also provide feature position coordinates. Multiple locate windows can provide coordinate positions of multiple features from which the number of pixels between these features can be determined. Examples of these measurement windowing techniques are given in Fig. 9.

Recognition. This may be defined as the identification of unknown objects. Of all MV tasks, the ability to recognize what is in the camera's field of view has one of the largest application potentials. The number of tasks in sorting, packaging, sensing equipment malfunctions, the presence of parts for machine loading, character identification, and even security surveillance to detect unrecognized conditions give the recognition task a wide range of opportunity.

The task of recognition is quite different from inspection, locating, or measuring in that there

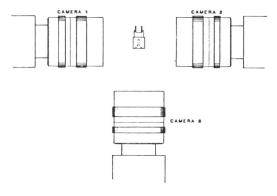

Fig. 11. Camera placement for connector assembly inspection.

100 Machine Vision

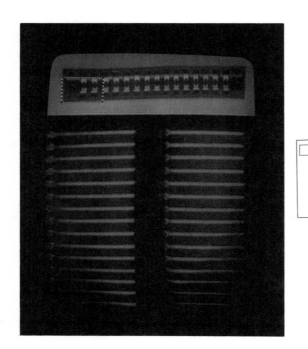

(a) Placement of locate windows used to precisely position the image for testing.

(b) Result of executing the locate command which positioned the test image.

Fig. 12. Steps in connector assembly inspection task. Facsimiles directly below photos of circuitry will assist in locating windows in photo. See (c), (d), (e), (f), (g), and (h) on following pages.

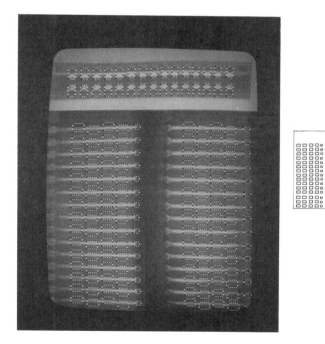

(c) Camera 1 inspects one row of male pins for bent, missing, and improperly seated pins. View shows the window placement for bent pins.

(d) An image of Camera 1 showing a bent pin condition.

102 Machine Vision

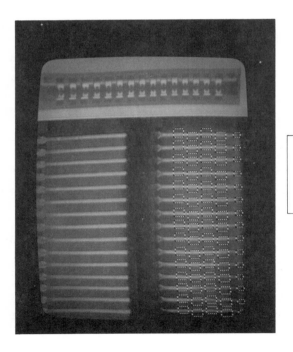

(e) Camera 2 inspects the other row of male pins for bent, missing, and improperly seated pins. View shows the window placement to test for missing and improperly seated pins.

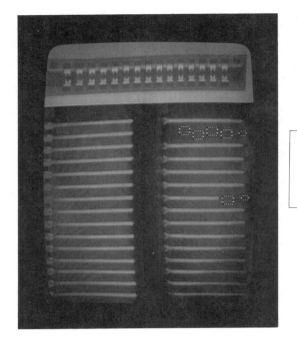

(f) Camera 2 image with failure. View shows an image with both a missing pin and an improperly seated pin.

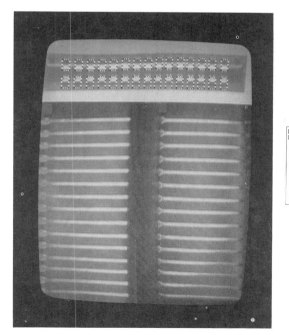

(g) Camera 3 inspects the row of female pins for missing pins. View shows the window placement to test for missing pins.

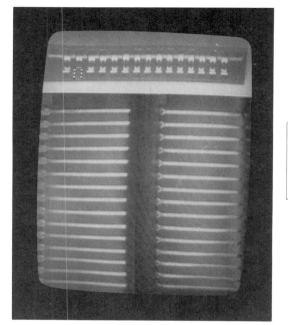

(h) Camera 3 image with failure. View shows an image with a missing pin.

usually exists a known set of objects (e.g., characters) and recognition occurs by identifying which one in the set is present.

An Associative Pattern Processor accomplishes a recognition task by comparing a test part against a set of reference windows that determines the criteria for recognition. The actual comparisons can be considered a shape verification.

Each of the shape verification templates is set up as one separate test in an overall procedure. There can be up to a hundred such tests in each procedure and up to twenty procedures active at one time. (See Fig. 10.)

Implementation of Contrast/Windowing Technique

When the contrast/windowing technique is applied to a functioning task, one appreciates how its simplicity and versatility produce demonstrable results. Examples shown here are both inspection tasks, but as shown earlier, the technique's implementation in other areas is a simple adaption of the same steps.

Example 1. This is for a contrast/windowing system to visually inspect a connector assembly presented in some semi-fixtured method. The inspection technique places the proper inspections template over each connector image and then checks for specified conditions. The resulting data can be used to prevent rejected parts from entering the production line, to provide feedback concerning failure types and rates to the assembly function for corrective action, and to provide an accounting log of the failures for long-term reporting purposes.

The camera requirements for this application necessitate rather detailed readings of close-up images. There are three cameras required with each viewing the part from three different sides. Camera placement for this application is shown in Fig. 11.

What is referred to as a semifixtured presentation is where one knows approximately where the part will be so that cameras can be properly set, but the exact location can vary a few mils.

To obtain an exact location, the system uses its locate command. This command uses a generally placed window around the area where an expected feature will occur. By executing the command, one can find the feature's precise location and use it to offset the test windows. During the test operation, parts are moved in front of the camera and momentarily stopped. That is when the inspection system acquires the test image. However, it is not necessary that the part stop to accomplish the inspection.

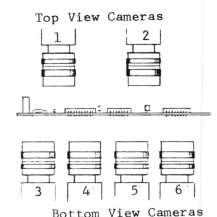

Fig. 13. Camera placement for printed circuit board inspection.

In a moving part application, a simple strobe light will stop the image long enough for the image acquisition. However, in this particular application, the assembly process has been designed to stop each part for inspection. The procedure is illustrated in some detail in Fig. 12.

Example 2. This example deals with inspection of the assembly of printed circuit boards. The inspection system here is a multi-camera dual-station system. The technique used is strategic placement of windows over an image of a part to detect "tested for" conditions.

The camera requirements are straightforward. Two cameras view the top side of the assembly and another four cameras are for viewing the bottom side. Multiple cameras are used for each station to increase resolution capacity. There will be a variety of printed circuit assemblies inspected over the life of the system and thus a six-camera setup adds the flexibility necessary to accommodate a larger variety of assembly sizes. Camera placement for this application is shown in Fig. 13. The inspection procedure is illustrated in some detail in Fig. 14.

In readying for the inspection of a specific part or subassembly, engineers are shown in Fig. 15 using two coins to determine desired window locations.

REFERENCES

Selected references on machine vision are included at the end of the next article in this *Handbook* section.

Machine Vision 105

(a) Top view locate window. View shows the placement of the locate window used to precisely position the window for testing.

(b) Top view locate window executed. View shows result of executing the locate command which positions the test image.

Fig. 14. Steps in printed circuit board inspection task. In these views, only a part of the printed circuit board assembly appears. This is because the precision of the inspection requires the assembly to be tested in sections by all six cameras in the inspection station. Facsimiles directly below photos of circuitry will assist in locating windows in photo. See (c), (d), (e), (f), (g), and (h) on following pages.

106 Machine Vision

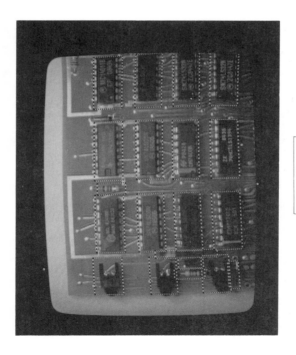

(c) Top view part presence windows. View shows the placement of part presence windows over an area of assembly.

(d) Top view part presence windows with errors. View shows an image with two components missing and the windows which found the error correction.

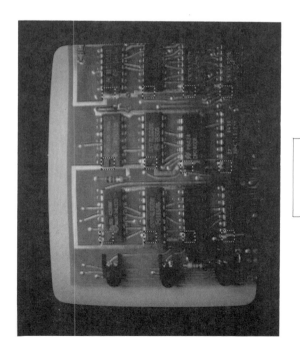
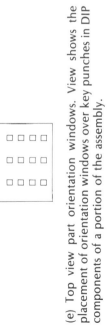

(e) Top view part orientation windows. View shows the placement of orientation windows over key punches in DIP components of a portion of the assembly.

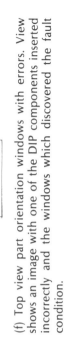

(f) Top view part orientation windows with errors. View shows an image with one of the DIP components inserted incorrectly and the windows which discovered the fault condition.

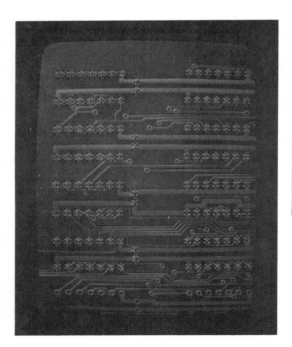

(g) Bottom view lead windows. View shows the placement of windows checking for the presence of component leads protruding through the board material.

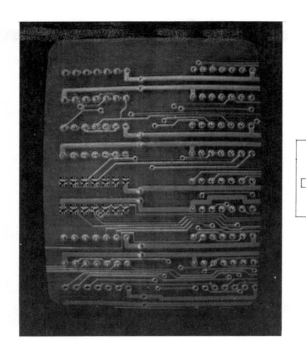

(h) Bottom view lead windows with errors. View shows an image with some of the leads not sufficiently through the printed circuit material for proper soldering.

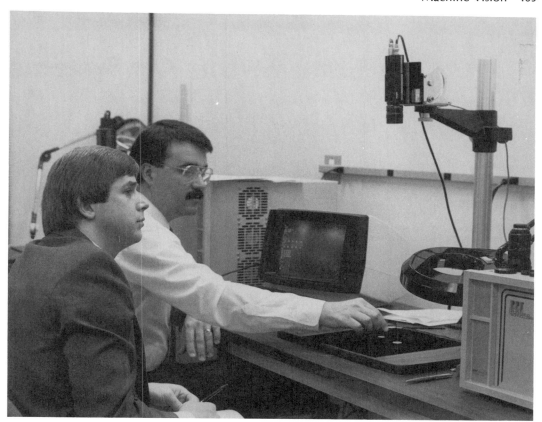

Fig. 15. Prior to inspecting a specific subassembly or part, engineers are shown using two coins to determine desired window locations. Camera is located directly above board with the coins. Images of coins are shown on the CRT to the left. (*Pattern Processing Technologies, Inc., Minneapolis, Minnesota.*)

Machine Vision:

State-of-the-Art Systems

Although much attention has been given in the literature to the application of machine vision systems in connection with their use for enhancing the ability and flexibility of robots (so-called *smart* robots), as of the latter half of the 1980s, statistics indicate that *non-robotic* applications for machine vision are in the majority. The general categories of such uses include: (1) identification, (2) classification, (3) sorting and measuring, (4) inspection, (5) verification, and (6) quality control. Machine vision systems are enjoying a rather wide acceptance in the electronics manufacturing industry. A number of robotic guidance applications of machine vision, in addition to those mentioned in this article, pertain to automated industrial operations, such as welding, machining, painting, assembly, materials handling, among others.

At the somewhat lower level of sophistication, machine vision systems sometimes compete with the more traditional technologies that provide identification, routine inspections, and related tasks. Such simpler methods include code readers and various types of mechanical, magnetic, and photoelectric object and position sensors, such as detecting the presence or absence of an object on a conveyor line. Such less expensive and simpler approaches will continue to suffice in applications where they are fully adequate. Such lower-cost methods are described in some detail in earlier articles in this *Handbook Section*. Although they are extremely fascinating from the standpoint of technology, it is fully evident that the more complex machine vision systems must be *cost-effective*.

GRAY-SCALE SYSTEMS

Illustrative of one system[1] are the following paragraphs and associated diagrams and photographs (see Figs. 1 through 6). The use of algorithms in certain machine vision systems is described in the preceding article on "*Machine Vision.*" The system is designed to provide quality control inspections at normal production line speeds. The system utilizes a binary vision system using a CCD (charge-coupled device) matrix array camera with 320 pixels. It can be used for determining part presence, absence, orientation, and alignment for sorting by size, shape, and pattern and for label position, fill level and closure inspection. Elements of the system include a microprocessor-based control unit, solid-state camera, light source, light-source power supply, and part present sensor.

The *template or pattern match algorithm* is best applied to applications where an *exact object position* is required. A typical rotary, roll-through, or applied labeling machine should offer exact positioning when a label is applied. To inspect this kind of exact location, the subject system must be mounted in or on the machine where such object positioning is repeatedly visible.

A known, good reference or pattern is stored in the controller. Each subsequent camera image is compared, pixel by pixel, with the reference pattern. In essence, an electronic template is overlaid over the camera image of each product passing the camera.

Any differences between the template and a new object image are "seen" and displayed on the LCD (liquid-crystal display) as a count of black versus white pixels. A count value above a set limit value will cause a reject signal output. As an example, a label is wrapped around a tuna-fish can (360°). The camera is placed so that the field of view is located horizontally across the area where the label ends touch. (See Fig. 1.)

To set up the reference template, the operator adjusts the horizontal/vertical axis of the camera by viewing the display on the controller, while, at the same time, adjusting the mounted camera until the display appears as shown in the top view of Fig. 2. The template algorithm looks at each pixel in each row and column to see if the present image is the same as the reference. If pixels change from black to white (or vice versa), the number of pixels that changed are counted and appear on the display as a count. If the count is not the same as the count on the reference, plus or minus a known and set variance,

---

[1]*Cutler-Hammer QR1000 Quality Recognition Monitor.*

Machine Vision—State-of-the-Art Systems 111

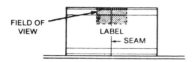

Fig. 1. Example of where a template or pattern match algorithm is used to check the accuracy of a label on a can of food. Referring to upper view, a label is wrapped around a tuna fish sized can (360°). The camera is placed so that the field of view is located horizontally across the area where the label ends touch. The camera's field of view should be located as shown in the lower diagram. To set up the reference template, the operator adjusts the horizontal/vertical axis of the camera by viewing a display on the controller, during which time the operator also adjusts the mounted camera to obtain the view shown in Fig. 2. (QR1000 System, Cutler-Hammer Products, Eaton Corporation.)

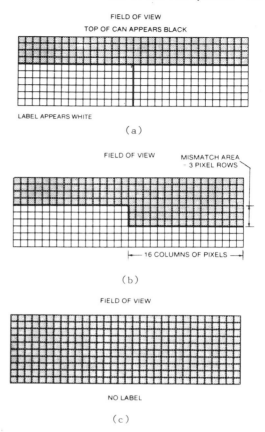

Fig. 2. Displays under three possible conditions: (a) Label is on can and is properly placed (can appears black; label, white); (b) a poorly placed label, indicating a mismatch area of three pixels; and (c) no label on can (screen appears black). (QR1000 System, Cutler-Hammer Products, Eaton Corporation.)

an output signal is energized. An example of a misplaced label and how it appears on the display is given in Fig. 2.

The *area algorithm* can be applied to applications in which the presence or absence of a label or object must be detected and where a photocell sensor is not adequate for the task. The area algorithm also can be used to look for similar objects that are not oriented the same (as, for example, upside-down). The area algorithm counts both black and white pixels within the field of view without reference to their specific location in the array.

To set up this application, the operator enters a high limit of pixel count for white area and a low limit for black area. These two limits allow rejecting missing objects, overlapping objects, or misshaped objects that may be rotated. The operator can adjust the camera to ignore changes in colors (of the object or the background), and by keeping the field of view small, can ignore the size of the background. An example of label presence and absence is given in Fig. 3. Examples of "no label," "two labels," or "presence or absence of nibs" are given in Fig. 4.

Y-Axis Movement. The Y-axis algorithm can be applied to checking a product-fill heights or other height-gaging tasks. It is useful for checking clear fluids as well as translucent or opaque materials or fluids. Tolerances of height changes may be as great as $\frac{1}{4}$ inch (6.4 mm) or as little as $\frac{1}{32}$ inch (0.8 mm), and are set by the operator. To set up the Y algorithm, a proper height is presented to the camera's view area middle. By touching the keypad "Reference," then "Enter," then "Run," the controller is given a reference height or template. By inputting two limits to the system, the instrument will reject any condition outside these limits. (See Fig. 5.)

For greater sophistication, a modular gray-scale system[2] can be used for automated inspection. The system, as shown in Fig. 6, consists of a multiple 16-bit microprocessor control unit, up to four cameras and light sources, with mounting hardware, a portable programmer, power supplies, and cables. The controller houses a master

[2] *Cutler-Hammer QR4000 Machine Vision Inspection System.*

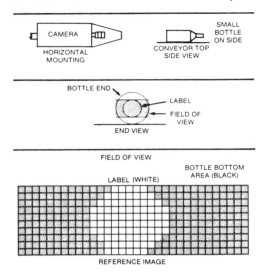

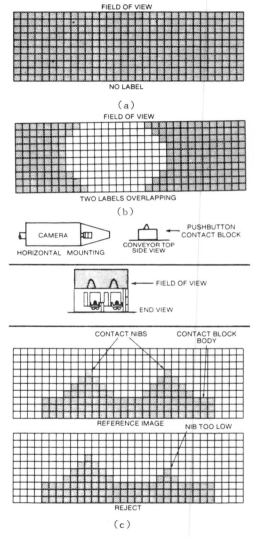

Fig. 3. Examples of where an area algorithm is used to check on the presence or absence of a label. Bottom view demonstrates that actual label coincides with the reference image. The latter has a high of 84 white pixels; a low of 236 black pixels. Under these conditions, the display will read zero (no change). (*QR1000 System, Cutler-Hammer Products, Eaton Corporation.*)

processor module and up to four co-processors plus video input modules selected for the application. A frame-display co-processor allows immediate displays of each target object and the registration and quality inspection windows used on them (using a video monitor). This allows convenient placement of the windows, dynamic viewing of the inspection process at any time, and assists in making adjustments to the production equipment for holding tolerances.

Each co-processor is connected to a camera, which can be used, singly or in combinations, for multiple views of a single object or for single views of four different objects. The controller thus can inspect and control rejection of up to four production lines simultaneously and at high speeds. A special architecture of the system eliminates signal bus contention problems often encountered with standard architecture. The cameras, light sources, and part-present sensors are located at the point of inspection on the line. Other elements of the system can be remotely located.

The sensor signals to the cameras to record a real-time image of the product to be inspected. Each image is digitized by the co-processor for comparison to the reference image stored in memory. If the on-line images deviate from the reference images by more than a specified amount, an output signal is generated for energizing a reject or alarm mechanism. Accept/reject data are stored for periodic data collection.

The comparison is achieved with a software program (algorithm) that specifies the pass-reject limits. A library of standard algorithms allows selection for given application needs.

A close-up of an installation utilizing four cameras on a containerized product production line is shown in Fig. 7. The selected inspection

Fig. 4. Displays under three possible conditions: (a) There is no label present and all screen pixels are black; (b) two overlapping labels are present. In this case, the display reads +32. (c) There is the presence or absence of nibs on pushbutton (DPDT) contact block. (*QR1000 System, Cutler-Hammer Products, Eaton Corporation.*)

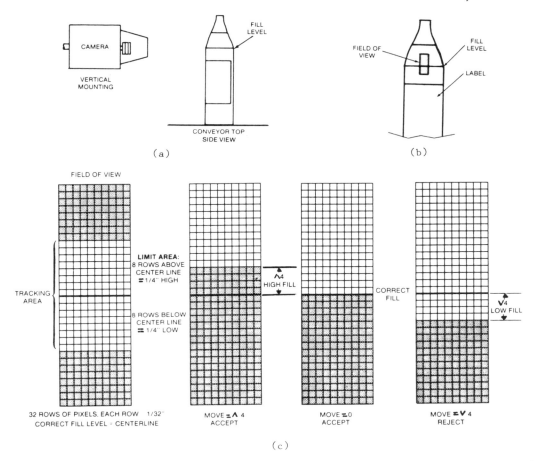

Fig. 5. Examples of where a Y-axis algorithm is used to check the accuracy of filling: (a) arrangement of camera and conveyor juxtapositions, (b) location of field of view for given high and low levels, and (c) reference images, showing high fill, correct fill, and low fill. (*QR1000 System, Cutler-Hammer Products, Eaton Corporation.*)

window portion of a bottled product is shown in Fig. 8; that of a connecting rod, in Fig. 9. A portable program, which is a part of the system, is shown in Fig. 10; and six modules in the master processor unit are shown in Fig. 11. The camera is equipped with a 25-mm "C" mount lens. A number of lens options are available.

The controller has three modes of operation: (1) run, (2) off-line (outputs disabled, and (3) program mode. The latter mode is used to install, set up, or change the system. While in this mode, the operator can obtain reference images or inspection parameters and may load a new processing algorithm, generate a new reference image, and perform installation/adjustment procedures by using the programmer and frame display coprocessor with monitor.

The system shown in Fig. 12 features 256 gray-scale processing[3] and finds application in inspection, robotic guidance, gaging, and data acquisition situations. Elements of the system include a video camera, system processor, terminal (including keyboard and monitor), and video monitor. The system processor, or master unit, includes the frame grabber, central processing unit (CPU), and mass storage devices. The frame grabber converts the analog video signal generated by the video camera to an 8-bit digital signal, which is stored in image buffers on the frame-grabber board and made available to other system components. One component is the video monitor, which displays the acquired image.

[3] *IVS-100, Analog Devices, Inc.*

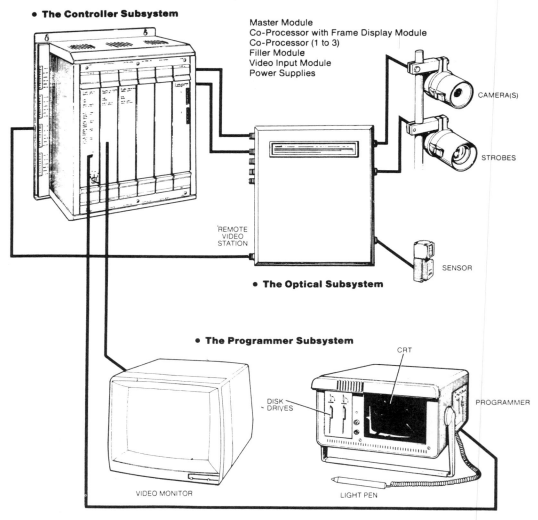

Fig. 6. Principal elements of modular gray-scale machine vision system. (*System QR4000, Cutler-Hammer Products, Eaton Corporation.*)

Another is the CPU, which can process, analyze, and interpret the image using a combination of software routines available from the manufacturer. The results of image analysis can be logged onto one of the mass storage devices (hard and floppy disk drives) or routed through the CPU's serial and parallel ports to external devices, such as host computers, programmable controllers, data-logging printers, and production equipment.

Quantization of the incoming video signal to an 8-bit-wide digital signal produces a range in gray level from 0 to 255. This wide range of gray scale provides the ability to identify subtle variations within images and to make reliable calculations (as, for example, compared with narrower gray scales). The wide gray scale is an asset in situations where illumination may be poor or fluctuating, where images are of poor contrast, and where there are problems with noisy data. (See Fig. 13.)

A hardware block diagram of the system is given in Fig. 14. Software for the system can be divided into three areas: (1) *Concurrent CP/M*^{TM} operating system, (2) programming languages, and (3) machine vision software. *Concurrent CP/M* is a multitasking, real-time operating system and includes the operating system proper plus a number of utilities. An added tool, enhancing productivity, is a full-function screen editor (*Pmate*^{TM}). This is a text editor for software

Fig. 7. Four cameras are used in this production-line inspection system to check labeling and other container characteristics. (*System QR4000, Cutler-Hammer Products, Eaton Corporation.*)

symbols. Based on the application of artificial intelligence techniques, experts studied how human inspectors judge printed characters. From this study, a specialized print evaluation software was developed. For example, the system "knows" that a break in a character is more objectionable than a thinning of the stroke width, even though both problems manifest themselves as similar amounts of missing ink.

A significant step of processing in early machine vision systems was *binary thresholding*; that is, systems that were sensitive to many levels of gray nevertheless set an intensity threshold during processing, calling all shades of gray lighter than the threshold *white*, and all shades of gray darker than the threshold *black*. Binary thresholding throws away much data and makes the system less accurate and consistent in all but very simple tasks. As a result, the trend, as reflected by this system, has been toward operating on all shades of gray throughout computing.

development and includes a number of features to help during program editing. Among these are multiple text buffers and an extensive macro-command language useful for a variety of editing tasks, such as source-code translations. The editor includes automatic indenting for structured languages, such as C and Pascal.

The machine vision software, produced by the system manufacturer, is designed as a hierarchical structure of software building blocks and embraces a number of routines, such as machine vision primitives, image acquisition and display routines, imaging processing routines, image analysis routines, image interpretation routines, and communication and control routines.

A system widely used in the electronics industry for inspection also has numerous uses for recognition, gaging, and robot guidance[4]. Data on 64 shades of gray at each point in an image are processed at the rate of up to 90 images per second. The processing speed for each application depends on the complexity of the inspection scene and on the decisions the user wants the system to make. In Fig. 15 the system is used for checking the quality of printed characters and

Fig. 8. Several inspection windows (up to a total of six) can be used to check various parts of the observed object. In the case of this bottle of shampoo: (A) Cap presence/absence, (B) product spillover, (C) container surface quality, (D) excessive adhesive, and (e) overall label quality. The additional registration window can be used for selected feature and skewed label detection. Windows can be of any size and positioned in any location within the field of view. (*System QR4000, Cutler-Hammer Products, Eaton Corporation.*)

[4]*Checkpoint 1100, Cognex Corporation.*

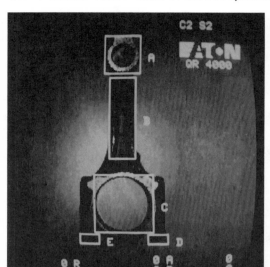

Fig. 9. Example of windowing in connection with the inspection of a connecting rod. For parts and sub-assemblies, measurement as small as 0.001-inch (0.025 mm) can be recognized and scaled up by use of optional camera lenses. The video monitor display focuses on: (A) piston pin hole presence/quality, (B) shaft configuration, (C) crank shaft rod hole presence/quality, and (D) and (E) nut presence/absence. The monitor also displays report-type data at the screen corners. (*System QR4000, Cutler-Hammer Products, Eaton Corporation.*)

Fig. 11. Main controller incorporates up to six modules—a master processor unit with EEPROM/RAM storage capacity, up to four camera co-processor modules, and a video input module. (*System QR4000, Cutler-Hammer Products, Eaton Corporation.*)

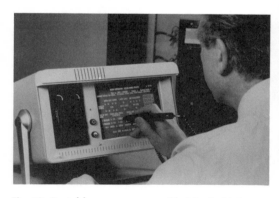

Fig. 10. Portable programmer with 9-inch, high-resolution screen also accommodates two disc drives that (1) download application programs, (2) upload/download application set-ups, and (3) provide output report statistics and data. A light pen serves to make selections and input data. An optional keyboard is available for manual data entry. (*System QR4000, Cutler-Hammer Products, Eaton Corporation.*)

Fig. 12. Machine vision system that features 256 grayscale processing and used for a variety of inspection, robotic guidance, gaging, and data acquisition applications. (*IVS-100, Analog Devices, Inc.*)

Machine Vision—State-of-the-Art Systems   117

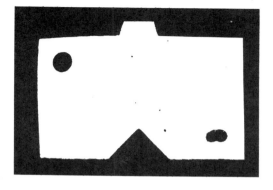
64 GRAY SCALE

8 GRAY SCALE

4 GRAY SCALE

BINARY

Fig. 13. Series of photographs that show the advantages of high-intensity resolution. With 64 levels of gray, for example, part boundaries are distinguishable from background. Fine features, such as the shadow cast by the post in the lower right-hand corner of the part, are discernible. Using fewer levels of gray (8, 4, and 2 as shown) renders the scene more subject to the effects of shading. The slightly shaded portion of the part in the binary system has deteriorated, as have other subtle features. (*Analog Devices, Inc.*)

An example of the use of a similar system in the semiconductor industry is shown in Fig. 16. Machine vision assists the automated operation by reading the alphanumeric identification codes that are laser-etched into the wafers.

The system shown in Fig. 17 inspects the dimensions, position, contrast, and shape of package elements. These include the cap, label, and neck seal placement as well as the fill level of pharmaceutical and beverage containers.

For many years, the food industry has depended mainly on human inspectors to visually check on the quality of its products. Where products flow past a human inspector at rather high speeds, the task soon becomes tiring and boring and, toward the end of a shift, can become less objective than what is required to ensure consistently high quality. With the system illustrated[5] in Fig. 18, inspections at the rate of 320 items/minute with a conveyor system moving at 150 feet (46 meters) per minute are possible. In one installation, a quarter million pizzas per day (three shifts) are inspected. (See Fig. 19.) Typical defects detected include incorrect size, incorrect shape, edge defects, and burnt spots. Based upon preselected criteria, the system makes accept/reject decisions and communicates corrective action to process controllers, reject mechanisms, or other down-line equipment. Relay, switch, or other signals are set using a menu and keypad.

[5]*Eye-Q*™, *Octek Inc.*, A Foxboro Company.

118 Machine Vision—State-of-the-Art Systems

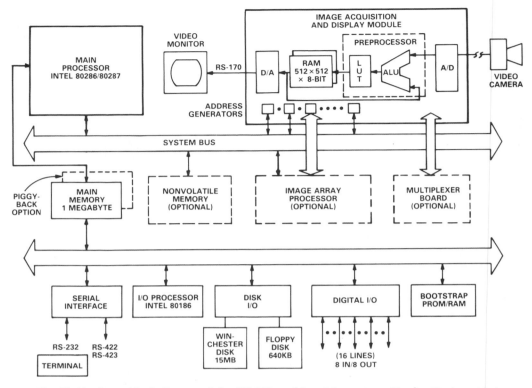

Fig. 14. Hardware block diagram of the *IVS-100* machine vision system. (*Analog Devices, Inc.*)

Fig. 15. Printers use machine vision to ensure work quality. The system shown here judges the shape, contrast, position, and skew of printed characters. The illustration shows a void in a magnetic-ink code character. (*Checkpoint 1200, Cognex Corporation.*)

Fig. 16. The semiconductor industry is automating rapidly in order to remove people from highly sterile manufacturing environments. The system shown here automatically reads the alphanumeric identification codes that are laser-etched into the wafers. (*DataMan 1000 Data Acquisition System, Cognex Corporation.* Positioning is by a wafer handler, *ADE Corporation.*)

BIN-OF-PARTS PROBLEM

As mentioned in the article on "Robot Technology Fundamentals" in *Section 4* of this *Handbook*, the "bin of parts" problem is portrayed as one of the most intriguing and unsolved problems facing robot designers for many years. A partial solution has been achieved by employing a combination of machine vision and tactile sensing. The system[6] shown in Fig. 20 and diagrammed in Fig. 21 operates as follows: The operational sequence (see Fig. 22) starts with the robot arm moving to a "home" position, allowing an overhead camera an unobstructed view of the entire scene. If a part is held in the gripper from previous operations, it is placed on the receiving station, which may be a moving conveyor, a rotating tabletop, a stationary chute, or any other mechanism for handling separated objects while retaining their alignment. Then the camera acquires an image of the scene, and the vision controller processes the gray-scale image inside a software-controlled rectangular window. At this point, a visual test is made to determine if the bin is empty, and if it is, the process stops until new workpieces are loaded in the bin, or a new bin with jumbled parts is moved into place under the camera. In case the bin is not empty and the robot is ready to accept new data, the part location and orientation data are communicated to the robot. The robot arm moves its end-effector toward the part and orients it to grasp the part, avoiding collision. The arm slowly descends into the bin until the gripper senses the part as it interrupts the photo beam between the tips of the jaws and grasps the part. As the jaws close, pressure sensors on the gripper indicate firmness of grasp between the jaws. The arm then transports the part to the receiving station, orients it as required, and releases the part. As soon as the robot arm moves out of the field of view, the entire cycle is repeated by activating the vision system with a new image and a new window covering a new area of interest. If, however, the collision sensor was activated during the acquisition process, an inquiry is made as to the availability of any other suitable candidates inside the same window. If other workpiece locations and orientations are available, a new grasping attempt is made to acquire the most appropriate candidate among them, based on their overall areas and elongations. If no candidates are available, a new image-processing cycle is initiated with a new image, and this process goes on until all objects have been acquired from the field of view inside the bin.

The basic system consists of a vision controller, a gripper controller with a pneumatic end-effector, and an imaging system that includes a solid-state camera. The vision controller (Fig. 22) includes a high-speed frame grabber with 320 × 240 pixels of 64 shades each, a 16-bit Intel 8086 microprocessor, 256K bytes of dynamic random access memory, 2K bytes of nonvolatile random access memory, 64K bytes of image-processing firmware, a hardware mathematics processor, an I/O expander, a display controller, and a 5-inch black-and-white video monitor.

[6]*ORS-i-bot*™, ORS Automation, Inc.

Fig. 17. In the pharmaceutical, food, cosmetic, and similar industries, machine vision inspects product packaging. In addition to inspecting packaging quality, the vision system also verifies expiration dates and lot codes. (*Checkpoint 1300, Package Inspection System, Cognex Corporation.*)

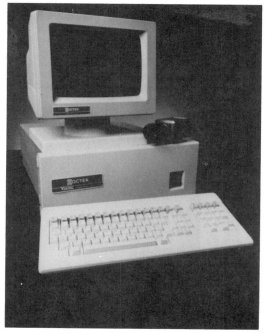

Fig. 18. Screen and keyboard of machine vision inspection system. (*Eye-Q*™, *Octek, Inc., a Foxboro Company.*)

120   Machine Vision—State-of-the-Art Systems

Fig. 19. Close-up of pizza production line showing machine vision inspection of products up to speeds of 320 items per minute. (Eye-Q™, Octek, Inc., a Foxboro Company.)

The system also includes a hardware histogram processor and dual floppy disk drives for optional peripheral storage. The system software is written in Pascal in the operating environment of CPM-86. The integrated system architecture consists of a multiaxis robot similar to a *PUMA 550/560* robot (*Unimation-Westinghouse*) with the parallel-jaw end-effector attached to the wrist of the robot, a rectangular bin of approximately 30 × 30 × 12 inches (76 × 76 × 30 cm) and an overhead camera (charge-coupled device) with a 35 mm "C" mount lens about 5 feet (1.5 meters) above the bottom of the bin. To provide col-

Fig. 20. Robotic system involving both machine vision and end-effector pressure sensors is designed for unloading randomly positioned parts from a storage bin. (ORS-i-bot™, ORS Automation, Inc.)

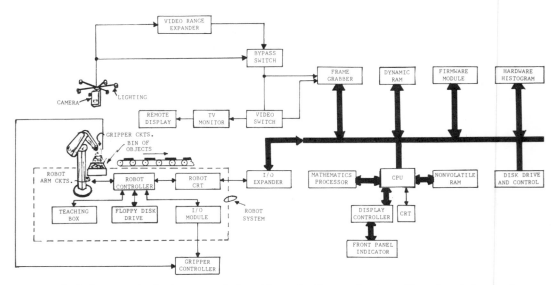

Fig. 21. Block diagram of robotic system designed for unloading randomly positioned parts from a storage bin. (ORS-i-bot™, ORS Automation, Inc.)

Machine Vision—State-of-the-Art Systems   121

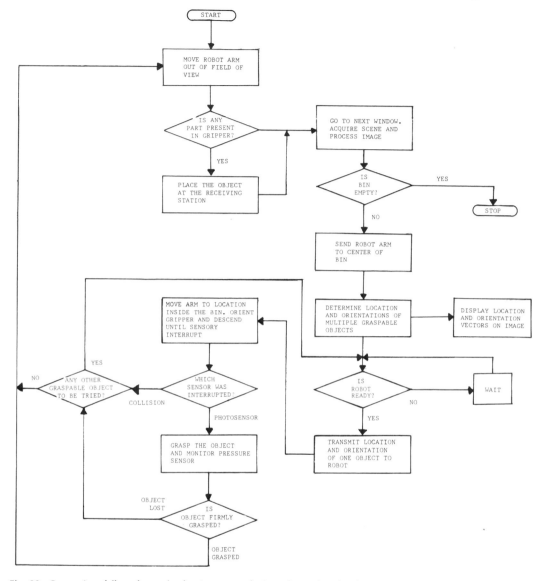

Fig. 22. Operational flowchart of robotic system designed to solve the 'bin of parts' problem. (*ORS Automation, Inc.*)

lision-free access to all workpieces, the field of view is made smaller than the cross section of the bin.

More detailed information on this system can be found in *Optical Engineering*, September/October 1984.

POST-OBJECTIVE SCANNING

Enlarging a camera's two-dimensional field of view has traditionally been an engineering task calling for complex mechanical devices and techniques. While applications vary from low-resolution verification of relatively simple two-dimensional objects to high-resolution measurements and gaging of complex three-dimensional objects, the designer must answer the question of how to acquire the images upon which inspection decisions are to be based.

Traditional techniques of image acquisition include: (1) X-Y tables, (2) the robot arm, and (3) multiple cameras. The first two techniques are mainly mechanical positioners. For example, the X-Y table either moves the target object beneath the camera, or the camera over the target object, but such movements are restricted to the X-Y plane. The robot arm provides more freedom of movement by positioning itself over the target field with a camera attached to its wrist. In multiple-camera systems, stationary cameras are placed above the target object, one above each frame. All of these techniques have proved to be expensive, mechanically complex and often not cost-effective.

A more recent alternative and perhaps the simplest mechanical configuration involves two galvanometer scanners that position the optical image before the camera. As shown in Fig. 23, the mirrors are placed orthogonally. The lower scanner has a smaller mirror and provides the X-axis scan. The upper scanner has a larger mirror that reflects the X-axis in the Y- direction, producing the Y-axis scan.

In any fixed-focus postobjective system, the general rule for system design must allow for a scan radius of at least two to four times the target field's diagonal dimensions for focus requirements. (See Fig. 24.) In this postobjective configuration, lens selections for image requirements are commercially available in most instances. When design restraints require a shorter scan radius, excessive scanning distortion, such as pincushioning, may occur. Standard correction hardware that digitally computes corrected scanning drive coordinates for the X-Y deflection system is available.

There are, of course, other configurations of laser scanners for two-axis scanning. One such configuration (Fig. 25) wraps the imaging field

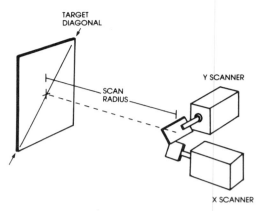

Fig. 24. Target diagonal and scan radius relationship in post-objective scanning. (*General Scanning, Inc.*)

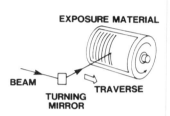

Fig. 25. External drum scanner.

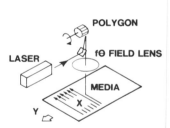

Fig. 26. Flatfield scanning with a rotating polygon scanner.

around a drum and then moves the drum in one axis while scanning the laser in the other axis. In another configuration (Fig. 26), a flat imaging field is moved in one axis while scanning the laser in the other. These systems tend to be more costly and mechanically more complex than the two-galvanometer system.

In one commercial system employing the aforementioned principles (two galvanometers), it is claimed that this technology offers the only method of XY random access motion with the dynamic focus that replicates the movement of the human eye[7]. The system is designed for ac-

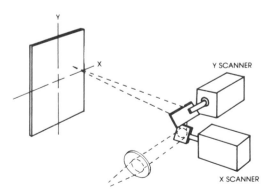

Fig. 23. Scanner position in post-objective scanning. (*General Scanning, Inc.*)

[7]gs/*EYE*™, *General Scanning Inc.*

curate positioning for both optical images or laser beams.

The system is illustrated in Fig. 27; a block diagram is given in Fig. 28. More details can be found in Vol. 390 (*Society of Photo-Optical Instrumentation Engineers*, Bellingham, Washington).

See also other articles on position, motion, and displacement measurement earlier in this *Handbook Section*.

ELECTRONIC INNOVATIONS FOR MACHINE VISION SYSTEMS

In the preceding article on "*Machine Vision*," the author explains the massive amounts of data that must be processed and hence demand much computer power with a majority of contemporary machine vision systems. Efforts have been underway during the last few years to better adapt the digital computer to these large demands.

Pipelined Image Processing Engine

The Sensory-Interactive Robotics Group of the U. S. National Bureau of Standards designed a Pipelined Image Processing Engine (*PIPE*)[8] as an experimental, multistage, multi-pipelined image-processing device for research in low-level machine vision. The device can acquire images from a variety of sources, such an analog or digital TV cameras, ranging devices, and conformal mapping arrays. The system can process sequences of images in real time, through a series of local neighborhood and point operations, under the control of a host device. The output can be presented to such devices as monitors, robot vision systems, iconic to symbolic mapping devices, and image-processing computers. (See Fig. 29.)

In addition to a forward flow of images through successive stages of operations as in a traditional pipeline, other paths between the stages of the device permit concurrent, interacting pipelining of image flow in other directions. In particular, recursive operations returning images into each stage and feedback of the results of operations from each stage to the preceding stage are supported in this manner. This architecture facilitates a variety of relaxation operations, interactions of images over time, and other interesting functions. Numerous operations are supported; within each stage they include arithmetic and Boolean neighborhood operations on images. Between-stage operations on each pixel include thresholding, Boolean and arithmetic operations, functional mappings, and a variety of functions for combining pixel data and converging via the multiple pipelined image paths.

PIPE was designed as a preprocessor for iconic (spatially indexed) images. It is intended to serve as a "front end" for the vision portion of a multi-modal sensory processing system also being developed for real-time robot guidance applications at the National Bureau of Standards. It is appropriate, however, for any sort of iconic image, such as those produced by tactile sensor arrays or range-image systems. The processed images are intended for a host machine that will perform global operations and/or image-understanding procedures, such as labeling, on the result. Thus, *PIPE* itself accepts iconic data images and typically produces iconic images whose pixel values are Boolean vectors describing local properties of the pixel neighborhood. *PIPE* relieves the host of costly low-level processing that must be performed over the entire image space.

PIPE is an attempt to seek a compromise between a fixed function, hardware image processor and a fully general-purpose, parallel computer. Through its design, it facilitates a variety of common and important image-processing techniques as well as several experimental approaches. Within the broad limits of the processes it supports, *PIPE* is an extremely fast and flexible device. On the other hand, it is not a general-purpose computer, and it is not possible to program arbitrary algorithms on it or at least not in an efficient manner.

The prototype version of *PIPE* consists of a sequence of identical processors (Fig. 30) sandwiched between a special input processor and a special output processor. The input processor accepts an image from any device that encodes two-dimensional images. It serves as a

Fig. 27. Commercial version of a post-objective scanning system. (*gs-EYE™, General Scanning, Inc.*)

---

[8]Now a trademark of Digital/Analog Design, Inc.

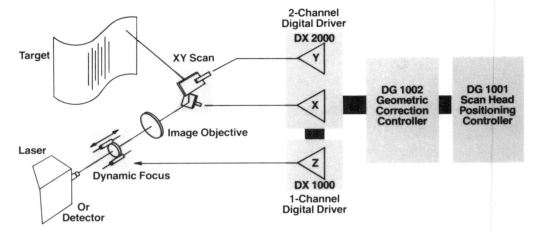

Fig. 28. Principal elements of the post-objective scanner. The DX components give the power interface for the analog galvanometers to digital portions of the system. The DG controllers are designed to simplify control and interface the scanning system to the user host terminal or computer. There is a series of XY scan heads. Major differences are in speed and pupil diameter (size of clear mirror surfaces for imaging at all angles). Small pupils move very quickly; large pupils are slower. (gsEYE™, General Scanning, Inc.)

buffer between the rest of the processors and the outside world. Each successive processing stage receives image data in an identical format, operates on it, and passes it on to the next stage for further processing. This sequence is repeated every television field-time. When an image emerges at the far end of the sequence, it is processed by the special output stage and presented to a host device, such as a robot vision system or a serial computer. The processors between the input and output stages are all identical and interchangeable, but each can perform different operations on the image sequences that it encounters. The internal architecture of a processing stage is shown in Fig. 31.

Further details are given in a paper by James M. Herriman et al. (Digital/Analog Design Associates, Inc.) and Ernest W. Kent et al. (U.S. National Bureau of Standards).

Geometric Arithmetic Parallel Processor

The development of the geometric arithmetic parallel processor ($GAPP^{TM}$) is best understood when explained against a background of earlier advancements in computer architecture that have been targeted toward improving processing power and speed. As Davis and Thomas (*NCR Corporation*) pointed out[9], "For machine vision systems and automated inspection systems, the power of the $GAPP^{TM}$ lies in the fact that a single chip can process 72 pixels in parallel. The sample inputs and outputs shown in Figs. 32 and

[9]Ronald Davis and Dave Thomas, NCR Corporation, Fort Collins, Colorado. One account of GAPP was given in *Integrated Circuits*, 28 (March 1985). *GAPP* is a trademark of the NCR Corporation.

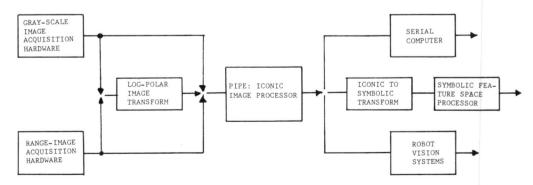

Fig. 29. The *PIPE*™ processor and its relations to other elements of the NBS Image-Processing configuration. (*National Bureau of Standards*.)

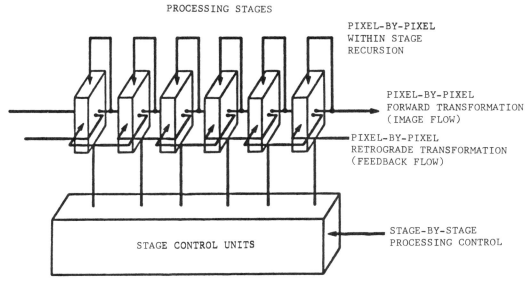

Fig. 30. Major connections between processing stages of *PIPE*™. (*National Bureau of Standards*.)

33 are taken from a 12 by 12 array of processors, or two *GAPP* chips. In 457 cycles, the image was processed. The high degree of parallelism in the *GAPP*'s architecture particularly qualifies it for automated inspection image manipulation tasks, such as template matching, edge detection, edge enhancement, binary correlation, convolution, and pattern recognition. In digital signal processing tasks such as these, the chip's ability to handle entire images, operating concurrently instead of one pixel at a time, greatly speeds throughput. In addition to robot vision and image enhancement applications, the *GAPP* is finding application in teleconferencing (video signal

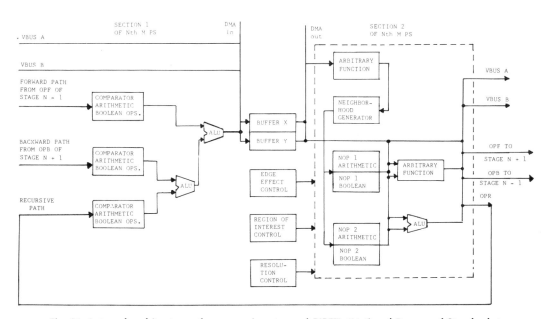

Fig. 31. Internal architecture of a processing stage of *PIPE*™. (*National Bureau of Standards*.)

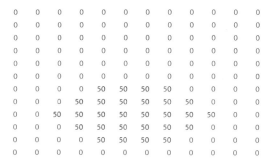

Fig. 32. A hypothetical edge extraction example, using the Sobel transform, processes this input data. (*NCR Corporation*.)

image compression for telephone line transmission); associative processor (content addressable memory; parallel logic machine); database machine (parallel search and sort); array processor (matrix arithmetic); digital signal processing for speech, seismic, and ultrasound systems; and in artificial intelligence (AI) systems."

Traditionally, computer architectures were based on the von Neumann model. (Fig. 34.) A more recent concept used to speed program execution is pipelining. (See Fig. 35). Another variation on the von Neumann scheme is known as Harvard architecture. (See Fig. 36.) A number of parallel processor architectures have been developed and are in two categories: (1) Single instruction multiple data (SIMD), and (2) multiple-instruction multiple-data (MIMD) processors (commonly called multiprocessors). These systems are comprised of several coupled processors that independently operate on different data in parallel. (See Fig. 37.) As distinct from MIMD systems, SIMD architectures are interconnected processors that work in conjunction. An example

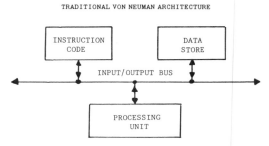

Fig. 34. Sequential fetch/execute cycles in a von Neuman processor result in slow operation. An instruction must be fetched from memory and decoded; then an operand must be fetched, followed by instruction execution. Programs implemented are performed in a step-by-step sequence of fetch/execute cycles. For digital signal processing (DSP), this can be a relatively slow process since each memory fetch can take 50 nanoseconds or more. (*NCR Corporation*.)

of an SIMD architecture is the systolic array[10], which is an array of simple, identical processor elements with a nearest neighbor interconnection pattern. (See Fig. 38.)

The systolic array has five characteristic features: (1) *Synchrony*—data are rhythmically computed and passed through a network paced by a global system clock, (2) *regularity* implies that the array is locally interconnected, consisting of a regular array of identical, mesh-connected processor elements, (3) *modularity* allows the array to be extended arbitrarily by connecting processor element modules, (4) *temporal locality* suggests that signal communications from one processor element node to a neighboring node require one unit delay, (5) *linear speed-up* pertains to a processing speed that increases linearly with the number of processing elements in the system. A measure of a systolic array processor system as compared with a single-element processor system is this speed-up factor. The latter is defined as the processing time for a single-element processor divided by the processing time for a systolic array processor.

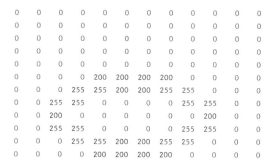

Fig. 33. The inputs shown in Fig. 32 produce this output pattern in a *GAPP*™ system. All values above 255 are set to 255. (*NCR Corporation*.)

[10]The term *systole*, originally used by physiologists, refers to the recurrent contractions of the heart and arteries that pulse blood through the body. Similarly, systolic computations are characterized by a "pumping" of data through an array of processor elements. A systolic system may be regarded as a circulatory system in which the processor is analogous to the heart. Each processor rhythmically pumps data—in and out of the processor element with some operation being performed on the data during each cycle. This action maintains a regular flow (circulation) of data within the network.

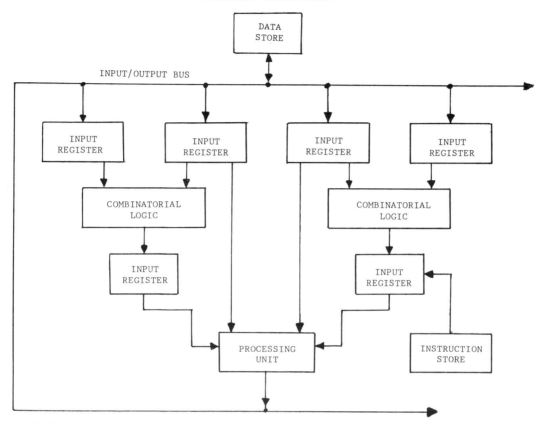

Fig. 35. Pipelining speeds program execution by breaking down a procedure into a set of independent steps that are processed at each state of the pipeline. The pipeline is an 'assembly line' in which each stage consists of combinatorial logic driven by an input register. The output of each stage is captured by the input register of the following stage. The scheme is to break down a procedure into a set of independent steps that can be processed at each stage of the pipeline. (NCR Corporation.)

The *GAPP* which has these five characteristics, consists of a 6-by-12 arrangement[11] of bit serial processor cells, packaged in an 84-pin plastic J-lead or 84-pin ceramic pin grid array. (See Fig. 39.) Each of the 72 processors on the chip contains an arithmetic and logic unit (ALU), 128 bits of random access memory (RAM), and bidirec-

[11]Research on reduced-instruction-set computers (RISCs) dates back to the mid-1970s when IBM began studies on instructions that were executed on computers. The research conducted there and subsequent research at universities, such as Stanford and the Univ. of California at Berkeley, indicated not only that simple instructions, such as LOAD, STORE, ADD, and SUB, are the most frequently used instructions but also that more complex instructions are executed up to 60% faster when implemented using several simple instructions.

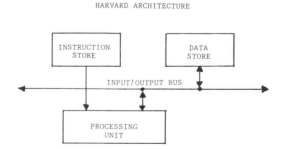

Fig. 36. A variation of the von Neumann scheme, Harvard architecture speeds computations by feeding data and instructions simultaneously into a processing unit. (NCR Corporation.)

128 Machine Vision—State-of-the-Art Systems

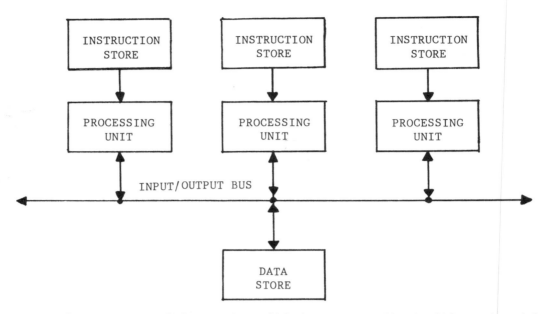

Fig. 37. Multiprocessors are multiple-instruction, multiple-data (MIMD) machines in which several coupled processors operate independently on different data in parallel. The overhead associated with interprocessor communication and synchronization frequently limits the number of independent processors that can be used in a system. (*NCR Corporation*.)

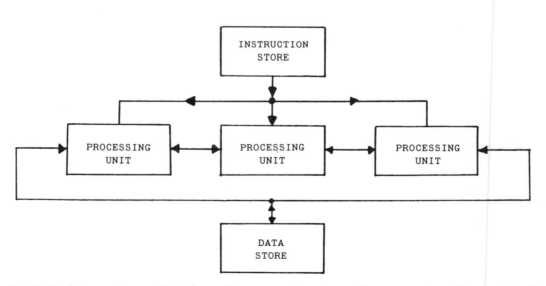

Fig. 38. Single-instruction, multiple-data architecture, such as the systolic array, consists of interconnected processors which simultaneously execute the same instruction on different data. (*NCR Corporation*.)

tional communication lines connecting each processor cell to its four nearest neighbors, one to the North, East, South, and West. Also, data can be input along the North/South axis without interfering with processor computation or interprocessor communication on a separate communication central memory (CM) line.

The *GAPP* is a RISC (reduced-instruction-set computer) whose instruction set is composed of a group of simple operations that are executed in one clock cycle. Because the execute up to five instructions concurrently, one on each of its four registers and one on its RAM. Horizontal microcoding results in nearly 6000 unique instructions that can be executed from the *GAPP*'s instruction set. Furthermore, because the *GAPP* is an SIMD architecture, the instructions are simultaneously performed by all processors in the system; that is, the same instruction is executed by multiple processors each operating on different data. (See Figs. 40 and 41.)

The cascadability of *GAPP* chips permits one to construct systems containing arbitrary numbers of processors (in multiples of 6 by 12). Further, since the *GAPP* is a single-instruction, multiple-data (SIMD) architecture, all processors can execute the same instructions concurrently. Thus, software compatibility is maintained as one modifies an existing system to use additional processors.

Image Processing. In highly parallel tasks, such as image processing, execution time does not depend upon the size of the image if the number of pixels in that image does not exceed the number of processors in the array. This contrasts with von Neumann architecture, for which execution time increases with the size of the image. The image may still be processed by the *GAPP* if the number of pixels exceeds the number of processors in the array by windowing portions of the image through the array. However, execution time will depend on the number of windows being processed. Programming of a *GAPP* system can be effected by a hardware board that is compatible with the IBM-PC I/O bus and contains a 12-by-12 array of processor elements implemented with two *GAPP* devices and a software package that allows the user to program the *GAPP* array in a high-level language and interactively debug a program. (See Figs. 42 and 43.)

Further details on *GAPP* may be found in data sheets NCR45GDS1, NCR45CG72, and NCR45CM16 (NCR Corporation, 1985) and in *Electronic Design* (October 31, 1984).

Associative Pattern Processing

Rather than increasing the capability of processing tremendous amounts of data in a shorter time span through the route of improving computer and data-handling techniques (hardware and software) as just described, a fundamentally different approach to the vision process has been taken by at least one supplier[12]. An associative pattern processing system (*APP*™) has three functional sections. An image analyzer contains all the pixel data for the current image on the camera and converts the data into a *unique statistical fingerprint*, which is a unique set of response memory addresses. The response memory contains the statistical fingerprints for all trained images. A matching histogram compares the fingerprints of the current image with those stored in the response memory. The theory of operation for the *APP*™ is based on a statistical phenomenon that reduces an image's data into a precise fingerprint. The fingerprint is a set of values for an image that is unique from other images and is a minute fraction of the total data in the image. Instead of one image being trained, all the images needed are trained and no programming is involved. One simply shows an image to the camera and allows the fingerprint memory locations drive the response memory and loads in

Fig. 39. The NCR45CG72 is a two-dimensional systolic array processor chip. It is a mesh-connected 6 × 12 arrangement of 1-bit processor elements. Each processor element can communicate with four neighbors: N., E., S., and W. (*NCR Corporation.*)

[12] *APP*™ pattern processor, Pattern Processing Technologies, Inc., Minneapolis, Minnesota.

130 Machine Vision—State-of-the-Art Systems

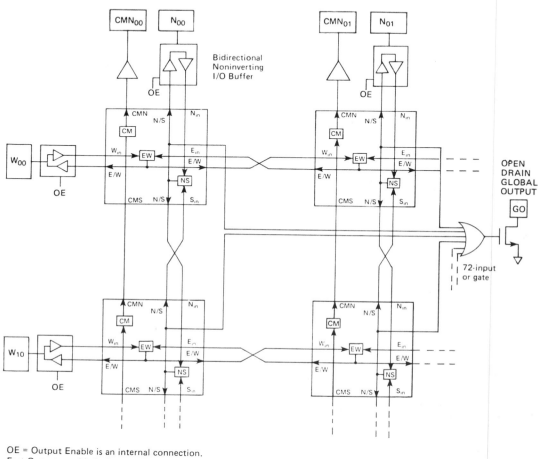

OE = Output Enable is an internal connection.
East Outputs enabled whenever EW:=W
West Outputs enabled whenever EW:=E
North Outputs enabled whenever NS:=S
South Outputs enabled whenever NS:=N
GO is pulled low whenever any NS register contains 1

Fig. 40. Block diagram of connections between four processor elements of the NCR45CG72 chip. (*NCR Corporation.*)

Machine Vision—State-of-the-Art Systems 131

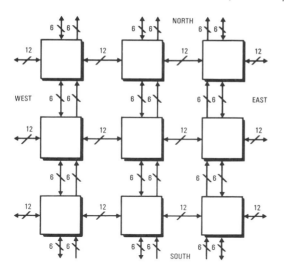

Fig. 41. The modularity inherent in the *GAPP*™ design allows an ensemble of discrete *GAPP* chips to be connected in a configuration that is almost topologically identical to that shown in Fig. 40. (*NCR Corporation.*)

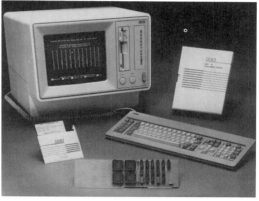

Fig. 42. Principal elements of the *GAPP*™ PC Development System. (*NCR Corporation.*)

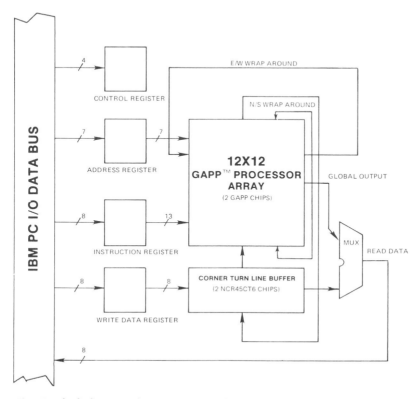

Fig. 43. Block diagram of *GAPP*™ PC Development System. (*NCR Corporation.*)

132 Machine Vision—State-of-the-Art Systems

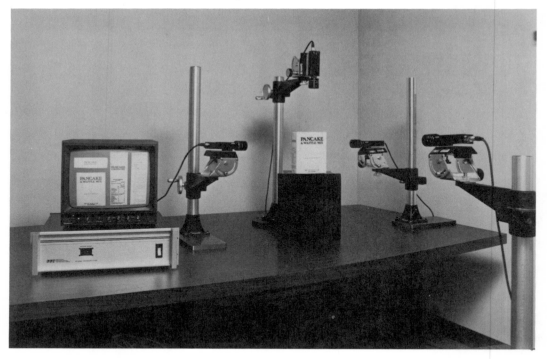

Fig. 44. Application of associative pattern processing technique (APP™) using four cameras to inspect front, two sides, and top of a pancake waffle mix box. Processor and video are shown at left. (*Pattern Processing Technologies, Inc.*)

a label for each image. This operation is explained in more detail in the prior article in this *Handbook* section.

The APP^{TM} processor may be desk or rack mounted. An application using the associative pattern processor is shown in Fig. 44.

ADDITIONAL READING

Aleksander, I., Ed.: "Artificial Vision for Robots," Chapman & Hall, New York, 1983.
Chester, M.: "Surveying the Array-Processor Landscape," *Electronic Products,* 42–47, July 1986.
Hunt, V. D.: "Smart Robots," Chapman & Hall, New York, 1985.

Morris, H. M.: "Industry Begins to Apply Vision Systems Widely," *Cont. Eng.,* 68–70, January 1985.
Morris, H. M.: "Robots See in 3-D," *Cont. Eng.,* 72–75, January 1986.
Poggio, T.: "Vision by Man and Machine," *Sci. Amer.,* 106–116, April 1984.

SECTION 3

Control Systems

Section Contents

Control System Architecture	136
Control System Hierarchy	136
Process Measurement and Data I/O	136
Direct Process Control	136
Process Monitoring	136
Process Management	138
Key Factors that Determine System Architecture (Table 1)	140
Reliability	140
Database Configuration	140
Protection, Isolation, Safety, and Security	140
Global (Systemwide) Attributes	140
Steps in the Evolution of Control System Architecture (Table 2)	141
Direct Digital Control (DDC)	141
Supervisory Control	141
Distributed Systems	141
Hybrid Distributed Control Systems	141
Single-Loop Controllers	141
State-of-the-Art Control Systems	139
Data Highways	139
Microprocessor-based Controllers	141
CRT Operator Stations	143
Higher-Level Process Control	144
Problems/Opportunities of State-of-the-Art Systems	145
Key Limitations in State-of-the-Art DCS (Table 3)	146
Key Limitations in State-of-the-Art Host Computer Systems (Table 4)	147
Trends in Distributed Control Systems (Table 5)	147
Broader Targets for Enhancing Distributed Control Systems	145
Programmable Controllers	152
Early Developments	152
Characteristic Functions of a Programmable Controller	152
Block Diagram of a Programmable Controller	153
The Processor Station	153
Power Supply	153
Memory	154
Central Processing Unit (CPU)	155
Processor Software	155
Executive Software	155
Multi-Tasking	156
User Software	157
Configuration	157
Languages	157
Ladder	158
Boolean	158
High-Level Languages	158
Grafcet	158
Programming	158
Program Loaders	158
I/O Systems	160
Parallel	160
Serial	161
I/O Circuits	161
Packaging	163
Communications	164
Point-to-Point	164
Network	164
Reliability	165
Noise Immunity	165
Availability	166
Programmable Controllers—State-of-the-Art Systems	169
Major Characteristics of Programmable Controllers	169
Micro Programmable Controllers	169
Small Programmable Controllers	174
Midsize Programmable Controllers	177
Large Programmable Controllers	182
Cell Controller	186
MAP Gateway	188
Programming the Programmable Controller	189
Languages	189
Natural Conversational Language	189
Natural Problem-Oriented Languages	189
Problem-Oriented Programming Languages	190
Machine-Oriented Programming Languages	190

Assembler Languages	190		Minimum-Variance Control	224
Narrowing the Communication Gap	190		Restricted-Complexity Control	224
Control System Flowchart (CSF)	191		Identification Models	225
Ladder Diagram	192		Explicit Model	225
Statement List (STL)	192		Implicit Model	226
Control System Flowchart (CSF)	192		Identifiers	227
Comparison of Programming Methods	193		Least-Weighted Squares	227
Structured Programming Enables Software Rationalization	196		Recursive Least-Weighted Squares	227
			Kalman Filter	228
Sequence Controllers—Hardware/ Software Trends	201		Restricted-Complexity Delay Identification	228
Graph Language "Graph 5"	201		**Numerical Control and Computerized Numerical Control**	230
The Ladder Diagram (LAD)	204		NC Fundamentals	230
Interaction of Hardware and Software	208		Types of NC	231
Program Processing	213		Computerized Numerical Control	232
Structure of a Stored Program	213		Implementation of CNC	233
			Designing for NC Production	233
Expert System and Model-Based Self-Tuning Controllers	216		**Computer Numerical Control— State-of-the-Art Systems**	235
Expert System Control	217		Two-axis CNC System for Turning Machine Applications	244
Pattern Recognition	217			
Knowledge-Based Rules	218		Three/Four Axis CNC System	249
Model-based Adaptive Control	220		Microprocessor-based CNC Systems	253
Controller Design	220			
Open-Loop Control	222			
Open-Loop Control with Feedback	222			

BERNATH, M. S., Industrial Automation Systems, Gould Inc., Andover, Massachusetts. (*Cell Control*)

BLAESER, J. A., Industrial Automation Systems, Gould Inc., Andover, Massachusetts. (*Cell Control*)

BOYLE, J., Giddings & Lewis Electronics Company, Fond Du Lac, Wisconsin. (*Numerical Control and Computerized Numerical Control*)

CHESTER, G. L., Divelbiss Corporation, Fredericktown, Ohio. (*Programmable Controllers*)

DOBSON, V. J., DynaPath Systems Inc., Detroit, Michigan. (*Computer Numerical Control*)

FLACK, T., Numa-Logic Department, Industry Electronics Division, Westinghouse Electric Corporation, Madison Heights, Michigan. (*Programmable Controllers*)

FOXX, D., Texas Instruments Incorporated, Johnson City, Tennessee. (*Programmable Controllers*)

HANSEN, P. D., The Foxboro Company, Foxboro, Massachusetts. (*Expert System and Model-Based Self-Tuning Controllers*)

HENDRICKS, H. V., Programmable Control Division, Gould Inc., Andover, Massachusetts. (*Programmable Controllers*)

*Persons who authored complete articles or subsections of articles or who otherwise cooperated in an outstanding manner in furnishing information and helpful counsel to the editorial staff.

KIMBALL, K. E., Siemans Capital Corporation, Iselin, New Jersey. (*Programming the Programmable Controller, Sequence Controllers*)

KRAUS, T. W., Formerly, The Foxboro Company, Foxboro, Massachusetts. (*Expert System and Model-Based Self-Tuning Controllers*)

MACALONEY, B., Industrial Automation Systems, Gould Inc., Andover, Massachusetts. (*Networking and Communications*)

MACKIEWICZ, R. E., Sisco, Inc., Warren, Michigan. (*Programmable Controllers*)

MARCOULLIER, J., Mitsubishi Electric Sales America, Inc., Mount Prospect, Illinois. (*Programmable Controllers*)

MEHAN, G., Siemens Energy & Automation, Inc., Peabody, Massachusetts. (*Programming the Programmable Controller, Sequence Controllers*)

MELLISH, M. T., Industrial Automation Systems, Gould Inc., Andover, Massachusetts. (*Cell Control*)

MOORE, M., Industrial Computer Group, Allen-Bradley Company, Highland Heights, Ohio. (*Programmable Controllers*)

SABINASH, E. R., Industrial Control and Power Distribution Division, Cutler-Hammer Products, Eaton Corporation, Milwaukee, Wisconsin. (*Programmable Controllers*)

SCAFARO, A. E., Reliance Electric Company, Cleveland, Ohio. (*Programmable Controllers*)

SIEMENS ENGINEERING STAFF, Siemens Aktiengesellschaft, Erlangen, West Germany. (*Programming the Programmable Controller*)

SULZER, E., Siemens Energy & Automation, Inc., Programmable Controls Division, Peabody, Massachusetts. (*Programming the Programmable Controller*)

VARNEY, T., Siemans Capital Corporation, Iselin, New Jersey. (*Programming the Programmable Controller, Sequence Controllers*)

ZIMMERMAN, C. K., Instrumentation and Control, Engineering Service Division, E. I. Du Pont de Nemours & Co. (Inc.), Wilmington, Delaware. (*Control System Architecture*)

Control System Architecture

By Carl K. Zimmermann[1]

Control system functionality and architectures have been evolving steadily for many years. Advances in semiconductor technology have allowed far more processing power to be applied at all levels of plant control than was believed possible only a few years ago. This power will be needed to achieve plantwide automatic control, which includes full integration of process control and plant business systems.

Control System Hierarchy

Process plant control has often been described as a hierarchy or layering of functions. Although this method is somewhat imprecise, it will be used here as an organizational aid. Four major layers, starting from the bottom up of Fig. 1 are: (1) process measurement and data input/output (I/O); (2) regulatory and other direct process control; (3) process monitoring; and (4) process and plant management.

Process Measurement and Data I/O

1. *Data acquisition*—sensors, actuators, contact inputs and outputs that are physically connected to the control system and converted to data in the control system's database. Speed and precision of conversion range rather widely and depend on the nature of specific points in the I/O.

2. *Remote information collection*—from points remotely located from the control center.

3. *Foreign device interfaces*—required, for example, with microprocessor-based devices. Here ASCII serial or other standard communications will be used. Such devices may include programmable controllers, motor speed controls, process analyzers, among others.

Direct Process Control

1. *Regulatory control*—closed-loop feedback systems for controlling variables, such as pressure, temperature, flow, et al. The fundamental building block of regulatory control is the PID (proportional, integral, derivative) control function. To this can be added many other dynamic functions, such as *lead/lag* and *deadtime*, as well as calculations to accomplish control of difficult situations that often involve a number of measurements and final control elements. Regulatory control execution rates vary with the type of loop. Pressure and flow require faster up-date speeds than temperature. Typical execution rates of one to three times per second are adequate for most loops.

2. *Advanced (nonconventional) regulatory control*—includes closed-loop control algorithms beyond the PID used for most loops and the conventional complex strategies, such as feedforward, overrides, and Smith Predictor. Examples of these nonconventional forms include (a) internal model control (IMC), (b) inferential control, and (c) dynamic matrix control (DMC). These approaches are usually based on process models and are used on processes that are not well regulated, such as those with intermittent observations (analytical measurements, for example), significant pure delays (deadtime), and other difficult-to-handle behavior, such as inverse response.

3. *Sequence control*—is related to the logic control of equipment having discrete states, such as motor start/stop, valve open/close, etc. For complex processes, this can involve many devices that must function in the proper order for process startup, shutdown, and other situations. Response times of about 10 milliseconds are typical. Operator interaction in the form of messages with operator responses are desirable.

4. *Sequence-of-events recording*—to assist in problem solving. In upset situations, for example, it is highly desirable to know what events occurred and in what order. Speed is critical since events can occur within a few milliseconds of each other. Events must be time stamped, stored, and available for recall for expert analysis. It is also desirable to have high-speed process measurement data for comparison with discrete events.

Process Monitoring

1. *CRT operator stations*—are used to interface the operator with the control system. From CRT screens and keyboards, the operator can look at values of input data, manipulate the process outputs, monitor alarms, and change process parameters. For batch processes, recipes may be called up from a master library and scheduled for execution by the system. CRTs can graphically display controller faceplates and process flow-

---

[1] E. I. Du Pont de Nemours & Co. (Inc.), Wilmington, Delaware.
Editor's Note: This article condenses a wealth of experience in automatic control that can be used by control, process, and production engineers and technical managers who are considering the development of specifications for future designs, procurements, and retrofits.

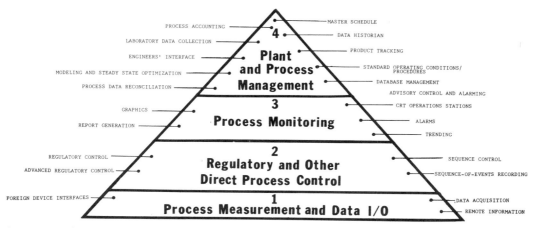

Fig. 1. Control system hierarchy. (1) At the process measurement and data I/O level is the myriad of sensors and actuators that allow a control system to function. (2) Regulatory control at the second level is the control of process parameters, such as flow, temperature, and numerous other variables. Also, at this level, is the first level of control logic used for equipment start/stop and cycling through simple operations. The combination of regulatory control and sequence control is termed direct process control. (3) At the third level, process monitoring includes the presentation of all the process-related data to the operators, giving them the ability to make changes as required. (4) At the highest level, process management includes a higher-level functionality, which will facilitate process diagnostic work, alter plant operating conditions for maximum profit, and schedule production based on demand for the product.

It should be noted that this pyramidal concept of control not only includes those elements (sensors, measurement, decision making, actuation and feedback) that make control of a process or machine possible, but includes numerous support functions, particularly the data display and analysis functions which assist both in the making of human as well as automatic decisions.

sheet graphics. The operator station also can recall stored data for trend and historical plots as well as accessing the many other functions in the system.

2. *Alarms*—to tell the operator when a process variable goes into an alarm condition. The location of the problem and the means for correcting the condition must be easily available and with means for the operator to correct the condition easily and rapidly. Operator comprehension of the alarm system cannot be overstressed.

3. *Trending*—is the procedure of plotting process variables versus time on CRT graphics, thus providing the operator with a useful tool for determining loop stability. Where several variables are visible on the same set of axes, the operator can determine the cause of a disturbance that can ripple through a process upsetting many loops. This type of disturbance can extend well beyond a single piece of equipment. Update times faster than once every five seconds are used where fast-responding loops are involved, such as in flow and pressure control. Plotting discrete points also can be helpful where equipment start/stop or change of state can trigger upsets. In advanced systems, laboratory data also can be plotted (based on when the sample was taken) on the same axes with continuous and discrete event data.

4. *Graphics*—to convey information quickly and efficiently to the process operator. CRT graphics with considerable capability and flexibility are provided at the operator station in many installations. Because many control systems are used as replacements to panelboard controls, the CRT graphics nevertheless must mimic panelboard functions. This not only can reduce the difficulty in learning the control system when it is first introduced, but also can provide a useful structure for organizing controls and alarm displays. For example, a CRT display showing multiple controller faceplates is far less sophisticated than pie chart or other pattern recognition technique, but it can be learned by the operator with very little effort and requires no display development or design effort. Process flowsheet graphics, while a useful training tool, often provide more information than an operator who is familiar with the process requires. They are, however, helpful in displaying batch progress.

5. *Report generation*—the presentation of process data is closely associated with its usefulness. Reports should have access to as much information in the system as possible. Time of day and date should also be available for key events. Arrangements of data or report formats must be easy to change so that reports can be custom-tailored to their users. For batch systems,

a batch-end report is desirable. This report should include start and end times for a batch (in time of day), the process parameters for the batch, and times of key events during the batch.

Process Management

These functions are related to the overall way a process is run. For example, these functions may adjust operating conditions to improve process yield or product quality. The loss of any of these functions will rarely lead to equipment shutdown. Normal plant operation can usually continue for several hours without serious product degradation, but over significant time spans they are important.

1. *Data historian*—this provides the long-term storage and retrieval of all available process information with respect to time. With this, problems in product quality, physical properties, or operational problems can be linked with specific equipment conditions or control problems. Storage of data is arranged so that process excursions are not lost, even if they are of very short duration. Since there is no way to predict where a process problem will develop, as many variables as possible are stored. The data historian also can be useful for looking back as far as a year when analyzing problems that are slow to be discovered, that is, of a seasonal or subtle nature. Data maintained by the historian may include all process parameters, event data, and laboratory data stored according to when the process sample was taken.

2. *Product tracking*—is an extension of the data historian, allowing specific batch or lot identifications to be associated with stored process parameters. This function is useful in connection with problems in a specific customer shipment, which can be compared with other production runs made under similar conditions.

3. *Standard operating conditions/procedures*—are the target parameters required to manufacture a particular product. For a continuous process, they may be setpoints for various process variables. For a batch process, they are more complex and will include times and rates of change. Such data are stored in the system by product type and grade. Associated with these conditions are procedures (automatic and manual) for transitioning from one product to another. Sometimes this is accomplished by slowly ramping setpoints to their new values. Equipment cleanout and startup procedures also may be included. In batch processes, the information may be termed recipe parameters and recipe files.

4. *Database management*—is particularly important when changes are made in plant equipment or procedures for manufacturing a product. Alterations beyond the routine operator changes in the control system may be involved. To minimize the disturbance to the manufacturing process, good database maintenance is essential. The system may have built-in protection against access by unauthorized personnel. Isolation between operating areas is also used to protect against inadvertently altering the wrong data. Some protection is also provided to ensure that changes can be put into use without unexpected disturbances in operating equipment. Automatic logging of the changes on a disk or printer can be helpful in establishing the cause of a disturbance should one develop. Provision also should be made for identifying alarm limits or other process parameters that are outside the desired or standard operating conditions. In such cases advising the system manager of the discrepancy may be the most appropriate system response. The system must be capable of documenting the database. For continuous control loop strategies, a graphical self-documenting function is desirable.

5. *Process accounting*—involves the collection of total materials and energy consumption by the system reported on an hourly, shift, weekly, calendar-month, and accounting-month basis. A report is generated at the end of the specified interval or on demand.

6. *Laboratory data collection*—involves the reporting of process samples that are analyzed in a laboratory (measurements such as composition and physical properties—variables that sometimes cannot be determined on-line). The results of lab data are delayed by the time required to complete the analysis. To facilitate the comparison of such data with other process data, it may be accessible on the basis of when the sample was taken and not when the lab work was completed. Statistical noise filters are often used in conjunction with lab data to determine when a change has occured since lab data tend to have considerable variations due to not only the analysis per se but also to the person performing the analysis. Lab data may be accessible by the trend and historian functions, previously mentioned, for reports and operator station plots.

7. *Engineers' interface*—can be used to make process data available to the engineer in a graphical format, enabling statistical analysis in the form of a library of statistical functions as well as providing high-level programming languages and tools. Flexible graphical plotting routines can also be provided. For example, historical plots may be easily expanded to show much detail or contracted to provide the "big picture." Variables also can be plotted against other variables in the form of an *X-Y* graph to establish sources of interaction.

8. *Modeling and steady state optimization*—are important functions for process management. Process models can be used to supplement unreliable measurements, compensate for slow or noisy lab analysis, enhance control strategies, and allow optimization of product quality, process energy consumption, and profit. Several modeling techniques are used, including steady state and dynamic first principle models, linear-dynamic models, and statistical steady state models. The choice of these depends upon the modeling func-

tion performed and the allowable computer time available. A valuable model-based function is steady state optimization. With this technique, model parameters are varied to find the set that yields the process optimum. Then actual process operating conditions can be changed to these values for improving the product quality, lowering energy consumption, and enhancing profit. Sophisiticated search techniques, such as linear and nonlinear programming, are usually used to change model parameters and find the optimal conditions with the least number of trials.

9. *Process data reconciliation*—is a function that uses material and energy balance calculations in conjunction with statistical data analysis to detect process measurement problems and to improve the quality of process data. Sometimes, the approximate values of measurements are predicted mathematically while a measurement device is under repair. Similarly, errors in accounting reports are mathematically corrected for measurement errors. This function is also valuable when process optimizers are in use to avoid problems due to measurement errors.

10. *Master schedule*—is a function that uses information on product inventory and anticipated demands so that a production schedule for a process unit or area can be created. Although this function is most common in connection with multistream batch processes, the approach can also be used for continuous processes to drive product transitions.

11. *Advisory control and alarming*—is a method of process control that is used when corrective action is not well defined and when operator intervention is desirable. In advisory control the digital computer is used to monitor the process, decide the most probable cause of a problem, and determine the best course of action to take. This information is then conveyed to the operator—to be combined with human insight into the operation of the process. The operator then will determine if (a) a change should be made manually, or (b) the system should be allowed to make the needed change automatically. This type of control has been in use for some time and may be a good candidate for the incorporation of advances in artificial intelligence (AI) as they develop (particularly expert systems).

All of the foregoing functions in the hierarchy of control affect the choice of control system architecture for a given application. Needless to say, these functions also have played a major role in molding the system architectures that are available today. Previously shown Fig. 1 can serve as a convenient checklist of these functions. Other important factors that define system architecture are summarized in Table 1.

Background of Control System Architecture

For an explanation of how and why control system architecture evolved into the configurations currently available and used, see Table 2.

STATE-OF-THE-ART CONTROL SYSTEMS

Having just provided a background on the development of control system architecture over the past several years, control systems are now assessed in terms of current times. Distributed control systems (DCS) and host process computers are reviewed. As mentioned earlier, DCSs have three primary constituents: (1) data highway, (2) microprocessor-based controllers, and (3) CRT operator stations. Many other modules can be added on to the data highway, such as calculation and application modules, trend memory units, analog multiplexers, and computer interface units. Information given here will be helpful in developing specifications.

Data Highways

The data highway is fundamental to the operation of the system. Through it, the control functions may be physically separated from each other and from the operator's console. The highway is also important in establishing functional isolation. For this, multiple processors can be dedicated to the control and operator interface functions. This is essential to achieving high reliability inasmuch as it minimizes the impact of any single failure in the system. This also helps to prevent user-induced failures by keeping each processor within its loading limitations. Functional isolation also can be helpful in preventing the ripple effect common among single processor systems—where a change in one function can impact the operation of another unrelated function.

Data highway communications must be able to handle heavy traffic loads and must be reliable. For example, response within one or two seconds is needed in reporting alarms to the operator, in displaying fast-moving process data for manual control, and in maintaining tight control where several controllers are involved. Loading requirements become more severe as system size increases. Communications speeds must be guaranteed even during abnormal operating conditions. During process upsets, for example, many alarm conditions may suddenly change and must be reported to the entire system quickly so that remedial action may be taken. Process startup and shutdown are other examples that cause heavy communications loads.

Because system dependence on data highway operation is increasing, there is a large need to maintain secure communications at all times.

140 Control System Architecture

TABLE 1. Key Factors that Determine System Architecture

RELIABILITY

Direct control functions require higher reliability than process management functions. In general, the downtime of a control system should be small with respect to the overall plant downtime. If a plant runs continuously for months, this should be reflected in the control system reliability. If a failure in the control system can have catastrophic effects, this must be reflected in control system design. Where the latter condition may exist, backup systems (or even separate safety systems) may be indicated.

TIMING

The timing requirements of control functions range from a few ms (sequence of events, interlocking) to several hours (optimization). Control of process variables (flow, temperature, pressure, etc.) usually requires between 1 and 5 executions per second. Effective transfer of information between processors over data highways often requires more than a second. To achieve the speed requirements for functions faster than one second, it is desirable to execute the function within one module. Regulatory control loop timing requirements frequently can be met by keeping the base-level control functions and related I/Os in one controller. Higher levels of control may require data from other controllers. Usually this can be accomplished without degrading the control because of timing. However, reliability may decrease because more processors are involved, increasing the likelihood of a failure. The data highway must be operational in such case for full control.

DATABASE CONFIGURATION

In the case of well-understood and common plant control functions, such as simple regulatory control, sequencing, and report generation, the database can be relatively simple, easy-to-use setup techniques that require little or no user programming. Fill-in-the-blanks or menu-driven selection techniques are available and require much less sophistication and training than do the programming/software-related functions.

When a programming environment is provided, several user tools should be provided, including text editors and compilers that pinpoint errors and provide English language error messages.

It is desirable to separate these two working environments so that (a) system support can be accomplished with minimum-required expertise, and (b) the difficulty of exchanging data between the two environments will not increase.

PROTECTION, ISOLATION, SAFETY, AND SECURITY

The question may be asked, "How many and which control functions should be handled by a single processor?" This involves system reliability and ultimately can impact on the ease of implementation and support. Generally, direct control functions do not belong in an environment shared with process management functions. Doing so subjects the control functions to loading effects which can adversely impact their critical timing requirements. Direct control functions should be isolated.

CRT operator stations should be used only for levels of regulatory control where timing is not critical. Regulatory control assigned to one processor should be limited so that timing can be guaranteed for all operating conditions and will not slip or vary with process upset or other abnormal operating conditions.

Control function isolation improves protection against errors made in changing the database inasmuch as problems induced within one processor's executable database are normally limited to that processor. Maintenance and overall support are also simplified. By isolating functions, the impact of hard failures in a processor can be minimized.

GLOBAL (SYSTEMWIDE) ATTRIBUTES

Even though system functions may vary in terms of the reliability needed, timing, and sophistication of working environment, they must work together if they report in some fashion to a network, that is, are part of an integrated system. In such cases, an alarm limit, for example, should be made only once for that change to be complete for the entire system. Conventions for naming devices and components should be consistent throughout the system. All identification of a particular measurement should be similar throughout the system regardless of the level or working environment. For example, having a 12-character tag name in one area of the system and a numbering system for processors and inputs in another area is undesirable, especially if cross referencing must be done manually.

For simple systems with few control system functions, deciding on a system architecture is usually straightforward. For complex systems, the selection process can be difficult and can present many tradeoffs.

The first step is to decide which functions are to be included in the system, followed by developing a "functional description." This should be based on the need and potential benefit to the operating process. Although planning an architecture for continued growth is excellent practice, system design should avoid unnecessary functions and keep the system as simple as possible. Having an overview of the entire system is essential. Available space and environmental conditions and other constraining factors are also needed, against which candidate system architectures can be tested.

Additionally, some fundamental decisions must be made about the system architecture. Will a host computer be needed? Will a distributed control system (DCS) be used? Will programmable logic controllers be used? Will maximum integration be sought?

If regulatory control with CRT operation stations is required, a DCS will probably be used. Sequencing operations may elect a programmable controller, or it may be desired to include this functionality in the DCS. If process management functions are required, a host computer may also be needed. When a DCS and a host computer are both used, considerable overlap in functionality may exist. In these cases, functions must be assigned to a system based upon reliability requirements, speed of execution, relative cost, and ease of support.

After analysis of the foregoing restraints and targets, the basic architecture best suited will largely have been selected.

TABLE 2. Steps in the Evolution of Control System Architecture

ENTRY OF COMPUTERS INTO CONTROL SYSTEMS

As early as 1959, computers were introduced in a serious way for regulatory process control. A few years earlier, computers were used to monitor processes.

Direct Digital Control (DDC). In early systems a single mainframe computer accomplished data acquisition, control, reporting to the operator, and higher-level computation. In the first systems the computer directly controlled the process—without intervening controllers. This architecture was called direct digital control. (See Fig. 2.) Only one computer was used in the first systems because of the high cost per processor and because of the general absence of computer-to-computer communications. However, in the DDC system shown, an auxiliary (redundant) computer was proposed.

Early DDC system applications had a number of drawbacks. For example, if only one processor was used, a single failure could affect a large number of controlled variables and possibly disable an entire process. Although redundant processors were introduced to provide backup for hardware failure, additional processors for functional distribution did not appear until much later.

Supervisory Control. This control concept, shown schematically in Fig. 3, developed over a period of years in an effort to use the many benefits available from the computer, but without the many drawbacks of DDC, as previously mentioned. This concept had several advantages; in particular, it preserved the traditional panelboard control room while adding the capability of a digital computer with associated CRT displays. This minimized the disruption to operations that could accompany a transition to DDC. It also offered a buffer between the computer and the process—so that the operator had the option of selectively disabling the computer for troublesome loops. Although the impact of a computer failure was substantially reduced, it, too, used few processors. Thus, considerable functionality could still be disabled by a single failure.

Distributed Systems. This concept was introduced during the mid-1970s and combines three technologies, most of which either appeared for the first time or were much improved during the interim that dated back to DDC. (See Fig. 4.) These technologies are: (1) microprocessors, (2) data communications, and (3) CRT displays. Multiple microprocessors were used in one system as dedicated control loop processors, CRT operating stations, and a variety of other control and communication functions. Although these systems could not always accomplish all the function of the previous computer-based architectures, they achieved considerable fault tolerance by minimizing the effect of a single failure.

Hybrid Distributed Control Systems. Because distributed control systems are limited in computer functionality and in logic or sequential control, they often are combined with supervisory host computers and with programmable controllers in "hybrid" architecture, as shown in Fig. 5. These systems use data communications channels between the host computer and both the programmable controllers and the distributed control system. Data communication between the programmable controllers and the DCS is also used even though it is not always required. Although these systems offer the strengths of each subsystem, it is difficult to establish communication between the subsystems and often confusing to provide software inasmuch as each subsystem uses different programming techniques and organizes data in separate databases.

Single-loop Controllers Replacing the DCS. In another version of the hybrid approach, the DCS is replaced by single-loop digital controllers. These offer some advantages in the precision of digital control, but still use a conventional panelboard as the primary operator interface rather than CRTs. This approach has limitations similar to the DCS hybrid approach except that the databases are inherently simpler because of the elimination of the CRT-based primary operator console.

Besides immunity from internal effects, such as sudden (burst) load increases, highways (and control systems in general) must be able to withstand the power surges in an industrial plant and electrical storms. The latter are particularly severe where data highways are run outside over long distances. Dual (redundant) data highways are commonly used to diminish the risk of communications loss should one cable be damaged. The two highways can be routed over different paths to minimize the possibility of both being subject to the same source of damage.

Much more information on data highways is given in *Section 5* of this *Handbook*.

Microprocessor-Based Controllers

There are two classes of controllers available in DCS systems: (1) the *multi-loop* controller, and (2) the *single-loop* controller. The multiloop controller can use one or more microprocessors (MPUs) to accomplish the control function on several (usually no less than eight) modulating control loops. Each controller includes dedicated I/O modules to interface to process measurement and control equipment. The controller can accommodate analog (continuous I/O), discrete (digital), and other field I/O. In some cases, serial ASCII interfaces to "smart" field analyzers, weigh scales, among others, are provided. The controller also manages communications to and from the data highway. This communication is primarily to report process information to the operator's console, but also can include data transactions to and from other process controllers (peer-to-peer) as well as to historian and other higher-level modules, programmable controllers, and mainframe computers.

The control functions available in the multi-loop controller are usually limited to those in the algorithm library. These include regulatory feedback control algorithms, dynamic function blocks, logic processing and sequencing blocks, and some calculating capability. The entire algorithm library

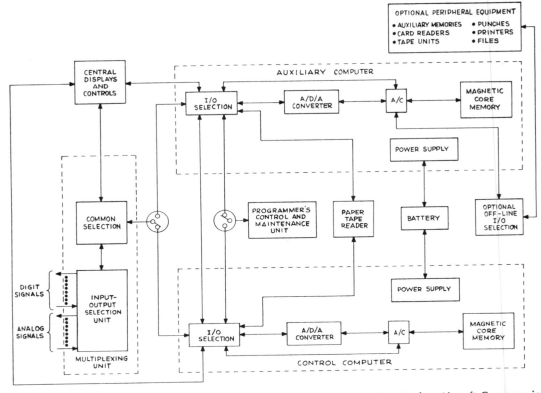

Fig. 2. Direct digital control system (with auxiliary computer) developed by Hughes Aircraft Company in 1965.

is generally included in read-only memory (ROM) in each controller—whether or not all algorithms are used. This is done by the manufacturer by using re-entrant (subroutine) programming so that the user simply provides the changeable process data (e.g., scaling, alarm limits, and tuning) and the block linkages (the sources and destinations of data passed between blocks). This approach to data organization greatly decreases

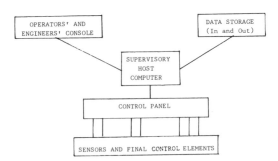

Fig. 3. Schematic diagram of a supervisory control system.

the user's potential for errors in generating the controller database and is well suited for easily implementing control strategy changes.

The single-loop controller is similar to the multiloop controller in both functionality and physical components. These controllers often can acquire many analog and discrete inputs, but are usually limited to about four analog outputs and some correspondingly small number of discrete outputs. The number of components is considerably smaller than multiloop controllers, and there is usually only one or two printed circuit boards.

Multiloop controllers offer advantages in system cost, communications, and packaging density. Because there are more shared components, multiloop controllers are less expensive to manufacture per control function. These cost advantages are passed on to the buyer and result in a lower hardware cost per loop. The cost savings in using multiloop versus single-loop controllers can be substantial in large (over 500 loops) applications. Communications for multiloop controllers also can be simpler where many process inputs must be used for a coordinated control strategy, or where one variable is used

Control System Architecture

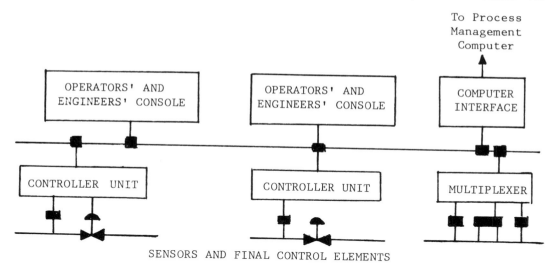

Fig. 4. Schematic diagram of an early distributed control system.

repeatedly by many control loops. This can occur in feedforward control or in advanced control strategies. In single-loop controllers, the multiple use of a single variable may entail communication over the data highway, setup procedures that are often complex, and additional highway loading. On large applications (especially on retrofits), space for controller cabinets is often limited. With multiloop controllers, the space required can be significantly less because of the higher packaging density and corresponding fewer number of cabinets.

The use of multiloop controllers, however, is not without drawbacks. A single component failure can disable every loop within a given controller. In some controllers, these can number into the hundreds. For this reason, they are available with redundant (backup) processors. In most cases, these will automatically take over in the event of a failure without any human intervention and with minimal disturbance to the process under control. In many cases redundant (auxiliary) I/O is also available to minimize the risk of an I/O failure. Both processor and I/O redundancy may be provided as either one-on-one or one-on-several to allow some flexibility for varying process criticality. Single-loop controllers are also available with redundancy, although this is not required for most applications.

CRT Operator Stations

The operator station provides the primary window to the process and to the performance of the controller modules. Each station has a keyboard and one or more CRTs for display. Although more than one CRT is usually desirable per operator, usually there are additional keyboards provided, with a usual ratio of one keyboard per CRT. A printer or other hard-copy device also is provided for permanent records.

The operator station allows process information to be displayed, parameters to be changed, automatic control sequences to be run, and, in most cases, the full system database to be maintained. The display of information has several formats.

A *system overview display* provides deviation and alarm information for a large number of points. From this, the appropriate *group display* can be easily accessed. This display shows controller faceplate bar graphs of eight to sixteen loops and allows controller parameters, such as alarm limits, setpoints, controller mode, and controller output to be changed. A *detail display* of a single controller is often provided to allow additional parameters (e.g., controller tuning) to be changed. *Trend displays* are also available to plot process variable versus time to facilitate controller tuning and analysis of control upsets. *Process graphics* or *flowsheet displays* can be constructed to pictorially show process vessels, piping, and current operating conditions. Provi-

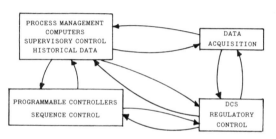

Fig. 5. Hybrid control system.

sions are made to allow controller adjustment without leaving the flowsheet display. *Alarm annunciation displays* provide a graphic representation of conventional alarm annunciator panels. These provide information on all alarms in the area on dedicated displays.

Keyboard designs vary widely among systems. Many are supplemented with methods of coordinating the operator's hand and eyes with *touchscreens* or *joysticks*, for example. Some suppliers offer touchscreen keyboards for highly flexible keyboard designs. Keyboard design must allow the operator quick access to any location for which there is operator responsibility. After an alarm is displayed at the operator station, a series of interactions or keystrokes allows access to the loop in alarm. By adjusting the setpoint, manually changing the valve position, or other actions, the operator returns the process to normal operation. When this happens, the alarm will clear itself. Even if the process returns to normal without operator intervention, the alarm condition will remain in the system until it is acknowledged by the operator.

Two major considerations in selecting consoles are (1) prestructuring, and (2) backup. Console software as it is produced by the manufacturer should provide a complete set of prestructured displays and keyboard functions for loop controller display and operator actions, as well as for alarm display and acknowledgment. Absence of these features can add significantly to the implementation effort of the system user. In the event of a console failure, one console should be able to back up another. This seemingly simple requirement is not trivial in many systems. Not only must access be provided to all control points for both consoles, but alarm/acknowledge must be fully operational (without dual acknowledgment) for both the normal and failed console conditions.

Operator consoles can be microcomputer- or minicomputer-based and usually can handle between 500 and 10,000 loops. The number of operator stations varies from one to four per computer processor. A disk is usually provided for additional display capability and when the console is used as a configuration station. In many cases the console will remain fully functional (minus some displays) in the event of a disk failure.

Higher-Level Process Control

Process management functions traditionally have been handled by higher-level supervisory host computers. As "second generation" systems add to their capability, many process management functions may also be handled in the distributed control system (DCS). These functions, some of which are overlapping, are described in the next several paragraphs.

Historian. Data historian and event-recording functions traditionally have been the domain of the host or supervisory computer. Large amounts of data require the storage capacity of large computer disks and tape drives. Retrieving the data in short time requires the speed of a larger disk and significant computer power. Distributed control systems traditionally have been limited to floppy disks and low-end microprocessors. This has resulted in considerable shortage of storage capacity and often a slow speed of retrieval. Placing this function in the host has meant severe loading increases on the DCS highway and computer interface for large systems. Database and software of the host system must also be maintained separately, which contributes to the overall database maintenance effort.

Language-based Application Processor. Programming language capability of the type required for process modeling, complex calculations, and optimization has also been primarily the domain of the supervisory host computer. Support for these functions can require considerable memory and processor power. Furthermore, existing process models and optimizing packages often exist in FORTRAN language. Distributed control systems traditionally have either lacked higher-level languages or have been limited to proprietary BASIC-like languages or (slow) interpretive BASIC. In many cases language-based applications must be squeezed into already heavily loaded operator station processors.

Gateway. The control system must often exchange data with other plant subsystems, such as programmable controllers, weigh system, gaging systems, and "smart" process analyzers. Although some standard gateways were developed, any unusual communications link could not be accomplished by the distributed control system (DCS). In some cases the DCS will have the electrical interface and the language required to complete the interface but will lack the internal data communication power (throughput) to effectively pass this information through the system for use as process data in a control scheme or as historical data. Host computers traditionally have provided much better flexibility for interfacing with other devices. However, sharing large amounts of this information with the DCS can become a loading problem, both for the computer and the DCS. If the data are regulatory control related, the reliability of the host also can become an issue and the redundancy and functional isolation of a DCS may be an advantage.

Report Writers. Reports generated on demand or on a regular schedule are a key part of process supervisory functions. While most general-purpose computers have comprehensive "forms" or other report-generating utilities, many DCSs have little capability. Packages should offer screen-oriented text editors with live data fields that can be updated when the report is generated.

Engineers' Workstation. A separate CRT workstation can be provided to analyze process data. Easy access to on-line and historical data is essential. The CRT should facilitate plotting data versus time and plotting one variable against another to establish interaction. A library of statistical functions and higher-level programming language capability should also be provided. The independence and security of this workstation must be assured so that its loading will have no effect on the operating process. The foregoing functions have required a supervisory computer in the past. In the future, personal computers and other microprocessor modules can be provided for this function as integrated components of a DCS.

Sequence-of-Events Processor. Although not a process management function, this function has been the province of the general-purpose computer for some time. Considerable speed and processing power are required to resolve the order in which process events occur (often to one millisecond).

Recipe Batch Unit. For batch control, records of product formulations and sequences of operation for their manufacture should be maintained and available as files in the control system. In the past, this required a host processor inasmuch as the products manufactured can number in the hundreds, to be further complicated with numerous grades of product and other manufacturing variations for each product. File structures of the host computer operating systems are well suited to product master recipe file handling. However, since these recipes must be directly transferable to the direct control processors for manufacturing, this function traditionally had computer interface and database limitations.

Problems/Opportunities of State-of-the-Art Systems

The foregoing description of systems represents a composite of what is generally available from a variety of suppliers. The following explores the development of these systems.

Distributed control systems can be considered in two classes: (1) those systems that reflect a predominant panel-board background, and (2) those systems that have a computer or "systems" background. The former class tends to be very structured in nature, with little computer sophistication required by the user and with many protective features to help the user avoid inadvertent mistakes and corruption of data handled by the system. The latter class tends to be more difficult to use, with fewer amenities to protect the user and, therefore, requires a higher degree of computer literacy. These systems are sometimes called "hard" and "soft" systems, respectively.

What is needed in DCS systems is to combine the favorable attributes of both classes and eliminate the undesirable qualities. In general, this would yield the most flexibility—providing the utilities of soft systems, while preserving the structure, user protection, and simplicity of the hard systems.

The limited functionality of the DCS often necessitates the use of separate subsystems in a *hybrid* approach, previously described. This creates a large burden on the user, both in developing communication interfaces between systems and in maintaining separate databases for each system.

Some of the key limitations in state-of-the-art DCS systems are summarized in Table 3. Limitations of current host computer systems are summarized in Table 4. Trends for overcoming the limitations of present distributed control systems are summarized in Table 5. DCS system development is a slow, costly process. Decisions on new developments are based not only on new control functions but also on new and emerging microprocessor and communication technology—and, of course, on reducing manufacturing costs.

In general, developments in DCS systems will allow a transition from the *hybrid* systems of the past to more fully integrated systems. (See Figs. 6 and 7.) In the hybrid approach, the DCS was one of several subsystems in the control hierarchy at the direct control level, including programmable controllers and the supervisory computer, which had considerable regulatory control functionality. DCS systems are now progressing toward inclusion of most of the direct control functions and numerous process management functions.

BROADER TARGETS FOR ENHANCING DISTRIBUTED CONTROL SYSTEMS

In addition to the factors described in Tables 3 and 5, there are a few broad targets that should be sought within the next few years.

Reliability

Through redundant processors and limited functionality per processor (functional separation), digital technology has been well established as having reliability as good as or better than the traditional analog control which was essentially the only control approach available for many years. For critical functions of safety interlocking, where personnel or equipment are endangered, additional sophistication in design will be required. Two examples: (1) two-out-of-three voting, and (2) fail-safe electronics.

TABLE 3. Key Limitations in State-of-the-Art Distributed Control Systems

SEQUENCING

Programmable controllers are required to provide sequence control. These can do the logic functions at higher speed and have superior I/O hardware as compared with DCSs. The cost per discrete point generally is much lower in the programmable controller than in the DCS. Distributed control systems in the past have provided only limited sequencing capability, most of which has been language-based and much slower (usually one second) to execute.

CALCULATION

Many DCSs have very little number processing capability. Math functions are often limited to a few very basic algorithms. Languages are frequently required for anything beyond the most simple functions and may not be available in every controller. Substantial calculating power should be provided within the configurable function blocks.

COMMUNICATIONS

Linking the DCS with host computers and other intelligent devices (e.g., programmable controllers and smart analyzers) has been difficult. In most cases extensive user software has been required. Interface capability to common host computer and other process equipment should be provided by the system manufacturers.

CAPACITY

Distributed control systems have been severely limited in the number of points that can be handled by a single highway and by a single console. Generally, it is an advantage to the system designer if all area points can be handled by the same highway and, therefore, accessed at a single point. Likewise, consoles should be capable of handling all highway points. By doing this, differences in console databases can be minimized, and one console can be used to back up others in the event of a console failure. Some care must be given to ensure that only those alarms that are within an operator's domain be activated on a given operator's console.

PROGRAMMING

Although fill-in-the-blanks configuration is desirable for most direct control functions, higher-level language is desirable for many complex functions, such as process modeling. For these needs, FORTRAN or other standard languages are desirable, inasmuch as the models in many cases exist before installation of the DCS. Direct transportability from one system to another is important to minimize the cost of implementing these functions. Proprietary languages are undesirable since they require special training for programmers and do not allow transport of software between systems.

LACK OF INTEGRATION

Past attempts to add sequencing and other functions have led to separate databases configured from separate workstations. As functions are added, they should appear to the user as having a *unified* database structure. All points in the system should be defined from the same workstation. Separate workstations and processors for continuous and sequence control, for example, are undesirable. Console points also must employ a unified structure. When a tag name or transmitter range is changed for one console, it should be automatically reflected in all other consoles. In general, the duplication of databases within the system should be avoided and, where required, should be made transparent to the user.

USER ORIENTATION

The limited need for both user software and unstructured computerlike environments should be continued. Database setup of control systems should remain highly structured. Direct control functions should require very little computer literacy. Entering and modifying routine control should be fully protected against corruption of unrelated system functions and unidentified errors in compilation or downloading. A comprehensive, error-free set of functions for both continuous and sequential operations should be maintained with easy-to-use and well-defined procedures for application and modification.

Two-out-of-Three Voting. In this scheme signals are input to the system on three separate paths to each of three separate processors. Through an elaborate algorithm, the processors decide the validity of the data. Each processor then issues a corresponding output, which is processed by the output logic to decide what the valid output is. By requiring that two signals must match in both of these cases, the probability of an error in processing due to a single failure is eliminated.

Fail-Safe Electronics. This approach addresses the unpredictability of electronic output circuits. Normally, it has been impossible to predict whether an electronic circuit will fail when "on" or "off" (high or low, etc.). With this type of circuit, the failure state can be accurately determined. The technique currently is becoming available on a limited basis. It is interesting to note that control engineers, particularly those in the process industries, over the years became accustomed to fail-safe devices as, for example, a control valve purposely configured to fail open or fail closed.

System Self-Diagnosis. Considerable improvement is also possible in the area of system self-diagnosis. When a system fails, a full English language description of the failure and how to handle it should be almost instantly available to the operator.

TABLE 4. Key Limitations in State-of-the-Art Host Computer Systems

PROGRAMMING

In the past host computers have required a high degree of programming and operating system expertise by the user. Although the addition of high-level, user-oriented packages for reports and database maintenance has reduced this need in some areas, these systems require considerable improvement to achieve "user friendliness."

LACK OF INTEGRATION

Like other subsystems in the control hierarchy, most computers suffer the effects of poor integration with the rest of the system. Not only must a separate database be defined and maintained for the host computer, but also independent operator stations, often with unique operating procedures, must be used to access the host data.

SPEED

Because of the limited number of processors at the host computer level, many points and functions are usually processed together. Because of this, the speed of response can be a serious problem. Data acquisition operator and engineer workstation, and database access can all experience significant delays in a heavily loaded system.

RELIABILITY/SECURITY

These have been key limitations of host computers. Because they are not functionally distributed, the operating environment can be very complex, increasing the probability of user error. In addition, there usually are large amounts of process data in the host computer. These systems traditionally have been vulnerable to downtime because of user errors and scheduled or unscheduled maintenance.

COMMUNICATIONS

Although host computers have good flexibility to accommodate data links to other computers and intelligent devices, they have been limited in capacity for large amounts of data. Slow-speed data links in particular have contributed to processor loading and have been low in efficiency. Special software interfaces have been required on many devices because few comprehensive standards have existed for communications between systems. See also *Section 5* of this *Handbook*.

TABLE 5. Trends in Distributed Control Systems

FASTER AND IMPROVED COMMUNICATION HIGHWAYS

As data communication technology advances, control system manufacturers are moving to take advantage of these changes. Data highway speeds are increasing from several hundred thousand bits/second to several million bits/second. Many systems now use *Ethernet* (CSMA/CD), token passing, baseband, and broadband techniques and will most likely employ other standard developments, such as chips, to convert to IEEE standard protocols as they become available.

In order to realize true plantwide control, system communication capability must improve considerably. Not only must standards (e.g., General Motors Manufacturing Application Protocol—MAP) be developed, but also system manufacturers must offer gateways to their systems that accept these standards. For systems to be truly compatible, they must adopt and offer compatibility at higher communication layers, including the highest or application layer. It is to be noted that the ISO (International Organization for Standardization) now describes seven layers of communication: (1) physical, (2) data link, (3) network, (4) transport, (5) session, (6) presentation, and (7) application. This organization can ensure that data in different systems can be structured similarly and thus eliminate the need for much of the translation and reorganization presently needed for exchanging data between systems.

Since control systems must accomplish real-time tasks, there also is a need for communication throughputs of high speed that permit transfers of dozens (if not hundreds) of points between systems within a one-second time frame. This speed requirement often can complicate system design inasmuch as faster access is often acquired through local data tables. Such tables then must be changed with the primary database when changes are made. See *Section 5* of this *Handbook*.

MORE POWERFUL MICROPROCESSORS

Advanced digital microprocessor technology now offers considerably enhanced processor capability for systems. Higher clock speeds, more memory addressing capability, and lower memory costs permit far greater capability in each system module.

SEQUENCING AND PROGRAMMABLE CONTROLLER CAPABILITY

Although most systems have offered interface "black boxes" to programmable controllers for some time, fully integrated sequence capability has been limited in most systems. High-speed (10- to 20-millisecond response) sequencing modules are now offered by several systems. They offer relay ladder logic and fill-in-the-blanks function block configuration for simple tasks and programming for complex operations. Some sequence modules are also available with rugged I/O electronics, similar to that of commercial programmable controllers.

MORE FLEXIBLE OPERATOR CRTs

Graphics capability on operator color CRTs is improving at a fast rate. Character-oriented graphics suppliers are being replaced by vector-oriented graphics. Operator keyboards are becoming more flexible, incorporating special function keys to prove quick access to main displays and a variety of devices (e.g., touchscreen, trackball, joystick, etc.) to aid eye-hand coordination and to improve the speed and accuracy of operator actions. Keyboard design is an active area of improvement. CRT touchscreens with full graphics are currently used as supplemental keyboards to afford maximum flexibility for user-modified keyboard designs. Alarm capability is also improving. Alarm annunciator CRT displays now have become available as standard equipment. Logic for combinatorial alarm processing is also becoming available.

TABLE 5. Continued

OPERATOR STATIONS

The CRT-based operator stations of the present have numerous advantages over former traditional panelboard instrumentation. For control rooms with many control loops and monitored variables, they can eliminate the continuous travel of an operator along a panelboard during the course of running a process. In addition, the intelligence within the control system can be used to provide improved information to the operator. CRTs do have the drawback of providing only a limited view of some of the variables. With a panelboard, the operator can see a large portion of the process at one time. Similar observations on a CRT may require selecting several different displays with corresponding keystrokes and delays in the system. Changes to the process are also complicated by the narrow CRT window. With a panelboard, the operator can do several things simultaneously. Numerous keystrokes and the need for different displays can prevent simultaneous operations and decrease the speed of response to some process upsets. In general, larger screens allowing multiple display segments or windows would improve this situation. Joystick and other ergonomic aids can be used to improve operator response time. Operator stations still are limited in their communications with host computers. To provide the most useful window to the process, one station should access both direct control data and the data resident on the host processor—while using the same keyboard and procedures.

DATABASE MAINTENANCE

This task is still far too difficult in both current distributed control systems and host computer systems. Single database changes (e.g., change in alarm limits) should be made only once to be reflected everywhere necessary throughout the system. The ideal way to accomplish this is to have a single data item exist in only one place in the system. For distributed databases, however, this can become a speed constraint if many distributed data items are needed at high speed. For this reason, local duplication of data in host processors and other places in the system is almost unavoidable.

Initial database entry in diagram form should be allowed for continuous control functions. This should be automatically translated to a tabular English language form and also into machine executable code. Self-documenting systems of this type are beginning to appear.

Considerable improvement is needed in the area of complex batch control. Although some work has been done to standardize the terminology for software structure, better process analysis techniques are needed. Presently, a very large part of the batch control effort is spent in defining the batch operation. Better analysis methodologies can greatly facilitate this and help to manage what is currently a high-risk, human-power-intensive enterprise. Process control systems used for sequencing and complex batching must include the three features that now exist in programmable controllers: (1) high speed, (2) rugged yet universal I/O hardware, and (3) self-documenting logic configuration that provides real-time displays of the logic, as do relay ladder diagrams.

DATA STORAGE AND RECALL

Distributed control systems are generally limited to several days of data storage. Longer periods are usually available only by using longer averages, which can lose information on short-term disturbances. Larger capacity mass storage devices can partially remedy this situation, but still may fall short of what is needed. Process computers have long used *data compression* techniques to better utilize available mass storage and extend the length of storage time without losing short-term excursion data. Data compression should be offered when long-term storage is required. For short-term data storage (e.g., for tuning), data compression may be undesirable because it can obscure process dynamics. Considerable improvement in DCS storage of process event data and batch tracking is also desirable.

INTERFACE TO FOREIGN DEVICES

As digital technology is incorporated into various subsystems throughout a plant, it becomes highly desirable to pass data to and from the control system. Analyzers, weigh systems, gage systems, and motor speed controllers are among the devices now using digital data communication. Control systems are presently offering communication ports and gateways to their data highways. They include standard communications driver programs as well as user programming so that an interface to a foreign digital device can be established.

INTEGRATED MEASUREMENTS

Sensor technology is rapidly moving in the direction of microprocessor intelligence in nearly every process transmitter. This opens up many new possibilities for integrated measurements. Diagnostic capabilities of the MPU can be used to perform self-checks on transmitters and report to the operator through the control system. A network of transmitters, possibly using fiber optics, can be used to significantly reduce field wiring in a plant. The difficulties existing in present systems, when, for example, a transmitter range or tag name is changed, can be largely reduced by integrating the intelligence of the transmitter into the control system. If a transmitter parameter is changed and the information resides only in the transmitter, it will no longer be necessary to reflect the change throughout the system, as is current practice.

PROGRAMMING LANGUAGE CAPABILITY

Many systems currently offer programming language capability to supplement their structured fill-in-the-blanks control environment. Although proprietary languages (often enhanced BASIC) still predominate, FORTRAN, PASCAL, and C language capability should become available in the relatively near future.

DATA STORAGE CAPABILITY

With the advent of inexpensive Winchester Disk technology, the data storage and recall capability of control systems has undergone a quantum leap. Floppy disks, which always have been limited for this function, no longer must be used for historical data storage.

TABLE 5. Continued

CONFIGURABILITY

Presently, a wide range of features exists to provide assistance in control system configuration. This is generally associated with the fill-in-the-blanks function block setup procedure required by many DCS system for continuous control. There should be a similar high-level environment for batch control. Use of programming should be kept at a minimum. Also, there should be an operational hierarchy organized around recipes, units, phases, steps, et al., which will ensure an organized structure for the user. This will reduce the effort needed to change and support the system.

Standard displays and prestructured keyboards must remain an integral part of control systems. For continuous control, groups of standard controller faceplate with dedicated function keys must be provided. For batch control, standard displays for recipe maintenance, selection, and operation, as well as unit status displays, must be provided.

Designing for Plantwide Control

Most of this article is directed toward the application of control systems limited to particular units or areas. For plantwide control, numerous additional functions are required. There are additional layers to the hierarchy required, and this increases the overall communication and database access task. As described in greater detail in *Handbook Section 5*, factory floor automation is moving along a path toward standardized communication (example: MAP developed by General Motors) and toward *open-system* architectures. In this type of architecture, all subsystems on a network can uniformly access data from other subsystems. No similar direction has been established for process control systems (e.g., distributed control systems). A number of manufacturers have agreed to provide GM MAP gateways, but it is likely that most suppliers will continue for a while to use proprietary communication schemes within their own systems. To the system user, all points of data in the system should be accessible in a uniform manner without concern for where it is located within the system. This is sometimes referred to as a *global database* concept. To achieve plantwide control, fully integrated systems with truly global database access must be achieved.

Potential of Artificial Intelligence

Although control system features with some artificial intelligence (AI) capability are relatively new, it is expected that many improvements to system performance will be in these areas, particularly in the "expert system" area. In the relatively near future, all distributed control systems should offer self-tuning on their regulatory control algorithms. Advanced alarming should also be offered. This capability will allow alarm grouping and sequence of occurrence to be used to let the system recommend corrective action to the operator that is based on programmed "intelligence" about the process. Although AI will require significantly more memory, speed, and processing power than is currently available in control systems, this type of feature could begin to capture human experience (from past incidents, for example) that can be used in running a process and that can be available to operators who may be less familiar with how to respond to process problems and emergencies. Before there is widespread use of AI techniques, much more development is required.

See article on expert systems later in this *Handbook* section.

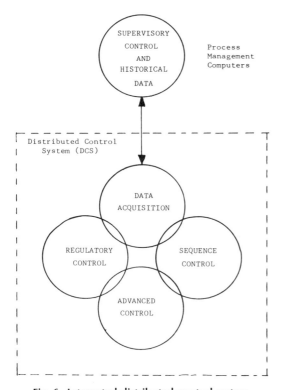

Fig. 6. Integrated distributed control system.

Personal Computers

Several personal computers currently available are enhancing control system capability, as in recipe handlers for batch control, configuration database generation and management, and configuration self-documentation. These new capabilities (sometimes called *second-generation*) will allow the DCS to take on more central role

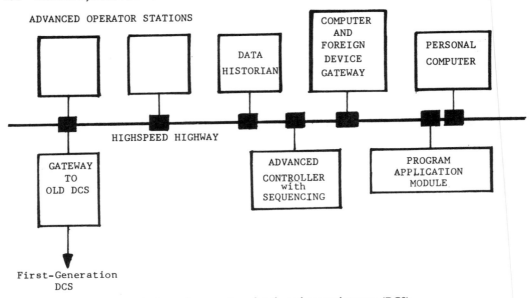

Fig. 7. Second-generation distributed control system (DCS).

in assuming a more complete set of direct control and process management functions. In doing this, it will become a larger portion of the control system hierarchy. In practice, the added capability of the first-level system will mean the migration of functions previously relegated to higher-level systems to first-level DCS. This will provide additional choices for system architecture. Application programs previously possible only in a higher-level computer will run at the lower level, too. This will mean easier integration with the direct control functions, but it may be more expensive because of the higher reliability present at the lower level. Programming environments and languages also may be more limited at the lower level. Similar choices will exist elsewhere in the system. More powerful sequencing, for example, may be available in programmable controllers from specialized vendors. This must be weighed against the ease of implementing and supporting a sequencer that is fully integrated within the system.

Changing Role of System Suppliers

The increased functionality of the DCS is also changing the business of the system suppliers. DCS development has become far more software intensive. Currently as much as 80% of the development investment may reside in software alone. Development costs in excess of $50 million for a complete system are not uncommon. The preponderance of software means that risk of errors in a new system is increasing dramatically. Achieving flawless operation may require several years from when a new system is first available to users.

In general, greater capability in the first-level system will be a major advantage to users. System support will be simplified because fewer interfaces to subsystems will exist. The number of separate databases will also be reduced as more functions are integrated into the first-level system. This, too, will simplify support and facilitate initial system implementation. Operator stations with broader capability will reduce the number of CRTs/keyboards that an operator will need. Instead, there will be a window to all functions within the system, thus eliminating the need for separate CRTs that are concerned with programmable controllers, distributed control system, and host computer. In turn, training can be reduced and efficiency in responding to process problems will improve. Complex actions involving multiple CRTs will be directed from a single CRT.

The newer, more flexible systems envisioned for the future will increase user responsibility. Critical functions must be properly isolated and distributed in the system, and they will be executed by processors with appropriate reliability. For very critical functions, dedicated processors with full backup and redundant I/O may be required. The choice between flexibility and simplicity will be more difficult with second-generation systems. Considerable judgment will be required to assure that systems to be installed will have achieved a degree of simplicity appropriate to the operating environment at a given plant site.

Editor's Note: During the late 1980s, the acronym SCADA (Supervisory Control and Data Acquisition) has gained some prominence in the industrial control literature. Still to be acceptably defined, a SCADA system originally was assumed to incorporate: (a) a hundred or more processors located at remote sites that transmit information at the instance of an event or upon demand; (b) a host computer that furnishes high-level computations (optimization); (c) a radio or microwave transmission system between the remote sites and a centrally-located supervisor; and (d) for security and reliability, a coded system that provided these assurances. Once mainly identified with long-distance situations, such as utilities and pipelines, SCADA is expanding in concept and the term is now being used increasingly for describing similarly functioning systems in manufacturing and processing plants where distance is not a criterion.

ADDITIONAL READING

Blickley, G. J.: "SCADA Systems Affected by Distributed Control," *Cont. Eng.,* **32** (3) 79–81, March 1985.

Farmer, E. J.: "Modernizing Control Systems," Van Nostrand Reinhold, New York, 1984.

Laduzinsky, A. J.: "Would SCADA By Any Other Name Still Be the Same?" *Cont. Eng.,* **33** (2), 72–75, February 1986.

Manuel, T.: "What's Holding Back Expert Systems," *Electronics,* **59** (28) 59–63, August 7, 1986.

Morris, H. M., Sr.: "Batch Applications Proliferate: Control System Manufacturers Respond," *Cont. Eng.,* **33,** 54–57, July 1986.

Skrokov, N. R.: "Mini- and microcomputer Control in Industrial Processes," (Planning and executing the computer project, pp. 9–58), Van Nostrand Reinhold, New York, 1980.

Williams, T. J.: "The Use of Digital Computers in Process Control" (a short history of the field, pp. 11–30; distribution of duties, pp. 31–48), Instrument Society of America, Research Triangle Park, North Carolina, 1984.

Zimmerman, C. K.: "Evaluating Distributed Control Systems," *Cont. Eng.,* **31** (11) 109–112 (1984).

Programmable Controllers

By Ralph E. Mackiewicz[1]

As recently as the early 1960s, industrial control systems had been constructed from traditional electromechanical devices, such as relays, drum switches, and paper tape readers[2]. The control relay was the most widely used device for controlling discrete manufacturing processes. Although these earlier devices are still used today, and many of the problems associated with using them have been eliminated due to technological advances in their design, such approaches continue to suffer from some inherent problems. Relays were susceptible to mechanical failure, they required large amounts of energy to operate, and they generated large amounts of electrical noise. Extreme care had to be taken in the design of relay-based control systems because it was not uncommon for the outputs to "chatter," that is, turn on and off rapidly when they changed states. The logic of the circuit was dictated by the wiring of contacts and coils, and in order to make changes, more time was required to rewire the logic than was needed in the first place.

In the late 1960s, the need to design more reliable and more flexible control systems became apparent. For example, the automotive industry was spending millions of dollars for rewiring control panels in order to make relatively minor changes to the control systems at the time of the yearly model changeover. In 1968 a team of automotive engineers[3] wrote a specification for what they called a "programmable logic controller." What they specified was a solid-state replacement for the relay logic. Instead of wires there would be bits inside of a memory circuit that would dictate the logic. The machine would use solid-state outputs and inputs, instead of control relays, to control the motor starters and sense pushbuttons and limit switches. The first commercially successful programmable controller was introduced in 1969. By today's standards, it was a massive machine containing thousands of electronic parts. It should be stressed that this machine was designed at a time long before the availability of microprocessors. The early PC used a magnetic core memory to store a program that was written in a graphical language (Relay Ladder), a scheme long established in connection with relay logic. (See Figs. 1 and 2.)

In the late 1970s, the microprocessor became a reality and greatly enhanced the role of the PC, permitting it to evolve from simply replacing relays to the sophisticated control system component that it has become today. Programmable controllers now have the ability to manipulate large amounts of data, perform mathematical calculations, and communicate with other intelligent devices, such as robots and computers. Concurrent with the increased capability and flexibility of the PC was the expansion into many other industrial applications, including the control of machine tools, material handling systems, food-processing operations, and use in the continuous process control field.

A programmable controller is currently defined by NEMA (National Electrical Manufacturers' Association) as a "Digital electronic device that uses a programmable memory to store instructions and to implement specific functions such as logic, sequence, timing, counting, and arithmetic to control machines and processes."

Characteristic Functions of a Programmable Controller

Seven of the most important characteristics of a programmable controller include:
1. *Field-programmable by the user*. This characteristic allows the user to write and change programs in the field without rewiring or sending the unit back to the manufacturer for this purpose.
2. *Contains preprogrammed functions*. PCs, when procured, are already programmed with at least logic, timing, counting, and memory functions that the user can access through some type of control-oriented language.
3. *Scans memory and input/output (I/O) in a deterministic manner*. This feature allows the control engineer to precisely determine how the machine or process will respond to the program.
4. *Provides error checking and diagnostics*. A PC will periodically run internal tests on its memory, processor, and I/O systems to ensure that what it is doing to the machine or process is what it was programmed to do.

---

[1] SISCO, Inc., Warren, Michigan.
[2] This general observation applies mainly to the discrete-piece manufacturing industries.
[3] General Motors Corporation, Hydramatic Division.

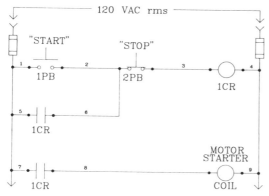

Fig. 1. Typical motor control circuit. When the pushbutton labeled START (1PB) is pressed, the control relay (1CR) is energized. A contact from 1CR is then closed and is used to 'seal' 1CR 'on' after 1PB is released. Another contact from 1CR is used to energize the motor starter coil, turning the motor on. When the STOP pushbutton (2PB) is pressed, it de-energizes 1CR, which 'unseals' 1CR, and deenergizes the motor starter coil which stops the motor. Implementing this motor control circuit requires nine wires, not counting the power supply. The equivalent PC ladder diagram program is shown in Fig. 2.

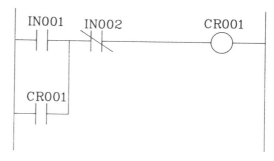

Fig. 2. Ladder diagram program used to control the motor circuit shown in Fig. 1. In this case, all the inputs and outputs are assigned variable names, such as IN001 for the START input and CR001 in place of the control relay 1CR. This diagram is then drawn on a program loader and entered into the programmable controller's user memory. The PC's processor then solves the logic that is stored in memory. Only six wires are needed between the PC's output and the motor starter coil and between the PC's inputs and the pushbuttons (not counting power supply wiring).

5. *Can be monitored.* A PC will provide some form of monitoring capability, either through indicating lights that show the status of inputs and outputs, or by an external device that can display program execution status.
6. *Packaged appropriately.* Modern PCs are designed to withstand the temperature, humidity, vibration, and noise found in most factory environments.
7. *General-purpose suitability.* Generally, a PC is not designed for a specific application, but it can handle a wide variety of control tasks effectively.

Block Diagram of a Programmable Controller

A simplified model of a PC is shown in Fig. 3. The input converters convert the *high-level signals* that come from the field devices to *logic-level signals* that the PC can read directly. The logic solver reads these inputs and decides what the outputs should be based on the user's program logic. The output converters take the logic-level signals output from the logic solver and convert them into the high-level signals that are needed by the various field devices. The program loader is used to enter and/or change the user's program into the memory and to monitor the execution of the program.

THE PROCESSOR SECTION

A detailed block diagram of the processor section of a PC is shown in Fig. 4. This section consists of four major elements: (1) power supply, (2) central processing unit (CPU), (3) memory, and (4) I/O interface. This interface is discussed later.

Power Supply

The basic function of the power supply is to convert the field power into a form more suitable for the electronic devices that comprise the PC (typically +5 V dc and/or +/− 12 V dc). The power supply is one of the most critical components of a PC for several reasons: (1) Unless well designed, the power supply can be one of the most unreliable components of a PC because of the large amounts of power consumed—hence heat-removal problems, (2) typically the power supply is nonredundant—hence its failure may cause the entire PC to fail, and (3) the noise immunity of the processor's power supply can dictate the noise immunity of the entire control system. Useful guidelines when considering the power supply include:

1. The power supply should be packaged properly so that the heat can be removed for

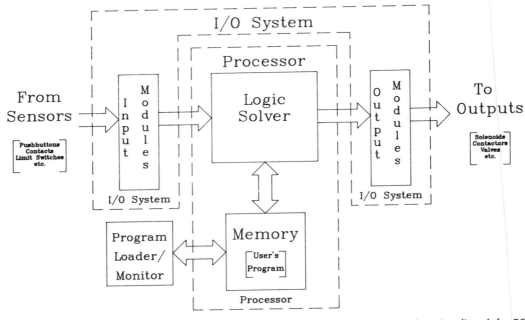

Fig. 3. Simplified block diagram of a programmable controller illustrates the basic functionality of the PC. The control engineer (user) enters the control program on the program loader. The program loader writes this program into the memory of the processor. The logic solver reads the states of the sensors through the input modules, then uses this information to solve the logic stored in the user memory (program) and also writes the resulting output states to the output devices through the output modules.

reliable operation. Packaging is discussed later in this article.

2. The power supply should meet at least one reputable standard for noise immunity. Two of the most popular are NEMA ICS2-230 and IEEE472. The power supply also should be capable of withstanding line voltage dropouts, common in industrial facilities.

Memory

A PC's memory can be of two different types—volatile or nonvolatile. Volatile memory loses its contents when power is removed, whereas a nonvolatile memory does not. PCs will use nonvolatile memory for a majority of the user's memory because the program must be retained during a power-down cycle, meaning that the user will not have to reload the program every time power is lost.

It is important that all nonvolatile memory in a PC use some form of error checking in order to assure that the memory has indeed not changed. This error checking should be done "on-line" or while the PC is controlling the machine or process so as to ensure safe execution of the user's program.

Battery-backed-Up CMOS RAM. This is probably the most widely used type of memory. Although most RAMs (Random Access Memory) are inherently volatile, the CMOS (Complementary Metal-Oxide Semiconductor) variety consumes so little power that a small battery will retain the memory during power losses. The batteries used vary from short-life primary cells (alkaline and mercury batteries) that require periodic replacement every six months to a year to long-life cells (such as lithium, which may last up to ten years) to rechargeable secondary cells, (such as NiCad and lead-acid). The latter may also last several years.

EPROM Memory. An EPROM (Electrically Programmable Read Only Memory) is programmed using electrical pulses and can only be erased by exposing the circuit to ultraviolet light (also called UV-EPROM).

EEPROM Memory. This memory is like the EPROM, but with the exception that it can be erased by using electrical pulses. In some PCs, use of EEPROM only alleviates the need for an ultraviolet source; while in other PCs EEPROM is the only type of memory used. This allows for the flexibility of reprogramming like the CMOS RAM, but without the disadvantage of having to provide battery maintenance.

For applications where the end user cannot reload programs or provide battery maintenance, it is considered better to use EPROM or EEPROM.

Central Processing Unit (CPU)

How the CPU is constructed will determine the flexibility of the PC (whether or not the PC can be expanded and modified for future enhancement) as well as the overall speed of the PC. The speed is expressed in terms of how fast the PC will scan a given amount of memory. This measure, called the *scan rate*, is typically expressed in milliseconds per thousand words of memory. Faster PCs will typically cost more than slower PCs. Thus, it is important to choose a PC with a scan time appropriate to the application.

It is important to note that many of the commercially available PCs specify their scan time using contacts and coils only. A real program that uses other functions, such as timers, counters, and mathematical functions, may take considerably longer to execute. Also in procuring a PC, one should not forget to include the scan time of the I/O, the scan time of the memory, and any additional time overheads the processor requires when one attempts to predict the overall scan time for a given application.

Processor Software

The hardware of the PC is not too different from that of a lot of computers. What makes the PC special is the software. The executive software is the program that the PC manufacturer provides internal to the PC that executes the users' program. The executive software determines what functions are available to the user's program, how the program is solved, how the I/O is serviced, and what the PC does during power up/down and fault conditions.

Executive Software. A simplified model of what the executive software does is shown in Fig. 5. Specific PC designs perform the basic functions shown somewhat differently. This can make a big difference in program execution. For example, some PCs may perform diagnostics only at a single point in the executive program, while

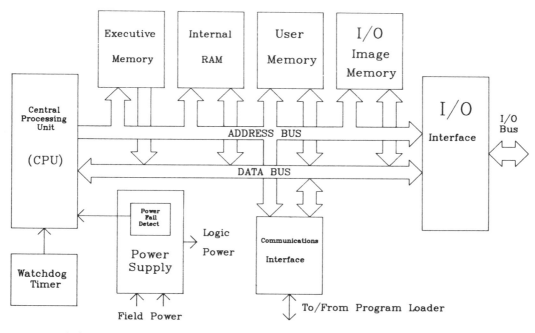

Fig. 4. Detailed block diagram of the processor section of a programmable controller. The central processing unit (CPU), typically a microprocessor, executes a program written by the manufacturer of the PC that is stored in the Executive Memory. This executive program that the CPU executes gives the CPU the ability to interpret the user's program. The CPU does not operate on the I/O directly. Rather, it works with an image of the I/O that is stored in the I/O image memory. The I/O interface is responsible for transferring the image outputs to the I/O system and reading the inputs from the I/O system and writing them into the image memory. A 'watchdog' timer is provided to time how long it takes the CPU to execute the user's program. If this time exceeds a predetermined value, the watchdog timer causes the processor to fault. If the CPU fails and does not execute the user's program, the watchdog timer will ensure that at least a fault will be indicated and that the processor will shut down in a safe manner.

156 Programmable Controllers

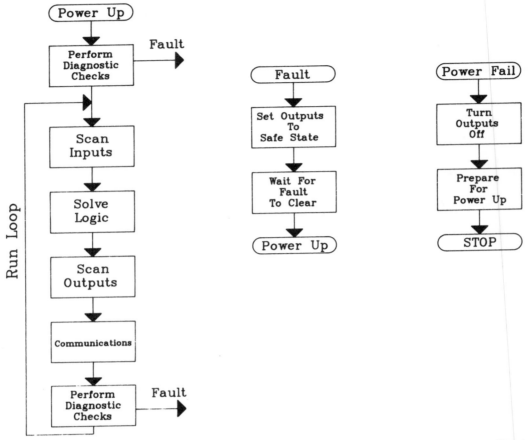

Fig. 5. The Executive Program shown here controls the functionality of the programmable controller. It controls the actions of the CPU to perform the indicated actions. Diagnostic checks must be run at power-up as well as during the run loop, which is executed while the PC is controlling the process. When faults are detected, the outputs must be set to a predetermined 'safe' state. The user usually has the choice of turning all of the outputs off or leaving them in their last state. When advance warning of power failure is given by the power supply, the executive program shuts down the CPU in a controlled manner. During the run loop, inputs are only scanned once. This allows the entire user program to operate from a consistent set of inputs because they are only determined prior to executing the user's program and do not change state in the middle of the run loop.

others may perform diagnostics "on-line," that is, while the user's program is being solved.

Close attention must be given to how the PC runs diagnostic tests and what it does during failures. Ignoring this aspect of the PC can result in an unsafe system.

Multi-tasking. In a later development, PCs capable of executing multiple tasks with a single processor appeared. Multi-tasking takes several forms, two of which are: (1) time-driven, and (2) event-driven. In a time-driven multi-tasking system, the user writes programs and assigns I/O for each task. The user will then configure the processor to run each task on periodic time intervals. This type of system is shown graphically in Fig. 6. This feature allows the *time-critical* portion of the control system, such as the portion that controls high-speed motions or machine fault detection, to run many times per second, while allowing the *noncritical* portions, such as servicing indicator lights, to run much slower. Because only the time-critical logic and I/O need quick solutions, versus the entire user's program, faster throughput can be achieved.

Event-driven multi-tasking (also called interrupt-driven) is similar. In this case the user defines a particular event, such as an input changing state or an output turning off, that causes each task to run.

In either the time-driven or event-driven case,

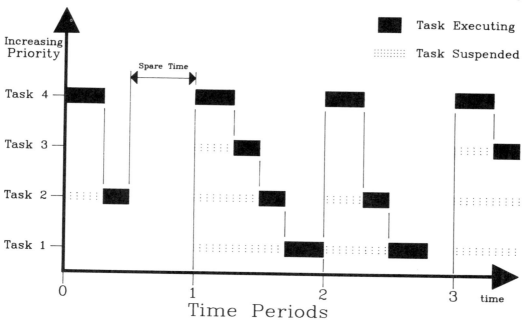

Fig. 6. In a time-driven multi-tasking system, tasks are scheduled to run on predetermined time intervals. In the example shown here, tasks 4 and 2 are scheduled to run every period while tasks 3 and 1 are scheduled to run every other time period. The higher-priority tasks always execute before the low-priority tasks. During period #1, all four tasks are scheduled to run. However, there is not enough time for task 4 to finish executing. Its execution is suspended until the spare time in period #2 is available. Care must be taken to ensure that there is enough spare time for all the tasks to execute. Some multi-tasking systems will provide an indication that not enough time exists to execute all the programmed tasks, thus making it easier to debug the programs.

it is important to recognize the priority of tasks. For example, if Task A has a higher priority than Task B and if Task A is currently active, then Task B will not execute until after Task A is complete, even though Task B may be scheduled to start because of an event occurring or because the specified time interval has elapsed. If Task B, on the other hand, has the higher priority, then Task A will be temporarily suspended until Task B is completed.

Some multi-tasking systems allow any task to access any variable, such as an I/O point. Thus, caution must be used when programming multiple tasks that access the same variables. It may be difficult to determine which task is writing which variable while trying to debug a program.

User Software

This is the software that the control engineer writes and stores in user memory in order to perform the required control over the machine or process. User software can contain both configuration data and language programs. The configuration data contain information that tells the processor what its environment is and how it should execute the language problem.

Configuration. The configuration process typically consists of assigning I/O points to particular I/O racks, telling the processor how much memory and I/O it has, assigning specific memory for tasks, determining fatal versus nonfatal faults, and many other items interactively on a program loader. This is covered in more detail later in this article under "Communications." Not all PCs require that they be configured, but being able to configure the processor can enhance the efficiency of the PC. For example, a PC that scans only the memory and I/O that is configured will run faster than one that scans all memory and I/O whether it is used or not. The user should thoroughly investigate how the configuration is done on a given programmable controller. Some PCs require reprogramming if the configuration changes.

Languages. Inasmuch as the modern PC is required to do more and more in terms of operator interfacing, communications, data acquisition, and supervisory control, more is required of the language that implements these functions. Therefore, it is crucial that the various aspects of the language, as described in the following several paragraphs, be considered when making decisions.

1. *Variables*—these are the way the language allows the user to access the I/O and internal data. Variables typically take the form of:

```
X YYYY
!   !
!   !_____Reference Number
!
!_____Type Designator
```

The *type designator* will indicate the type of variable, such as Input bit, Output word, Internal bit, Internal word, and so on. The *reference number* indicates which specific variable of the specified type is being accessed. Some PCs allow the user to represent data in many different forms, including Bits, Words, Tables of Bits, Tables of Words, File, Bytes, and so on. Languages that allow the same data to be represented in many different ways are more flexible to use and allow the user to write the programs more readily and efficiently. Some languages allow the user to use variable names instead of reference-number-based variables. This feature can further enhance the readability of the program.

2. *On-line versus Off-line*. An on-line language is one that can be programmed interactively with the PC. An off-line program requires that the programs be generated separately and then "down-loaded" to the PC. The user must be careful when using an on-line language while the PC is controlling the process. Inadvertently changing the wrong thing in an active program can wreak havoc on the process.

3. *Flow control*. This feature dictates how easy it is to make decisions and control the program's execution. Some languages only give conditional jump instructions, while others may allow subroutines and loop functions.

4. *Functions*. The functions provided by the language vary widely from one PC to the next. Some languages only provide the minimum set of logic, timing, counting and memory functions, while others may provide many additional instructions, such as drum controllers, matrix operations, shift registers, mathematical functions, and many others.

5. *Maintainability*. This refers to how easy the language is to debug, modify, and teach to others. Some computer languages, which offer the ultimate in flexibility, are not suited to applications where the people who have to maintain the machine cannot easily be trained to understand the language.

6. *Speed*. Some languages take longer to execute than others. Choosing a slow but powerful language in time-critical applications can make the programming much more difficult than using a fast, but simple language.

7. *Efficiency*. This is a measure of how much memory is required for a language to implement a given function. Typically measured in words (16 bits) of memory per function, this can vary widely from PC to PC. Since PCs are priced according to memory size, a more efficient PC can result in significant cost savings in terms of the hardware required for a given application.

Examples of some Languages Used

LADDER—this is still the premier language of the PC. The language has many advantages: (1) it is easy for users to understand and simplifies training, (2) the language provides graphical display of program execution by showing power flow through the diagram, (3) it is fast, (4) it generates more readable programs for sequence control, and (5) the language has generally good functionality. However, ladder generally lacks good flow control instructions.

BOOLEAN—this language is generally used in very small PCs. Boolean uses AND, OR, NOT, STORE, and RECALL instructions in order to describe the program logic. This language is not easily debuggable unless a single-stepping feature is provided that slows down program execution so that the effects on the I/O can be observed visually. Although fast, Boolean languages typically lack good functionality and flow control and can be difficult to maintain for large programs.

HIGH-LEVEL LANGUAGES—examples of which include *BASIC*, *Pascal*, and *C*. These languages can be very powerful and are identical to the languages used by computers. They exhibit excellent flow control and functionality and provide for the accessing of many different types of variables. They also offer reasonable speed. Most plant-floor maintenance personnel, however, do not understand these languages, and it is difficult to monitor program execution in real time.

GRAFCET—is a flow chart language that allows the flow of program logic to be described graphically. Although more recently considered in the United States, the language has been applied in Europe for several years. GRAFCET can be a valuable aid in designing large control programs by allowing the engineer to describe the process graphically before using another language, such as Ladder or Pascal, to implement the control actions. (See Fig. 7.)

Programming

How the user's program is entered is dictated by the programming equipment provided. How this programming equipment operates can determine how long it will take to develop a program. Virtually all the time spent programming and debugging a PC is spent in interacting with a *program loader*, not with the PC directly. A program loader can either help or hinder the user in this programming effort—thus it is important that the user be familiar with the various func-

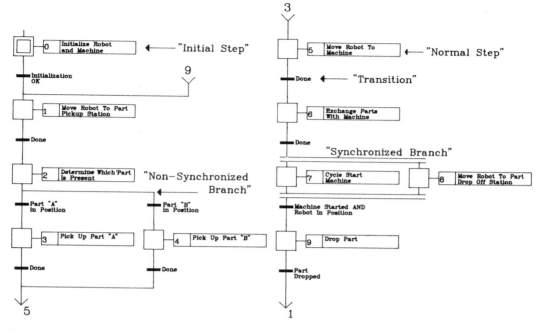

Fig. 7. GRAFCET is a flowchart language that allows the process to be described graphically. A GRAFCET chart consists of 'steps' and 'transitions.' A step contains a program that implements an action, such as "Move Robot to Part Pickup Station" (Step 1). A transition contains a program that determines whether or not a particular condition has occurred, such as "Part A in Position" (Transition between Steps 2 and 3). An 'initial step' is the step that the processor executes first. A "Non-Synchronized Branch" (between Step 2 and Steps 3 or 4) is used to decide between mutually exclusive actions. A "Synchronized Branch" (between Step 6 and Steps 7 and 8) is used when multiple actions are to be executed simultaneously. This language offers an easily readable method of describing a control program's action.

tions of the program loader and how they operate. These include:

1. *Programming*. Of primary importance, the program loader provides an environment for entering programs. This function must be studied closely because it can determine the amount of time required to write a program.

2. *Monitoring*. The program loader also provides an environment for monitoring program execution in real time. Some loaders provide displays that show "power flow" through a relay ladder diagram, while others only allow the user to see the I/O points turn on and off. The speed at which a program loader can monitor varies widely from one loader to the next.

3. *Program storage*. This function allows the program to be stored in some machine-readable format that is separate from the PC itself. Typically, this is done on some form of magnetic media, such as floppy disks or cassette tapes.

4. *Documentation*. Most program loaders allow the user to obtain a hard copy or printout (non-machine-readable) of the program. Some program loaders allow the user to get cross references of variable usage, enter comments in the program, and define names for all the variables. These features can greatly improve the readability and maintainability of the program. Some PC manufacturers do not supply these functions. In such cases a documentation software package is usually available from a third party.

5. *System maintenance*. This feature, not available on all loaders, allows for limited access to the PC for the purpose of maintenance. With this feature, maintenance personnel can gain limited access to do monitoring and change presets for counters and timers, but it will not allow program changes without a password or key.

The user is well advised to pay close attention to the operating characteristics of the program loader. Some loaders are menu or soft-key driven and are simple to learn. Others may use hidden keys and difficult key sequences that need to be memorized. If the user cannot understand how a loader works within a short period of time, this indicates a poor ergonomic design.

Types of Program Loaders. These include:

1. *Hand-held loaders*—these are small but low-cost loaders that typically use liquid crystal (LED)

displays. These units are primarily for applications where cost is of the utmost importance or for small, simple applications because such loaders typically are limited in the amount of information that they can display.

2. *Specific-purpose graphic-based loaders*—these typically use CRT or large LED displays. They are capable of displaying a wide variety of data and can monitor power flow through the relay ladder programs. Although more expensive than the hand-held loader, they offer a vastly improved environment for program development and debugging.

3. *General-purpose graphic-based loaders*—these typically use personal computers (typically IBM-PC (R) compatible) for their hardware. The PC manufacturer provides a software program that makes the personal computer look like a program loader. Although the cost is about the same as for a specific-purpose graphic-based loader, the general-purpose variety can be a valuable asset because it can also run the multitude of other programs that are available for the personal computer. These can be advantageous to the control engineer who needs a personal computer for other purposes but cannot justify the purchase of both a personal computer and a program loader.

I/O SYSTEMS

Direct I/O, as the name implies, is the brute force way of getting I/O to and from the PC's processor. There is one input signal and one output signal corresponding to the number of inputs and outputs the processor supports. This approach is typically used in the very small PCs that have all the I/O circuits in the same package as the processor (sometimes called internal I/O). Cost is the principal advantage of internal direct I/O. Some flexibility, however, is lost because the processor must be changed in order to change the I/O.

Direct I/O systems are typically more cost effective when the number of I/O points is low (less than 128 I/O). When the number of I/O points is large, bus-oriented systems (parallel or serial) offer a better cost/performance ratio.

Parallel I/O Systems. In a parallel I/O system, a parallel I/O bus emanates from the processor's I/O interface and individual I/O modules are plugged into this bus. The I/O module contains the necessary circuitry to decode the bus signals and convert these signals into voltage levels that can drive the necessary loads. I/O modules will typically drive multiple loads. This multiplicity of I/O points is called the modularity of the I/O system. Most commercially available I/O systems have modularities of 2, 4, 8, 16, or 32 I/O points per module. Adding more I/O points on a module will commonly reduce the cost per I/O point and reduce the amount of space required to install a given number of I/O points. (See Fig. 8.)

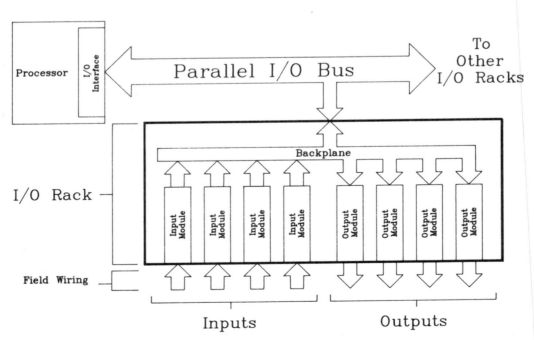

Fig. 8. Block diagram of a parallel I/O system.

The failure of one I/O module with many points of I/O on it can be disastrous if it controls many critical devices, such as those causing motions or controlling emergency stops. Thus, it is good practice to split up the critical I/O between the high-density modules.

Serial I/O Systems. Parallel systems are limited in the distance over which one can extend the I/O bus, typically less than 50 feet (15 meters). If the machine should be 100 feet (31 meters) long, one would have to use two PCs. Serial I/O systems solve this problem by transmitting the I/O information over a serial data link capable of being extended over longer distances (1,000—10,000 feet; 305—3050 meters). A serial bus emanates from the processor and is connected to a parallel bus through a serial-to-parallel converter. Since a single serial bus contains fewer wires than does the wiring to the loads, large wiring cost savings can be realized by using serial systems. (See Fig. 9.)

Care must be taken when using serial I/O systems in time-critical applications because two I/O buses have to be scanned—both the serial and parallel bus—instead of one, thus making them slower than straight parallel systems. Some, but not all, serial systems may "desynchronize" themselves from the logic scanning, thus making it more difficult to predict I/O responses to fast-changing signals.

I/O Circuits

An I/O module performs signal conversion and isolation between the internal logic-level signals inside the PC and the field's high-level signals. There are many different types of I/O circuits available that are capable of driving almost any conceivable load and sensing the status of a wide variety of sensors. Most of these I/O circuits fall into one of five categories. (See Fig. 10.)

1. *Pilot duty outputs.* Outputs of this type are typically used to drive high-current electromagnetic loads, such as solenoids, relays, valves, and motor starters. These loads are highly inductive and exhibit a large inrush current. Pilot duty outputs should be capable of withstanding an inrush current of ten times the rated load for a short period of time without failure. They also should include some form of noise suppression because of the electrical noise that many pilot duty loads generate.

The isolation on a pilot duty output is a critical safety feature. If the isolation fails, the PC could become "hot" and thus an electric shock hazard. The isolation on a pilot duty output should withstand a dielectric test voltage of 1000 VAC RMS plus twice the working voltage of the circuit (minimum) for at least one minute. See UL reference at end of article.

2. *General-purpose outputs.* These are usually low voltage and low current and are used to drive indicating lights and other noninductive loads. Noise suppression may or may not be included on these types of modules.

3. *Discrete inputs.* Circuits of this type are used to sense the status of limit switches, pushbuttons, and other discrete sensors. Again, noise suppression is of great importance in preventing false indication of inputs turning on and off because of noise. The more noise immune an input is, the slower it will be because of all the

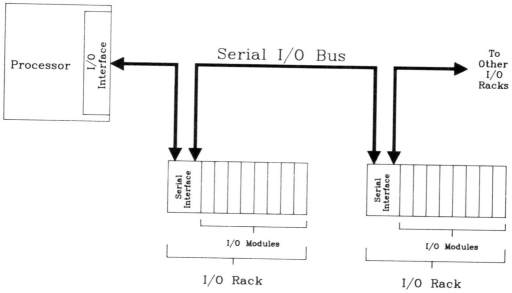

Fig. 9. Block diagram of a serial I/O system.

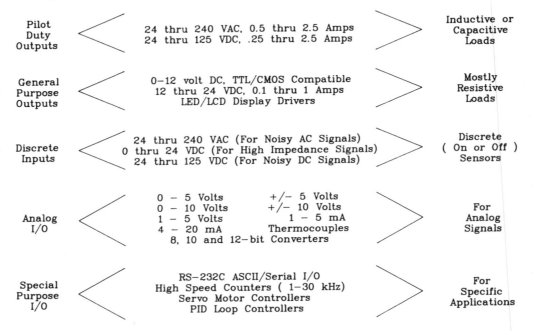

Fig. 10. Examples of commonly available I/O circuits.

filtering required to reject noise signals. The user should examine an application closely to make certain that an I/O module is selected with the appropriate amount of noise rejection.

4. *Analog I/O.* Circuits of this type sense and/or drive analog signals. Analog inputs come from devices such as thermocouples, strain gages, pressure sensors, among others, that provide a signal voltage or current that is derived from some process variable. Analog outputs can be used to drive devices, such as voltmeters, X-Y recorders, servomotor drives, and so on. Analog I/O circuits consist of either digital-to-analog (D-A) converters for outputs or analog-to-digital (A-D) converters for inputs. These analog converters transform the digital signals of the I/O bus to/from the analog signals of the devices.

Four major characteristics are of interest concerning analog I/O circuits besides the voltage and/or current levels. These factors include: (1) conversion time, (2) accuracy, (3) drift, and (4) resolution. The conversion time is a measure of how long it takes the analog converters to perform their operation. The faster the signal from the end device can change, the faster must be the converter that is needed to sense or control this change. The accuracy is a measure of how precisely the analog converter will sense or drive the analog signal. This is typically expressed in terms of a percentage of the full-scale resolution (FSR). A 1-5 V analog signal with an accuracy of $\pm 1\%$ will be $\pm (5-1) \times 1\%/100$, or ± 0.04 V. The drift is a measure of how the accuracy of the converter changes with time. This is critical in control systems with a long lifetime. The resolution is a measure of how many different voltages the converter is capable of representing and is specified in terms of bits. An 8-bit converter can represent an analog signal in one of 256 ways; a 12-bit converter can represent 4096 different voltages.

Although not obvious, the accuracy and resolution are independent of each other. It is possible to construct analog circuits with very high accuracy, but with very low resolution. For example, a 1-bit high precision analog converter is called a *threshold detector*. A threshold detector will turn its single bit on or off when the voltage reaches a specified value.

5. *Special-purpose I/O.* Circuits of this type are used to interface PCs to very specific types of circuits, such as servomotors, stepping motors, PID (proportional integral derivative) loops, high-speed pulse counting, resolver and encoder inputs, multiplexed displays, and keyboards, among others. Using these special-purpose modules makes for very efficient control systems because they relieve some of the burden of control from the PC. This has several advantages: (1) the speed of the PC increases because it has less logic to solve to control the device, (2) the PC can fail, and it still may be possible to maintain control of the device because the special-purpose I/O module has not failed, and (3) a wider variety of devices can be interfaced with the PC.

PACKAGING

The manner in which a PC and its I/O system is packaged is critical in determining whether or not a particular PC is feasible for a given application. If the PC needs to be mounted on the factory floor directly next to a machine or process, it must be packaged differently from, for example, a personal computer designed for desktop use. The constant vibration, electrical noise, and dirt found in most industrial manufacturing environments will adversely affect a PC that is not packaged with such environmental conditions in mind.

Several of the factors involved in designing the package for a PC include:

1. Heat removal. All electronic devices generate heat when power is applied to them. Some means must be provided to remove this heat from the PC and move it to the surrounding environment. If the heat is not effectively removed, the PC will operate at temperatures too high for proper performance of its electronic devices, thereby causing a drastic reduction in the lifetime of the PC or an outright failure. Reducing the temperatures internal to the PC can dramatically improve its reliability.

Venting is commonly used to transfer heat from the PC to the environment. For proper operation, the vents should be near the top of the package. Additional venting of the package at the bottom allows airflow through the enclosure, thereby increasing the efficiency of heat removal. Vents should be designed so that items like screws, nuts, et al. cannot inadvertently fall inside the equipment. PCs with vents should be mounted upright for proper operation.

Forced-air cooling is a technique for heat removal that takes advantage of the better heat removal by a moving flow of air as contrasted with essentially still air that applies to vents only. But, as mechanical devices, fans and blowers are prone to failure and require filtering that may clog with time. Thus, it is important that a PC with fans or blowers can still operate reliably without them in case of failure.

Heat sinking is primarily used to remove the heat of specific devices, such as power supply transistors. In some power supplies, these transistors may consume 40% of all the power consumed by the PC, and thus such devices should be physically attached to a heat sink. Obviously, a heat-sink material should be a material of high thermal conductivity, such as aluminum. A heat sink will work better if it has a radiating surface outside the PC. The greater the area on the outside compared with the area inside, the more effective the heat sink will be.

2. Mounting. The packaging should be constructed so that it mounts appropriately where called for by the application. Almost all PCs are intended to be mounted inside some other enclosure that carries a NEMA rating, such as oil-tight, dust-proof, RFI shielded, etc. Such enclosures must be chosen wisely. The top-of-the-line enclosures are very expensive, whereas the lower-cost enclosures may not provide the proper amount of protection. These enclosures should *not* be vented.

The mounting of the PC is done by either rack or panel mounting. In panel mounting, the PC is mounted to a flat piece of metal (panel). This panel is then mounted inside an enclosure. Inside the enclosure there is a large space that is only partially consumed by the PC. The remaining space absorbs heat rejected by the PC as previously discussed. Currently, this is the most popular method of mounting PCs in the United States.

Rack mounting offers a much higher density, but this means that there is less space for heat rejection. Rack-mounted installations will often require forced air for cooling. Rack mounting is much more popular in lower-temperature applications and where density is of the utmost importance.

3. Wiring. The impact of wiring on installed cost cannot be minimized. Wiring costs can easily exceed the cost for the hardware and cause large maintenance problems if not implemented properly. Two types of wiring systems are commonly used:

Fixed wiring is typically used on the PC's power supply and on small PCs with direct I/O. Obviously, in order to remove something with fixed wiring, one must first *unwire* it.

Removable wiring is typically used for I/O modules. In this case, the field wiring is done to a removable terminal block. In order to remove an I/O module for repair or fuse replacement, one simply detaches the terminal block and unplugs the module from the I/O rack.

In some systems, the terminals are fixed, but the I/O module is removable. However, not all PCs use removable wiring in their I/O systems.

Because wiring is so important to safety, one must make certain that the terminals are suitable for a given application. 120V pilot duty outputs and 120V inputs should use 300V, 10A terminals. The wiring system should not be "too removable" and thus be unsafe where vibration is present or where a slight pull on the wiring will cause a disconnection, particularly at a time when the PC is controlling a machine or process.

COMMUNICATIONS

The communications aspects of a PC can severely limit or greatly enhance the applicability of these devices.

Point-to-Point Communications

Most PCs have at least one communication port built in, that is, the program loader interface. However, only a few manufacturers release the information needed in order to communicate over this interface. Even so, these ports typically use unusual protocols that can require considerable effort to implement. Some manufacturers of peripheral equipment, such as color graphics displays, have converted their equipment to talk to some PCs directly, thereby saving the expense of writing specific communications software.

Most PCs also provide some form of ASCII communications. Some PCs have separate I/O modules for this purpose, while some others allow the user to reconfigure the program loader port for this purpose. With ASCII communications, it is possible to talk to a wide variety of devices, such as color graphics terminals, intelligent pushbutton stations, barcode readers, servomotor controllers, ASCII terminals, etc. Usually these ASCII ports use an RS232 interface, making it easy to communicate over telephone lines as well.

The instruction set of the PC can impact on how ASCII information is processed. PCs with few instructions can be difficult to program for ASCII information processing. Functions, such as search, compare, ASCII-to-binary, binary-to-ASCII, and flexible conditional jumps are crucial for the efficient handling of ASCII information.

Network Communications

Most PC manufacturers provide some type of network allowing for communication between their own PCs. With these networks, it is possible to distribute PCs physically, but yet have them work in unison by using the network's communication functions. Most of these networks provide three basic functions: (1) reading variables, (2) writing variables, and (3) program upload and download. However, because these networks are designed to provide communication functions, not necessarily control functions, using a network inside of a control loop requires careful planning and evaluation. Some networks have difficulty transporting information from two points on the network in such a manner as to allow the use of this information in a time-critical control application. The following points should be considered when one is evaluating a network for a control application.

1. Response Time. The length of time required from an input changing state on one node of a network to a second node receiving notice that the input has changed is a critical parameter when trying to implement control of a process over a network. Some networks give only a probabilistic response time based on some hypothetical installation. However, one should always know the precise response time limits before putting control information on a network. Not all networks are intended for control and thus must be evaluated carefully. A MAP version 2.1 network does not guarantee delivery of all messages, and a CSMA/CD network, such as Ethernet, has a probabilistic access time that increases with the number of active points on the network, making reliable prediction of response times nearly impossible.

2. Error Checking. Any network that is used for transferring control information should utilize extensive error checking on the information sent over it. Both ends of the network, the sender and the receiver, should be capable of detecting errors and should also perform specific and known error recovery mechanisms, such as retransmission or, at a minimum, be able to notify both the sender and receiver that there was an error so the control engineers can program their own recovery scheme.

3. Access Mechanisms. Because a network usually contains only one channel over which all PCs must talk, some method for determining who has access to the network at any given time must be used. Two of the more popular access mechanisms are master/slave and peer-to-peer. (See Fig. 11.)

On a *master/slave system*, there is only one master PC. The master sends commands out to the other slave PCs, and they respond appropriately. The slaves on the network never initiate their own commands. They always respond to what the master commands them to do.

The *peer-to-peer* mechanism allows any PC on the network to initiate messages. However, as in the case of humans talking, if everybody talks at once, nothing intelligible can be heard. Peer-to-peer networks need some mechanism for determining access between all the PCs—not just between the master and the slave. Various mechanisms for determining access have been implemented, such as token-passing and carrier-sense multiple-access with collision detection (CSMA/CD). More details on these methods can be found in *Section 5* of this *Handbook*.

The PC user should make certain that the access mechanism used is deterministic; that is, the access time is known and not probabilistic. Control signals, such as end-of-travel and emergency stop, should be hardwired. Depending on a network to send safety interlocks and related information without hardwired backups can result in a hazardous system should the network fail or if access times degrade.

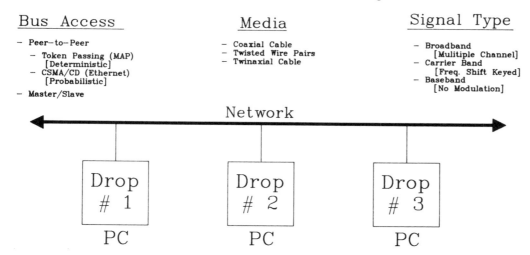

Fig. 11. Simplified network block diagram with typical features. In master/slave networks, only the master can initiate communications. In peer-to-peer networks, any drop on the network can initiate communications. Although peer-to-peer networks typically offer faster communications, they are sometimes difficult to use for control because they require that a large number of variables be sent between a large number of drops. This causes the number of communication paths to increase exponentially. A network that is used for control should have guaranteed response times and known error recovery methods. A master/slave system can be easier to maintain if variables in the drops change—because only the master and the drop in which the variable changes need be updated. Some networks alleviate this problem by allowing variables to be accessed by names instead of addresses. The software tools that are provided to communicate with can be the most important factor to consider. The media and signal type affect the noise immunity of the network. Coax and twinaxial based networks offer good noise immunity, but may be more expensive than some twisted wire networks. Broadband allows for multiple communication channels on the same network (much like a TV has multiple channels), but is very expensive. Baseband is the lowest cost, but does not offer the noise immunity that modulated signals provide.

RELIABILITY

Of utmost concern in any control system is the reliability of the components in that system. PCs have gained a reputation for being very reliable, and that partly accounts for their widespread use. However, not all aspects of reliability are strictly a function of how well the PC manufacturer designs and builds the equipment. Significant improvements in reliability can be gained by the proper installation and maintenance of the equipment. Improving noise immunity and improving availability are key factors.

Noise Immunity

Many failures of control systems can be traced to noise conditions that cause intermittent faults resulting in lower reliability. Fortunately, noise problems can usually be prevented by proper application of equipment. Although most PCs are designed to withstand certain levels of noise, the user will find it advantageous to go beyond the inherent protection provided by the PC manufacturer and install the equipment in a manner that reduces the likelihood of noise problems. Among excellent techniques to consider are:

1. Grounding. This is one of the least costly and most effective means for improving noise immunity. All electronic equipment requires good grounding to operate reliably and safely. The manufacturer's recommended grounding procedures should be followed carefully so that the inherent noise immunity of the equipment can be utilized to the fullest. In addition, it is good practice to ground all metal chassis, use large wires for grounding to minimize impedance for high-frequency noise, provide solid earth grounds for all electronic equipment, and avoid putting noisy devices, such as arc welders, on the same grounding system as the electronic control equipment.

2. Isolation. Noise immunity can be improved by separating the noise-generating devices from the noise-susceptible devices. Isolation transformers should be used on all power supplies, and field wiring should be kept separate from logic wiring (e.g., I/O bus and communication cables).

3. Suppression An effective way of improving noise immunity is by reducing the noise itself. Zero-crossing AC outputs should be used instead of phase-controlled AC outputs. Zero-crossing AC outputs generate less noise because they turn on only when the voltage across them is zero. Phase-controlled AC outputs can turn on at any point in the AC waveform, resulting in signals with fast rise times, which generate high-frequency noise. Another method involves the use of noise suppressors on the noise-generating devices, such as DC solenoids and relays, or by putting noise suppressors on the input to noise-susceptible devices, such as power supplies. (See Fig. 12.)

Availability

Even though reliability gets the most attention, it is *availability* (the percentage of total time that the system operates reliably and satisfactorily) that is of larger concern. For instance, suppose that System A fails twice a year while System B fails only once every five years. Obviously, System B is more reliable. However, suppose that it takes five minutes to repair System A, while it takes one day to repair System B. Over the course of five years, System A will fail ten times, thus requiring a total time of fifty minutes to repair, while over the same period of time, System B will be out of service for repair a whole day because of only a single failure. Availability is typically expressed as a percentage of uptime and is calculated by dividing the mean time between failures (MTBF) by the sum of the MTBF and the mean time to repair (MTTR) and multiplying the result by 100%.

$$\text{Availability} = \frac{\text{MTBF}}{\text{MTBF} + \text{MTTR}} \times 100\%$$

Many schemes can improve the availability of a system. Some of the more popular means are improving (1) serviceability, (2) fault tolerance, and (3) redundancy.

1. Serviceability. By improving the serviceability of a system, dramatic improvements in availability can be achieved. Choosing equipment with removable wiring and modular design can significantly decrease MTTR because a failed module can be replaced easily with removable wiring. Keeping spares of processor modules, I/O modules and power supplies close at hand avoids lengthy repair cycles, particularly if the equipment must be returned to the manufacturer for repair.

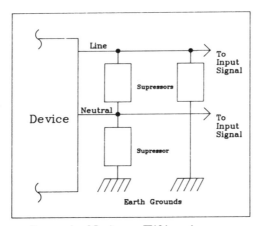

Input Noise Filtering

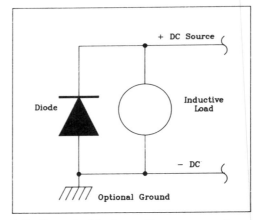

DC Load Noise Suppression

Fig. 12. Examples of common noise suppression techniques. Although most programmable controllers are designed to be relatively noise immune, some applications require additional precautions. Input noise filtering will increase the immunity of any input circuit or power supply. The suppressors can be capacitors, metal-oxide-varistors or *Tranzorbs*™ (General Semiconductor Industries). The suppressor across the line (or +DC) and neutral (or −DC) will improve the differential or normal mode noise immunity, while the other suppressors, from each line to ground, will improve the common mode noise immunity. In many cases, it is more effective to eliminate the noise at the source, particularly with inductive DC loads. In this case, a diode across the load, as shown, will limit the voltage surges that occur when the load is turned off. Some output circuits will not withstand the current pulses that will occur with the diode in the circuit or cannot have their negative terminal connected to ground. In these cases, the diode alone should be connected to a good earth ground.

2. **Fault tolerance.** Although they are expensive, fault tolerance techniques, such as graceful degradation and error detection and correction (EDC), can significantly improve availability.

Graceful degradation means that the equipment still can function partially in the presence of faults. For instance, most PCs will continue to update I/O for good modules, even though some failed ones may be present. *Distributed control* is also a form of graceful degradation. In a distributed control system, the control is split up among many PCs that are distributed over a network. In this case, if any one PC fails, control to only a small part of the process will be lost.

Error detection and correction is typically applied to memories. By inserting certain codes (i.e., Hamming codes) as well as the user data into the user memory, not only does it become possible to detect errors, but the codes allow for the correction of errors as well. Even though a memory fault occurs, it can be corrected before it causes a system failure.

3. **Redundancy.** Some PC manufacturers offer redundancy for their larger PCs. In these cases two processors can be tied into one I/O system, and some means is provided that switches control from the failed PC to the backup when a failure occurs. (See Fig. 13.) In addition, some PCs can use redundant cabling. In this case the PC will try to transmit its I/O information over the primary cable. If it detects that it cannot do so, it can automatically switch over to a redundant cable. I/O circuits can be wired so that the failure of any one point will not prevent the PC from controlling the output. (See Fig. 14.)

The user never should disregard what happens when a PC fails. Despite its reliability, a PC, like all electronic equipment, will fail from time to time. Failure to consider this in a system design can result in an unsafe system.

Acknowledgments: The author wishes to thank Jon Martin, President of Sisco, Inc., and Angelo Vinch, Westinghouse Electric Corporation, Numa-Logic Dept., for their assistance in the preparation of this article.

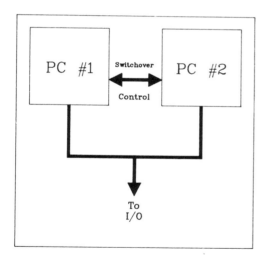

I/O Bus Sharing

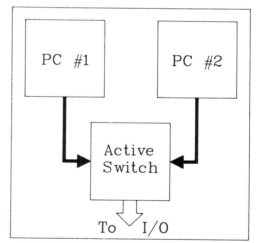

Active Switch

Fig. 13. Common methods for processor redundancy, particularly popular for large programmable controllers. Two common methods are shown here. In an I/O Bus Sharing system, both processors connect to the same I/O system. They are constantly exchanging information concerning condition and status while both processors are running. When the active PC fails, it disconnects from the I/O bus and the backup connects itself to the I/O bus. This type of system could be susceptible to failures in the I/O interfaces of the processors. In an active switch system, both processors 'think' they are in control. An active switch monitors the I/O link of both processors and determines which one has actual control of the I/O. The active switch must be relatively reliable. If it is less than or equal to the reliability of the PC itself, then the availability will decrease rather than increase.

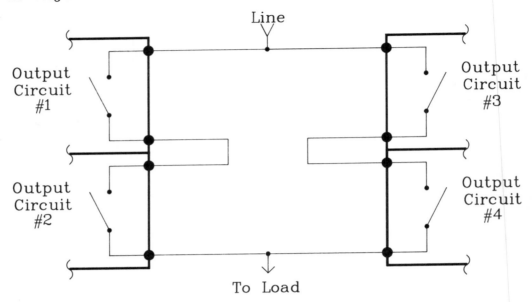

Fig. 14. Redundant I/O circuits. Redundant output control can be achieved by wiring together four output circuits as shown here. In this case, control over the output can be maintained in spite of the failure of any one output circuit. In addition, control can be maintained with two output failures depending on how the output circuits fail (i.e., fail to energize or fail to deenergize). The major disadvantage of using redundant I/O circuits is cost. It requires four output circuits for every load.

REFERENCES

Chilton: "I&CS Buyers Guide," published annually by the Chilton Company, Philadelphis, Pennsylvania.

Flynn, W. R.: "1985 Programmable Controller Update," *Cont. Eng.*, **32**, 1, 79–83, (January 1985).

Flynn, W. R.: "1986 Programmable Controller Update," *Cont. Eng.*, **33**, 1, 50–54, (January 1986).

NEMA: "NEMA Product Statistical Bulletin for Systems Group (4-IS-2), National Electrical Manufacturers Association, Washington, D.C. (February 9, 1981 and February 21, 1985.)

Quatse, J. T.: "Programmable Controllers of the Future," *Cont. Eng.*, **33**, 1, 59–62 (January 1986).

UL: "Standard for Safety—Industrial Control Equipment, UL508," 13th Edition, Underwriters Laboratories, Northbrook, Illinois. (Revised periodically)

Programmable Controllers:

State-of-the-Art-Systems

Because of the very large number of quality manufacturers of programmable controllers (PCs) in the United States, Europe, and Japan, most of whom do not offer a single product but rather a family of related products designed to satisfy the requirements of a large range of applications, selecting a system best suited to a given application has become increasingly difficult during the past several years. Consequently, efforts have been made to construct tabular comparisons of the capabilities and features of PCs. In one tabular system[1], the following criteria are listed:

Total system I/O	Motion control
Maximum discrete I/O	Documentation
Maximum analog I/O	PC data highway
Relay ladder logic	Type of interface
High-level language	Scan rate
PID capabilities	Type and size of memory

A unique graphical comparison has been developed by G. L. Chester (copyright 1984) and is illustrated in Fig. 1. In what is termed a *Spidergraph*™, Chester bases system comparisons on eight generalized criteria, namely:

Maximum input/ output (I/O)	Special internal functions
Input/output types	Peripherals and program loaders
Maximum scanning time	Communications
User program memory	Diagnostics

Other important considerations suggested by Chester include price, size, programming compatibility, standard I/O, and internal features, such as auxiliary relays, timers/counters, latching relays, shift/data register, real-time clock, and sequencers.

In the prior article of this *Handbook*, author Mackiewicz describes the fundamental operating principles and applications for PCs and provides considerable insight concerning factors that the control engineer must consider when procuring a PC for a given application. This immediate article represents an attempt to present the hardware and some of the software aspects of currently available PCs (early 1986). The devices described have been selected as being reasonably representative of the current inventory of commercially available PCs, but of course their inclusion does not indicate a preference or endorsement on the part of the editors of this *Handbook*. Inasmuch as there are well over 200 models of PCs, emanating from scores of manufacturers, there are more omissions than inclusions in this gallery. The order of appearance is roughly in increasing size of memory.

MICRO PROGRAMMABLE CONTROLLERS

The PLC-4 *Microtrol*, produced by Allen-Bradley[2], is described as a self-contained, solid-state control device. All processor, input, communication, and power supply is contained in the controller module. The controller's twenty-input and twelve-output points, along with over $\frac{1}{2}$K of memory, provide access to machine and process functions beyond simple on/off functions. The controller provides timing, up/down counting, and a sequencer instruction that permits the control of complex sequential operations with a simple instruction. The memory is EEPROM/RAM (640 programmed instructions; 32 timers/counters, 99 storage bits). RAM with lithium battery backup is also available. A summary of specifications is given in Table 1.

Mitsubishi[3] offers a *Series F* line of small PCs for handling from 20 to 80 I/Os. The two models share a common programmer as well as a ROM writer, program loader, and other peripherals. By adding an extension unit to the basic unit, the number of I/Os can be doubled. The *F-20M* basic unit handles 20 I/Os and is expandable to 40 I/Os; the *F-40M* basic unit handles 40 I/Os

---

[1] Used by the editors of *Control Engineering* in their annual *Programmable Controller Updates*.
[2] Allen-Bradley Company, Systems Division, Highland Heights, Ohio.
[3] Mitsubishi Electric Corporation, Mt. Prospect, Illinois and Rancho Dominguez, California.

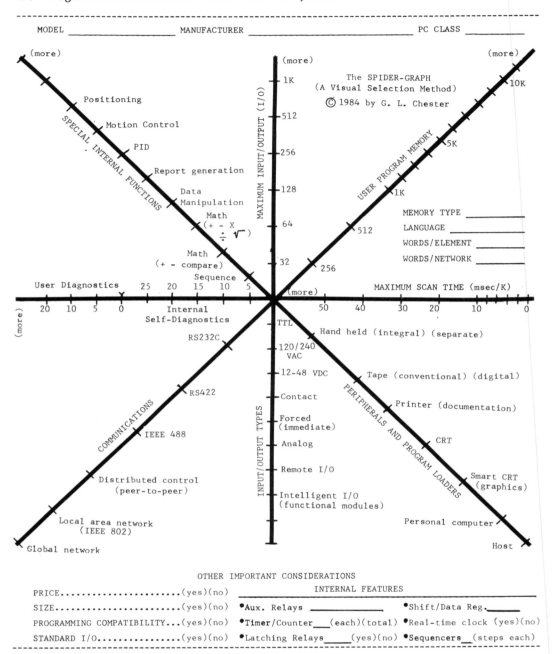

Fig. 1. Chart designed to assist engineer in specifying and purchasing a programmable controller with features best suited for a given application. Eight major parameters are considered: (1) Maximum input/output; (2) types of input/output; (3) user program memory; (4) maximum scan time; (5) special internal functions; (6) communications; (7) diagnostics; and (8) peripherals and program loaders. Copyrighted under name of *SPIDERGRAPH*, this form was designed by G. L. Chester. (*Divelbiss Corporation, Fredericktown, Ohio*.) (used here with permission.)

TABLE 1.   General Specifications—Allen-Bradley *PLC-4 Microtrol*

PLC-4 PROCESSOR

Relay and sequencer instructions . . . . . . Cost-effective equivalent to relays and drum controllers. Provides solid-state reliability and reduces panel space.
Self-monitoring diagnostics . . . . . . Verifies system operations through diagnostic messages and LED indicators. Assists in troubleshooting.
Hardware and memory tested automatically . . . . . . Increases integrity of data that are sent to outputs to minimize system damage.
Memory backup with lithium battery or EEPROM . . . . . . Long-term battery life. EEPROM provides nonvolatile storage and easily accommodates program changes. EEPROM can be programmed without extra hardware.
System available in a variety of voltage inputs . . . . . . Readily satisfies a variety of user requirements, such as European (220/240 AC), United States (120V AC), and off-shore and remote-area applications (24V DC).
Proximity switch input capability . . . . . . Allows direct input from a variety of standard proximity switches with available adapters (120V AC, 220/240 AC, and 24V DC).
Input simulator . . . . . . Facilitates system programming, testing, and debugging.
Removable user wiring strip . . . . . . Can modify processor/system and perform routine maintenance without disturbing user wiring.
System expansion capabilities . . . . . . Up to 4K memory, 256 I/O.
Processor-to-processor communication . . . . . . Provides effective distributed control in small applications with up to eight *PLC-4*s on a 4,000-foot (1220 meters) loop.
Data highway compatibility . . . . . . Integrates *PLC-4*s into a plant-wide communication network.
Environmental ratings . . . . . . 32–140°F (0–60°C); 5–95% humidity (no condensation.)

PROGRAMMER

Full ladder diagram display; sealed keypad; full programming/monitoring capability with password protection; cassette tape storage and ladder diagram printout.
Specifications are subject to change.

and is expandable to 80 I/Os. Execution speed is seven microseconds per instruction, comparing favorably with some of the larger PCs. Specifications are summarized in Table 2.

Mitsubishi has introduced a "Handy Graphic Programming Unit" to meet the needs for a low-cost, compact, multifunctional graphic programmer (*GP-80F*) that can be used with the *F-20M* and *F-40M* PCs. The programmer incorporates off-line programming wherein the program can be stored in the HGP RAM by either instruction code or relay ladder diagram programming. The program can be transferred to and from the PC. The PC operation can be monitored on the display. A cassette magnetic tape recorder can be connected for the purpose of program storage or retrieval. The display incorporates an advanced and large LCD matrix (480 × 128 dots). The relay diagram display allows eleven contacts plus one coil wide × 7 lines high plus two message lines in a screen. One circuit can extend over three screens. The screen can be scrolled up and down, and the contrast and view angle can be adjusted.

Applications for which the *F-20M* and *F-40M* controllers have been designed include printing, finishing, papermaking, packing and wrapping, woodworking, transfer, textile, mold-injection, and assembly machines as well as paint-spraying and finishing lines.

The smallest PCs currently offered by Gould[4] are the *PC 0085* and *PC 0185*, the design objectives of which were directly largely toward original equipment manufacturers (OEMs). This market calls for low-cost PCs that are easy to design into a machine and that have the functionality to do just what is necessary for a specific application. A common requirement of OEMs who manufacture a standard product in high volume is space efficiency. To the OEM, reductions in machine size usually translate into lower costs. From a packaging standpoint, the *PC 0085* is one of the world's smallest PCs. Miniaturized through the use of custom LSI, a standard unit includes a 24VDC power supply, the CPU, a high-speed 5 kHz counter, and 24 input/output points. Build-

Fig. 2. Gould *PC 0085* programmable controller shown at left; the *PC 0185* is shown at right. The programmer, which can be used with either PC model, is in lower portion of the photo. (42Gould Inc.)

[4]Gould Inc., Programmable Control Division, Andover, Massachusetts.

TABLE 2. General Specifications—Mitsubishi F-20M and F-40M

	F-20M	F-40M
Power supply	90 ~ 132 VAC, 50/60 Hz	90 ~ 132 or 220/240 VAC +10% −15%, 50/60 Hz
Consumption	<11VA (basic unit)	<25VA (basic unit)
Power failure compensation	If failure duration is 20 ms	If failure duration is 25 Ms
Ambient temp./storage temp.	0 to +55°C/−15 to +60°C 32 to 131°F/+3 to +140°F	0 to +55°C/−15 to +60°C 32 to 131°F/+3 to +140°F
Ambient humidity	95% (no condensation)	95% (no condensation)
Vibration resistance	10—55 Hz; 0.5 mm (max.2G)	10—55 Hz; 0.5mm (max.2G)
Insulation resistance	5 megohms (500VDC)	5 megohms (500VDC)
Insulation	1500VAC, 1 min	1500VAC, 1 min
Noise immunity	1000V, 1 microsec	1000V, 1 microsec
Noise spike	NEMA-1CS2-230	NEMA-1CS2-230
CPU	μP 8049	μP 8039
Self-diagnostics	Watchdog, sum check	Watchdog, sum check
Battery	Lithium, 5-yr backup	Lithium, 5-yr backup
Memory	320 words C-MOS RAM/EPROM	890 words, C-MOS RAM/EPROM
Speed	Average 45 microsec	Average 45 microsec
Input	12 + 12 points	24 + 24 points
Output	8 + 8 points	16 + 16 points
Timers	8 (0.1 to 99 sec)	16 (0.1 to 999 sec)
Counters	8 (1 ~ 99) power-failure compensated	16 (1 ~ 999) power-failure compensated
Auxiliary relays (marker)	64 (16) power-failure compensated	192 (64) power-failure compensated
IN/OUTPUT LOADS Input voltage	24VDC (sink)	24VDC (sink)
Input current	6mA	7mA
Output relay voltage	24VDC, 220/240VAC	24VDC, 220/240VAC
Output relay current	max. 2A/Output (resistance load)	max. 2A/Output (resistance load)
Output transistor voltage	24VDC	24VDC
Output transistor current	max. 1A/Output (resistance load)	max. 1A/Output (resistance load)
Output voltage solid-state relay	220/240VAC	220/240VAC
Output current solid-state relay	max. 1A/Output (resistance load)	max. 1A/Output (resistance load)

Specifications are subject to change.

ing in only the I/O necessary for the application reduces costs and keeps size to a minimum. The designers reasoned that a PC with a lot of tools to simplify frequent reprogramming usually has little value to an OEM. Once the PC is programmed, the OEM copies it into other machines. Improvements in ease-of-use or time savings are important only when the task is frequently repeated.

The *PC 0085* is expandable to 120 I/O points and contains a 928-word instruction memory. The unit is housed in an enclosure that measures less than 35 inches square (~ 1 meter square) and fits into small spaces for stand-alone machine control.

The *PC 0185* was designed for applications that require more control functionality. However, its small size is also a key design feature. using remote I/O, the OEM can place I/O very close to the actual process, near the motors, solenoids, and limit switches. Up to seven remote locations can be utilized as far as 1500 feet (457 meters).

Abridged specifications for both the *PC 0085* and *PC 0185* are given in Table 3.

A universal programmer will accommodate both models. The programmer can be attached to the mainframe, mounted on a panel, placed on a desk, or held in the hand. Programs are stored on standard cassette tapes, with the cassette interface in the program panel. The *PC 0185* has an electroluminescent panel for viewing in dimly lighted locations.

The *D100* micro PCs comprise a series of controllers offered by Cutler-Hammer[5] which

---
[5]Cutler-Hammer Products, Eaton Corporation, Milwaukee, Wisconsin.

TABLE 3. Abridged Specifications—Gould PC 0085 and PC 0185 Controllers

Item	PC 0085	PC 0185
I/O points (standard)	24	192
I/O points (extended)	120	512
Remote I/O points	—	Up to 1500 feet (520 meters)
Memory (words)	1K	3.5K
PROM memory	Optional	Optional
Timers and counters	48	128
Instructions	23	54
Logic scan time	6 ms/K	2 ms/K
Programming panel	Portable	Universal
Mounting	On panel or DIN rail	Various options

Specifications are subject to change.

range from 20 I/O (12 input/8 output) to 120 I/O (72 input/48 output). There are three basic units (20, 40, and 40 high-performance I/O) and two expansion units (20 and 40 I/O). The controllers can further be obtained with relay, triac, or transistor output. By selection of suitable basic and expandable controllers, a mix of outputs can be obtained. The basic unit is, in itself, a standalone programmable controller. (A 20 I/O controller is shown in Fig. 3.) Also shown in Fig. 3 is the *D100* programmer, which will program all basic and expansion units in the *D100* family. The programmer is plugged into the basic unit and receives its power from the unit. The program can be written into the RAM memory (2K) of the basic unit or can be written into a PROM module.

The programmer has an LCD display and keyboard. The keyboard uses membrane keys and features an audible feedback. The LCD displays eight rungs of a ladder diagram at one time; each rung can contain nine elements (contacts) and a coil function. The ladder diagram can be paged through in sequence or in reverse sequence. The screens appear in the order in which they are scanned. The programmer is used to write a program, to troubleshoot the system operation, and to write PROMs. PROMs are used as backup to the controller RAM memory or to program other similar systems located locally or remote. Once a system has been programmed and debugged, the controller can operate independently of the programmer.

One programmer can be used with a number of systems. The programmer is not used to operate the system after it has been programmed and is running. The programmer can be used to monitor a system for diagnostic purposes or to modify or revise the program. If a number of similar systems are used in remote locations, the programmer is used to write the revised program onto PROMs—these can then be transported to remote locations to update other controllers.

Cutler-Hammer offers the *MPC1* programmable controller shown in Fig. 4. There are several versions in the *MPC1* family of PCs. The most recent additions to this family are two low-cost PC processors capable of supporting intelligent, analog I/O modules along with an enhanced instruction set. Identified as the *MPC1C12* (with 32 I/Os) and the *MPC1C22* (with 128 I/Os), these processors incorporate register I/O commands and double precision multiply/divide math functions that accommodate new programmable and microprocessor-based I/O modules. For simplicity, parameters are down-loaded to the I/O modules under familiar ladder logic control. Thus, functions like signal range selection, scaling to and from engineering units (reading in °F, rpm, psi, et al.), input signal filtering, output signal conditioning (with linear ramp or logarithmic response) are module imple-

Fig. 3. The *D100* micro programmable controller for serving twenty input/outputs (12 input/8 output). Shown below is the hand-held programmer. (*Cutler-Hammer Products, Eaton Corporation.*)

mented. This eliminates manual adjustments on modules, inventorying of various modules for different applications, and ladder logic exercises.

LEDs for each of two isolated input channels and discrete inputs to the PCs are also provided by the analog I/O modules.

SMALL PROGRAMMABLE CONTROLLERS

Designed as a relay replacer, the *AutoMate 15* offered by Reliance Electric[6] finds application in small control systems requiring sequencing, timing, counting, and shift register functions. The unit is designed for installation on the production floor. The processor is integrated with the power supply and memory and is designed to control a maximum of 64 digital I/O points. Maximum scan time for a fully loaded memory with maximum I/O is 7 milliseconds. Average scan time is 3–4 milliseconds. An optional *R-NET* processor allows the unit to communicate with other like units and with other members of the overall *Automate* family. The *Automate 15* memory is a nonvolatile R/W (1K) protected against loss of program and status without battery backup. The memory can store up to 1000 ladder elements.

An expanded version of the *AutoMate 15* is the *AutoMate 20*, with capacity for larger applications. (See Fig. 5.) Although relay replacement is the primary function of this controller, it can serve the more complex functions of timing,

[6]Programmable Controls Div., Reliance Electric Co., Stone Mountain, Georgia.

Fig. 4. The *MPC1*, a low-cost programmable controller capable of supporting intelligent, analog I/O modules. The *MPC1C22* (with 128 I/O) is shown at top, with an analog input and an analog output module shown at bottom of view. (*Cutler-Hammer Products, Eaton Corporation.*)

Fig. 5. Programmable controller 4245C220 in the *AutoMate 20* family of PCs. Processor capability: 256 I/O points, 768 internal coils, 512 data registers, remote I/O, networking via *R-NET*. Memory is 2K × 16 non-volatile R/W. Battery backup is not required. (*Programmable Controls Division, Reliance Electric Company.*)

Programmable Controllers—State-of-the-Art Systems 175

counting, shift registers, and other enhanced functions. The capacity of the *AutoMate 20* has been expanded to control up to 256 digital I/O. There are 256 programmable internals and 512 registers. The controller is also capable of communication on the *R-NET* local area network. The memory (2K), the same as that used in the *AutoMate 15*, has been expanded to the equivalent of 2000 ladder elements.

This family of PCs also includes the *AutoMate 30* (8K memory) and the *AutoMate 40* (104K memory). These are, of course, classified as mid-sized and large PCs, respectively. (See Figs. 6 and 7.) All *AutoMates* are MAP compatible with gateway.

R-NET (Reliance Electric's Local Area Network) provides a real-time communications scheme for distributed control and centralized data-gathering applications. *R-NET* allows up to 255 *AutoMate* PCs and/or other intelligent devices to communicate with each other. *R-NET*,

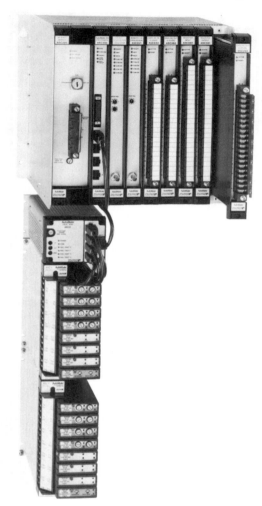

Fig. 6. Built around the *Motorola MC68008*™ microprocessor and a custom logic processor, the *AutoMate 30* programmable controller shown here is an expanded and sophisticated version of the *AutoMate 20* PC. Execution speed is 1.6 millisecond/K. The controller handles up to 512 digital I/O, 512 internal relays, 1,024 data registers, 123 analog/register I/O, and from 2K to 8K $\times$ 16 application memory. The memory is non-volatile R/W and does not require battery backup. Enhanced ladder diagram programming is used. The controller communicates via *R-NET* (local area network). Diagnostics detect any single point failure and report it. On-line memory parity, checksums, processor checks and bus checking are designed to detect failures that could cause loss of production or unsafe operation. (*Programmable Control Division, Reliance Electric Company.*)

Fig. 7. The *AutoMate 40* programmable controller shown here is built around the *Motorola MC68008*™ microprocessor and a custom logic system. The architecture of the *AutoMate 40* is based on the *Intel Multibus*™. This bus allows the user to integrate *Multibus*-compatible cards into the *AutoMate* control system. The execution speed is 0.8 millisecond/K. The controller handles up to 8,192 digital I/O, 8,192 internal relays, 7,168 data registers, 2,048 analog/register I/O, and from 8K to 104K $\times$ 16 application memory. The controller communicates via *R-NET* (local area network). On-line memory parity, checksums, processor checks and bus checking are designed to detect failures that could cause loss of production or unsafe operation. The memory is non-volatile R/W and does not require battery backup. Enhanced ladder diagram programming is used. (*Programmable Control Division, Reliance Electric Company.*)

a baseband coaxial cable system, is arranged in a multidrop configuration, as shown in Fig. 8. The network connects devices up to 12,000 feet (3658 meters) apart. Utilizing a token-passing process, each network node controls the network in turn and transmits for a certain maximum time. When the transaction is complete, the token is passed to the next node in sequence (specified by card front switches), Because of the deterministic nature of the token-passing scheme, a data rate of 800K baud and a typical token holding time of 2–3 milliseconds, this speed usually exceeds the need of real-time control applications. The *R-NET* is based on the *Motorola 68000* and a high-speed serial controller. Communications are discussed in more detail in *Section 5* of this *Handbook*.

The *PLC-5*, shown in Fig. 9 and offered by Allen-Bradley, is a single-slot processor that incorporates a broad range of functions, including: (1) PID (proportional, integral, derivative) capability for closed-loop operation of multiple continuous processes, (2) advanced math functions, including floating point, signed integer, and square root for flow control, (3) sequential function chart programming, in addition to standard ladder logic programming, to cut scan times by partitioning programs into blocks and executing them on a selective basis, (4) a built-in interface for peer-to-peer communications, remote programming, uploading and downloading and networking functions without additional modules, and (5) built-in remote I/O scanner capability for distributing control of 512 I/O points, among one local and three remote racks along a 10,000 foot (3050 meters) distance. The processor features 6K CMOS RAM memory that is expandable to 10K with a 4K RAM memory expansion module. This supports 512 discrete I/O in remote configurations.

The *Director 4001* programmable controller, offered by Struthers-Dunn[7], was designed as a modular system, both in hardware and programming. A leading design factor was simplicity of operation and programming. The processor is capable of controlling up to 256 discrete I/O with any mixture of inputs and outputs. It provides 240 discrete internal storage points (relay coils) and 64 timers or counters that may be programmed with 4-digit presets (0.1 to 999.9 seconds or 1 to 9,999 counts, up or down). User program memory is CMOS RAM or nonvolatile EAROM and is available in 1, 2, 3, or 4K increments. In addition to its prime task of examining the I/O status, the user program, and calculating the proper output response, the processor also provides continuous fault detection monitoring of itself and the I/O modules and tracks to which it is connected. This self-diagnosing function encompasses the processor's executive program, watchdog timer and the correct interpretation of the user program code. Errors in any of those areas result in illumination of the CPU fault LED and disablement of the outputs. If a communication error exists among various

[7]Systems Division, Struthers-Dunn, Inc., Bettendorf, Iowa.

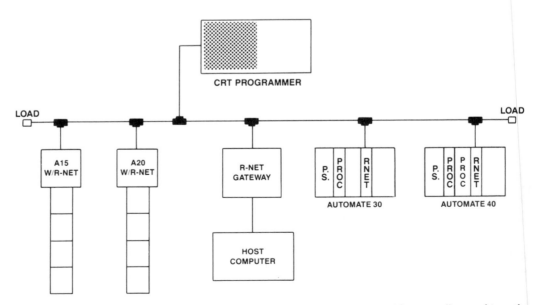

Fig. 8. Block diagram of *R-NET* which allows up to 255 *AutoMate* programmable controllers and/or other intelligent devices to communicate with each other. (*Programmable Controls Division, Reliance Electric Company.*)

I/O tracks, the Track Fault LED will illuminate. An eight-position thumbwheel switch on the processor front panel can then be used to determine the location of a faulted track by dialing through the 1-8 positions.

As an option, the *Director 4001* can have the capability of data handling. This provides the addition of increased hardware and software for more sophisticated applications. Data Handling provides an additional 32 input/output ports, each capable of 16 bits of parallel data. Various modules may be inserted in these 32 positions to provide functions of A/D and D/A conversions, 16-bit parallel data interfacing, high-speed counting, et al. The processor also provides 512 internal 16-bit storage locations that can be used for solving mathematical formulas, storing data, recipes, et al.

The computer interface option allows the Director *4001* to directly communicate with a host computer for total two-way communications. The computer interface provides either RS-232-C or RS-422 serial transmission with selectable baud rates of 110 to 56K. Each CPU with computer interface has a selectable identification number from 1 to 32, which means 32 CPUs can be distributed throughout a single communication network. The host computer can monitor the status of all functions in the CPU as well as update all data and force the condition of I/O. User programs may be uploaded and downloaded to the host computer. When the computer interface option is selected, a real-time clock is also provided.

A smaller version, the *Director 4002*, shown in Fig. 10, utilizes the same I/O modules as the *Director 4001* as well as the same program loader. The processor is designed around a *6502* microprocessor with 2, 4, or 6K of user memory. The memory is CMOS RAM with battery backup or an optional EAROM backup. The processor can handle 64 discrete I/O points or 256 I/O using 5-30VDC I/O modules.

Fig. 9. *PLC-5* programmable controller designed to meet the more complex needs of small-system users. The processor features 6K CMOS RAM memory that is expandable to 10K, with a 4K RAM memory expansion module. (*Systems Division, Allen-Bradley Company.*)

MID-SIZE PROGRAMMABLE CONTROLLERS

The *Gould 884* programmable controller was designed for small-to-mid-size PC needs. The system features modular, integrated construction. System components are designed as separate, detachable modules. (See Fig. 11.) In addition to the Gould *P190* programmer, the IBM Personal Computer or IBM-Compatible Computer can be used to program the 884 PC. Incorporating the same editing configurations and programming features as the *P190*, the software used in the IBM personal computer allows on-line development of *884* programs and storage on 5.25-in. floppy disks or hard disks. A remote I/O system, operating with an *884*, allows up to five racks of remote I/O to be located as far away as 7000 feet (2134 meters) from the *884* controller mainframe. Some of the design features are shown in Fig. 12. Abridged specifications are given in Table 4.

TI Model 530, shown in Fig. 13, is one of Texas Instruments'[8] *500 Series* programmable controllers. This controller fills the traditional PC requirements of sequencing, timing, and counting, as well as added capabilities because of its special I/O bus structure. This unit handles up to 1023 I/O points. Memory size is 5K-7K,

---

[8]Texas Instruments Incorporated, Dallas, Texas.

Fig. 10. *Director 4002* programmable controller. The unit can replace just a few relays or perform complex controls using PID loops, computer interfacing, data handling, et al. The processor, with 2K, 4K, or 6K of user memory, can handle 64 discrete I/O points, or 256 I/O using 5-30 VDC I/O modules. (*Systems Division, Struthers-Dunn, Inc.*)

Fig. 11. Components of Gould *884* programmable controller system, include the *884* controller, a *P800* power supply, and an *800 Series* of I/O modules. System uses CMOS RAM with lithium battery backup. Logic program storage is 2K, 3.5K, or 8K words. Total memory (logic plus 1,040 4-digit registers and 1,280 discrete variables) is 3.5K, 5K, or 9.5K words. (*Gould Inc.*)

Keylock, power supply machine

800 Series I/O Module

Rack mounted, two housing system

Front mounted fuse

Fig. 12. Some of the design features of the Gould *884* programmable controller. (*Gould Inc.*)

expandable to 11K. There is VPU, TIWAY I[9] compatibility. Distributed I/O—placement up to 1000 feet (300 meters) from the CPU permits configuration to almost any application requirement. The I/O base accepts a variety of modules in any configuration. Diagnostic capability is provided. The unit will handle 511 control relays, 128 timers/counters, or, in the extended version, up to 512 timers/counters. It is available with video programming units and used for programming, troubleshooting, and documentation. It employs ladder logic language. The CRT is 9 inches. The programmer allows an immediate hard-copy printout of the program.

A smaller model in the *TI 500 Series* is the *TI 510*, which has a memory size of 256 words and is designed to replace up to 20 relays, timers and counters, or multiple drum timers. There are 12 inputs, 8 outputs; expandable to 40 I/O points.

[9] VPU=Video Programming Unit; TIWAY=Texas Instrument Network.

TABLE 4. Abridged Specifications—Gould 884 Programmable Controller

I/O Capacity
512 discrete and 32 analog channels per scan.
Up to 32 total I/O modules

Memory
CMOS RAM with lithium battery backup (Battery shelf life, 5 yrs)
2, 3.5, or 8K words of user logic program storage, plus 1040 4-digit registers, plus 1280 discrete variables.
Total memory: 3.5, 5, or 9.5K words

Communications Capability
Number of ports............Two (integral)
Port type..................RS-232C
Protocol:
   Modbus (Hierarchical Master/Slave)
   Baud rate, 50 to 19,200
   Modes: ASCII or RTU
   Parity: None, odd, or even
   Stop bits: One or two
Port parameter setting........Hardware (dip switch) and software (RAM)

Environment
P190 Programmer...........41 to 104°F (15 to 40°C) (when operating)
884 Modules................32 to 140°F (0 to 60°C) (when operating)
Humidity..................0—95% (noncondensing) for modules
                          20—80% (noncondensing) for P120 programmer

Specifications are subject to change.

Fig. 13. Texas Instruments *Model 530* programmable controller handles up to 1,023 I/O. Memory size is 5K to 7K, expandable to 11K. Memory is RAM, EEPROM. The controller uses ladder logic language. The programmer incorporates a 9-inch CRT and facility for making immediate hard copy of program. (*Texas Instruments Incorporated*.)

It is designed for use in severe industrial environments. It includes a powerful instruction set, including four drum timers. (See Fig. 14.)

The Allen-Bradley *PLC-3/10* programmable controller was designed specifically for mid-size manufacturing and process control applications. The controller features an integral power supply, and 16K or 32K word memory with single-bit error detection and correction (EDC) and 2048 input/output points. The unit utilizes the Allen-Bradley Universal I/O system, which allows location up to 10,000 cable feet (3480 meters) from the processor. The unit incorporates floating-point math, double-precision integers, and a powerful instruction set. (See Fig. 15.)

The *Gould 984* is a mid-size programmable controller. Designed to meet increasing user demand for expanded capabilities, this unit combines large PC functionality with increased logic-solve capacity. The designers recognized the need to provide large PC power in a modular design—with expandability, functionality, and speed constituting major design objectives. (See Fig. 16.) The CPU module utilizes a four-bit-slice microprocessor configured in a 16-bit architecture. The CPU incorporates a high-speed logic solver (200 nanoseconds per instruction) and a time-of-day clock with separate 5-year battery backup. The processor uses RAM and PROM. Of the 16K memory, users can program up to 8K of PROM to ensure controller reliability and prevent logic tampering. Up to 2048 I/O points can be configured. The I/O processor supports 2048 discrete I/O points and 1920 holding registers for analog I/O. All I/O points can be configured in a continually expanding network up to 15,000 feet (4572 meters) from the controller. Other I/O processor features include single coax cable multi-drop configuration (up to 16 I/O drops

Programmable Controllers—State-of-the-Art Systems 181

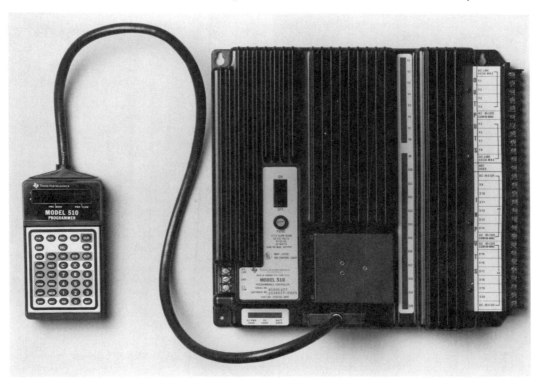

Fig. 14. The smallest in the Texas Instruments *500 Series* of programmable controllers, this unit was designed mainly for the OEM (original equipment manufacturer) market. The unit addresses applications that require 20, 28, 40, or 64 I/O. Memory size is 1K words. There are 12 to 40 inputs; 8 to 24 outputs; 68 central relays; 20 timers, and 8 counters. Programming is by detachable unit or EPROM (optional). (*Texas Instruments Incorporated.*)

Fig. 15. The Allen-Bradley *PLC-3/10* programmable controller designed for mid-size manufacturing and process control applications. The controller features an integral power supply, and 16K or 32K word memory with single-bit error detection and correction. The unit handles 2,048 input/output points. (*Systems Division, Allen-Bradley Company.*)

Fig. 16. The Gould *984* modular programmable controller. Shown at left is the mainframe, which includes power supply and high-speed CPU, communications processor, mixable PROM/CMOS memory module and I/O processor. At right is the data access panel which operates as register access panel and simple ASCII message display. (*Gould Inc.*)

with 512 I/O points per drop), a 1.544 MBaud communication rate, and a run-status indicator. A data access panel constitutes a low-cost operator/machine interface. A handheld data access panel (DAP), plugs directly into the front panel of the programmer and allows users to access register values in the PC and to alter data, such as timer presets or PID loop parameters at the controller site without using a programming panel. The DAP port will also act as a simple ASCII port capable of generating 64-character messages on a four-line LED readout display.

LARGE PROGRAMMABLE CONTROLLERS

The *TI 560/565* models are currently the largest of the Texas Instruments *Series 500* line of programmable controllers. (See Fig. 17.) The designers particularly had expandability and upgradability in mind. For example, memory boards can be added in the field. The chassis can accept three memory boards, added in increments of 65K, 128K, or 208K, up to 256K. Memory can be added without disturbing the existing program.

The *560/565* controllers incorporate the complex control capabilities of the *TI 530* model previously described. The systems were conceived so that future user control problems, as manufacturing operations expand, can be handled in terms of manageable bits. It is claimed, for example, that the user may elect to reautomate just discrete processes first with the *560* and when that has gone through a shakedown period, other discrete operations, such as materials handling and robot-assisted assembly, can be added. Later the system can be upgraded to a *565*, using the parallel special function CPU to handle batch operations, for example. The *560/565* permits allocation of memory to match new application requirements as they arise. For example, an assembly-line application may require 80% relay ladder logic (RLL) and only 20% variable, constant, et al.; a batch weighing or blending operation may require only 20% RRL, but 80% constant memory for recipe storage, or data acquisition may require 80% variable memory. The system permits any mix of discrete, analog word, or intelligent I/O with up to 8192 I/O points.

A hot backup option is offered with the system. A dedicated hot backup card is installed in

Fig. 17. The *TI 560/565* programmable controllers are currently the largest of the Texas Instruments *Series 500* line of PCs. Designed for expandability, memory boards can be added in the field. The chassis accepts three memory boards, added in increments of 65K, 128K, or 208K up to a total of 256K. Memory can be added without disturbing the existing program. The system permits any mix of discrete, analog word, or intelligent I/O with up to 8,192 I/O points. (*Texas Instruments Incorporated*.)

each chassis, communicating by a dedicated fiber optic link. When an error is detected, control is automatically transferred. The *560/565* can handle remote I/Os as far as 15,000 feet (4572 meters) away without scan latency. The controller can be used with a *TI CVU 5000* (see Fig. 18) operator interface. This provides up to 25 configurable pages of operator displays. The unit uses a 256 RAM, a single floppy disk drive, and eight-color high-resolution graphics.

A hosted industrial local area network (LAN), known as *TIWAY I*, provides the means to connect a mix of up to 254 PCs, including the *560/565* and other units in the *TI 500 Series*. Associated with this is the *Universal Command Language* (UCL) of high-level commands, permitting all PCs to be addressed in a similar way, including PCs from most other manufacturers. *TIWAY* host adapters permit the use of mini- and microcomputers, including IBM and DEC (Digital Equipment Corp.), as network hosts. Software packages are available for DEC PDPs, running RSX11 and IBM Series I computers and for some other major minicomputers and popular personal computers.

The Gould *584L* programmable controller is a microprocessor-based system that is specifically designed for industrial control. The control logic is entered in a form similar to a relay ladder diagram. The controller is available in a variety of memory sizes (CMOS type). Additionally, any controller can be equipped with a variety of intelligence levels and reference sizes. The CMOS memory is protected by lithium batteries that will retain memory for nine months in envisioned worst conditions. The batteries have a shelf life of five years and are *not* rechargeable. So long as DC power is maintained to a CMOS memory system (i.e., AC power on), the batteries can be removed and replaced. The power supply for the *584L* is included in the mainframe. System operation is not interrupted when the door is opened for access to internal components. As shown by Fig. 19, on the front of the mainframe unit are three indicators: A keylock, an interface connector, and a keypad/display unit. The three indicators show the satisfactory output of the batteries, the presence of all internal DC power (proper operation of the power supply), and the operation of the controller (RUN light). AC power is required to operate any LED, including battery OK. The keylock switch controls Memory Protect logic.

The *584L* controller can communicate with a wide variety of I/O devices and up to 32 channels of I/O. Each channel of I/O can contain up to 128 input points and 128 output points. Controllers can operate with four channels of local I/O; two of these channels are powered from the main power supply contained within the *584L* mainframe, and two require auxiliary power supplies. The user has a choice of four local and 28 remote channels or 32 remote I/O channels. Modules can be installed or removed without disturbing the controller's scan or the field wiring.

A Gould *J211* redundancy supervisor, used in conjunction with two *584Ls*, provides a fault-tolerant, high-availability control for applications where no interruption of the process can be tolerated. The *J211* monitors the readiness of both *584Ls*. Transfer of control to the standby PC is initiated the instant a fault in the active PC is detected. High-speed data transfer passes all status information data from the active to the standby PC at the end of every scan. No master/slave relationship exists between the two PCs.

In 1984, the memory of the *584L* was expanded up to 128K. Optional memory capacities in excess of 32K are available in increments of 48K, 64K, and 128K. Increased memory capacity provides mass storage for data stored in BCD, 16-bit binary, hexadecimal, or ASCII format. The data are stored in files made up of storage registers. A *584L* controller can have up to ten files, each with 10,000 registers. Extended memory can be loaded into the *584L* with a data transfer function or from the peripheral port using either a *Modbus* master or the program panel. For memory sizes over 32K, a *Modbus* communication system allows extended memory access using general read/write capabilities. Also, an additional parity-error detection scheme is incor-

Fig. 18. The *TI CVU 5000* operator interface, shown at left, provides up to 25 configurable pages of operator displays. The unit uses a 256K RAM, a single floppy disk drive, and eight-color high-resolution graphics. A loop-tuning page permits on-line tuning of up to three loops in bar-graph format, and reports can be generated by an accompanying printer. (*Texas Instruments Incorporated*.)

porated as a security feature. A parity error detected in the logic area of memory will be considered catastrophic and will shut down the *584L*. A parity error detected while performing an extended memory read/write function will allow the user to select one of two states: (1) detect and log the parity error but not interrupt the operation of the *584L* or (2) scan the *584L* and stop at the end of the segment.

Modbus communications capability provides the means to integrate multiple controllers into a communication network—for (1) monitoring controller operation, (2) collecting data for management reports, (3) supervisory control of multiple

Fig. 19. Gould *584* programmable controller. Optional memory is available in increments of 48K, 64K, 96K, and 128K words (maximum). (*Gould Inc.*)

controller operation, and (4) initializing large systems. As shown by Fig. 20, the *Modbus* communication system consists of: (1) a protocol for communicating that allows slave addresses up to value 247, ten operational functions and sophisticated error detection, (2) a smart device as the master, which can be a computer, a programmable controller, or the *P190* programmer, (3) multiple slaves (up to 32) on one *Modbus* link. These slaves can be any mix of several Gould controllers with their appropriate interfaces. A communication link can be either locally installed twisted pair cable or a telephone network. The total cable length can be up to 15,000 feet (3 miles or 4.5 kilometers) and allows data rates up to 19,200 baud. Telephone links can be established by using commercial modems or data sets at the master and each slave. RS-232-C ports are provided on the appropriate model interface.

Numa-Logic programmable controllers available from Westinghouse[10] include the *PC-900* (RAM memory; 1024 words minimum and 2560 words maximum) with 128 discrete inputs and 128 discrete outputs; and 16 register inputs and 16 register outputs.

[10] Numa-Logic Department, Industry Electronics Division, Westinghouse Electric Corp., Madison Heights, Michigan.

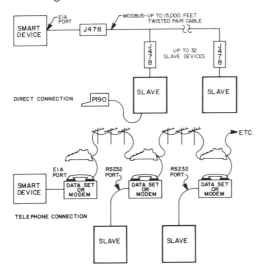

Fig. 20. *Modbus* communication system used with programmable controllers. (*Gould Inc.*)

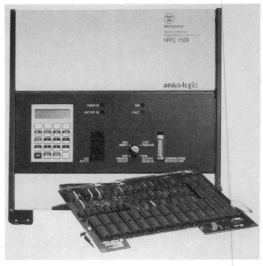

Fig. 21. Westinghouse *HPPC 1500* programmable controller with a RAM memory of 8K words (minimum) and 224K words (maximum), with 8,192 I/O points of any mix. The controller is programmed with relay ladder logic, using proprietary programmers or an IBM personal computer. Communications involving numerous PCs can be accomplished via the *WESTNET II* data highway. (*Numa-Logic Department, Industry Electronics Division, Westinghouse Electric Corp.*)

The *PC-1100* has a RAM memory (512 words minimum and 3584 words maximum) with 64 discrete inputs, 64 discrete outputs, 8 register inputs, and 8 register outputs.

The *HPPC-1500*, shown in Fig. 21, has a RAM memory (8K words minimum and 64K words maximum) with 8192 I/O points in any mix of discrete or analog inputs or outputs.

The *HPPC-1700* has a RAM memory (8K words minimum and 224K words maximum) with 8192 I/O points of any mix.

All of the foregoing controllers are programmed with relay ladder logic, using Westinghouse programmers or an IBM personal computer. Communications involving numerous PCs can be accomplished via the *WESTNET II*[11] data highway.

Westinghouse also offers several smaller PCs for less-demanding applications, including the *PC-100* (320 words memory; 30 I/O); the *PC-110* (1024 words; 112 I/O); the *PC-1100* (3584 words; 350 I/O); and the *PC-700* (8192 words; 1500 I/O).

CELL CONTROLLER

Very late in 1985, Gould introduced the *FM 1800*, a factory-hardened controller that combines the features of a programmable controller with those of a minicomputer. The cell controller is designed for the coordination, control, and communication of numerous devices and networks within islands of automation. In particular, the flexible manufacturing system (FMS) was an application objective. Up to four microprocessors, running in parallel, provide the cell controller, shown in Fig. 22, with the power required to handle high volumes of real-time data, typical of the FMS cell. Up to 16 ports and a library of communication protocols allow the user to connect the cell controller directly to PCs, robots, CNC equipment, gages, I/O, and ASCII equipment, such as bar code readers, terminals, and printers. A simple *flow chart and function plan programming* permit the coordination of multiple devices. Using the built-in MAP[12] (Version 2.1) capabilities, the user can upload, download, and verify cell data with standard vendor-independent MAP commands.

The principal advantages claimed for a cell controller of this type include:

1. Provides a means of manufacturing multiple parts or chemical batches from one cell or group of machines. For example, an automotive

---

[11] WESTNET = Westinghouse Network.

[12] MAP = Manufacturing Automation Protocol.

Fig. 22. Gould *Cell Controller FM 1800* is designed as a factory-hardened controller that combines the features of a programmable controller with those of a minicomputer. The cell controller is conceived for the coordination, control, and communication of numerous devices and networks within islands of automation. (*Gould Inc.*)

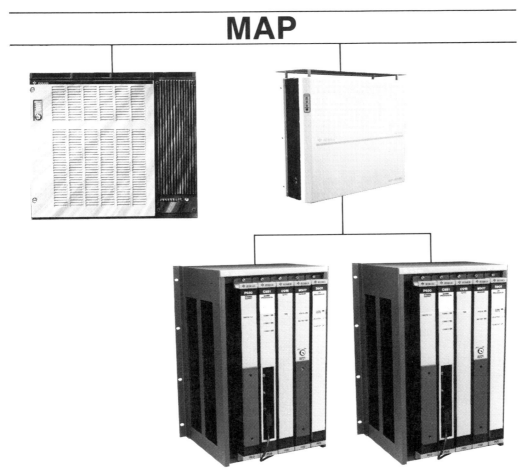

Fig. 23. The Gould *NW 0200 MAP Gateway Unit* is shown in upper right-hand portion of diagram. The *FM 1800* cell controller is shown in upper left-hand portion. Modular programmable controllers are shown at bottom. The *Gateway Unit* provides a cost-effective means of integrating diverse, segmented manufacturing systems along MAP backbone networks without replacing existing equipment. (*Gould Inc.*)

manufacturer can produce camshafts for four-cylinder cars on the same equipment used to make camshafts for six-cylinder cars.

2. A cell controller ensures higher quality through local statistical process control. The available statistical process control enables users to take an actual reading of the conditions of equipment during the production cycle. These readings are then fed into a statistical process and compared with the ideal conditions for making an ideal part. Where the data do not match, an operator can be immediately notified by way of an interface device or alarm.

3. A cell controller permits more efficient use of capital equipment—mainly because the controller can monitor the activities of its own cell and because a cell controller can communicate over a MAP network to other cell controllers and MAP nodes. Thus, the overall production scheduling process can become automated in real time.

4. A cell controller assists in reducing cell downtime through comprehensive cell diagnostics, detailed fault enunciation, and ability to program "around a fault."

5. A cell controller provides increased availability of information over the MAP network. By connecting the controller to a MAP network, it is possible to channel actual production data to host computers, running a variety of applications, such as programmable device support, materials requirement planning, inventory, purchasing, among other manufacturing functions.

The Gould *Cell Controller* is designed to operate in a factory environment, up to 140° F (60° C) without need for cooling fans. From one to four 32-bit microprocessors are used. There are up to sixteen RS232 or RS422 serial communication ports. The controller supports a variety of device protocols, including *Modbus*, DDCMP, Marposs gage, and GMF robots. The controller is addressable to 8M bytes of dynamic RAM. There are up to 3M bytes of battery-backed RAM file systems. The controller can be connected directly to MAP, Version 2.1 (ten megabit standard) local area network (LAN). The unit is a real-time operating system with built-in power-up diagnostics.

The *FM 1800* cell controller supports programming in two different languages, including *flow chart and function plan*, used extensively in Europe.

MAP Gateway

Another late-1985 introduction and closely associated with the cell controller just described is the Gould *NW 0200 MAP Gateway*, shown in Fig. 23. This unit provides a cost-effective means of integrating diverse, segmented manufacturing systems along MAP backbone networks, without replacing existing equipment. The unit permits users to incorporate various stand-alone devices, such as robots and PCs, in a multi-vendor factory-floor environment. This is a multi-*68010* microprocessor-based system. The system incorporates ability of 10 MBPS raw data rate; 512K byes of memory; VMSE-based bus connection to all boards within the system. Protocols include: DDCMP versions 3 and 4; CCITT X .25 *Telenet* certification; RS 232/RS 422 ports (up to 19.2 Kbps Async./56 Kbps Sync.); expandable protocols without additional hardware.

REFERENCES

Please refer to list of references at end of prior article in this *Handbook*.

Programming the Programmable Controller

By Erich Sulzer[1] and Engineering Staff, Siemens Aktiengesellschaft[2]

In connection with manufacturing automation, software may be defined as the informal conversion of a technical problem into a sequence of instructions (program) that are understood by the programmable controller. Software preparation involves the analysis of problems and the generation of the program by the user. A relatively recent addition to the definition of software is the inclusion of program documentation.

As related to programmable controllers (PCs), frequently called *memory programmable controllers* (MPCs) in Europe, the concept of artificial intelligence (AI) is of questionable nature—because the intelligence-requiring tasks performed by the programmable controller are the intellectual input of the human programmer. The programmer's "pre-thought" instruction sequences are "frozen" into the memory of the control system and implemented by the processing procedures performed in the system's central processing unit (CPU). During the processing of the instruction sequence, the CPU simply reads a sequence of stored instructions, which appear to human operators as nondescript, monotonous combinations of zeros and ones.

The *machine-oriented language*, also known as machine code (MC), is the basic language of the PC and is also the *final form* for any other higher and, for humans, more comprehensible language. However, the more comprehensible such a language is to humans, the higher are the expenditures of time required to convert a given higher language into a machine-oriented language. As shown in Fig. 1, a large communication gap exists between humans and machines, since the *conversational language* for the verbal description of the problem as well as its programming in machine code require in each of them a maximum expenditure of time. In this connection, the basic problem and the tasks of the programming language come to light—*unequivocal and simple communication*.

[1]Siemens Energy & Automation, Inc., Programmable Controls Division, Peabody, Massachusetts.
[2]Erlangen, West Germany. Systems described here are marketed in the United States by Siemens Capital Corporation, Iselin, New Jersey.

Languages may be subdivided into four categories: (1) natural conversational languages, (2) natural problem-oriented languages, (3) problem-oriented programming languages, and (4) machine-oriented programming languages. (See Fig. 2.)

Natural Conversational Languages. These languages evolve from their respective cultural and historical environment and are used to describe a technical problem. They require a relatively large number of words with a relatively low information content. Usually additional explanations and limitations are required for providing a clear (not ambiguous) description of the problem.

Natural Problem-oriented Language. These are the languages used *between experts* and usually require relatively few words with a relatively high information content to describe a technical problem. Here a single technical term can often be used to describe an entire complex issue.

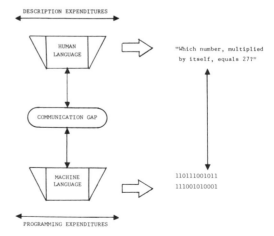

Fig. 1. Communication gap between humans and machines.

190 Programming the Programmable Controller

Fig. 2. Technical languages and programming languages.

Problem-oriented Programming Languages. In these languages instructions are modeled in terms of problem-solving technical terminology. Languages in this category that are directed to scientific problems are, for example, *FORTRAN* or *ALGOL*; for business-related problems, *COBOL*; and for industrial control tasks, *STEP® 5*[3].

Relatively few words with a relatively high information content are used with problem-oriented programming languages to describe specific technical problems. For example, the mathematical problem, "Find the square root of 27" is represented in the programming language *ALGOL* as SQRT (27).

Machine-oriented Programming Languages. These are the basic languages of programmable controllers. The term "machine-oriented" defines a language that is modeled on signals "understood" by the machine, that is, the binary signals "0" or "1."

In its most machine-oriented form, the instructions are written down as combinations of binary signals (bit pattern) in the sequence of the machine processing. In the machine code, each structural part of a bit pattern (such as a group of four bits) is assigned a specific meaning during the execution of the instruction. Thus, instructions may appear as:

Instruction No. 1:   1100 0011 1001 0110
Instruction No. 2:   0011 1100 1010 1001

The foregoing instructions are in machine code.

The advantage of applying machine-oriented programming languages is that the program can be designed for optimal utilization of the memory and, therefore, for optimal scan time. The disadvantage is that the programs require a great deal of writing and are difficult to test. A relatively

---

[3] Language developed by Siemens AG, Bereich Verabeitende Industine, Erlangen, West Germany, and used on the *SIMATIC® STEP®5* control system. This particular system is used as the basis of descriptions throughout this article, although the principles discussed will apply to any system of this type. Differences in abbreviations and symbols used in Europe and in English-speaking countries are resolved at the end of the article.

large number of words is required (bit pattern) with very low information content to describe a technical problem.

Assembler Languages. More advanced are the so-called *assembler* languages, where symbolic addresses or instruction groups consolidated into macro-instructions can be applied, such as MOVE A,B. (Transfer the content of memory cell identified by A into memory cell B, where A and B are symbolic addresses.) Since the controller has no direct understanding of this language, a special *conversion program* must be used to translate the assembler language into the machine code (bit pattern).

Narrowing the Communication Gap

When comparing the expenditures required to describe a specific technical problem with the various natural and artificial languages, the results when graphically displayed appear roughly as an "hour glass." (See Fig. 3.) It will be noted that the formulation of a specific issue in either a natural conversational language or in a machine-oriented language involves larger efforts than its formulation in a natural technical language or problem-oriented programming language, and the narrower the communication gap the result will be a lower programming expenditure and efforts in writing a program. In the most optimal case, the technical language and the problem-oriented programming language are identical. In this instance, the highest level of "programming friendliness" would have been achieved. The programming language, *STEP®5*, was created with this particular goal in mind.

A major problem of such highly sophisticated problem-oriented programming languages is their incomprehensibility for the programmable controller. The language has to be converted with a so-called *compiler* into the associated bit pattern of a machine-oriented programming language. (See Fig. 4[a].) The translation of the compiler is redundant; this means the programs are no longer optimal with respect to memory space and scan time. This disadvantage is unacceptable with modern, high-speed processing controllers. The problem of memory space can be solved by adapting the structure of the machine to the respective programming language, (See Fig. 4[b].) As a result, the insertion of extensive code-generating conversions is largely eliminated.

At the same time, it becomes evident that the advantages of "programming friendliness" and those of optimal memory space and scan time can be fully utilized only when (1) the problem-oriented programming language has been largely adapted to the respective technical language, and (2) the structure of the programmable controller has been adapted to the programming language. (See Fig. 4.)

To allow an immediate interpretation of the language *STEP®5* with the respective measures such as electronical circuits and microprogram,

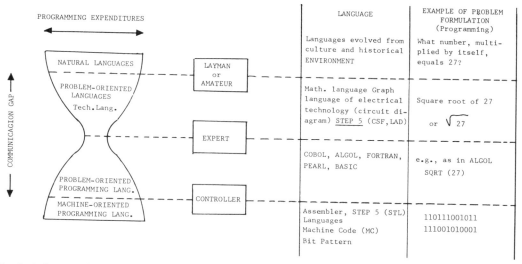

Fig. 3. Software-related communication problems.

microprocessors are implemented in S5 programmable controllers.

Control System Flow Chart (CSF)

By way of example, the near elimination of the communication gap between the problem-oriented technical language and the problem-oriented programming language STEP®5, used to program SIMATIC® S5 programmable controllers, is shown in Fig. 5.

When the limit switch b1 and the limit switch b2 are closed, the drive M1 will be running (AND operation). The language STEP®5 of the system SIMATIC® S5 has been designed to accept the symbols of this technical language (according to Deutsche Industrie Normenausschuss [DIN] and International Electromechanical Commission [IEC]) directly as program symbols. (See Fig. 5.) As can be seen on the basis of this example, the communication gap is practically nonexistent with this language.

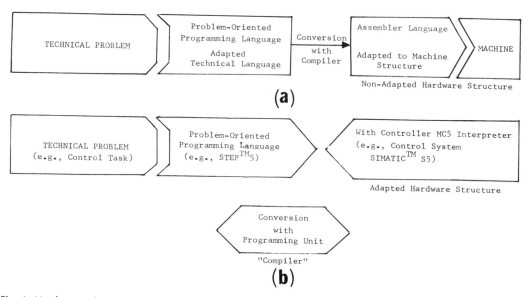

Fig. 4. Hardware adapted to software.

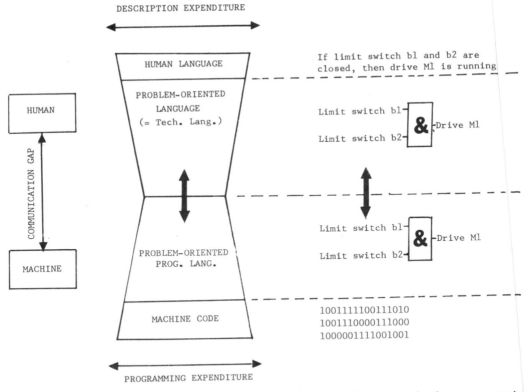

Fig. 5. Example solved in conjunction with control system flow chart (CSF), one of the three programming methods of STEP™5.

Ladder Diagram. There are several other descriptive methods used in conjunction with controller technology. For example, circuit diagrams are usually applied in relay/contactor technologies. So that all relay/contactor technicians need not change to the foregoing introduced programming method of the control system flow chart, a second graphic programming method has been provided in *STEP®5*, namely the ladder diagram (LAD).

The example in Fig. 6 demonstrates the close correlation between a problem-oriented technical language (circuit diagram) and the ladder diagram of the *STEP®5* language.

Statement List (STL)

A third programming possibility is based on a verbal description of the functions to be programmed. Mnemonic code (a pseudo code in which information, usually instructions, is represented by symbols or characters that are readily identified with the information) is used. Letter A represents AND, O represents OR, and N represents NOT. The equation sign (=) stands for ON/OFF according to the logic status. Programs generated in this manner are known as statement list (*STL*).

An example is shown below:

AND limit switch b1	AI	b1
AND limit switch b2	AI	b2
= drive M1 ON	=Q	M1

This list-like display of a program (shorthand of the descriptive text) is not as easily grasped as graphic methods. However, this approach provides for high-level program optimization and can include the entire instruction set of the *SIMATIC® S5*. Furthermore, this kind of language is not limited by the size of the display screen of the programming unit.

Options. To date no one programming method of the three methods described has succeeded in replacing the others. In view of the different educational levels and familiarity of users with programmable controllers (PCs), it seems unwise to insist on a preferred programming method. Thus, there is a distinct advantage when a PC can be programmed with all three of the described programming methods. Which type of display may prove to be most optimal for the user depends on a variety of factors, among them

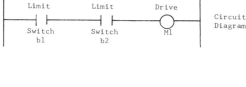
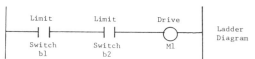

Fig. 6. Example solved with the ladder diagram (LAD), the second graphic programming method of STEP™5.

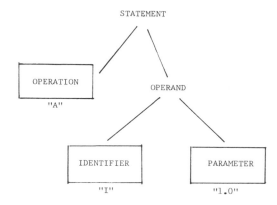

Fig. 7. STEP™5 instruction based on example AI 1.0 (AND input 1.0).

the educational and experience levels of the users. For users who have had years of experience in programming contactor (relay) controllers or who maintain on-site installations, the ladder diagram is probably the most acceptable method. It should be noted that in the U.S., the ladder diagram currently enjoys a high preference.

For other users who are experienced in handling process computers or microcomputers and who have had opportunity to write basic programs, the computer-oriented program [statement list (STL)] approach may be the most preferred method.

The function diagram (FD) is frequently used since it has proven itself as a design and planning tool, especially in Europe. It is common practice to document controller tasks in "CSF" to provide nontechnical personnel with a completely commonly understood language medium. The advantage of the "CSF" is that the design drawings can be used for the final program documentation as well as for the programming of the PC.

In general, a trend to use a graphic programming method is evident because it provides the possibility of easy training for third parties.

Comparison of Programming Methods. A summary of the advantages and disadvantages of the three programming methods previously described is given in Table 1.

Functions of the Programming Language STEP® 5

To facilitate an understanding of the essential functions in *STEP®5*, the smallest autonomous unit of the program will be introduced, namely, the statement. A statement is a command for the processor and consist of two parts (1) an operation part, (2) an operand consisting of an identifier and a parameter. (See Fig. 7.)

TABLE 1. Comparison of Programming Methods.

Method	Advantages	Disadvantages
Control System Flowchart (CSF)	Flow chart for programmable controllers. Display of complex functions. Processor-independent projection and programming language. Easy overview with programming units, including screen. Builds program in block diagrams	Amount of flow chart functions to be displayed is limited by size of screen. Displayed program no longer represents processing in the controller.
Ladder Diagram (LAD)	Use of already familiar program terms and symbols. Easy overview with programming units including screen. Can be supplemented with control system flow chart elements.	Does not allow complex functions; black boxes must be used. Amount of flow chart functions to be displayed is limited by size of screen. Displayed program no longer represents processing in the controller.
Statement List (STL)	Easy display of all instructions. Interlocking depth is not limited. Simple programming units. Optimal utilization of controller capabilities. Comments can be inserted in program flow.	Additional projection documents must be generated (e.g., circuit diagram, function diagrams). Different countries use different abbreviations for operation and identification. Programming. Badly arranged programming is possible.

194   Programming the Programmable Controller

The *operation* describes the function to be performed by the processor. The *operand* includes the information necessary to execute the operation. The programming language "knows" the following essential operand fields in connection with basic functions:

Inputs "I" . . . . These represent the interface from the process to the controller.
Outputs "Q" . . . . . . These represent the interface from the controller to the process.
Flags "F" . . . . These are used for storing intermediate binary results.
Data "D" . . . . These are used for storing intermediate digital results.
Timer "T" . . . . These are used to realize time functions.
Counter "C" . . . . . . These are used to realize counting functions.
Constants "K" . . . . These represent a constant number.

The operand consists of identifier and parameter. For addressing an operand, the respective parameter must be entered.

The parameter is the address for a particular identification. The operand inputs I, outputs Q, and flags F are divided in byte and bit address. A period (.) separates the bit address from the byte address. (See Fig. 8, upper left-hand margin.) Individual inputs/outputs (bit 0 to 7) are called with the bit address. Note: Two bytes (B) are considered a word (W), that is, a 16-bit wide data format.

An example illustrating *STEP®5* statement structure is given in Fig. 8. The typical functions are compared in the three language methods given in Fig. 9.

In the function diagram, the operation is stated by means of the description within the function symbols, by the arrangement of the function symbols, and in the ladder diagram display by the arrangement of the contacts. When using the statement list as display method, the controller instructions are programmed according to the sequence in which they are executed in the programmable controller.

Boolean operation output Q 1.1 has been dedicated in the example of the Boolean operations (AND operation), which means that the output is immediately disabled when the interlocking conditions (AND function) are not met. The enable/disable conditions are therefore checked during each program cycle, and the output is operated according to the operation result, that is, the logical binary result.

The example provided in regard to (flag) functions shows that the switch-on (set) status of F 2.0 remains constant until the presence of switch-off conditions (resetting), even though the setting conditions might have disappeared. During memory functions, one status will be dominant, if the setting as well as resetting requirements are simultaneously met. The dominance of a particular status is ensured in the instruction list by the sequence of program entry—the last instruction dominates. With the graphic display methods, the input of the memory (flag), arranged below, will dominate.

Timer Function

In the timer function example, output Q 2.1 is enabled 10 seconds after (load KT 10.2) if the switch-on conditions have been met. During the

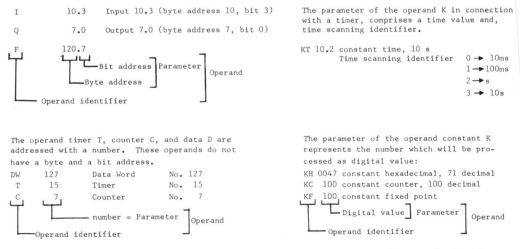

Fig. 8. Example illustrating *STEP™5* statement structure. The typical functions are shown in the three language methods. See Fig. 9.

Programming the Programmable Controller

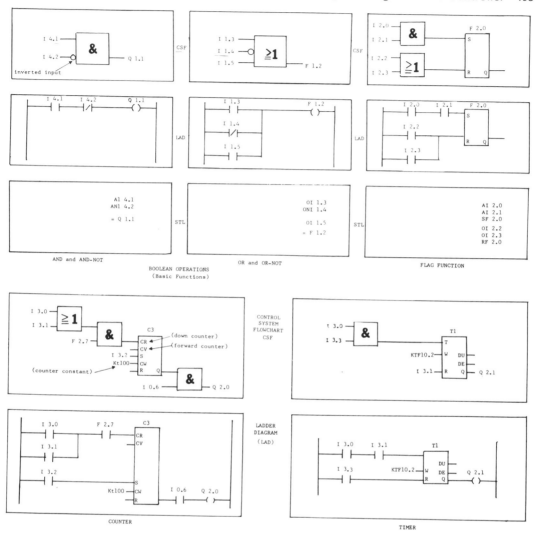

Fig. 9. Comparison of three display methods in *STEP*™5 programming. (Q = Output; I = Input; S = Set; R = Reset)

starting of the timer, the timing characteristics (e.g. delay time or inputs) have to be stated as well. The timing characteristics are shown in the graphic display methods. In the statement list method, the timing characteristics have to be explicitly stated when setting the timing function.

Counter Functions

In the counting function example, a down counter counts down to zero beginning with a loaded presetting (LKC 100, SC 3, load constant 100 into the counter 3), which provides a binary result at the counter output. The counter output is connected in series to the input I 0.6 for enabling output Q 2.0.

The following generally applies to counters: A counter can be set or reset with the counter value changing at the same time. The counter is set to the value allocated by means of a loading operation. The counter status is set to zero by means of resetting. The count-up and count-down operations evaluate the status change of the count inputs from "0" to "1." A static signal on these inputs does not lead to changes in the counter value.

Compare Functions. In the compare-function example, the input word No. 4 is compared with the fixed-point number 27. If they are compared, the result on output "Q" can be used together with input I 13.7 to activate F 12.3.

With the load instructions L IW 4 and L KF 27, both figures are loaded into the two accumulators of the CPU and compared. The instruction "= F" is used to determine the identity of two fixed-point numbers.

The compare operations also include comparisons with respect to (1) greater than, (2) greater than or equal to, (3) less than, and (4) less than or equal to.

CONCLUSION

Some of the most important programming functions in *STEP®5* have been displayed and explained and are now applied in a model for further illustration.

Structured Programming Enables Software Rationalization

The basic idea is that, for example, a 20,000-statement-long program should not be written down line by line beginning with statement 1; instead it should be subdivided into organization blocks (OB), program blocks (PB), function blocks (FB), and Data Blocks (DB) for easier manageability, transparency, and cost efficiency (See Fig. 10.) inasmuch as the cost for each programmed instruction increases considerably with the length of an unstructured program. The reasons for this cost increase of an unstructured program is the lack of transparency generated through too many unconditional jump operations. Also, in extreme cases, only the programmer of the program will be able to handle the program.

In sum, structured programming means the subdivision of the user program into easily manageable sections known as *program blocks* (PBs). By assigning one program block to each technological process, the generated program is a reconstruction of the technological structure of the controlled process.

How the individual program blocks are to be processed and in which sequence is determined by the organization block.

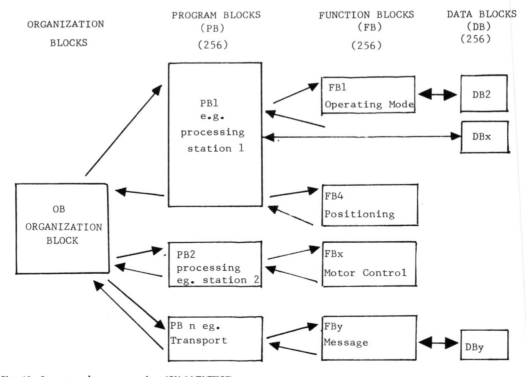

Fig. 10. Structured programming (SIMATIC™S5).

Function blocks are *standardized* program parts of *frequently occurring and universally applicable functions* and are compiled in a standard software library (e.g., divider, multiplier, code converter, shift register, message functions, etc.). These "parameterizable" function blocks can be inserted into the program flow and frequently called. These multi-addressable function blocks can be used to execute a specific function and called from different locations in the program. For example, fifty equivalent valves are to be controlled by a programmable controller. The control function to drive a valve is programmed once into a functional block. The functional block can be called fifty times and provided with different parameters.

Since trained users are able to program their own standard function blocks, this possibility becomes an essential prerequisite for a uniform software configuration on the basis of customer-specific standards.

These standard blocks are stored once in the memory and called and utilized by the individual program blocks an infinite number of times.

Aside from the cost-savings realized during programming, the fixed or permanent entry of frequently occurring functions enables the standardization of all controller programs used in a factory. This kind of standardization is the most important prerequisite for any type of rationalization. In the system described here (*Siemens*) a product spectrum of almost 250 function blocks for universally applicable standard functions is available and shown in a catalog. Thus, it is not necessary to "reinvent the wheel."

For example, the multiplication standard FB 8 (FB "MUL") multiplies the input word 24 with the flag word 17, and the result is stored in FW 18. (See Fig. 11.) If the numerical value is 0, output Q 7.3 is enabled.

The following statements are used to subdivide the program:

Jump functions (valid in FBs only)

Unconditional Jump   JU = symbolic address

Conditional Jump   JC = symbolic address

Block Calls

Unconditional Call   JU PBx, JU SBx, JU FBx

Conditional Call   JC PBx, JC SBx, JC FBx

Jump and Call Functions

During branching the linear processing of the program is interrupted. The new destination must be identified by a symbol address when the branching operation was started through jumps. (See Fig. 12.) Such structuring is possible only within a function block. Entire program sections can be jumped around. However, the user program can be structured through block calls only.

When calling a program block, the program is automatically continued at the instruction following the call, or after the transfer parameter of a function block. (See Fig. 13.)

When the status of the F 20.1 is a logical "1," the function FB1 is called. If the status of the F 20.1 is a logical "0," then the FB1 is bypassed.

The foregoing discussion illustrates software tools available to the user to design his program with the *SIMATIC*™ *S5* system in conjunction

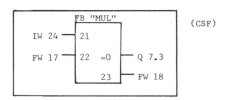

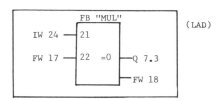

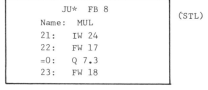

* unconditional jump

16 bit multiplication function block

Fig. 11. Example of saving programming costs through use of standard function program blocks. The multiplication FB 8 (FB "MUL") multiplies the input word 24 with the flag word 17 and the result is stored in FW 18. If the numerical value is 0, output Q 7.3 is enabled.

Fig. 12. Symbolic address. Jump function within a function block.

198 Programming the Programmable Controller

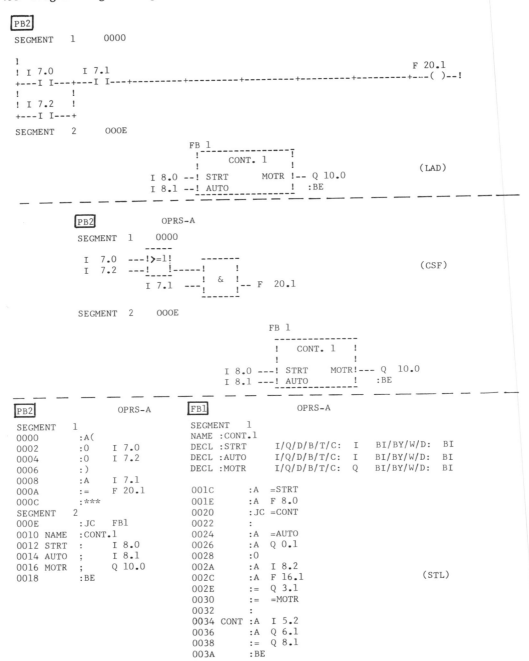

Fig. 13. Calling a function block (FB).

Programming the Programmable Controller 199

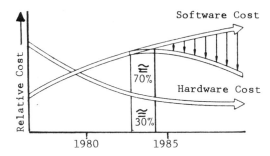

Fig. 14. Trends in cost to automate a process.

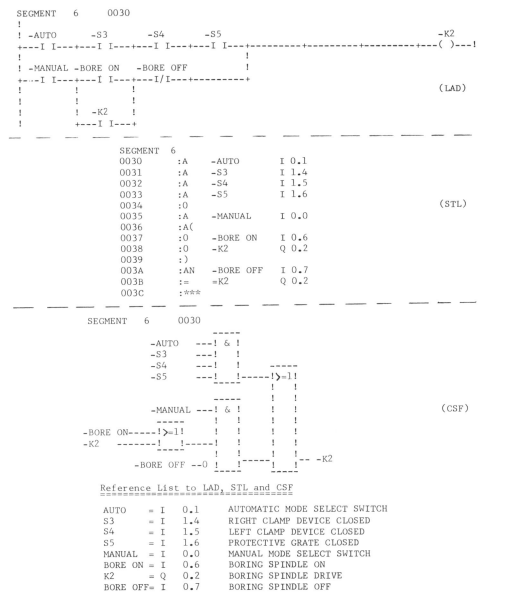

Fig. 15. Comment and symbolic identifier.

with *STEP*™*5*. Software and its generation are today an important topic due to the disproportional software and hardware costs for process automation. While hardware costs have decreased with the event of microelectronics, software costs have been increasing steadily. One reason for this is that software tasks are personnel intensive. Software costs have been rising in proportion to the commonly experienced increase in skilled personnel costs. Another reason, of course, is the growing trend to use automated equipment in lieu of the prior labor-intensive factory operations. (See Fig. 14.)

In addition to such functions as controlling, monitoring, and processing, the error diagnosis for automated processes and the operation and visual control of devices are becoming increasingly important. Currently, programs are generated where the programming portion required for the actual task literally disappears with respect to the costs incurred for diagnosis, operation, and visual control of the processes/machines to be automated. Therefore, the rationalization of software is a major concern.

Especially in the area of *documentation*, the supporting effect of performance-oriented software in connection with programming a system can be observed—explanatory texts with respect to individual networks and symbolic identification in place of absolute parameters. (See Fig. 15.)

Sequence Controllers—Hardware/Software Trends

By Erich Sulzer[1] and Engineering Staff, Siemens Aktiengesellschaft[2]

The application of the function diagram, as described in the preceding article in this *Handbook*, is further supported by the possibility of projecting and programming a sequence control simultaneously by means of supplemental graphic elements.

The logical control technology still widely used today is increasingly influenced by the need for providing rapid and transparent diagnostic means. Traditional controls based on relay technology require that all interrelated means for driving and interlocking actuators, drives, etc., are presented statically; that means the relation of temporal sequences and the connection and disconnection of control signals are inseparably tied together in this mode of presentation. In the case of logical control systems it is, therefore, difficult to rapidly picture and recognize the sequential or simultaneous conditions in the event of a malfunction. As a result, diagnostic programs that could be designed with a logical control through utilization of programmable controllers are very extensive.

The situation is fundamentally different in the case of sequence control. With the increasing utilization of sequence control, an additional possibility for the description of the process interlocking in an installation is described here—the *temporary sequence*.

In the event of a malfunction, the special structure of the step sequence provides rapid diagnosis inasmuch as only the defective step contains the continuing interlocking conditions in a temporary unequivocal manner. As a result, diagnostic programs that automatically display the cause of failure (as by a monitor or printer) can be added to sequence control programs written for programmable controllers. (See Fig. 1.)

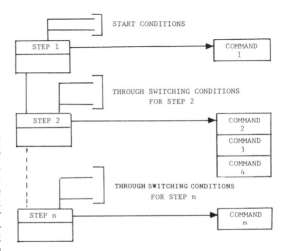

Fig. 1. Sequence control in an earlier method of presentation.

GRAPH LANGUAGE "GRAPH 5"[3]

GRAPH 5 is a graphic language that enables the engineer to design and program a sequence control rapidly. This language divides the sequence control into two methods of display: (1) the *overview presentation* of the structure of the steps, and (2) the *detailed* or *magnified display* of the transition or step (See Fig. 2.) The actions are defined in the steps that present the processing status during program processing. A programmable interlocking capability is assigned to each step—called *transition*. This concept has been adopted internationally. Transitions present the interlocking conditions to the assigned step. The

[1]Siemens Energy & Automation, Inc., Programmable Controls Division, Peabody, Massachusetts.
[2]Erlangen, West Germany. Systems described here are marketed in the United States by Siemens Capital Corporation, Iselin, New Jersey.
[3]Developed by Siemens.

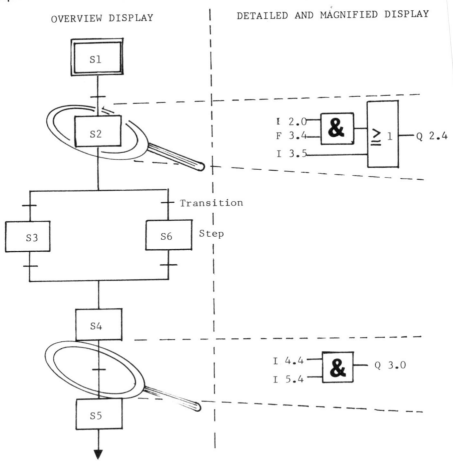

Fig. 2. Overview and detailed presentation of a sequence control in *GRAPH 5*.

new graphical language allows the numerous different structural possibilities, as outlined in the following paragraphs.

Linear Sequence. The chain (see Fig. 3) is the simplest basic structure. It is a series of steps with transitions between them. If a step (S1) is enabled and the next following transition (T1) is performed, the step (S1) is reset and the subsequent step (S2) is set.

Simultaneous Branching. Also called AND branching (see Fig. 4), simultaneous branching presents two or more branches that are started concurrently by a common transition (T3). The branches are simultaneously processed like linear chains.

Synchronization. Another common transition (T5) reconnects the independently running chains when all parallel branches have completed their last step and the transition (T5) has been performed. These steps are reset simultaneously, and a common subsequent step (S6) is set.

Alternative Branching. In the case of alternative branching, also called OR branching, a step (S7) is followed by a number of chains, each of them starting with a transition (T6, T8). (See Fig. 5.) Only one of these transitions can reset (S7) and set its subsequent step (S8 or S9). The left transition has priority in the event that *both are performed simultaneously*. At the end of the alternative branching, the chains are continued by a common step (S11) through various transitions (T7 or T10).

Jump Function. The *jump* (see Fig. 6) is a further important basic function. A transition (T14) can, without providing a graphical connec-

Sequence Controllers—Hardware/Software Trends 203

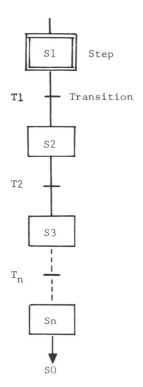

Fig. 3. Linear chain.

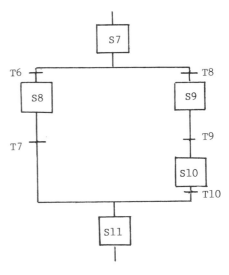

Fig. 5. Alternative branching.

tion in the diagram, lead to a specific step (12). Through the use of jumps, loops can be programmed.

Initial Step. The initial step is activated without process interlockings after the start of the sequence program.

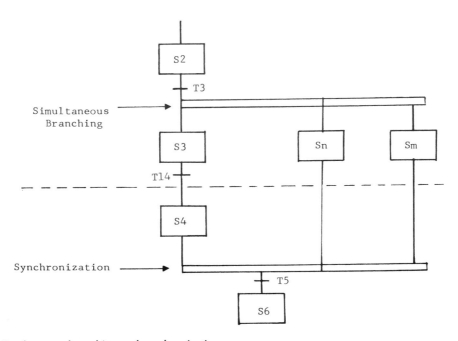

Fig. 4. Simultaneous branching and synchronization.

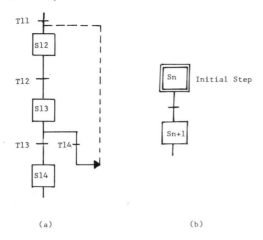

Fig. 6. (a) Jump function, (b) initial step.

Wait Time and Monitoring Time. Every step has a *wait time* and a *monitoring time*. The wait time is the minimum run time of the step and will be enabled even though the subsequent transition has been previously completed. The monitoring time checks whether the next switching condition will be performed within a specific time period. A comment can be presented for each step and each transition. These comments are presented in the overview as well as in the respective detailed diagram. (See Fig. 7.)

Advantages of *GRAPH 5* include: (1) The same method of presentation used for projecting and programming this method also provides comments and symbolic addresses and can be used as final documentation. The most important characteristic is the simplicity of the diagnostic capability, which provides rapid assistance during malfunctions. In addition, the structure facilitates the insertion of diagnostic software packages for the display of malfunction messages (as on a monitor

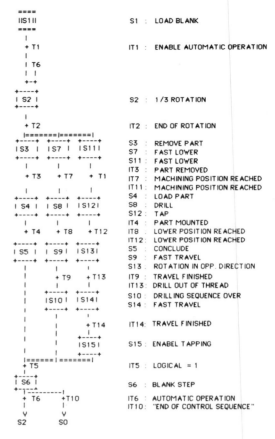

Fig. 7. Sequence control as document example.

or printer). (2) The language is able to present the three methods of the *STEP®5* language, such as "CSF," "LAD," and "STL." *GRAPH 5* is an example of a higher-level language. (See Fig. 8.)

THE LADDER DIAGRAM (LAD)

Since the ladder diagram provides for visual clarity and incorporates previously used documentation means, it is especially suitable for low-performance applications or for users who are equipping small installations with programmable controllers (PCs) for the first time.

In these areas of application it is advantageous to offer a programming method that makes the widest possible use of circuit-diagram know-how within the user's facility. The ladder diagram only partially performs this requirement inasmuch as it constitutes a graphical interpretation of the statement list (STL) of the *STEP 5*[4]

[4] A specific programming language developed for Siemens PCs.

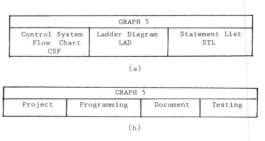

Fig. 8. (a) *GRAPH 5* as a comprehensive programming method, (b) as an advanced project design language.

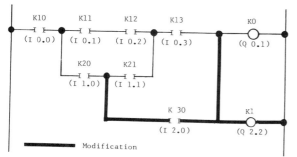

Fig. 9. Initial network of circuit diagram with the indicated changes. The absolute addresses are shown in parentheses.

language, which contains certain language rules that ultimately relate back to Boolean algebra. For example, OR functions in a combination of OR before AND functions must be the various logic possibilities and must be solved by auxiliary flags. In the application of PCs, these language rules present a barrier to understanding. This is eliminated when using the language ladder diagram (LAD).

The following example illustrates this by way of circuit-diagram modifications and realizations in the ladder diagram as compared with the relay circuit diagram (R-LD).

The network shown in Fig. 9 is completed with a modification. The network and the modification is shown in ladder diagrams (LAD) and in the actual statement list (STL). (See Fig. 10.) This illustrates the essence and characteristics of the relay ladder diagram (R-LD) for this program modification.

When simulating the modification (dotted line) as graphic display in the ladder diagram, the modification is rejected by the PC. This is because the language (LAD) and also the graphical interpreter must correspond to Boolean language rules.

In practical terms, this means that the undefined combinations in the statement list (STL) to be presented in the graphic mode must be eliminated by the application of *parenthesized commands* or *flags* as intermediate results. Thus, it is made certain that a network can be created in LAD or that it can be converted into LAD.

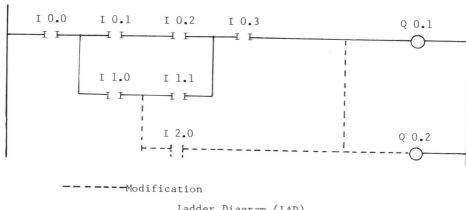

Ladder Diagram (LAD)

```
Statement List (STL)
(without the modification)
    :A      I 0.0
    :A(
    :A      I 0.1
    :A      I 0.2
    :O
    :A      I 1.0
    :A      I 1.1
    :)
    :A      I 0.3
    :=      Q 0.1
```

Fig. 10. Initial structure as ladder diagram instruction list (modification shown as dotted line).

```
!                                                                               !   Q 0.1   !
!   I 0.0     I 0.1     I 0.2     I 0.3                                         !           !
+---I I---+---I I---+---I I---+---I I---+---------+---------+---------+---( )---!
!         !         !         !                                                 !           !
!         !         !         !                                                 !   Q 0.2   !
!         !   I 1.0     I 1.1     I 0.3                                         !           !
!         +---I I---+---I I---+---I I---+                                       +---( )---!
!                   !                   !                                                   !
!                   !    I 2.0          !                                                   !
!                   +---I I---+---------+                                                   !
!                                                                                           !
```

Ladder Diagram (LAD)

```
Statement List (STL)
    :A     I 0.0
    :A(
    :A     I 0.1
    :A     I 0.2
    :A     I 0.3
    :O
    :A     I 1.0
    :A(
    :A     I 1.1
    :A     I 0.3
    :O     I 2.0
    :)
    :)
    :=     Q 0.1
    :=     Q 0.2
```

Fig. 11. Solution utilizing the language rules and the standard command set.

The transparency of graphical languages leads to a situation in which certain "grammatical rules" in the initial language would have been violated.

The solution with the standard command set of STL and its presentation in LAD is shown in Fig. 11.

In the graphical solution it becomes immediately evident that the input I 0.3 is displayed twice as opposed to what preceded, which of course is the result of the language rules. As can be seen from the highly complex configuration, the STL solution already contains the term I 0.3 twice to facilitate a clear definition.

Owing to practical considerations, however, often the complete and highly efficient range of the language is not utilized. Instead, a program problem is divided into a number of small, rapidly visualizable segments. Here the versatility and flexibility of the ladder diagram (LAD) offers a series of helpful constructions that assist in solving the problem in a highly graphical manner.

Figure 12 demonstrates the system of segmentation of the problem by usage of flag functions for storing intermediate binary results. The flags are sequentially used to solve the problem, with a final instruction and graphic instruction sequence. This method is quite simple and offers numerous similar variations, depending only on the creativity of the programmer. With a second method, all single rungs are identified that can energize the outputs. The rungs are in parallel mode, as shown in Fig. 13.

```
SEGMENT  1
!
!   I 0.0      I 0.1      I 0.2      I 0.3                                              F 1.0  !
+---I I---+---I I---+---I I---+---I I---+----------+----------+----------+----------+---( )--!
!
SEGMENT  2
!
!   I 0.0      I 1.0                                                                    F 1.1  !
+---I I---+---I I---+----------+----------+----------+----------+----------+----------+---( )--!
!
SEGMENT  3
!
!   F 1.1      I 2.0                                                                    F 2.0  !
+---I I---+---I I---+----------+----------+----------+----------+----------+----------+---( )--!
!
SEGMENT  4
!
!   F 1.1      I 1.1      I 0.3                                                         F 3.0  !
+---I I---+---I I---+---I I---+----------+----------+----------+----------+----------+---( )--!
!
SEGMENT  5
!
!   F 1.0                                                                               Q 0.4  !
+---I I---+----------+----------+----------+----------+----------+----------+----------+---( )--!
!         !                                                                             !
!   F 2.0 !                                                                             !  Q 0.2 !
+---I I---+                                                                             +---( )--!
!         !                                                                             !
!   F 3.0 !                                                                             !
+---I I---+                                                                             !
```

Ladder Diagram (LAD)

```
        Statement List (STL)        STL (continued)
    SEGMENT 1                   SEGMENT 4
            :A    I 0.0                 :A    F 1.1
            :A    I 0.1                 :A    I 1.1
            :A    I 0.2                 :A    I 0.3
            :A    I 0.3                 :=    F 3.0
            :=    F 1.0                 :***
            :***                    SEGMENT 5
    SEGMENT 2                           :O    F 1.0
            :A    I 0.0                 :O    F 2.0
            :A    I 1.0                 :O    F 3.0
            :=    F 1.1                 :=    Q 0.4
            :***                        :=    Q 0.2
    SEGMENT 3
            :A    F 1.1
            :A    I 2.0
            :=    F 2.0
```

Fig. 12. Solution in ladder diagram (LAD) with flag functions.

```
! I 0.0      I 0.1      I 0.2      I 0.3                                          Q 0.1 !
+---I I---+---I I---+---I I---+---I I---+----------+----------+---------+---( )--!
!   I 0.0      I 1.0      I 1.1      I 0.3   !                           !         !
+---I I---+---I I---+---I I---+---I I---+                                +---( )--!
!                                        !                                         !
!   I 0.0      I 1.0      I 2.0          !                                         !
+---I I---+---I I---+---I I---+----------+                                         !
!
```

Ladder Diagram (LAD)

```
Statement List (STL)
    :A      I 0.0
    :A      I 0.1
    :A      I 0.2
    :A      I 0.3
    :O
    :A      I 0.0
    :A      I 1.0
    :A      I 1.1
    :A      I 0.3
    :O
    :A      I 0.0
    :A      I 1.0
    :A      I 2.0
    :=      Q 0.1
    :=      Q 0.2
```

Fig. 13. Solution in ladder diagram (LAD) in parallel mode.

INTERACTION OF HARDWARE AND SOFTWARE

The conversion of a circuit diagram into a ladder diagram is subject to a number of critical considerations because a direct conversion is not possible. The following conversion rules should be considered to ensure a problem-free conversion of circuit diagrams into programmable controllers.

When converting a circuit diagram (see Fig. 14) into a ladder diagram, the logical signal processing of the relay control is replaced by a PC. The external signals are retained, and connected to the inputs of the PC. The relays and timers are replaced by internal software functions integrated within the PC. This becomes an important factor in deciding to increase the usage of PCs to reduce costs. Of course, relays will continue to be required when the current of the digital output is not adequate to drive starters or valves, etc. directly.

The example shown in Fig. 14(a) was converted into separate functional areas. Figure 14(b) provides a better overview of the external hardware circuitry and the internal signal processing by usage of flags and timers.

The implementation of signal processing does not pose a problem for a PC because program languages support the implementation of relay/contactor controllers and provide additional functions, such as flags, timers, and counters, that provide hardware cost reduction.

Conversion rules for internal signals into internal processing are shown in Fig. 15. In summary, the following assignments may be made:

Relays ....are replaced with flags

Hardware timers......are replaced with software timers

Contactors, etc. ......are replaced with outputs

On the hardware side, the cost reduction due to the elimination of relays becomes immediately evident, while on the external signal side, all signals are retained.

The situation is different with external signals (inputs), i.e., pushbuttons, contacts, etc. For the inputs of the PC, it makes no difference in terms

Sequence Controllers—Hardware/Software Trends   209

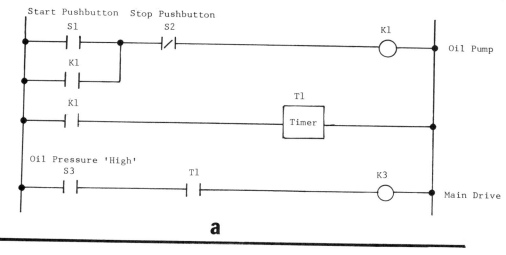

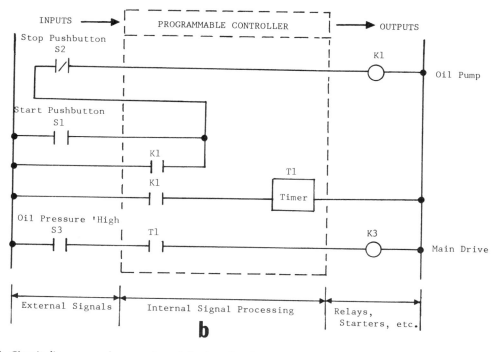

Fig. 14. Circuit diagrams and conversion of the functional areas.

of hardware, whether the signal status "1" is activated from a normally open contact or from a normally closed contact. However, in terms of the software function, the logical purpose of this condition has its own meaning. This circumstance also involves conversion rules, which are explained by the following description. In order to explain what has been covered, Fig. 16(a) and (b) show a separation of the hardware wiring of the PC and its associated software. In order to illustrate the conversion of the example in Fig. 14, the ladder diagram (LAD) is used. In order to ensure the same function as shown in Fig. 14, the seal-in function (parallel branch contact K1 to the start pushbutton) has to be converted into a set and reset flag function (F1 because it uses a

210 Sequence Controllers—Hardware/Software Trends

INTERNAL SIGNALS		INTERNAL PROCESSING	
Explanation	Symbolic	LADDER DIAGRAM (LAD) Symbolic	STATEMENT LIST (STL)
Normal Open Contact	* ─/─ ─┤ ├─	─┤ ├─	A I 2.2 O I 2.2
Normal Closed Contact	* ─/─ ─┤/├─	─┤/├─	A I 2.3 O I 2.3

* European format

Fig. 15. Conversion rules for internal signals into internal processing.

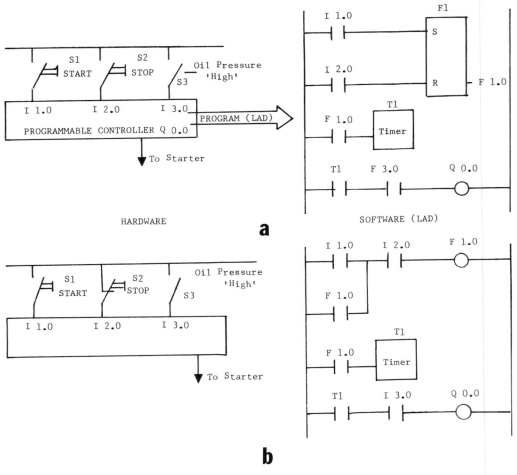

Fig. 16. Conversion of example of Fig. 14(a) with stop pushbutton as normally open contact; (b) as normally closed contact.

normal open contact for the stop pushbutton (See Fig. 16[a].)

On the software side, the similarity to the original diagram is evident. See Fig. 16(b). However, the STOP contact I 2.0 does not appear as the original normal closed contact but as a normally open contact instead. (See Fig. 16[b].) How does this discrepancy arise that prevents the direct conversion circuit diagrams into ladder diagrams?

In order to explain this discrepancy, an example with all possibilities of the usage of external signals and their various applications is given here. A technical schematic is shown in Fig. 17, which shows a hydraulic pusher. To measure the temperature of the hydraulic oil, a temperature limit switch is used. The temperature must exceed a minimum to avoid overloading the pump motor. The control of the hydraulic system is not shown. To reach the maximum position on the right, a position switch (S2) is provided.

To ensure proper control of the hydraulic pusher, operation interlocking, start-up interlocking, and protective interlocking functions have to be used. The operation interlocking provides protection of the hydraulic pusher during the entire operation even if a cable disconnection occurs. For this purpose, a "normally open contact" of the external signal has to be used. The start-up interlocking provides protection of the pump motor until a minimum oil temperature is reached. For this purpose and to protect the pump motor against broken cables, a "normally open contact" of the external signal has to be used.

The protective interlocking provides protection during the entire operation even if a cable disconnection occurs. For this purpose, a "normally closed contact" of the external signal has to be used. (See Fig. 18.)

In normal operation, the object of the control is to actuate the contactor K2; if the position switch is not reached, minimum temperature is exceeded and no STOP actuated.

In Fig. 19(a), the position switch S2 would switch off the contactor K2 if the end position is

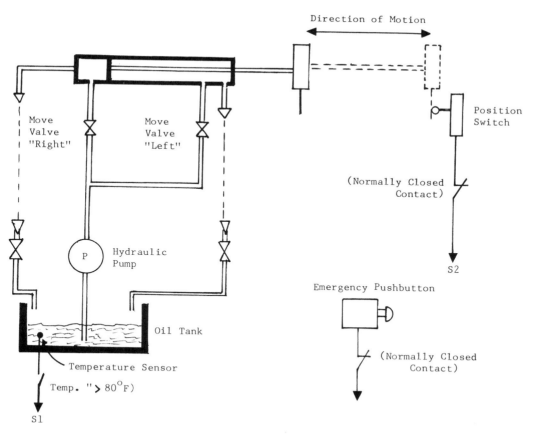

Fig. 17. Technical schematic for the example of a hydraulic pusher.

212 Sequence Controllers—Hardware/Software Trends

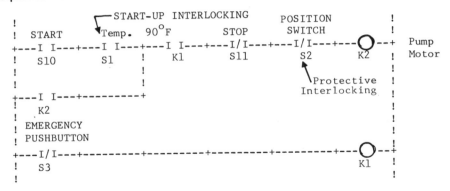

Fig. 18. Circuit diagram for technical schematic of Fig. 17.

reached. As long as the minimum temperature S1 has not been reached, the contact S1 will not release the contactor K2.

In Fig. 19(b), the position switch S2 upon reaching the end position will switch off the contactor K2, or hold until the minimum temperature is not reached.

The conversion rules used are not apparent inasmuch as the conversion of the external signal S3 in LAD is, unexpectedly, carried out with the symbol. In fact, S3 is shown differently. (See Fig. 19 [hardware].)

The meaning of this conversion rulse is contained in the signal function and in the hardware execution as shown in Fig. 20. The *aim of the ladder diagram* is to provide a positive logical status signal to energize the output and to carry out the function of the hydraulic pusher. Consequently, both conditions shown in Fig. 20 can be programmed in positive logic.

Generalizing the table and using the following definitions:

External Condition   Closed: +1
                                              Opened: −1
Logic   Function Enabled: +1
                                            Function Disabled: −1

Table 1 illustrates the general conversion rules for external signals for the programmable controller.

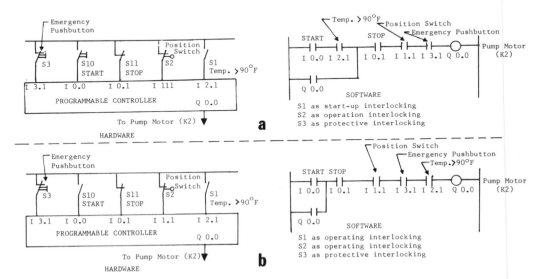

Fig. 19. Representation of circuit diagram of Fig. 18 as a ladder diagram (LAD). (a) With start-up interlocking, (b) without start-up interlocking.

Sequence Controllers—Hardware/Software Trends 213

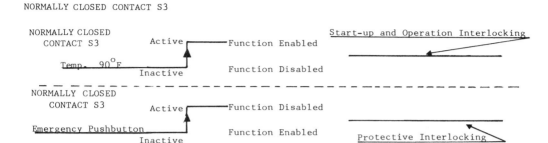

Fig. 20. Overview diagram for the various external signal conditions in connection with the signal function.

PROGRAM PROCESSING

The cycle time is used as a criterion to measure the reaction of a PC to a rapidly changing input signal. The basic cycle time provides information on how fast 1K (1024) binary statements (AI, OI, AF, AQ, etc.) are being processed. The cycle time evidences a common characteristic of PCs—the serial processing of statements which are cyclically repeated. (See Fig. 21.)

STRUCTURE OF A STORED PROGRAM

A stored program control system consists of a PC together with externally connected activating contacts or sensors on the input side and actuating elements or annunciators on the output side.

The PC comprises several units essentially consisting of a central processing unit (CPU), which includes the processor and the program storage module, a number of input/output (I/O) modules, and a power supply.

The external activating contact or sensor voltages are applied to the input modules of the PC (input terminal connections). The processor in the CPU then executes the program, which is stored in the memory and thereby scans the individual inputs of the PC for the presence or absence of input voltage. Depending on the scan result obtained and the program in the memory, the processor then issues command signals to certain output modules, causing voltages to appear (or vanish) at the respective output terminals of the PC. The actuating elements or annunciators connected to these output terminals are, therefore, energized (or deenergized) as their output module is either switched on or off.

The power supply converts the main voltage to a voltage level of 5V DC required by the electronic modules that make up the PC.

The external I/O devices (activating contacts or sensors and actuating elements or annunciators) require a higher voltage level (e.g., 24V DC or 115V AC), which is provided by a separate power supply unit or isolating transformer.

Program Counter. During program processing by the CPU, each individual memory location is addressed sequentially via the program counter. The programmed statement, which is located in the memory location addressed by the program counter, appears instantaneously at the memory output gates to be transferred to an intermediate register known as the statement register. The statement contained in this register is then executed by the processor. Subsequently, the program counter is incremented so that the immediate following memory location is addressed. (See Fig. 22.)

Detecting Errors. The duration of the cycle time is monitored by the processor of the CPU (watch dog). Should the processing cycle extend

TABLE 1. Logic (multiplication) Table for the Conversion of External Signals for a Program Controller

Switching condition			Signal function		PC processing
Normal open contact	Close (+1)	●	Function Enabled (+1)	=	+1 Positive
	Open (−1)	●	Function disabled (−1)		⊣⊢
Normal closed contact	Close (+1)	●	Function Disabled (−1)	=	−1 Negative
	Open (−1)	●	Function enabled (+1)		⊣⊢

214 Sequence Controllers—Hardware/Software Trends

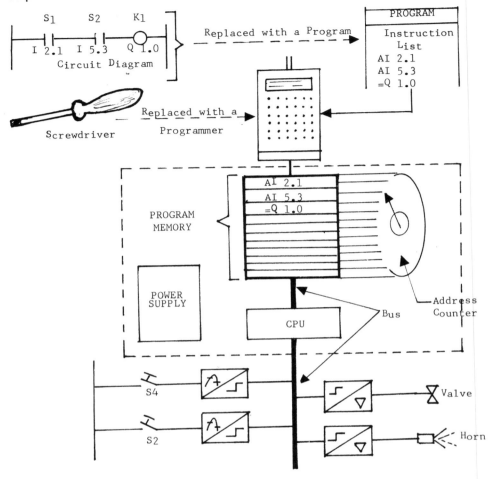

Fig. 21. Principle of a programmable controller (PC) with cyclic processing.

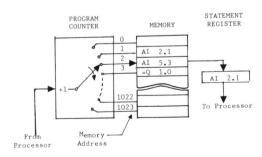

Fig. 22. During program processing by the central processing unit (CPU), each individual memory location is addressed sequentially via the program counter.

beyond a certain time interval, then either a programming error (software) or a machine error (hardware) may be the cause. This has the effect of stopping program processing and switching off all output modules of the PC.

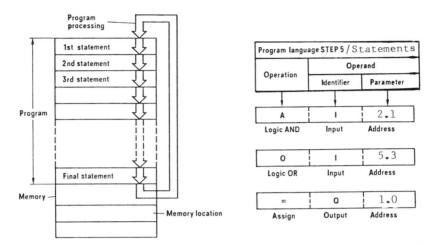

Fig. 23. The description of the control task to be performed by a stored-program control system needs to be broken down into individual statements.

Expert System and Model-Based Self-Tuning Controllers

By Peter D. Hansen[1] and Thomas W. Kraus[2]

The necessity of self-tuning[3] controllers is best illustrated by an example. Consider a heat exchanger that uses saturated steam to heat water that flows through its tube bundle. A simple control scheme senses the outlet water temperature and attempts to position the steam valve so that the actual water temperature equals the desired water temperature. Effects of both nonlinearities in the steam valve and changing steam pressure can be reduced by using a second control loop to control the steam flow.[4] The slower-acting temperature controller now adjusts the set point of the faster-acting steam flow controller. Unfortunately, a fixed parameter temperature controller has difficulty because of the nonlinear, time-varying behavior of the process. A change in the water flow rate changes the effective delay time and heat transfer characteristics of the process. Gradual fouling of the heat exchanger tubes also changes the process dynamics over time. Good control performance at one operating condition can give way to very poor performance (overdamped or unstable response) at another operating condition.

In general, time-varying process dynamics, variable operating conditions, nonlinear process dynamics, and lack of expertise during control loop commissioning have all led to an interest in *self-tuning* controllers. Economic incentives are ultimately behind the thrust since control loop performance directly affects product quality, energy consumption, product yield, production rate, pollution control, and plant safety. The meaning of the term "self-tuning" is somewhat nebulous, especially when it is combined with the term "adaptive." Self-tuning or self-adaptive control implies the ability to learn about the closed loop process in real time. The controller initially tunes itself and remains tuned as the process dynamics and operating conditions change. The implied assumption underlying the theory is that the best present control strategy can be based upon past closed loop observations. Many schemes—such as dead-time compensation, gain scheduling, and feedforward—that have been labeled adaptive are actually preprogrammed adjustments, based upon measurable quantities. Such techniques are well suited to compensate for nonlinearities caused by actuator characteristics or easily modeled physical phenomena, and good design practice dictates that they should be used wherever possible. There is nothing to be gained by employing the complexity of self-adaptation to learn what is already known. There are many instances, however, where the causes of behavioral changes are either unknown, unobservable, or difficult to model. These are the situations to which self-tuning techniques are best applied.

Two entirely different self-tuning approaches have evolved and are commercially available, an expert system and a process model approach. The expert system uses heuristics or rules of thumb. Its goal is to achieve a desired control loop response by incorporating tuning rules used by control engineers to manually tune controllers plus additional rules discovered during field tests. Tuning changes result directly from the process response without any need to mathematically model the process. The control loop response is expressed in terms that describe its pattern. Examples of such pattern features are peak heights, period, slopes, frequency content, integrated areas, zero crossings, and other shape information. These are the same features that the eye detects when it scans a strip chart recording of the reference input (set point), measurement, and controller output. The discrete nature of the pattern characteristics easily combines with the tuning rules that are expressed in the IF-THEN-ELSE format. The control structure usually chosen is a discrete form of the Proportional+Integral+Derivative, PID, controller because of widespread industrial acceptance and a long history of tuning experience.

The model-based, self-tuning approach depends entirely upon a process model. Its goal is to achieve a desired control loop response by updating

---

[1] The Foxboro Company, Foxboro, Massachusetts.
[2] Intec Controls Corp., Foxboro, Mass.
[3] Also called *self-adaptive* control and described later in this article.
[4] This type of *cascade control* is commonly used in process control.

coefficients in a process model and using the coefficients to calculate the control parameters. If the model is appropriately constructed, the calculation is simple, since the model coefficients can be directly used in the controller. The desired response is expressed as a transfer function that relates the measurement to the reference input and process load disturbance. The identified process model coefficients minimize the mismatch between the actual process response and the model response to measured inputs. This model-based approach is flexible enough to accommodate a wide variety of parameter identification techniques and controller design strategies. The complexity of the chosen control structure often necessitates self-tuning, since it is almost impossible to manually tune a controller having more than three adjustments.

Both self-tuning approaches have their own advantages and disadvantages. Although it is usually difficult to get universal agreement because of subjective reasons, the comparative issues are presented here to allow a look into the future. The advantages of the expert system arise since it is extremely robust and additional rules can be easily added. It is also easy to apply because it is designed to mimic manual tuning rules that are understood by control engineers and technicians. Furthermore, it responds directly to the quality of the closed-loop process response and does not require a process model. On the negative side, the expert system is engineered for a particular controller structure, and its rule base makes it impossible to analyze mathematically. It prefers single-event, unmeasured loop disturbances and may be more dependent upon the nature of an arbitrary unmeasured disturbance.

The model-based approach, on the other hand, has a rigorously defined performance criterion, and mathematical analysis is possible. Also, the process model allows the flexibility of implementing different controller structures and may be used for process diagnostics, such as detection of sensor or actuator failure. On the negative side, the model structure may not allow a match with the process, and final actuator dead zone or backlash causes underestimation of process gain. An estimate of the process dead time is needed, and the model sampling time is critical. Too short a sampling interval leads to a model with too many parameters, and too long an interval results in sluggish control behavior. Rapid process changes and unmeasured load disturbances, even single-event disturbances, may give problems. Finally, it is difficult to develop a weighting scheme for past errors and to factor the model into its controllable and uncontrollable (i.e., delay time and nonminimum phase zero) parts.

Not one of these problems is insurmountable, however, since many successful applications are installed and running. The controller of the future will not be a plug-in black box because new processes tend to continually require more sophisticated controls, and, at the same time, specifications become tighter.

The controller of the future will, however, incorporate both the expert system technology and the model-based technology. The two approaches will become one as their unique characteristics are blended for the common goal of process control.

EXPERT SYSTEM CONTROL

The first method of achieving self-tuning control is to formalize the rules used to manually tune controllers. The art of manually tuning controllers based on pattern recognition, has evolved over a number of decades. The control engineer either disturbs the closed loop by making a set point change or a load change, or else he will wait for a natural disturbance. Based upon the closed-loop response pattern to the upset, the controller tuning parameters are adjusted. This manual tuning procedure does not utilize a process model because, in general, processes are nonlinear, time varying, and noisy, and they often have discontinuous regimes. As a result, processes are almost impossible to describe mathematically.

Figure 1 shows a block diagram of an expert system approach to self-tuning. The knowledge-based rules are the formal implementation of the manual rules just described. The inputs for these rules are the set point, measurement, controller output, controller output limits, and the current tuning parameters. Since a process model is not used, the tuning rules must be tailored to a particular controller structure, such as the common proportional + integral + derivative (PID) controller.

Pattern Recognition

There are two hurdles that must be overcome in building such an expert system. First, the adapter must recognize the pattern of the closed-loop response. Second, after recognizing this pattern, it must be able to take proper control action. Figure 2 shows a transient response of error versus time for a load change. The dominant pattern features of this response are the peak heights shown as E1, E2,

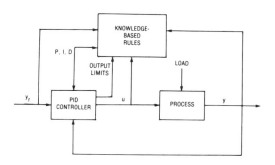

Fig. 1. An expert system approach to self-tuning.

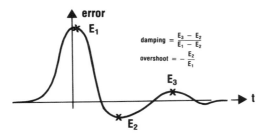

Fig. 2. Pattern recognition characteristics.

and E3. In a specific expert system configuration, these peak heights are normalized to define two variables, the overshoot and the damping (Refs. 1, 2, and 3). Note that the value of 1.0 for both these parameters indicates neutral stability, and the value of 0.0 represents an overdamped response. The damping is independent of the zero error line, wheras the overshoot depends strongly on this line. As shown in Fig. 3, another parameter is necessary to completely define the response pattern. A single process with different controller settings can produce two responses that are very similar in terms of overshoot and damping, but different in timing. The response period is the third pattern feature needed to distinguish the response. The period is the time between the first and third peaks, and it is normalized with the integral term of the controller and with the derivative term of the controller to produce the ratios, integral/period and derivative/period.

These two time parameters are similar to the Ziegler and Nichols tuning ratios, but there are two important differences. First, this period includes the effect of the controller integral and derivative terms. That is significant because the integral and derivative terms of the controller change the closed-loop period. The second difference is that the optimal integral and derivative ratios change depending upon the process characteristics. Processes having a large amount of dead time require smaller ratios whereas processes having more lag require larger ratios. The optimal integral and derivative ratios are adjusted automatically based upon the pattern of the process response.

Figures 4, 5, and 6 show contour plots of damping, overshoot, and integral/period for one specific process. The axes of these plots are the proportional term and the proportional term multiplexed by the integral term. The contours represent lines of constant damping and overshoot. The similarity of the damping and overshoot contour shapes indicates that these two parameters are not independent and that the same shape can be achieved from a number of different controller tuning settings. The knee in Figs. 4 and 5 represents the good tuning area. Note that the damping and integral/period contour plots are orthogonal in this area, which suggests rapid convergence and unique controller tuning parameters produce the desired response pattern.

Knowledge-Based Rules

Up to this point, the response pattern is defined in terms of peak heights and period. These pattern features are normalized to produce damping, overshoot, integral/period, and derivative/period. This information is built into an expert system via knowledge-based rules, as shown in the state diagram of Fig. 7. The quiet state corresponds to a mode where the measurement is close to the set point and new process information is not available. The adapter begins to watch the closed-loop response when a process disturbance causes the measurement to drift away from the set point. Simultaneously, the controller, with its parameters fixed, takes action based upon the changing measurement. The adapter locates and verifies Peak 1, locates and verifies Peak 2, and locates and verifies Peak 3. The time it takes to observe these three peaks is highly process dependent. However, once the three peaks are observed, it quickly moves into the adapt state and adjusts the tuning parameters, if necessary.

The new tuning parameters are adjusted in two steps. First, the integral and derivative terms are set based upon the desired integral/period and derivative/period ratios. Because of parameter interaction, the proportional term is adjusted to compensate for the integral and derivative changes. Second, the observed damping and overshoot are compared with the desired values. If distinct peaks have occurred and both damping and overshoot are less than the desired values, the proportional term is decreased. The amount of decrease depends upon either the difference between the desired and actual damping, or the difference between the desired and actual overshoot. Since damping and overshoot are not independent, the smallest (or most negative) difference is used. If distinct peaks

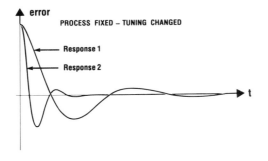

Fig. 3. Damping and overshoot are not independent.

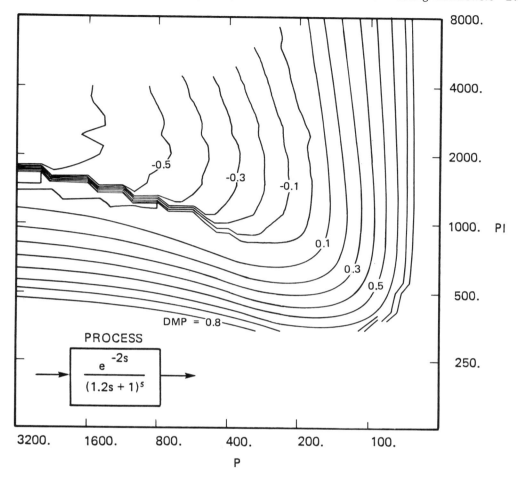

Fig. 4. Damping mapped into control space.

are not detected, all tuning parameters are decreased by an amount that depends upon the desired damping or overshoot.

After the adapt state, the process is controlled with new tuning parameters. The adapter then goes through a transient state called settle and back to the quiet state, where it is ready to repeat this procedure again as required. When the response is overdamped and peaks are not present, the rule base recognizes the lack of peaks and goes directly to the adapt state, as shown in Fig. 7. Also shown in Fig. 7, the dotted lines represent a large set point change that occurs during the response to a previous control loop disturbance. The set point change alters the response error pattern and causes the control algorithm to abort its current pattern search.

The vast majority of tuning rules lie within the adapt state. This state contains kernels of knowledge in the IF, THEN, ELSE, format. The rules represent small blocks of knowledge that work together to produce a very robust adapter. There are no hidden assumptions about the process or the controller. Stability is ensured by the direct performance feedback aspect of this design. If the loop is initially unstable, the adapter watches the oscillations and adjusts the controller's tuning parameters until the oscillation decays. The adapter moves the integral and derivative based on the period and primarily moves the proportional band based upon the peak heights and the desired response shape. Once proper tuning is achieved, subsequent process disturbances produce the desired response.

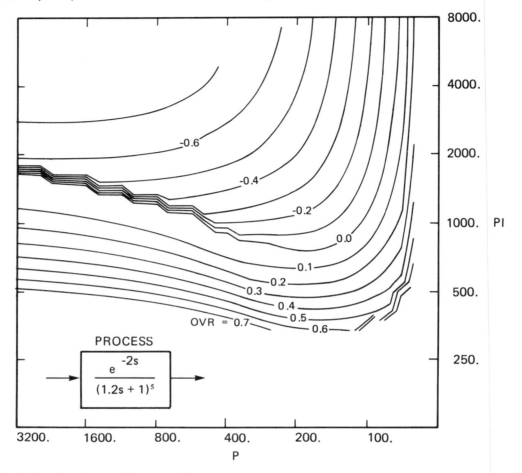

Fig. 5. Overshoot mapped into control space.

MODEL-BASED ADAPTIVE CONTROL

Automatic tuning of a controller can be based upon a process model whose parameters are continually updated to match the input-output behavior of the actual process. The structure of such an adaptive system is shown in Fig. 8. Both the Self-Tuning Regulator (Ref. 4) and some Model Reference Adaptive Systems (Ref. 5) are examples. In the next three sections, the issues relating to model-based adaptive control are discussed. First, the issues affecting controller design and loop performance are addressed and related to model structure. Next, various model structures for identification of the controller parameters are presented. Finally, the mechanics behind a number of identifiers are discussed.

Controller Design

Assume a linear process model with fixed but unknown parameters. These parameters are updated periodically and frequently so that the model may be assumed a good representation of the process. Several conditions must be met for these assumptions to be valid. First, for a nonlinear process, large changes in input quantities must occur much more slowly than small changes. Second, the process must not exhibit significant small signal nonlinearity such as is caused by a sticking-valve or dead-zone.[5]

As seen by a digital controller, the process model (including sample, hold, and anti-aliasing filter) is written using the time delay operator z^{-1} as:

$$A(z^{-1})y = z^{-k} B(z^{-1}) (u - u_0) + C(z^{-1}) e. \quad (1)$$

A, B, and C are polynomial operators with lead

---

[5] Often the process may be made to appear much more linear to the controller by employing local feedback loops and gain scheduling through valve positioners and cascaded ratio set flow loops. These techniques, although often crucial to good performance, will not be addressed here.

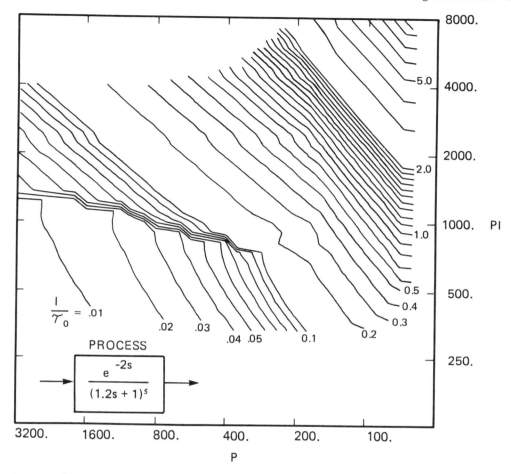

Fig. 6. 1/Period mapped into control space.

coefficients of $a_0 = 1$, $b_0 \neq 0$ and $c_0 = 1$. The process dead-time is represented as an integer number of sampling intervals, k. The measured process output is y, and the controller output is u. The bias, u_0, is the value of the controller output that would cause y to be zero in steady state. The unmeasured disturbance, e, is usually considered to be white noise in stochastic design and an impulse in minimum variance deterministic design. A bias term could have been included in e in lieu of u_0.

For the present, assume A, B, and C as known polynomials and consider various strategies for designing a linear controller of the form:

$$N(z^{-1})(u - u_0) = T(z^{-1})r - S(z^{-1})y. \quad (2)$$

The reference input (set point) is r. The polynomial $N(z^{-1})$ must have a stable inverse in order that the controller output, u, be finite for finite r and y. The closed-loop behavior is expressed by eliminating u from eqn. (1) and eqn. (2). The z^{-1} notation expressing dependence of the polynomial operators on the unit time delay operator is suppressed.

$$y = \frac{z^{-k} BT r + CNe}{AN + z^{-k} BS} \quad (3)$$

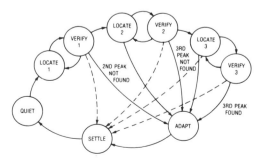

Fig. 7. State diagram.

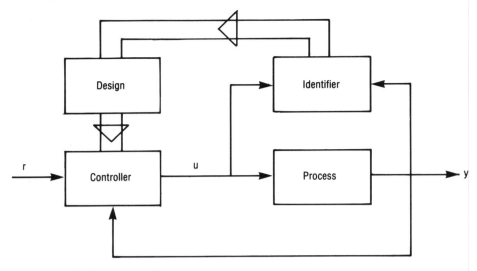

Fig. 8. Model-based adaptive control.

The general form of the control law, eqn. (2), does not automatically include integral action. However, steady state error can be removed without explicit integral action through the identifier-design feedback path of Fig. 8. A bias term may be included in the identified process model and then used to adjust the controller output bias, u_0. The effective integral time of this method depends on the identifier characteristics and may be far from optimally matched and not adaptively tuned to the process. Furthermore, this integral function is lost entirely should the identification be frozen.

Up to this point, the general process model, eqn. (1), and the general controller, eqn. (2), have been combined to form the closed-loop response, eqn. (3). A, B, C, and k are assumed to be known, but controller design dictates that N, S, and T must be calculated so that the actual closed-loop response, eqn. (3), matches the desired response of eqn. (4).

$$y = \frac{H(z^{-1})}{E(z^{-1})}(z^{-k}r + F(z^{-1})e) \qquad (4)$$

In this form H and E are specified and must have the same value in steady state to assure y approaches r. The process delay, z^{-k}, cannot be overcome and is therefore included in the desired response. A unique solution for N, S, and T cannot be found without a further restriction on F, N, S, or T. In the next five subsections, various options for this restriction are explored. All the restrictions that prespecify H and E can be considered model reference or pole-zero placement approaches.

1. *Open-Loop Control.* The first approach to controller design uses an open-loop control restriction ($S = 0$) to avoid any loop stability problems. Matching the reference portion of the responses in eqn. (3) and (4) yields $T/N = AH/BE$ which, when substituted into eqn. (2), yields a control law of the form:

$$u = \frac{AH}{BE} r + u_0 \qquad (5)$$

The zeroes of the target function, H, should be chosen to include the nonminimum phase (noninvertible) factors of B to ensure stability. The predicted process output response is obtained by combining eqn. (5) and (1) to produce:

$$y = z^{-k} \frac{H}{E} r + \frac{C}{A} e \qquad (6)$$

Note that the process must be open-loop stable, since the predicted process output disturbance response has A in its denominator. Reference (set-point) tracking can be very good if the controller output does not bound. However, recovery from a step load disturbance, which requires the bias, u_0, to be reidentified, may be very slow.

2. *Open-Loop Control with Feedback.* The second approach to controller design incorporates explicit integral action in a model reference control system by forcing $S = T$. The previous requirements that H include the noninvertible factors of B and that the process be open-loop stable apply.

Matching the reference portion of the responses in eqn. (3) and (4) yields $T/N = AH/(E - z^{-k}H)B$ which, when substituted into eqn. (2), yields an error-driven control law of the form:

$$u = \frac{AH}{B(E - z^{-k}H)}(r - y) + u_0 \qquad (7)$$

The steady state gain on $(r - y)$ is infinite since, as

mentioned earlier, $H = E$ in steady state. This allows u_0 to be omitted. The predicted process output response is obtained by combining eqn. (7) and (1) to produce:

$$y = z^{-k}\frac{H}{E}r + \left(1 - z^{-k}\frac{H}{E}\right)\frac{C}{A}e \qquad (8)$$

Again, reference tracking can be very good if the controller output does not bound, but recovery from a step load disturbance can be slow if the process polynomial A contributes a dominant lag uncancelled by a lead from polynomial C. Furthermore, the controller should incorporate an internal bound to prevent integrator "wind-up." Two equivalent control structures representing eqn. (7) and (8) are shown in Fig. 9(a) and (b). In the first, the positive feedback of the desired closed-loop response provides discrete integral action. In the second, the net feedback is the difference between process and open-loop model responses. In both, the forward path controller is the same as the open-loop controller. This is why we call this scheme open-loop control with feedback (it could also be termed pole cancellation or error-driven control). Others have called it or similar schemes Dahlin Control, Model Algorithmic Control, Dynamic Matrix Control, Inferential Control,

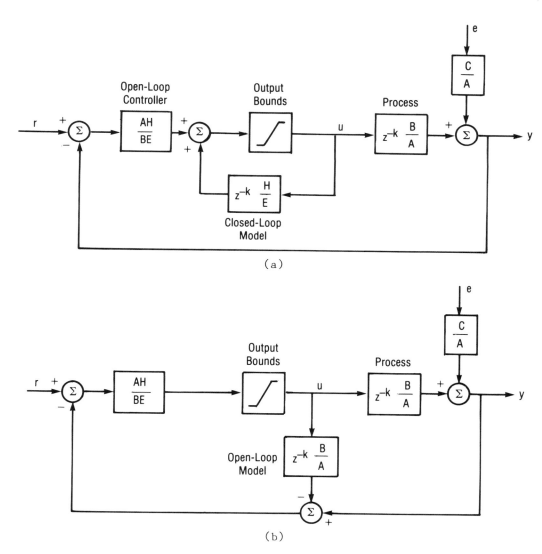

Fig. 9. Two equivalent open-loop control with feedback (or pole cancellation) structures: (a) Closed-loop model feedback; (b) Open-loop model feedback.

Internal Model Control, or Vogel-Edgar Control (Ref. 6). E is chosen to provide a low pass filter characteristic to improve the loop stability margin, with respect to high-frequency mismatch between the process and model and to reduce controller output activity due to process noise.

The Vogel-Edgar version of this scheme (Ref. 7), which makes $H = B$, is particularly interesting because it does not require factoring the model numerator, $z^{-k}B$, into delay, nonminimum phase (non-invertible) factors, and minimum phase (invertible) factors. No part of the model numerator is inverted. As a result, this method achieves implicit delay identification avoiding the known, fixed delay assumption inherent in many other model-based adaptive control systems. Because the delay is not explicitly identified, the method can be readily applied to multivariable situations that may have different delays between each input and output.

3. *Minimum Variance Control.* A third approach to controller design with explicit emphasis on disturbance rejection results in more feedback gain and faster recovery from bounds for a dominant lag process. This approach minimizes the expected variance, J, of the control loop error k time steps in the future (see eqn. 4) with respect to the present value of u, eqn. (1).

$$J = \underline{E}\left(\frac{E}{H}z^k y - r\right)^2 = \underline{E}\left(\frac{BE}{AH}(u - u_0) + z^k \frac{CE}{AH}e - r\right)^2 \quad (9)$$

($\underline{E}$ is the expectation function.) H/E is usually chosen to have a low-pass filter characteristic in order to compensate for measurement lag and to make the controller less sensitive to high-frequency noise and model mismatch.

The disturbance, e, is assumed to be white noise or a unit impulse. Its effect on the control loop error, CE/AH, is separated into two parts by performing polynomial division: one uncontrollable (the quotient, F) and the other controllable (the remainder, G)

$$\frac{CE}{AH} = F + z^{-k}\frac{G}{AH} \quad (10)$$

$F(z^{-1})$ is a finite polynomial function with terms up to $z^{-(k-1)}$. The criterion minimization with respect to u results when the controllable terms (everything except $z^k Fe$) are set equal to zero. This yields the control function:

$$u = \frac{CHr - Gy}{BFH} + u_0 \quad (11)$$

and the closed-loop process measurement:

$$y = \frac{H}{E}(z^{-k}r + Fe) \quad (12)$$

The self-tuning regulator and some model reference controllers use the minimum variance control strategy.

As with open-loop control, minimum-variance control relies on the identifier-design loop to provide effective integral action. Hence, steady state control will not be maintained when the adaptation is frozen. Furthermore, because G is not chosen arbitrarily, it cannot be assumed to contain any factors common to B. Therefore, if B contains a non-minimum phase factor, the control function is not stable. B is more likely to contain a non-minimum phase factor (resulting from sampling a continuous process) when the sampling interval is small (Ref. 8). Thus, it may be possible to avoid instability and at the same time accommodate variable-process delay by choosing a sampling interval so large that the model delay is always one sampling interval, $k = 1$. On the other hand, much tighter stable control might be achieved with a continuous or fast sampling suboptimal algorithm of the restricted complexity type described later.

4. *Minimum Variance with Detuning.* A fourth approach to controller design employs a manually adjusted detuning parameter, η, to stabilize a minimum variance controller and to reduce controller output activity (Ref. 9). This parameter appears in the modified minimum variance criterion function

$$J = E\left(\frac{E}{H}z^k y - r + \eta V u\right)^2 \quad (13)$$

where $V(z^{-1})$ is an arbitrary polynomial with a factor $(1 - z^{-1})$ to prevent steady state error. "Cautious" control results if V is chosen so that its magnitude becomes larger as the model uncertainty increases. The control and predicted closed-loop functions become:

$$u = \frac{CHr - Gy}{H(BF + \eta CV)} + u_0 \quad (14)$$

$$y = \frac{H(z^{-k}Br + (BF + \eta CV)e)}{BE + \eta AHV} \quad (15)$$

These same results can be shown to follow from a different criterion function that combines squared control error with a squared measure of controller output activity. Detuning may be capable of stabilizing a loop when the model delay, k, is incorrectly assigned and when B has a nonminimum phase factor. On the other hand, it may not be possible to achieve stability by adjusting η if the process is unstable open loop.

5. *Restricted Complexity Control.* Instead of designing the controller structure to fit the process model, the process model structure can be selected to accommodate a fixed controller structure such as the PI (proportional, integral) controller of Fig. 10 (Ref. 10). The minimum variance design approach is applied to the incremental ($\delta y_t = y_t - y_{t-1}$, etc.) restricted complexity model (α_0, β_0, k):

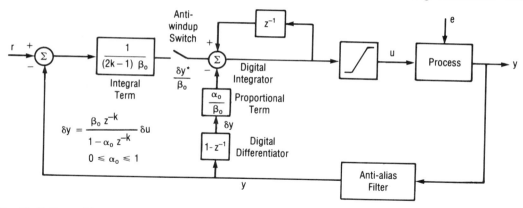

Fig. 10. PI Control loop design based on a restricted complexity model of the process.

$$\delta y_{t+k} = \beta_0 \, \delta u_t + \alpha_0 \, \delta y_t + \delta \epsilon_{t+k} \quad (16)$$

The term $\delta \epsilon_{t+k}$ is the incremental error caused by unmeasured disturbances and modeling mismatch. By setting the expected value of δy_{t+k} equal to the reference δy_t^*, an incremental proportional control law results

$$\delta u_t = \frac{1}{\beta_0} (\delta y_t^* - \alpha_0 \, \delta y_t) \quad (17)$$

To avoid positive feedback, α_0 must be positive. If the model actually matched the process, the closed inner loop would be a pure delay of k sampling intervals.

$$\delta y_t = \delta y_{t-k}^* + \delta \epsilon_t \quad (18)$$

An outer-loop controller can be designed, assuming that this inner-loop objective is achieved. For PI control, a robust outer-loop integral control strategy may be used.

$$\delta y_t^* = \frac{1}{2k-1} (r_t - y_t) \quad (19)$$

This approach yields a closed outer-loop reference response that approximates the response of a third-order Butterworth filter. If the model were an exact match for the process, the response takes the form:

$$y = \frac{z^{-k} r + (2k-1)(1 - z^{-1}) \epsilon}{z^{-k} + (2k-1)(1 - z^{-1})} \quad (20)$$

Because of the explicit integral action, provision (shown as a switch in Fig. 10) must be made to avoid integrator windup when the controller output bounds.

Eqn. (16) can be viewed as a digital model of a lag-delay process. Lumped with the delay are all of the process dynamics except the most dominant lag. By matching the low-frequency behavior of the model to a lag delay, the effective lag-to-delay ratio and the effective delay time are found to be:

$$\frac{\tau_{\text{lag}}}{\tau_{\text{delay}}} = \frac{\sqrt{\alpha_0}}{1 - \sqrt{\alpha_0}} \quad (21)$$

$$\tau_{\text{delay}} = \frac{k \, \delta t}{1 + \sqrt{\alpha_0}} \quad (22)$$

The computing interval is δt. For the model to be open-loop stable ($\tau_{\text{lag}} \geq 0$), $0 \leq \alpha_0 \leq 1$.

Identification Models

All of the controller designs of the previous section (except the restricted complexity type) were based on the process model of eqn. (1). An explicit (indirect) identification determines the parameter values of this model. If the model is restructured so that the controller parameters become the model parameters, the identification is called implicit (direct). In either case the model can be absolute or incremental. If it is incremental, it is not necessary to identify the bias, u_0.

1. *Explicit Model.* Explicit identification is most convenient for determining the parameters of the open-loop controllers (eqns. 5 and 7) because the H polynomial is not known in advance. H must include the non-invertible factors of B. For best performance, H should not include the invertible factors of B. If it is assumed that all of the mismatch between the model and the process is caused by a random white noise unmeasured disturbance, e, and that the process and unmeasured disturbances are stationary, the identification will be unbiased when the identifier inputs, y and u, are prefiltered by $1/C$. Usually C is not known and otherwise need not be identified since it is not involved in the open-loop controller functions. When the filter polynomial is D (with $d_0 = 1$) instead of C, convergence to the correct values for the A and B polynomials is still possible provided that a positive real (passivity) condition is satisfied in the frequency domain ($z^{-1} = e^{-j\omega\delta t}$) (Ref. 11).

$$\text{Re}\left(\frac{D}{C}(\omega)\right) > \frac{1}{2} \quad (23)$$

The explicit identification model can be put in the form:

$$\bar{\psi}_t = \phi_t^T \theta + \epsilon_t \quad (24)$$

where $\bar{\psi}_t$ is the filtered measurement,

$$\bar{\psi}_t = \bar{y}_t \quad (25)$$

ϕ_t is a vector of filtered process inputs and outputs

$$\phi_t^T = [\bar{u}_{t-k}, \bar{u}_{t-k-1}, --, -\bar{y}_{t-1}, -\bar{y}_{t-2}, --, 1] \quad (26)$$

and θ is a vector of parameters to be identified,

$$\theta^T = [b_0, b_1, --, a_1, a_2, --, e_0] \quad (27)$$

The bars indicate that the measured variables are filtered. If the identification error, ϵ, is entirely caused by the unmeasured disturbance, e, then:

$$e = \frac{D}{C}\epsilon \quad (28)$$

Note that the positive real condition, eqn. (23), requires that at all frequencies (ω) there be a component of the process noise, $e(\omega)$, in phase with and at least half as big as the identification error, $\epsilon(\omega)$.

The controller output bias is calculated from e_0 by comparing eqn. (24) with eqn. (1).

$$u_0 = -\frac{\Sigma d_i}{\Sigma b_i} e_0 \quad (29)$$

In summary, the explicit model, eqn. (24), allows identification of the A and B polynomials and the bias term, u_0. The latter is related to the identified bias, e_0, by eqn. (29). The identification process works best when the identifier is excited by white noise. This would be achieved by prefiltering the variables y and u by C, in effect, dividing each term in eqn. (1) by C to leave e, the white noise disturbance. Since C is not normally known, a prespecified D filter may be used instead.

2. *Implicit Model.* An implicit model form simplifies the design calculation for minimum variance types of controllers (eqn. 11 or 14). By defining

$$\psi_t = \frac{E}{H} y_t + \eta V u_{t-k} \quad (30)$$

$$\alpha(z^{-1}) = G \quad (31)$$

$$\beta(z^{-1}) = H(BF + \eta CV), \beta_0 \neq 0 \quad (32)$$

$$\gamma(z^{-1}) = CH - 1, \gamma_0 = 0 \quad (33)$$

and combining with eqns. (1) and (10) there results:

$$\psi = \frac{z^{-k}}{1+\gamma}(\beta(u - u_0) + \alpha y) + Fe \quad (34)$$

As mentioned earlier, V is assumed to contain the factor $(1 - z^{-1})$. In this form identification of γ is not easy because the model error, Fe, does not depend linearly on γ. In another model the α, β, and γ polynomials are linearly related to the model error.

$$\bar{\psi} = z^{-k}(\beta\bar{u} + \alpha\bar{y} - \gamma\bar{r} + e_0) + \epsilon \quad (35)$$

In this model, as in the explicit model, the process inputs and outputs are filtered by $1/D$ as indicated by the bar. $\hat{r}_{t-k}$ is the predicted value of $\bar{\psi}_t$ made at time $t - k$. This variable is used in place of $\bar{\psi}_t$ because it is uncorrelated with the disturbance term, Fe. Also, the bias is represented by e_0 instead of u_0. If the identification error ϵ were entirely caused by the unmeasured disturbance, then

$$\epsilon = \frac{1+\gamma}{D} Fe \quad (36)$$

Assuming a complete model (adequate parameterization), stationary process, and stationary unmeasured disturbance, the identifier can converge to the correct values for α, β, and γ, provided a positive real condition is satisfied (Ref. 11).

$$\text{Re}\left(\frac{D}{1+\gamma}(\omega)\right) > \frac{1}{2} \quad (37)$$

When D is set equal to the identified $1 + \gamma$, the identifier is called "maximum likelihood." If D is 1 and γ is identified, the identifier is called "extended," e.g., extended least squares. As with explicit identification,

$$\bar{\psi}_t = \phi_t^T \theta + \epsilon_t \quad (38)$$

where now:

$$\phi_t^T = [\bar{u}_{t-k}, \bar{u}_{t-k-1}, --, \bar{y}_{t-k}, \bar{y}_{t-k-1}, --, -\hat{r}_{t-k-1}, --, 1] \quad (39)$$

$$\theta^T = [\beta_0, \beta_1, --, \alpha_0, \alpha_1, --, \gamma_1, --, e_0] \quad (40)$$

The controller output bias, u_0, can be calculated from e_0.

$$u_0 = -\frac{\Sigma d_i}{\Sigma \beta_i} e_0 \quad (41)$$

Identification of the γ polynomial in the implicit model (eqn. [35]) requires that $\hat{r}$, the predicted values of $\bar{\psi}$, be active. When the controller (eqn. 11 or 14) is functioning, $\hat{r}$ is forced to track the reference r. Therefore, γ cannot be identified when the reference is constant. Furthermore, α and β should not have excess parameters. When the reference is constant, a common factor cannot be uniquely identified and might be unstable. Also, when the reference is constant, β_0 cannot be determined and must be assigned an arbitrary value. A poor choice for β_0 might cause the positive-real condition for convergence to be violated and result in an invalid identification (Ref. 4).

An incremental model can be used if the controller has an explicit integral action, such as is provided in the open loop with feedback and restricted complexity controllers. In these cases the model is not required to identify the steady-state bias, and the prefilter, $1/D$, can have a high-pass or band-pass characteristic.

In summary, the implicit model, eqn. (35), allows identification of the α, β, and γ polynomials, and the bias term, u_0 (through eqn. [41]). These terms are used directly in the minimum variance control eqns. (11) and (14).

Identifiers

The unknown coefficients of the chosen identification model can be found in a number of ways. Four identification procedures are described in this section.

1. *Least Weighted Squares.* The method of least weighted squares processes a block of observations and is, therefore, most often used for off-line identification. The model is not updated with each new observation, and the C or γ polynomial is not identified. The model is identified by minimizing the sum of weighted squared errors

$$J_t = \frac{1}{\nu_t} \sum_{i=1}^{t} \mu_i \epsilon_i^2 \tag{42}$$

with respect to the elements of the θ vector. The result can be shown to be:

$$\theta_t = \left(\frac{1}{\nu_t} \sum_{i=1}^{t} \mu_i \phi_i \phi_i^T\right)^{-1} \left(\frac{1}{\nu_t} \sum_{i=1}^{t} \mu_i \phi_i \bar{\psi}_i\right) \tag{43}$$

As a result, the identification error, ϵ, is uncorrelated with ϕ.

$$\frac{1}{\nu_t} \sum_{i=1}^{t} \mu_i \phi_i \epsilon_i = 0 \tag{44}$$

The positive real condition of the previous section ensures that ϕ is also uncorrelated with the unmeasured disturbance, e.

The factor, μ_i, that weights each error, generally is selected to increase with time so that the most recent errors are most heavily weighted. The factor ν_t is used to scale the sums so that all remain finite even as the number of observations approaches infinity.

The inverse P matrix is defined as:

$$P_t^{-1} = \frac{1}{\nu_t} \sum_{i=1}^{t} \mu_i \phi_i \phi_i^T \tag{45}$$

The P matrix appears in eqn. (43) and also in the recursive equations of the next section. In order that the P^{-1} matrix not be dominated by steady state components of ϕ, it is customary to choose the ϕ and ψ components so that they have nearly zero mean. If the mean were completely removed, the P^{-1} matrix would be a weighted covariance matrix for ϕ and the P matrix would be a weighted covariance for θ.

For θ to be calculable, the P^{-1} matrix must be non-singular. Non-singularity is difficult to guarantee. If any of the process inputs and outputs included in the ϕ vector were quiescent over the identification period, P^{-1} would be singular. This could happen if the controller output were bounded or if the controller were in its manual mode. When the P^{-1} matrix is non-singular, the process is said to be persistently excited. It may be necessary to add otherwise undesirable probing signals to the normal process inputs to achieve persistent excitation.

If the control loop were marginally stable, the measured process input and output might be oscillating sinusoidally. As a result, there would be enough information to identify only two parameters from gain and phase. Guaranteed recovery from instability of a model-based adaptive controller requires either a unique identification or constraints on the nonuniquely identified parameters in order that any allowed identification lead to stabilizing control action. This need for robustness conflicts with the completeness requirement for convergence: that the model have sufficient, well-chosen degrees of freedom to be able to match the process small-signal, noise-free behavior exactly.

Since the identification model error is a nonlinear function of the delay, k, its value is usually assigned rather than identified. If its value is incorrect, control based on the identified model may be very poor, possibly unstable (Ref. 12).

2. *Recursive Least Weighted Squares.* In practice, the least weighted squares identification is difficult to implement because it requires storage of a large number of past measurements and controller outputs. Also, it requires a great number of multiplications and does not identify the C polynomial. The recursive technique overcomes these difficulties.

The computation required to identify the process at each time step can be simplified substantially by updating the previous identification with the new observation. This is done with standard recursive equations that can be derived from eqn. (43), (Ref. 13),

$$\theta_t = \theta_{t-1} + K_t (\bar{\psi}_t - \phi_t^T \theta_{t-1}) \tag{46}$$

where

$$K_t = \frac{P_{t-1} \phi_t}{\frac{\nu_{t-1}}{\mu_t} + \phi_t^T P_{t-1} \phi_t} \tag{47}$$

and

$$P_t = \frac{\nu_t}{\nu_{t-1}} \left[1 - K_t \, \phi_t^T \right] P_{t-1} \quad (48)$$

Although it may be desirable that ν_t and μ_t approach infinity as t increases, the ratios involved in eqns. (47) and (48) can be made to be finite.

In place of the weighting factors, ν_t and μ_t, a forgetting factor, λ_t, may be used to exponentially deweight past history:

$$\lambda_t = \frac{\nu_{t-1}}{\nu_t} = \frac{\nu_{t-1}}{\mu_t} \quad (49)$$

The value of λ_t is usually selected to be between .95 and 1.0. Deweighting of past history is necessary to allow the identified model to track a time-varying or nonlinear process. With $\lambda_t = 1$, the P matrix will eventually approach zero. Then new observations could not affect θ. On the other hand, enough past history (redundant observations) must be remembered so that process noise or unavoidable mismatch between process and model can be distinguished from process variation. During a quiescent period, ($\phi_i = 0$, assuming the elements of ϕ have zero mean), P_i increases by the factor $1/\lambda_i$ with each time step. When a change in ϕ_t finally occurs, K_t may be large, causing a large change in the control parameters, θ_t. This may cause the loop to "burst" into instability. Balancing these considerations may require that the weighting coefficients be varied depending on the information content of each observation. A scheme for doing this is suggested in Ref. 14.

When γ or C is identified, this boot-strapped algorithm responds slowly to process changes because the predicted measurements, $\hat{r}$, are based on old identifications. Poor tracking of process changes may cause poor control loop performance. Furthermore, a single event disturbance may cause a mis-identification when γ or C is identified because the stationarity condition is violated.

3. *Kalman Filter.* When the process has time varying parameters, a Kalman filter can be used to identify the state variable θ_t from the observations $\bar{\psi}_t$. The model's unpredictable time-varying characteristics are represented as a random walk:

$$\theta_t = \theta_{t-1} + w_t \quad (50)$$

The random components of the vector, W_t, are considered to be zero mean white noise with known covariance, Q_t.

$$E(w_t \, w_t^T) = Q_t \quad (51)$$

The filtered measured variable, $\bar{\psi}_t$, is related to the state variable by:

$$\bar{\psi}_t = \phi_t^T \, \theta_t + \epsilon_t \quad (52)$$

where the random measurement error ϵ_t is also assumed to be zero mean white noise uncorrelated with w_t. Its known variance is R_t.

In application, the statistical properties of w_t and ϵ_t may not be known. Then Q_t and R_t may be treated as weighting factors and assigned arbitrary time-varying values. The ratio Q_t/R_t affects θ_t.

The Kalman filter recursively gives the optimal update for the parameter vector θ_t based on the new observation, ψ_t and ϕ_t (Ref. 13).

$$\theta_t = \theta_{t-1} + K_t \, (\bar{\psi}_t - \phi_t^T \, \theta_{t-1}) \quad (53)$$

$$K_t = \frac{P_{t-1} \, \phi_t}{R_t + \phi_t^T \, P_{t-1} \, \phi_t} \quad (54)$$

and

$$P_t = [1 - K_t \, \phi_t^T] \, [P_{t-1} + Q_t] \quad (55)$$

The recursive least weighted squares method is included as a special case if R_t and Q_t are chosen to satisfy:

$$R_t = \frac{\nu_{t-1}}{\mu_t} = \lambda_t \quad (56)$$

and

$$Q_t = \left(\frac{\nu_t}{\nu_{t-1}} - 1 \right) P_{t-1} = \left(\frac{1}{\lambda_t} - 1 \right) P_{t-1} \quad (57)$$

However, the Kalman filter approach allows greater flexibility in assigning weighting factors, since Q_t can be an arbitrary positive-definite symmetric matrix. When Q_t is constant and ϕ_t is zero, P_t increases linearly rather than geometrically. Furthermore, Q_t can be chosen so that P_t does not tend to become singular. This makes a "burst" instability less likely.

Another special case of the Kalman filter, the projection algorithm, results when it is assumed that there are no measurement errors, $R_t = 0$, and P_t is not updated (or is reinitialized each iteration). This algorithm updates the parameter vector by minimizing the sum of weighted squared parameter changes consistent with the new observation. Old information inconsistent with a new observation is completely forgotten. All consistent information is remembered. The "burst" problem is avoided because P_t is not updated. P_t is the a priori parameter covariance matrix usually chosen to be diagonal.

Orthogonal projection results from the Kalman filter when Q_t and R_t are both assumed to be zero and P_t is updated. This method completes its identification when the number of observations equals the number of parameters, provided each observation contains independent information. The P_t matrix, and therefore, K_t, become zero. No new information inconsistent with old information is taken into account at any time step.

4. *Restricted Complexity Delay Identification.* Identification of the value for the time delay, k, is difficult because the model error is a nonlinear function of k. The recursive methods used to determine θ do not apply. For this reason, the delay

is usually assumed to be known or calculable based on measured variables. To overcome these difficulties, a multi-model approach can be used, each model having a different k. A criterion, such as maximum β_0, can be used to judge which model would lead to the best control (Ref. 10). This approach is particularly attractive for identification of the restricted complexity model used to tune a PI controller. By time shifting, all of the prediction filter models can be made to depend on the same ϕ vector. The prediction error of the k time step (denoted with a superscript) filter is:

$$\delta\epsilon_t^k = \delta y_{i+k-k\max} - \phi_i^T \theta_t^k \quad (58)$$

where

$$\phi_i^T = [\delta u_{i-k\max}, \delta y_{i-k\max}] \quad (59)$$

$$\theta_t^k = [\beta_0^k, \alpha_0^k]_t \quad (60)$$

If the weighting factors ν_t and μ_t or R_t and Q_t are chosen to be independent of k, then the Kalman gain, K_t, and parameter covariance, P_t, are independent of k. Only the parameter update equation (46 or 53) depends on k.

$$\theta_t^k = \theta_{t-1}^k + K_t (\delta y_{i+k-k\max} - \phi_i^T \theta_{t-1}^k) \quad (61)$$

In order that a restricted complexity model provide useful information for stabilizing a control loop, it must reasonably match the process in a critical frequency range near the open-loop unity gain crossing. The frequencies where the process-model match will be best are those having significant harmonic content in the inputs. Therefore, it is important to employ a band-pass filter to precondition the process inputs and outputs before identification. Because the closed-loop time constant (and therefore the cross-over frequency) is related to k, the filter cut-off time constants can be related to $k_{\min}$ and $k_{\max}$.

Also, to prevent an invalid identification, it is useful to constrain the parameters. For example, to avoid negative controller gain

$$\alpha_0^k \geq 0, \beta_0^k > 0, k \geq 1 \quad (62)$$

For the model to be open loop stable:

$$\alpha_0^k \leq 1 \quad (63)$$

Even if the outer-loop reference were constant, the inner-loop reference and measurement will be active whenever the process is disturbed, allowing both α_0^k and β_0^k to be identified.

Normal changes in process inputs may be only an order of magnitude greater than resolution errors caused by analog to digital conversion. As a result, identification of more than one term in each of the α and β polynomials may be unreliable, particularly when the sampling interval is small, $k \gg 1$.

REFERENCES

1. Kraus, T. W., and T. J. Myron: "Self-Tuning PID Controller Uses Pattern Recognition Approach," *Cont. Eng.*, 106–111 (June 1984).
2. Kraus, T. W.: "Self-Tuning Control: An Expert System Approach," *I.S.A. Pcdgs.*, 685–704 (October 1984).
3. Kraus, T. W.: "Self-Tuning Control: A Technique for the '80s," *TAPPI Process Contro Symposium*, 81–88 (1985).
4. Astrom, K. J., and B. Wittenmark: "On Self-Tuning Regulators," *Automatica*, **9**, 195–199 (1973).
5. Landau, I. D.: "Adaptive Control—The Model Reference Approach," Marcel Dekker, New York, 1979.
6. Bristol, E. H., J. P. Gerry, and P. D. Hansen: "Modern Control and modern Control," *10th Annual Advanced Control Conf.*, sponsored by Cont. Eng. and the Purdue Laby. for Applied Industrial Control, Purdue University, West Lafayette, Indiana, 150–170 (1984).
7. Vogel, E. G., and T. F. Edgar: "Application of an Adaptive Pole-Zero Placement Controller to Chemical Processes with Variable Dead Time," *American Control Conf.*, Washington, D.C. (1982).
8. Harris, C. J., and S. A. Billings: "Self-Tuning and Adaptive Control: Theory and Applications," IEE (London and New York), and Peter Pergrinus Ltd., Stevanage, U.K. and New York, 39–44 (1981).
9. Clarke, D. W., and J. P. Gawthrop: "Self-Tuning Controller," *Pcdgs IEEE*, **122(a)**, 929–934 (1975).
10. Hansen, P. D.: "Robust Identification for a Self-Tuning Controller," *Pcdgs. Yale Workshop on Applications of Adaptive Systems Theory*, Becton Center, Yale University, New Haven, Connecticut (1983).
11. Ljung, L., and T. Soderstrom: "Theory and Practice of Recursive Identification," MIT Press, Cambridge, Massachusetts, 1982.
12. Bristol, E. H.: "Experimental Analysis of Adaptive Controllers," *Pcdgs. Yale Workshop on Applications of Adaptive Systems Theory*," Becton Center, Yale University, New Haven, Connecticut (1983).
13. Goodwin, G. C., and K. S. Sin: "Adaptive Filtering, Prediction, and Control," Prentice-Hall, Englewood Cliffs, New Jersey, 1983.
14. Fortescue, T. R., L. S. Kershenbaum, and B. E. Ydstie: "Implementation of Self-Tuning Regulators with Variable Forgetting Factors," *Automatica*, **17**, 831–835 (1981).

Numerical Control and Computerized Numerical Control

By John Boyle[1]

Numerical Control (NC) is the term universally applied to the flexible automation through electronics of general-purpose machine tools. From the late 1950s on, it has made increasing strides on all machine types, bringing with it the benefits of improved productivity and quality.

Since the 1970s, the use of computer techniques has led to a new generation of units known as Computer Numerical Controls (CNCs). This has allowed more features to be added to the controls, giving still greater productivity and quality.

NC FUNDAMENTALS

The basic property of a numerical control is its ability to accurately position the slides (called *axes*) of a machine and to control the cutting feeds and speeds from information created by the user. This information is called the *part program* and is put together for each workpiece by a trained methods and tooling engineer, commonly referred to as a *programmer*. The data are traditionally passed to the machine controller on a punched paper tape. The controller then processes this information and moves the machine axes accordingly to cut the workpiece. The role of the operator is to set up the part, start the control, carry out any manual interventions, such as tool changes, and resolve problems as they occur.

The control user has many alternatives, ranging from a complete do-it-yourself system to a turnkey package. Equipment manufacturers offer a wide range of assistance, from application notes to complete systems. In addition, suppliers[2] offer training schools where customers can learn more about their controls and other customers' applications. System houses and controls consultants can also be a valuable resource.

The benefits of NC are higher productivity and quality through such areas as greater use of the machine and reduction of human errors. These benefits have made NC an indispensable technology in all sizes of metalworking plants.

Machine Requirements

Successful application of NC requires certain features of a machine tool. NC is not a magic wand that converts an older manual machine to an accurate and reliable piece of equipment. NC may, in fact, increase the frequency of breakdown of older machines due to greater utilization. New machines are designed specifically for NC application with special consideration being given to stiffness, accuracy, and heat dissipation as well as to position measurement and drive location and capability.

Position Measurement

As stated earlier, the basis for NC is the accurate positioning of machine axes. Two types of devices have emerged as the standards for position measurement. They are both available in linear and rotary form. The linear devices are more accurate because they are a direct measure of axis position, but they are also more awkward to mount and protect and are more susceptible to damage. The rotary types are more compact, but are less accurate since they measure via lead screws or rack and pinion methods with gearing.

The first type relies on *electromagnetic induction*. The rotary version is known as a *resolver*, and one proprietary version of the linear type is known as the *Inductosyn*[R]. Both are absolute devices within one cycle of measurement and, as such, are less sensitive to electrical noise upsetting their accuracy.

---

[1] Manager—Quality Assurance, Giddings & Lewis Electronics Company, Fond Du Lac, Wisconsin.
[2] As, for example, Giddings & Lewis.

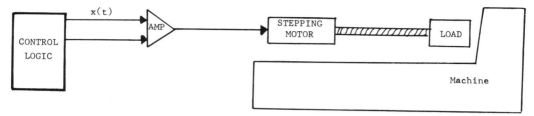

Fig. 1. Open-loop numerical control system.

The second type uses *optical coupling* and is known as an *encoder*. These devices are more sensitive to dirt interfering with their operation than are the magnetic devices.

Typical resolution of either type of device is .0001 inch.

Positioning devices are described in greater detail in *Section* 2 of this *Handbook*. Also consult alphabetical index at the end of book.

Axis Drives

Machine motion is provided by variable speed motors. Hydraulic motors were at one time in common usage but now tend to be replaced by DC electric units. Currently, AC variable frequency drives and motors are expected to gain wide acceptance in the future. For smaller machines, stepper motors are also used. They have the advantage of not requiring position feedback inasmuch as one step of the motor represents a known increment of position. However, if for any reason the motor fails to respond to a commanded step, accuracy will be lost. This latter configuration is known as *open-loop control* and is represented in Fig. 1. The more preferred *closed-loop control* is shown in Fig. 2.

Variable speed motor drives are described in more detail in *Handbook Section* 4.

TYPES OF NC

Many of the earlier distinctions used to classify types of numerical control have become obsolete with the advent of CNC. This is because CNC either incorporates all of the functions or has made the more preferred one available at a cost low enough to obsolete the alternative. For example, controls at one time worked in either inch or metric units; now one switch available to the operator allows the unit to measure in either mode. Another distinction was whether the unit was capable of contouring or capable only of point-to-point operation. Contouring is the ability to control two or more axes simultaneously to produce a shape, such as a circle. Point-to-point control was the name given to the movement of one or more axes into position independently

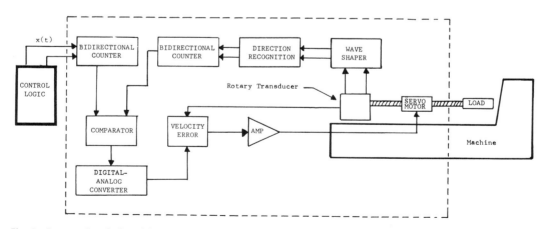

Fig. 2. Conventional closed-loop numerical control.

and then the actuation of a single axis to cut the part, as in the case of a drilling machine.

Drilling machines have tended to become machining centers that can both mill and drill; and hence contouring is almost a standard CNC feature. In many cases programmable controllers have taken over the simpler point-to-point applications.

COMPUTERIZED NUMERICAL CONTROL FEATURES

The flexibility of software has allowed a multitude of optional features to be offered. This makes comparison of different manufacturers' products a complex task. The potential user needs to take the time and trouble to analyze each feature and its value to the application being considered.

Number of Controlled Axes

The number of axes controlled can range from two in the simplest machine to ten or more in the most sophisticated applications. Rotary tables, auxiliary slides, tool changers, and work changers now have more sophisticated designs that qualify them as additional axes of controlled motion.

Operator Interface

The operator interface usually consists of a push-button panel and a display. The push buttons can range from a few buttons specific to the application to a full typewriter keyboard allowing greater flexibility. Displays can range from simple numeric readouts to color CRTs, in some cases with graphical representation of parts being machined.

Data Entry Devices

The one-inch-wide punched tape reader is still by far the most common data entry device. However, floppy disks, magnetic tape cartridges, and direct links to other computers are appearing at a growing rate.

Part Program Storage

The reduction in memory costs and the availability of new technologies allow the storage of one or more part programs in the CNC. This has two main advantages. It eliminates repeated reading of punched paper tape, which is often an unreliable operation. And, it allows changes to be made to part programs at the machine and subsequent parts to be run without the need to go back and forth producing new tapes. Semiconductor memory is still the most common, but bubble memory technology is also in use.

Fixed or Canned Cycles

Where groups of machine movements are used frequently, they are stored as a set of instructions in the control by the manufacturer. The user can then invoke the cycle by one command rather than several. Typical cycles include those for drilling and tapping.

Parametric Subroutines

These are groups of instructions that the user creates directly in the plant as may be determined by specific needs. Parametric subroutines are then invoked by a single instruction, and parameters, such as axis travel and feed rate, are passed for each case.

Macros

Macros are codes used for repeat operations, such as bolt hole circles surrounding a bore. By specifying radius of the circle and number of holes as well as start point to the macro, the programmer is able to minimize the programming task as well as the chance for errors.

Software End Limits

These allow the user to specify limits to the travel that each axis is permitted to make. End limits are used to prevent crashes from occurring on the machine due to programmer or operator error.

Probing

This feature is used in conjunction with a touch-type position measuring probe usually held in the spindle. Accurate measurement of surface or hole dimensions can be taken. They can be used to determine if further machining is needed or even if the correct part is in place.

Adaptive Control

By sensing such parameters as motor current and torque, the control can automatically adjust feed rates in order to meet optimum cutting conditions. Such variables as tool wear and varying material hardness are compensated in this way, and minimum cutting time is achieved without impacting such quality characteristics as surface finish.

Other control features, varying from one manufacturer to the next, are available in CNC systems. See immediately following article on "Computer Numerical Control-State-of-the-Art Systems."

IMPLEMENTATION OF CNC

Vendors of CNCs are either electronics companies that specialize in industrial controls or machine tool companies that believe they know their own requirements better and thus can provide a totally integrated package. Earlier approaches to CNC often took the form of a general-purpose minicomputer. This computer was housed in an air-conditioned industrial cabinet because it could not survive unprotected on the shop floor. The minicomputer was also equipped with special purpose interface boards for handling the machine devices, such as drives, solenoids, and position measurement equipment.

As with many other applications, microprocessors have taken over the field. Several microprocessors are used in a distributed processing setup to provide the computing power needed. CNC is a real-time computing application that places heavy demands on computing speed and power in order to effectively and safely control the machine tool. The industrial environment also places demands on packaging of the electronic equipment to ensure its operation in a satisfactory manner. In particular, the dirt and electrical noise factors must be considered. The equipment also must withstand more physical abuse, as compared with an office environment, for example.

Software

Tailoring of software is needed by the machine tool builder in order to match it to the specific machine tool. This is usually provided by the control builder in the form of a machine interface package. Simpler programming languages similar to those used in programmable controllers are included. This allows the general-purpose signals coming from the main executive programs of the CNC to be tailored to the different forms needed for individual machine tools without having to know the complexities of the main software per se. Machine languages for automated machines are described elsewhere in this *Handbook*. Consult alphabetical index at end of book.

Maintenance and Troubleshooting

CNCs are normally provided with diagnostic programs that run when power is applied to the unit, or it can be invoked as required. These programs isolate problems to the level of a printed circuit board, which if in fault can be exchanged. The CNC itself is also programmed to detect machine, operator, and part programming errors as they occur.

DESIGNING FOR NC PRODUCTION

Although CNC has brought about a revolution in the metalworking industry in terms of flexibility and capability, this does not mean that engineering groups have *carte blanche* in terms of designing parts without consideration of available manufacturing methods. Indeed, for the high levels of productivity and quality needed for a business to survive, this knowledge is now more important than ever before. While cells and flexible manufacturing systems add to the flexibility of a plant, they also have considerations that the designer needs to be aware of in order to make optimum use of these high-priced installations.

The size and shape of the part are among the most obvious and important criteria. Most manufacturers define their machine capabilities in terms of "cube" of workspace that can be accommodated. This does not mean that the overall part size must fit into this cube, but rather that machining operations must be limited to it. Workpiece weight is another consideration for table-type machines. Where floor-type machines are available, these problems are less critical. Cells and flexible manufacturing systems introduce the additional consideration of the material-handling equipment. Size and weight are particularly important where parts are to be moved around automatically, and these limitations must also be known by the designer.

The number and type of cutting tools required to machine a workpiece must also be in the forefront of the designer's mind. The capacity of automatic tool-change equipment ranges from about 10 to 100 tools. Significant increases in manufacturing efficiency can be achieved if these do not require frequent change. In particular, tool-change time and thus machine-cycle time is reduced and errors caused by incorrect tools being put into tool-changer sockets can be eliminated. These factors are particularly important in cells and flexible manufacturing systems where several machines operating on several parts or families of parts are simultaneously in production. The major saving in tool quantity is still probably to be achieved in the area of tapped holes that often require three tools per thread type. As a by-product of saving tool magazine space, tool-change time is also reduced, thereby cutting down on total machining time.

Another problem made more critical by the advent of multiple machine cells and flexible manufacturing systems is that of holding the workpieces for machining. Not only must the part be held while being cut, but also it must be transported between machines and the system or cell load and unload stations. Where designers can help is by keeping required machining operations to as few sides of the part as possible and by providing surfaces or other features to locate and clamp on quickly and effectively. In most systems all load and unload as well as refixturing operations must be done at special stations and not on the machine tools themselves. Thus, each refixturing requires the part leaving a machine, traveling to a special station, and then traveling back again to continue being machined. This involves significant time as well as making the material transporter unavailable for other operations and can have an important detrimental effect on the efficiency of the system as a whole.

The designer must also know the rotary axis capability of the manufacturing plant. Many rotary tables can only position in 15-degree steps, and any subdivisions beyond this angle can result in severe difficulties for the shop floor. The cost increment between tables with indexing as opposed to full positioning capability to any angle can be very large, and thus designs that call for this requirement in order to be manufactured may be very expensive.

The designer must also keep in mind the part programmer and the capabilities available to simplify the task of producing effective NC tapes. Features such as origin shift, subroutines, probing, etc., previously mentioned are targeted at making programming easier in certain commonly found manufacturing situations. By designing the part to take advantage of these features, the designer can enhance the success of the manufacturing process.

Outlook

The advances in CNCs have paved the way for more complex applications of factory automation in the form of flexible machining cells and systems. Cells are usually two or three machines equipped with CNCs and tied into a simple work-handling system. A master computer monitors and controls the work flow on all machines. The machine for a given part is chosen, and the material handling system takes the part there. The part program is transmitted over communication lines to the CNC, and the cycle is automatically started. When it is complete, the CNC signals the master computer and the part is removed and the process is repeated. Flexible manufacturing systems are the same process on a larger scale with twelve or more machines under control.

ADDITIONAL READING

Arnold, F., Jr.: "Microstepping a Step Motor for Greater Position Resolution," *Cont. Eng.*, 111–113 (January 1985).

Harrington, J., Jr.: "Designing for N/C Production," in "Manufacturing Automation Management" (R. W. Bolz, Ed.), Chapman & Hall, New York, 1985.

Morris, H. M.: "Computerized Numerical Control Evolves in Response to Market's Changing Needs," *Cont. Eng.*, 76–79 (February 1984).

Shaiken, H.: "The Automated Factory: The View from the Shop Floor," *Technology Review* (M.I.T), 17–21 (January 20, 1985).

Staff: "CNC System for Small Machine Tools Brings CAM to the Job Shop," *Cont. Eng.*, 60 (June 1984).

Computer Numerical Control—State-of-the-Art Systems

The application of computer numerical control (CNC) to machine tools is mainly guided by four factors: (1) the volume of work produced during a given period of time, (2) the variety of work produced, (3) the complexity of the workpieces produced, and (4) the desirability of making possible the implementation of the newer manufacturing philosophies, notably computer-integrated manufacturing (CIM) and flexible manufacturing systems (FMS).

CNC systems are generally marketed in two ways: (1) as a package from the machine tool builder (who may or may not make the CNC system), or (2) the CNC system may be procured separately and engineered into the system by the user. In either case, of course, close cooperation between the CNC maker, the machine tool builder, and the user is required to ensure excellent performance and a maximum return on investment. CNC may be present at the outset—that is, it may be part of a new machine tool—or it may be applied as a retrofit in updating an older machine equipped with plain NC or, less frequently, in retrofitting an older machine to which NC has not previously been applied. Of course, in reviewing automation against the broad backdrop of manufacturing situations, there still remain cases where neither NC or CNC can be justified. Some degree of automation may be achieved, such as robot loading of machines, where numerical control may not be involved. A variety of situations is described by Figs. 1 through 8.

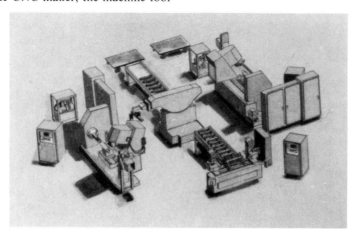

Fig. 1. Not all systems are large, multi-machine complexes. The situation shown here incorporates two CNC controlled lathes with automated parts changers. (*Giddings & Lewis.*)

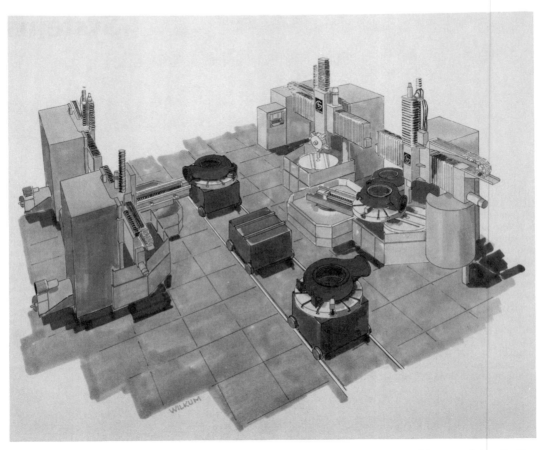

Fig. 2. Manufacturing systems can be relatively simple. Machining of large castings in this system is handled by four NC controlled vertical turning centers with workpiece and rotary pallet moved from transport to vertical turret lathe by sliding pallet shuttles. (*Giddings & Lewis.*)

CNC—State-of-the-Art Systems   237

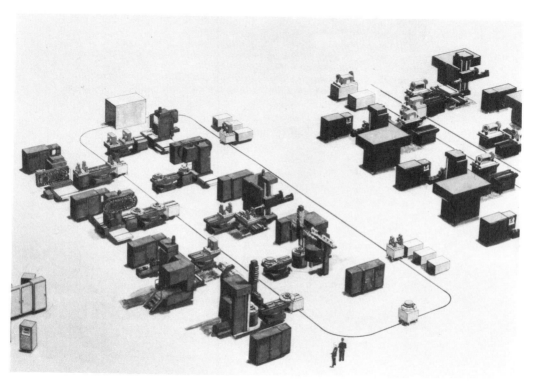

Fig. 3. Most manufacturing operations fall within the category of mid-volume, discrete batch lot production—not large enough for the economies of in-line transfer systems, but urgently needing improved efficiency of operation to substantially increase productivity. For example, a system can integrate turning, milling, drilling, tapping, and boring operations with automated parts handling. (*Giddings & Lewis.*)

238 CNC—State-of-the-Art Systems

Fig. 4. Mid-volume manufacturing is frequently best achieved with general-purpose equipment and controls arranged as production modules. Thus, familiar and proven components are kept in action by a master computer that has been preprogrammed for any number of parts. (*Giddings & Lewis.*)

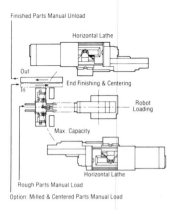

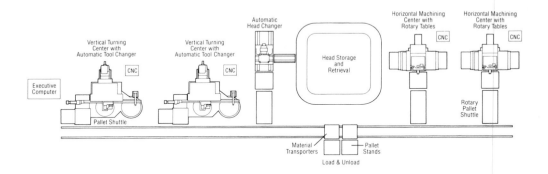

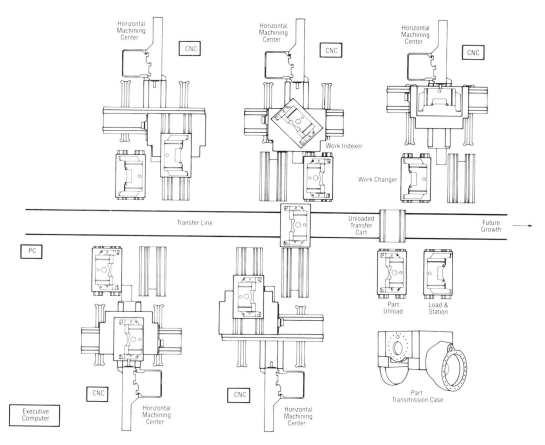

Fig. 5. Some manufacturing systems are basically machining centers equipped with work handling options and joined by transfer equipment. Each machining center in itself is a minisystem that processes parts through a single station rather than through sequential stations. PC = Programmable Controller. (*Giddings & Lewis.*)

240 CNC—State-of-the-Art Systems

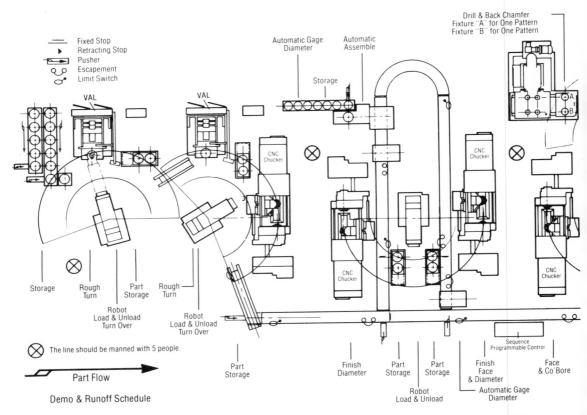

Fig. 6. A complex manufacturing system is essentially an organized arrangement of automated machining modules. The line rate, or output, determines the number of modules and a master controller schedules workpieces among series of parallel equipment. Computerized numerical control units or programmable controllers are identical from one machine to another so that operating personnel can gain familiarity with the system. Diagnostic capability is a standard feature of such systems, permitting rapid troubleshooting and routine

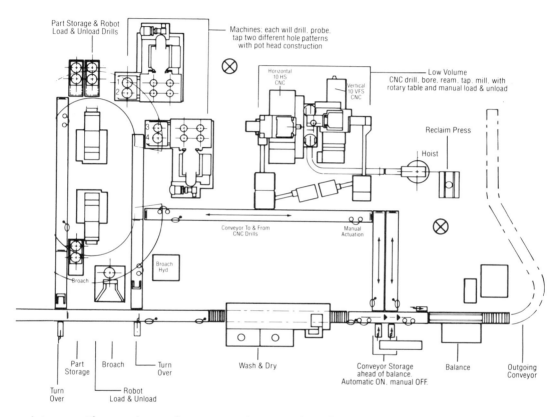

maintenance. The use of general-purpose equipment and standard controls permits manufacturing system flexibility. Machines can be added to, or subtracted from, the line whenever desired and the controlling equipment can be readily reprogrammed to include new parts, design changes, new techniques, or additional equipment. (*Giddings & Lewis.*)

242 CNC—State-of-the-Art Systems

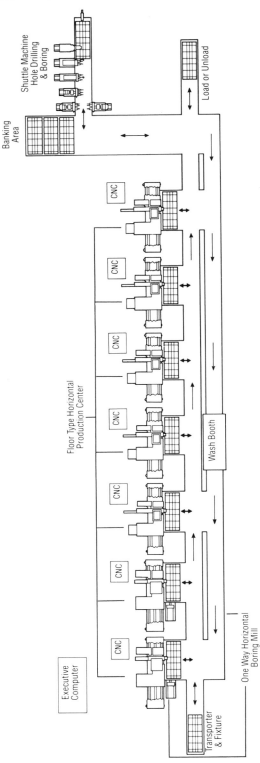

Fig. 7. System shown here is capable of handling and processing parts with envelope dimensions of 5 feet by 14 feet with a total weight for part, fixture, and transporter of more than 20 tons. (*Giddings & Lewis.*)

CNC—State-of-the-Art Systems 243

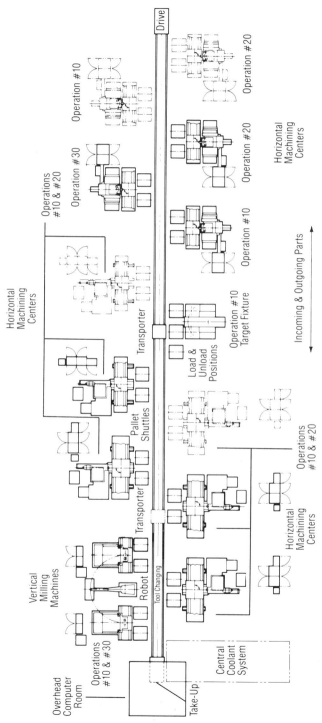

Fig. 8. Random machining system for large prismatic parts designed to accommodate planned future production increases. Operations in light outline were scheduled for installation in three steps to ultimately achieve maximum yearly production. Each machine in system is controlled by a CNC system with the entire system controlled by an executive computer. (*Giddings & Lewis.*)

TWO-AXIS CNC SYSTEM FOR TURNING MACHINE APPLICATIONS[1]

As shown in Fig. 9, this system is equipped with a 9-inch cathode ray tube (CRT) display and a laminated membrane "touch-tone" type panel. So-called conversational programming is designed to assist less experienced operators through step-by-step program generation via a menu of turning events, allowing part print programming in machining language familiar to the operator while retaining the capability for off-line or commercial time-sharing programming.

Fifteen types of turning events can be programmed via the CRT. (See Fig. 10.) In the set-up mode on the CRT, materials data for automatic entry in roughing and finishing events are available. The control contains tables for ten standard material types and eight special material types, which can be entered by the operator if required to accommodate special materials or tooling types not included in the standard tables.

[1] System 10T CNC, *Dynapath* Systems Inc.

The system also incorporates *tool nose radius compensation*, sometimes referred to as "work surface programming." This allows the operator to enter or display the values of tool radius, offset, and orientation. It permits part surface programming with the CNC calculating the actual tool path from the values entered for each tool. This feature also uses a "look ahead" capability to calculate the actual tool path for accurate cornering and intersection.

To generate a plot of the part profile, the *graphics mode* permits simple keyboard entries to define the display window and the portion of the program sequence for the plot desired. The X and Z part coordinates are entered to locate the display origination point, which becomes one corner of the window. The cursor can be jogged in graphics mode in order to determine the location of the origin for display of specific areas of the part. Also entered is either the horizontal or vertical size desired for the display window. The

(a)

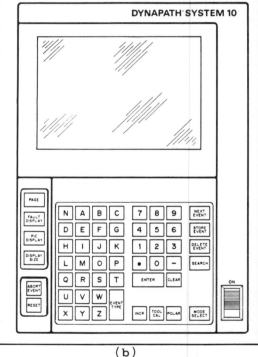

(b)

Fig. 9. (a) Nine-inch CRT and laminated membrane (touch-tone) panel for programming the System 10T CNC (*DynaPath*® Systems Inc.); (b) diagram of front of panel. This panel also obtainable in a horizontal format.

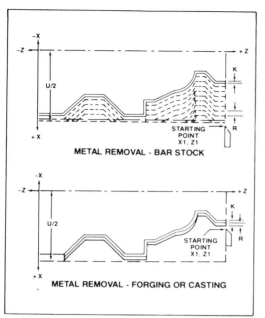
Rough and finish

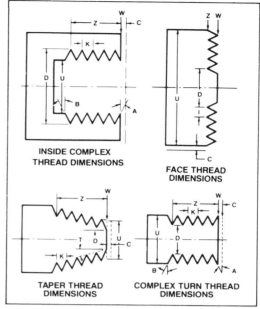
Threading

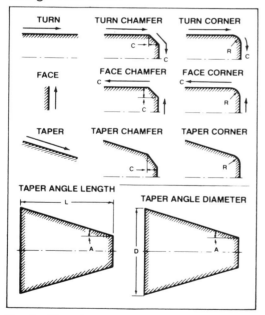
Basic Turning

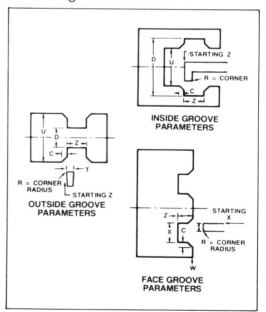
Grooving

Fig. 10. Fifteen types of turning events can be programmed via the CRT, including: Position, turn, face, taper, arc, groove, drill and tap, repeat, subroutines, M functions, set-up, roughing, finishing, text message, and threading. In threading, the operator can choose from among seven threading variables, including turn thread, face thread, taper thread, inside turn thread, inside taper thread, complex turn thread, and complex inside thread. (System 10T CNC, DynaPath® Systems Inc.)

system will then automatically scale both the height and width of the display. The part program (or portion of the program) to be displayed is then entered, and a push of the *cycle start* button begins the plot. The time required to cut the part is calculated by the control and displayed on the screen. (See Figs. 11 and 12.)

The display on the CRT of the 10T CNC system will plot each roughing pass, spindle, and tailstock safe zones and rapid moves at half-intensity while the finish pass and all moves at feedrate are shown at full intensity. The part is displayed in relation to the spindle centerline. If preferred, a single keystroke deletes the half-intensity display, leaving a simple display of the part outline. (See Fig. 13.)

Character information tells the operator the precise place where the program is in its sequence as well as the numeric position display of the tool tip, the plot scale, and the execution time. (See Fig. 14.) The system graphics also can depict vertical turning configurations in their proper perspective. (See Fig. 15).

A detailed summary of specifications for the 10T CNC System is given in Table 1.

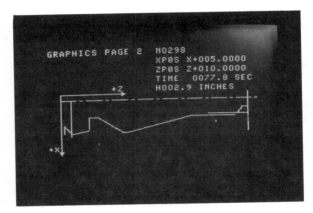

Fig. 11. Full intensity graphics display the part outline relative to the spindle centerline in the *graphics display mode*. The display origin is defined by the X and Z vectors in part coordinates. (System 10T CNC, DynaPath® Systems Inc.)

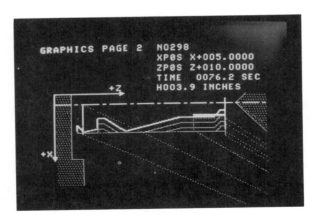

Fig. 12. A single keystroke adds half-itensity display information outlining threading as well as roughing process. The dotted line depicts rapid positioning moves. (System 10T CNC, DynaPath® Systems Inc.)

CNC—State-of-the-Art Systems 247

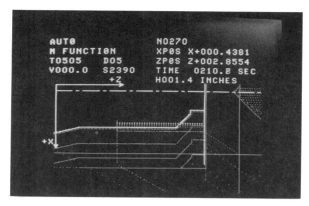

Fig. 13. Graphics may be viewed while the part is run in the *auto* or *single event* execution modes. The moving crosshair traces the tool tip as this display focuses on the area that is being machined. (System 10T CNC, *DynaPath*® Systems Inc.)

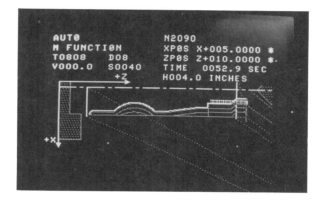

Fig. 14. As the graphics trace the part outline, text information above the display tells which event is being executed, the tool selected, spindle speed, and feed rate. (System 10T CNC, *DynaPath*® Systems Inc.)

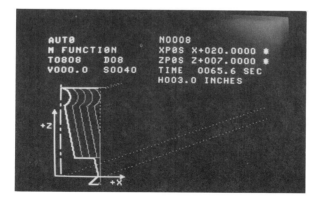

Fig. 15. The CRT display also can depict vertical turning configurations in their proper perspective. (System 10T CNC, *DynaPath*® Systems Inc.)

TABLE 1. Summary of Specifications for the *DynaPath®* 10T CNC System

MECHANICAL:
- Modular Configuration:
  Operator's Panel: 19"W X 12.25"H X 10"D
  System Module: 17.5"W X 15.75"H X 13.5"D
- Input Voltage: 115 VAC, ±10%, Single Phase, 47-63 HZ, 500 VA
- Ambient Temperature: 40°F to 120°F
- Humidity: 0-95%
- Mating Half of Interface Connectors
- Teletype Interface
- Tape Cassette Interface
- RS-232C Serial Port for Tape Reader, DNC, etc.
- Operating Program in Nonvolatile EPROM Memory
- CMOS Ram Memory for Retention of Part Programs and Compensations for One Year

MACHINE CONTROL CAPABILITY
- Two (X,Z) Axes of Control
- Departures up to ±999.9999 Inches or ±9999.999 mm
- Feedrates up to 400 IPM or 9999 MMPM
- Electric Servo Drive Outputs
- Interface for Digital Encoders Including Handwheel, inch or metric
- Spindle Speed Control: 4 Digits, ±10 Volt Analog
- Constant Surface Speed Control
- Stored Axis Travel Limits and Safe Zones
- Customer Initialized Machine Control Parameters
- Auxiliary Functions:
  M, S and T Codes
  BCD Outputs
  Decoded M Functions
  Strobe Signals
  Done Signals
- DC Inputs/Outputs

OPERATOR FEATURES
- Full Alphanumeric Keyboard with Tactile Feel and Audible Tone
- 9" CRT for Information Display
- Part Program Storage and Display 60,000 Characters
- Tool Set-Up
- Automatic Reference Zero
- Axis Home
- Axis Jog
- Test
- Program Inhibit
- Spindle Speed Display
- Fault Messages
- Text Messages
- On-Line Diagnostics
- Material Data Table for Standard and Special Materials Selected by Operator for Automatic Entry of Speeds and Feeds in Roughing and Finishing Events.

OPERATING MODES
- Jog Modes:

	Inch (IPM)	Metric (MM/MIN.)
Slow	1.0	25
Medium	10.0	254
Fast	100.0	2540
Rapid	Rapid Rate	Rapid Rate
Inc. Jog	0.0001	00.002
Inc. Jog	0.0010	00.020
Inc. Jog	0.0100	00.200
Inc. Jog	0.1000	02.000
Inc. Jog	1.0000	20.000
Zero Set	—	—
Home	—	—
Reference Zero	—	—

- Auto Mode
- Single Event Mode
- Program Modes:

Position	Turn
Face	Taper
Arc	Thread
Repeat	Subroutine
Drill	M Function
Roughing	Set-Up
Finishing	Text

- Display Program Mode (With Paging)
- Set-Up Mode:

Reversal Comp.	In/MM Selection
Program Allow/Inhibit	Travel Limits
Mirror Image	Test Modes
Safe Zones	Home Position Entry
Feedrate Override	Maximum Spindle RPM
Materials Table	

- Tool Offset Mode
- Part Surface Programming with Tool Nose Radius Compensation
- Record Program Mode
- Load Program Mode
- Delete Program Mode
- Learn Mode
- Catalog Mode for Storing up to 99 Part Programs

PART PROGRAMMING AIDS
- Decimal Point Programming
- Inch/Metric Input
- Reference Zero
- Absolute/Incremental
- Mirror Image
- Radius or Diameter Programming
- End Point Programming
- Angle-Length Programming
- Automatic Fillet Generation
- Automatic Chamfering Generation
- Special Events:
  (3) Types of Taper Programming
  (7) Types of Thread Programming
  (3) Types of Groove Programming
- Peck Drill Cycles
- Tapping Cycle
- Metal Removal Cycles:
  Roughing and Finishing Cycles to Automatically Cut Along a Contour Defined by Turn, Face, Taper and Arc Events. Inside, Outside or Face Cycles can be Programmed.
- Graphics to Display a Portion or All of a Part Shape on CRT (For more information see DynaPath® System 10T Graphics Data Sheet)

COMPENSATIONS
- Lead Screw Error Compensation
- Reversal Error Compensation
- (16) Tool Nose Radius Compensations—"Look-Ahead" Capability for Work Surface Programming Allows Storing Tool Tip Position and Tool Orientation
- (24) Tool Offsets

OPTIONS
- Integral Programmable Interface Control
  4000 Instructions
  EPROM Non-Volatile Memory
- Additional DC Inputs and Outputs

THREE/FOUR-AXIS CNC SYSTEM FOR MILLING MACHINE AND MACHING CENTER APPLICATIONS[2]

As with the 10T CNC System just described, this system also incorporates a 9-inch CRT display. The operator generates a program directly from the part print by selecting the appropriate operation and supplying dimensional information requested by the control. In this manual data input mode, the program is generated, recorded, and then proved by actually making a part or by displaying the programmed part shape on the cathode ray tube. The system also includes interfaces for loading and retrieving part programs via a variety of peripheral devices from teletype to cassette through full direct numerical control (DNC) systems. The system also offers a

[2]System 10M CNC, *Dynapath* Systems Inc.

Fig. 16. CRT display panel and control entry keyboard. One may plot the part profile with axis and spindle motion inhibited; or call on graphics while machining the part. One may select a view of the entire part outline or focus on the details. The information plotted is the running part program. The graphics uses the same information as the control system uses to machine the part. One can plane select to plot any single plane or two-plane views simultaneously. The visual display and text messages assist in locating part program errors. (System 10M CNC, *Dynapath*® Systems Inc.)

variety of "canned" cycles for such operations as cavity milling, frame milling, pocket milling, bolt circle generation, scaling, part rotation, and drilling, tapping, and boring cycles. The unit provides a large memory capacity for part programs. The system also incorporates multiple compensation capabilities for tool size, part misalignment, and lead screw/gearbox inaccuracies

The system has full feature graphics (Figs. 16 through 21). A detailed summary of specifications for the 10M CNC System is given in Table 2.

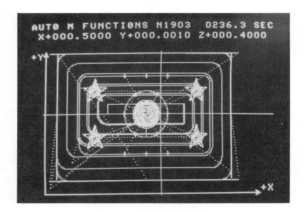

Fig. 17. Graphics display roughing passes and hole circumferences and centers at half density; rapid moves with dotted lines; and finish passes at full intensity. (System 10M CNC, *DynaPath®* Systems Inc.)

Fig. 18. A single keystroke deletes the half-intensity display leaving the finish part profile. (System 10M CNC, *DynaPath®* Systems Inc.)

CNC—State-of-the-Art Systems 251

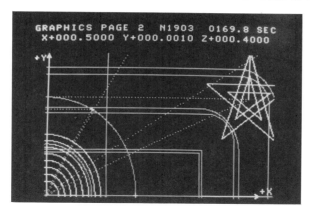

Fig. 19. A portion of the part profile can be displayed by simple keyboard entry. (System 10M CNC, *DynaPath*® Systems Inc.)

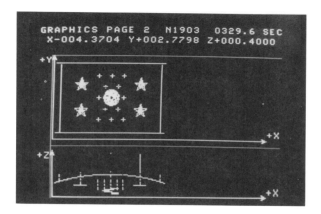

Fig. 20. Plane selection allows display of more than one plane. (System 10M CNC, *DynaPath*® Systems Inc.)

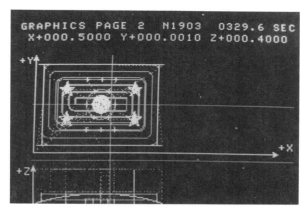

Fig. 21. Graphics may be viewed while the part is run in the *auto* or *single event* execution modes. The moving crosshair traces the tool tip with text information contained above the display on the cathode ray tube. (System 10M CNC, *DynaPath*®, Systems Inc.)

252 CNC—State-of-the-Art Systems

TABLE 2. Summary of Specifications for the *DynaPath®* 10M CNC System

MECHANICAL:
- Standard Enclosure:
  21"W X 14.5"H X 30"D
- Modular Configuration:
  Operator's Panel: 19"W X 12.25"H X 10"D
  System Module: 17.5"W X 15.75"H X 13.5"D
- Input Voltage: 115 VAC, ± 10%, Single Phase, 47-63 HZ, 500 VA
- Ambient Temperature: 40°F to 120°F
- Humidity: 0-95%
- Dust Tight
- Splash Proof
- Cooling: External Cooling Not Required
- Mating Half of Interface Connectors
- Teletype Interface
- Tape Cassette Interface
- RS-232C Serial Port for Tape Reader, Tape Punch, DNC, etc.
- Operating Program in Nonvolatile EPROM Memory
- CMOS Ram Memory for Retention of Part Programs and Compensations for One Year

MACHINE CONTROL CAPABILITIES
- Simultaneous Three (X, Y, Z) Axes of Control (4th simultaneous axis optional)
- Departures up to ± 999.9999 Inches or ± 9999.999 mm
- Feedrates up to 400 IPM or 9999 MMPM
- Electric Servo Drive Outputs ± 10V Max.
- Interface for Digital Encoders Including Handwheel, Inch or Metric
- Z Axis Tracking
- Spindle Speed Control: 4 Digits, ± 10 Volt Analog
- Stored Axis Travel Limits
- Customer Initialized Machine Control Parameters
- Auxiliary Functions:
  M, S and T Codes
  BCD Outputs
  Decoded M Functions
  Strobe Signals
  Done Signals

OPERATOR FEATURES
- Full Alphanumeric Keyboard with Tactile Feel and Audible Tone
- 9" CRT for Information Display
- Spindle Speed Display
- Part Program Storage and Display. 60,000 Characters
- Actual Feedrate Display
- Fault Messages
- Text Messages
- Grid Zero
- Program Zero
- Axis Home
- Axis Jog
- Test Modes
- Program Inhibit
- On-Line Diagnostics

OPERATING MODES
- Jog Modes:

	Inch (IPM)	Metric (MM/MIN.)
Slow	1.0	25
Medium	10.0	254
Fast	100.0	2540
Rapid	Rapid Rate	Rapid Rate
Inc. Jog	0.0001	00.002
Inc. Jog	0.0010	00.020
Inc. Jog	0.0100	00.200
Inc. Jog	0.1000	02.000
Inc. Jog	1.0000	20.000
Zero Set	—	—
Home	—	—
Machine Zero	—	—
Ref/Grid Zero	—	—

- Auto Mode
- Single Event Mode
- Program Modes:
  Position   Linear Mill
  Arc Mill   Frame Mill
  Circle Mill   Bolt Circle
  Repeat   Subroutine
  Dwell   M Function
  Rotate   Set-Up
  Cavity Mill (Option) Text
- Display Program Mode (With Paging)
- Set-Up Mode:
  Part Offsets   In/MM selection
  Reversal Comp.   Travel Limits
  Program Allow/Inhibit Test Modes
  Scale Factor   Home Position Entry
  Mirror Image   Axis Switching
- Tool Length Compensation
- Part Surface Programming with Cutter Diameter Compensation
- Record Program Mode
- Load Program Mode
- Delete Program Mode
- Learn Mode (Optional)
- Catalog Mode for Storing up to 99 Part Programs

PART PROGRAMMING AIDS
- Decimal Point Programming
- Inch/Metric Input
- Polar Coordinate Programming
- Cartesian Coordinate Programming
- Absolute/Incremental
- Mirror Image
- Part Rotation: 0.001° to 360°
- Scale Factor: 0.0001 to 99.0000
- Fixed Cycles:
  g0 - Cancel   g3 - Peck Drill
  g1 - Drill   g4 - Tap
  g2 - Counterbore   g5 - Bore
- Rectangular or Circular Pocket Milling Programmed as a Single Event
- Automatic Generation of a Radius Between Two Consecutive Linear Moves

COMPENSATIONS
- Lead Screw Error Compensation
- Reversal Error Compensation
- Part Offsets
- (48) Cutter Diameter Compensations — "Look-Ahead" Capability for Work Surface Programming
- (48) Tool Length Compensations
- Tool Calibration Feature for Automatically Storing Tool Length Compensations

OPTIONS
- Cavity Mill Event Enables Programming Concave or Convex Shapes Including Half Cylinders, Hemispheres, Toroids and Tapered Half Cylinders
- Learn Mode Enables the Operator to Manually Jog the Machine to the Desired Part Coordinates and the Control Automatically Records Part Dimensions
- Expanded Interface Provides for:
  Automatic Reference Zero,
  Axis Switching,
  Manual Handwheel Switching,
  Two-Range Spindle Speed Control
- Extension for Control of a 4th Axis (Linear or Rotary)
- Integral Programmable Interface Control
- Foreign Language CRT Display
- Graphics to Display a Portion or all of a Part Shape on CRT

MICROPROCESSOR-BASED CNC SYSTEM FOR METALWORKING APPLICATIONS[3]

Introduced in 1984, this system, based on microprocessor technology, was designed with a number of objectives in mind—simplicity of operation, use of advanced programming methods, and capability of being serviced by users. Typical applications of the control system are shown in Fig. 22. The cathode ray tube (CRT) serves as the visual interface between the control and the operator. Through simple menu selections, information such as actual and commanded position and feeds, speeds, and preparatory functions can be displayed. The operator may also use the CRT for program editing and for displaying messages.

For manual operation, a heavy-duty joystick enables single-axis or combined motions for both feed and rapid traverse. An infinitely variable potentiometer provides stepless feedrate control over the full range. The directional features and center button can be used for incremental feed modes.

The part program storage (PPS) provides memory capacity to store complete programs, a particularly helpful feature when repetitive long-run parts are scheduled. The complete part program is read into PPS from either the punched

[3]NUMERIPATH8000 System (Giddings & Lewis, Inc.)

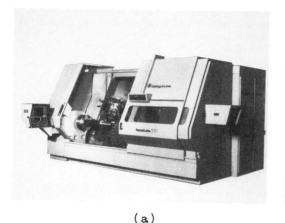

(a)

(b)

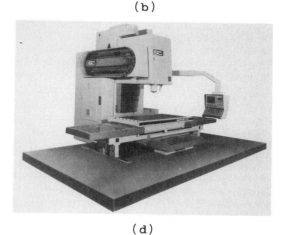

(c)   (d)

Fig. 22. Representative applications of the CNC 8000 *NumeriPath*® System in the metalworking industry: (a) Horizontal turning center; (b) double and single vertical chuckers; (c) vertical turning centers, machining cells and flexible manufacturing systems (FMS); (d) horizontal and vertical machining centers. Note that the control station is packaged in a hanging pendant or on a pedestal stand, which includes a drawing tray, built-in calculator, magnetic tape cartridge, and soft-touch membrane keys. (*Giddings & Lewis, Inc.*)

tape reader or the built-in magnetic cartridge unit. The program may be optimized or corrected directly at the machine in the PPS memory and then saved on a cartridge or on a paper tape punch. The keyboard and readout displays are used for modifying and optimizing the data. Entire blocks may be added or deleted. The system is programmed using the Electronics Industries Association Standards (EIA RS 274-D and EIA RS 447). Thus, programmers and operators have only a limited number of codes to learn. A new conversational programming language (Type II Data) can be used to create special functions and sequences for probing, subroutines, and software cycles. Tapes made for use in the prior generation of control system are compatible with the NumeriPath 8000 unit.

Programming in feed and speed terms enables the programmer to work directly from the part drawing, using common language familiar to the machinist. An inch/metric feature allows the operator to convert the system for handling data in the inch system to computing, handling, and displaying data in millimeters. Parts programmed in inches may be run in the metric mode and vice versa. Tool offsets are automatically converted to the measuring system in use.

For repetitive or similar operations on a part, a parametric subroutine may be programmed the first time the operation is encountered. Each additional time the operation is required, it is simply a matter of calling and executing the subroutine. Using such names as "Mill Square," when only major dimensions vary with each part, saves time by eliminating the need to write complete routines each time a subroutine is called. Resident subroutines (macros) can be used in any part program. Thus, special threading or probe sequences can be stored as macro to save time and memory space, and also to shorten tapes.

Looping is a means of repeating similar operations a specified number of times. This can use a spindle cycle in drilling a series of holes, for example. It can be used within a subroutine to simplify the subroutine program, or a subroutine may be part of the loop.

Where patterns or operations are repeated, the origin shift feature can be used. (See Fig. 23.) This procedure enables the programmer to shift the established part zero temporarily to any new point by moving x, y, and z axes with u, v, and p commands. Shifting the origin to the center of a bolt hole pattern simplifies calculations for the holes. For repeat patterns, only the reference point is shifted; the same x and y calculations previously made are used again.

An *auto reentry* feature allows for stopping in a cut, retracting the cutter, checking the work size, or modifying a tool offset, and then automatically reentering the tool path by pushing a button. The control system for lathes provides the ability to program directly in constant surface feet per minute. For example, c200 means 200 surface feet per minute (SFM). The control automatically increases or decreases table or chuck speed as the cutting tool moves in or out from the table or chuck centerline.

Other Programming Features

For those who desire to prepare and review part programs using advance tape preparation centers, color graphics can be interfaced with the system. Bulk storage of part programs is also possible. *Adaptive control* is available on new machines. This control function maintains selected horsepower by automatically adjusting the feedrates. The cut horsepower is continuously monitored and maintained to optimize production rates. Also a *recommended tool life* may be programmed for a cutting tool. When 90% of this time expires, the condition is indicated to the operator by the control. Automatic *probing* or in-process gaging is also available. Part surfaces may be accurately located and dimensioned. Probe results are displayed on the CRT or sent to a printer. Available on new machines, this *NumeriProbe®* system is described in more detail in *Section* 2 of this *Handbook*.

Diagnostics

When power is applied to the control, each printed circuit board is checked out by a resident diagnostic program. Board status is indicated by a light-emitting diode (LED) indicator. The control also monitors and records error conditions on the machine tool and displays them for operator action. The logic to control the machine is written in relay ladder diagram form. This can be displayed on the CRT with power flow indicated by the display intensity, thus enabling the maintenance worker to quickly isolate and repair problems.

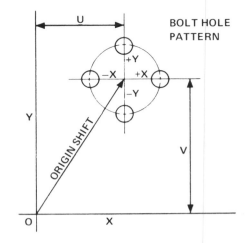

Fig. 23. Origin shift, a special feature available in the CNC 8000 *NumeriPath®* System. (*Giddings & Lewis, Inc.*)

Summary of Specifications

The general specifications of the NumeriPath® 8000 system are given in Table 3.

TABLE 3. Summary of Specifications for the *NumeriPath®* CNC 8000 System

Dynamic inch/metric switchable
999 tool offsets loaded by tape with a range of ±99.9999" (±2539.999 mm)
Shop talk programming of RPM. IPM (MMPM) and IPR (MMPR)
Parametric subroutine and looping for repetitive parts
3-axis simultaneous contouring in linear and circular interpolation (with quadrant crossing) g01, g02, g03
Magnetic tape cartridge input/output
Punched tape reader, 400 c.p.s., bi-directional
Tape format: Variable block word address, 8 channel punched tape
Multi-block buffer storage
Absolute/incremental dimensioning using g90/g91 tape codes
Plus-minus decimal point programming
Leading zero suppression
EIA/ASCII switchable tape format with automatic recognition
Block delete using slash code
Dwell programming (seconds or revolutions)
Parity detection
Full reference offset on all axis
Tool radius compensation: 999 values, range ±2.000" (±50.800 mm) (tangential paths)
Operational modes: manual, single block, automatic from tape or PPS.
Incremental feeds — 1.0, .1, .01, .001 .0001 inch (100.0, 10.0, 1.0, .1, .01, .001 mm)
Feedrate override (1 - 125%) infinitely variable
Spindle speed override (1 - 125%) infinitely variable
Programmable precent override for feeds and speeds
Automatic tool set and modify
Keyboard MDI using full alphanumeric characters
Sequence number search forward and reverse
Feed hold
Return-to-Path
Automatic acceleration and deceleration
Temperature rating 50°F. - 122°F., (10°C. - 50°C.), humidity to 95%
Power input 200, 230, 460, 575 volts, ±10% (select one), 3 phase, 60 herts, ±1 hertz; 220 or 380 volts, ±10% (select one), 3 phase, 50 hertz, ±1 hertz

Preparatory Function Codes
g00 - Point-to-point positioning
g01 - Linear interpolation
g02 - Circular interpolation, CW
g03 - Circular interpolation, CCW
g04 - Dwell
g09 - Non-model decelerate to zero
g17-24 - Plane select
g33 - Threadcutting, constant lead
g34 - Threadcutting, increasing lead
g35 - Threadcutting, decreasing lead
g36 - Threadcutting, segmented lead
g40 - Cutter compensation cancel
g41 - Cutter compensation left
g42 - Cutter compensation right
g50-59 - Adaptive control
g60 - Cancel envelopes
g61 - Enable envelope No. 1
g62 - Enable envelope No. 2
g70 - Inch
g71 - Metric
g80-89 - Canned cycles
g90 - Absolute dimension input
g91 - Incremental dimension input
g92 - Preload of part reference
g93 - Inverse time feedrate (velocity/distance)
g94 - Inches feedrate (millimeters) per minute
g95 - Inches feedrate (millimeters) per spindle revolution
g96 - Constant surface speed
g97 - Revolutions per minute

Miscellaneous Functions
m00 - Program stop
m01 - Optional stop
m02 - End of program
m03 - Spindle forward CW
m04 - Spindle reverse CCW
m05 - Spindle stop
m06 - Tool change
m07 - Mist coolant
m08 - Floor coolant
m09 - Coolant off
m12 - Synchronization code
m17 - Pallet change
m19 - Spindle orient
m30 - End of program
m31 - Interlock by-pass
m48 - Cancel override lockout (both feed and speed)
m49 - Override lockout (both feed and speed)
m58 - Cancel m59
m59 - Bypass constant surface speed
m90-m99 - Reserve for user

Maximum Programmable Data

	Inch	Metric
Coordinate	±9999.9999"	±99999.999 mm
Departure	±9999.9999"	±99999.999 mm
Arc Radius	±9999.9999"	±99999.999 mm
Resolution	.0001"	.001 mm

SECTION 4

Actuators and Materials-Transfer Systems*

Section Contents

Robots in Perspective	260
Robot Technology Fundamentals	262
Classification of Robots	262
Axes of Motion	262
Degrees of Freedom	262
Revolute (Jointed-Arm) Coordinates	262
Cartesian Coordinates	262
Cylindrical Coordinates	264
Spherical (Polar) Coordinates	264
Work Envelope	265
Robot Control Systems	265
Non-Servo-Controlled Robots	265
Servo-Controlled Robots	265
Point-to-Point Robots	267
Continuous-Path Robots	267
Programming Robots	267
Development of Programming Techniques	267
Playback Concept	267
Levels of Robot Programming	267
Load Capacity and Power Requirements of Robots	281
Electric Drives for Robots	281
Hydraulic Actuators for Robots	282
Pneumatic Drives for Robots	282
Electromechanical Drive Hardware	283
Dynamic Properties of Robots	283
Stability	283
Solutions to Oscillation Problems	284
Resolution	286
Differences Between Sliding and Rotary Joints	286
Repeatability	286
Compliance	287
End-Effectors	288
Mechanical Clamping	288
Firmness of Grasp	290
Vacuum Systems	290
Magnetic Pickups	292
Permanent Magnets	292
Electromagnetics	292
Fastening Tools to Robot Wrists	293
Standardized End-Effectors Available from Stock	293
Workplace Configuration and Environmental Factors	293
The Robot Setting	293
Work Is Around the Robot	293
Work Is Brought to the Robot	297
Robot Travels to the Work	298
Work Cells	298
Overview of Some Contemporary Robots and Applications	299
Evaluating Potential of Robots for a Given Application	318
Advanced Robotic Systems/Expanded Uses	318
A Robot Dynamics Simulator	321
Robot Model	321
Robot Dynamics	321
Actuator Dynamics	322
Complete Dynamic Model	323
Simulator	323
VAST Operation	324
Simulation Modes	325
Engineering Applications	326
Simulation of Robot Control Algorithms	326
Robot Control	326
Controller Performance	327
Control of Actuators in Multilegged Robots	330
Statement of the Problem	330
Kinematic Considerations	330
Equilibrium of the Structure	331
Servomotor and Servosystem Design Trends	336
Industrial Motors—A Perspective	336
Alternating Current (AC) Motors	336
Direct Current (DC) Motors	337
Brushless Motors	337
Microprocessors in Servo Control Systems	340
Stepper Motors and Controls	346
Variable Reluctance Stepper Motors	347
Permanent-Magnet Stepper Motors	347

Hybrid Stepper Motors	348
Stepper Motor Controls	353
Transition in Stepper Motor Technology	355
Microstepping	355
An Ironless Disc-Rotor Stepper Motor	360
Electrohydraulic Stepper Motors	362
Linear and Planar Motors	367
Types of Linear Motors	367
The Planar Motor	371
Dynamic Tests	373
Sawyer Linear Reluctance Motor	374
Solid-State Variable-Speed Drives	376
Development of Power-Switching Devices	376
Gate-turn-off Thyristors	376
Static Induction Thyristors	377
Variable-Speed Drive Hardware Development	377
DC Drives	378
Six-thyristor Full Converter	378
Dual Armature Converter	378
AC Drives	380
Slip-energy Recovery System	380
Load Commutated Inverter	380
Induction Motor Variable Speed Drives	381
Cycloconverter	381
Autosequentially Commutated Current-fed Inverter	382
Field-oriented Controls	382
Voltage-source Inverter	383
Transistors and GTOs	383
Pulse-width Modulated Drives	385
Materials Motion/Handling Systems	386
Overhead Movement of Materials	386
Power-and-free Conveyor Systems	386
Underhung Cranes and Monorail Systems	391
Floor-mounted Conveyors	396
Track-and-drive-tube Conveyors	396
Inverted Power and Free Conveyor Systems	404
Towlines	407
Automatic Guided Vehicles (AGVs)	412
Automated Storage and Retrieval Systems (ASRS)	412
Order Selection Systems for Cartoned Products	413
Order Selection Systems for Noncartoned Products	415
Large, High-density, High-rise AS/RS Systems	416

ARNOLD F., PMI Motion Technology, Division of Kollmorgen Corporation, Commack, New York. (*Stepper Motors and Controls*)

ARUM, H. R., Designatronics Inc., New Hyde Park, New York. (*Robot End-Effectors*)

BREEN, J. M., Adaptive Intelligence Corporation, Milpitas, California. (*Adaptive Assembly Robots*)

CONSTANTINO, P. J., Jervis B. Webb Company, Farmington Hills, Michigan. (*Power and Free Conveyors; Underhung Cranes and Monorail Systems; Inverted Power and Free Conveyors; Automatically Guided Vehicles*)

EBRAHIMI, N. D., Department of Mechanical Engineering, The University of New Mexico, Albuquerque, New Mexico. (*Control of Actuators in Multilegged Robots*)

FORD, R. E., Bodine Electric Company, Chicago, Illinois. (*Stepper Motors and Controls*)

HONCHELL, N., Cincinnati Milacron, Lebanon, Ohio. (*Industrial Robots*)

FLINN, P. A., GMF Robotics Corporation, Troy, Michigan. (*Robots for Automotive and Other Industrial Uses*)

MACKENZIE, P. C., Jervis B. Webb Company, Farmington Hills, Michigan. (*Power and Free Conveyors; Underhung Cranes and Monorail Systems; Inverted Power and Free Conveyors; Automatically Guided Vehicles*)

MAHONEY, W. A., Intelledex Corp., Corvallis, Oregon. (*Robots*)

MANDRILLO, V., Eaton-Kenway, a subsidiary of Eaton Corporation, Salt Lake City, Utah. (*Automatic Storage/Retrieval—Warehouse Systems*)

MAZURKIEWICZ, J., Pacific Scientific, Rockford, Illinois. (*Servomotor and Servosystem Design Trends—Microprocessors and Brushless Motors*)

MOHR, E. D., Unimation Incorporated (A Westinghouse Company), Danbury, Connecticut. (*Robotic Technology*)

MURRAY, J. J., The Superior Electric Company, Bristol, Connecticut. (*Stepper Motors and Controls*)

NEUMAN, C. P., Department of Electrical and Computer Engineering and the Robotics Institute, Carnegie-Mellon University, Pittsburgh, Pennsylvania. (*A Robot Dynamics Simulator*)

ORTEGA, D. G., Intelledex Incorporated, Corvallis, Oregon. (*Assembly Robots and Vision Systems*)

OSMAN, R. H., AC Drives, Robicon Corporation (A Barber-Colman Company), Pittsburgh, Pennsylvania. (*Solid-State Variable Speed Drives*)

*Persons who authored complete articles or subsections of articles or who otherwise cooperated in an outstanding manner in furnishing information and helpful counsel to the editorial staff.

RUSSELL, L., MTS Systems Corporation, Minneapolis, Minnesota. (*Electrohydraulic Stepping Cylinders and Motors*)

STORMS, T., Hitachi America, Ltd., Tarrytown, New York. (*Robots and Robot Vision Systems*)

THOMAS, J., SI Handling Systems, Inc., Easton, Pennsylvania. (*Automatically Guided Vehicles; Towline Conveyors: Track and Drive-tube Conveyors; Automatic Storage/Retrieval Systems*)

VOLPE, G. T., University of Bridgeport, Bridgeport, Connecticut. (*Linear and Planar Motors*)

WISMANN, D., PMI Motion Technology, Division of Kollmorgen Corporation, Commack, New York. (*Stepper Motors and Controls*)

Robots in Perspective

In the opinion of some authorities, robots and their application (robotics), as we approach the 1990s, command a somewhat disproportionately high share of the technical and public attention given to the total concept of manufacturing and processing automation. But, with many thousands of robots now installed, their presence is indeed impressive, and there are few sages who do not forecast a bright future for robotics during the remainder of this century.

For some industrial applications (welding of automobile and truck frames, painting of vehicle bodies, assembly and inspection of electronic gear, among other examples from the discrete-piece manufacturing industries), robots have already played a revolutionary and revitalizing role. To realize their full potential over a broader spectrum of large manufacturing firms and particularly at the medium-size and small-production levels will require a better and wider understanding of the basics of robot technology and economics. Very important will be the continued incorporation of advanced technology into robot systems by suppliers and users.

Notable progress has already been made toward the development and utilization of "smart" robots. These robots, according to some scientists and engineers, employ at least some of the elements of artificial intelligence, this observation based mainly on the design and use of technically effective and economically practical machine vision.

Before the comparatively recent niche for robots in manufacturing, the image of robots over past centuries has been burdened by an aura of mystique and romanticism that has resulted from an overworked comparison of robots with people, an image that still persists but to a much lesser degree.

During the late 1700s, there was much fascination with *androids*, which essentially were mechanical "people." For example, in 1774 Jaquet-droz exhibited three "mechanical marvels"—a musician, a writer, and a draftsman. These and other charmingly attired mechanical people were presented at court like visiting dignitaries. During that period Diderot in his *Encyclopédie* observed that in the construction of machines, engineers should look to monsters for inspiration, but that instead the eighteenth-century engineers looked to man and built beautiful automatons. (Ref. 1).

As recently as 1923, this general approach was carried on by a Czech playwright who used the Czech word *robot* (for worker) to describe humanoid creations in his play *Rossum's Universal Robots*. And in the 1940s the science-fictionist Isaac Asimov coined the word *robotics*. (Ref. 2)

Early industrial usage of robots, which resemble machines more than people, dates back some twenty to thirty years. Although these earlier robots performed well for certain applications (mainly for handling large and heavy loads under adverse conditions), their use did not commence to bloom until the 1970s. Their current wider acceptance, in the configurations with which we are familiar today, did not get well under way until the early 1980s.

In the discrete-piece manufacturing industries (metalworking and automotive as examples), many of the operations now performed or assisted by robots were formerly accomplished by what is referred to as *fixed* or *hard* automation. Such systems employed limit switches, relays, photoelectric sensors and other electromechanical and magnetic devices for controlling the motion- and position-related geometric variables. Fixed or hard automation, still best suited for some operations, is constrained by a lack of *flexibility*.

Transfer lines, for example, commonly used in the automotive and machinery industries for several years, were representative of fixed automation. With model changes or for a firm producing limited runs of many different products, the fixed system required costly retooling for going from one application to the next. The introduction of programmable relay logic systems (adjustable plug-board memories) and, in the late 1960s, the entry of the *programmable controller* brought a degree of flexibility and limited universality to prerobot systems. Programmable controllers (PCs) initially received their widest and most enthusiastic acceptance in the discrete-piece manufacturing industries. At a somewhat later date, PCs were found to be ideally suited for the control of batching operations in the process industries. Today, of course, PCs are used in thousands of applications in both of the major categories of manufacturing. PCs have become an integral part of many robotic systems.

As with so many of the technological advances of the last few decades, the concept of robotics benefited markedly from microelectronics, computer science, and improved design and performance of electrical, electromechanical, and hydromechanical servosystems.

In the relatively recent past, the highly selective applications for robots resulted in so-called *islands* of automation. With the development of more sophisticated automation concepts, as represented by computer-integrated manufacturing (CIM), users have learned that industrial operations are usually best automated through the integration of several robots with several machines into what is sometimes called a "work cell" scheme. In such configurations, the robots and machines that they serve are treated as a *unified system*.

This integration is difficult for the system designer today because of differences in protocols, languages, buses, and other *un*standardized hardware and software. But, progress is being made in all of these areas toward the establishment of acceptable standards. Successful integration results only when work cells can communicate plant-wide and when new tasks can be accommodated by designed-in flexibility, thus avoiding high retooling costs characteristic of fixed and special-purpose automation.

Just as computing technology is being absorbed into systems as the result of integration and thus losing its former separate identity, so it is predicted that the robot will ultimately be regarded as just one more part of an automated complex.

REFERENCES AND ADDITIONAL READING

1. Carrera, R., Fryer, D., and J. Paul: "Androids," *FMR Magazine* (American translation from Italian), 65-92 (November 1984). Franco Maria Ricci International, Inc., New York.
2. Engelberger, J. F.: "Robotics in Practice," Kogan Page Limited, London, 1980.
3. Horn, B. K. P., and K. Ikeuchi: "The Mechanical Manipulation of Randomly Oriented Parts," *Sci. Amer.*, 100-111 (August 1984).
4. Hunt, V. D.: "Smart Robots," Chapman and Hall, New York, 1985.
5. Aleksander, I.: "Computing Techniques for Robots," Chapman & Hall, New York, 1985.
6. Gardner, L. B.: "Automated Manufacturing," *ASTM STP 862*, American Society for Testing and Materials, Philadelphia, Pennsylvania, 1985.
7. Amber, G. H., and P. S. Amber: "Anatomy of Automation," Prentice-Hall, Englewood Cliffs, New Jersey, 1962. (out of print)
8. Todd, D. J.: "Walking Machines—An Introduction to Legged Robots," Chapman & Hall, New York, 1985.

Robot Technology Fundamentals

A widely accepted definition of robot is that of the Robot Institute of America: "A robot is a reprogrammable, multifunctional manipulator designed to move material, parts, tools, or specialized devices, through variable programmed motions for the performance of a variety of tasks."

CLASSIFICATION OF ROBOTS

Information categories for classifying robots include: (1) axes motion, including type of motion, number of axes, and the parameters of axis travel; (2) control system; (3) programming system; (4) load capacity and power required; (5) dynamic properties, such as stability, resolution, repeatability, and compliance; (6) end-effectors and grippers used; (7) workplace configuration; and (8) appropriate applications for a given robot design.

Axes Motion

A robot is movable from one factory location to another, as may be dictated by factory layout changes or by major alterations in job assignment, but for any given task that will be repeated over and over for long periods, a robot will be *firmly fastened* to the operating floor (in some cases the ceiling). This establishes a firm *geometric location of reference*, an unchangeable position that will geometrically relate precisely with associated machinery and, in the case of a work cell, with several machines and possibly other robots. For moderate changes in use, the average robot will incorporate within its design sufficient operating flexibility and adjustments obviating the need to alter the installed robot's reference location. In the case of a "smart" robot, final very small changes in the positioning of the arm will be made by outputs from a machine vision or tactile system.

Intentionally designed movable robots are discussed later in this article under "Workplace Configuration."

Degrees of Freedom. Once a robot is installed, its ability to move parts and materials will be established by the built-in axes of motion, sometimes called degrees of freedom. The axis of motion refers to the separate motion a robot has in its manipulator, wrist, and base. The designer will usually select one of four different systems of geometric coordinates for any given robot. The system of coordinates will define the geometric configuration of the robot at any given instant.

Four coordinate systems are found in robots: (1) revolute (jointed arm), (2) cartesian (X,Y,Z), (3) cylindrical (rectilinear), and (4) spherical (polar).

Revolute (Jointed-arm) Coordinates. In this system, the robot arm is constructed of several rigid members, which are connected by *rotary joints*. As shown by Fig. 1, three independent motions are permitted. In robots, these members are analogous to the human upper arm, forearm, and hand, while the joints are respectively equivalent to the human shoulder, elbow, and wrist. The arm incorporates a wrist assembly for orienting the end effector in accordance with the requirements of the workpiece. (See Fig. 2.) These three articulations are pitch (bend), yaw (swing), and roll (swivel). In some applications, fewer than six articulations may suffice, depending upon the orientation of the workpiece and the machine(s) that the robot is serving.

Cartesian Coordinates. In the cartesian system, all robot motions travel in right-angle lines to each other. There are no radial motions. Consequently, the profile of a cartesian robot's work envelope is a rectangular shape. (See Fig. 3.)

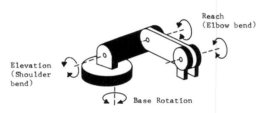

Fig. 1. A jointed-arm manipulator incorporating revolute coordinates.

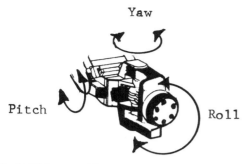

Fig. 2. Wrist assembly on robot arm for orienting the end-effector in accordance with requirements of the workpiece.

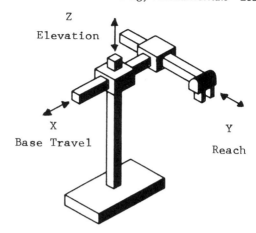

Fig. 3. A manipulator incorporating Cartesian coordinates.

Some systems utilize rotary actuators to control end-effector orientation. Robots of this type generally are limited to special applications. A robot may incorporate rectilinear-cartesian coordinates. In one example, a continuous-path extended-reach robot offers the versatility of multiple robots through the use of a bridge and trolley construction that enables it to have a large rectangular work envelope. When ceiling mounted, a device of this type may service many stations with several functions, thus leaving the floor clear. X and Y motions are performed by the bridge and trolley; the vertical motions are performed by telescoping tubes. Additional axes can be used. (Ref. 1)

In a cartesian coordinate system, the location of the center for the coordinate system is the center of the junction of the first two joints. Except for literally moving the robot to another factory location, this center does not move. In effect, it is tied to the "world" as if anchored in concrete. If the X measurement line points toward a column in the area where the robot is placed, the X line will always point toward that same column no matter what way the robot turns while performing its programs. These are known as the *world coordinates* for a given robot installation. (See Fig. 4.)

However, in the operation of a robot, having an origin for a measurement reference is not sufficient. We also need to know where we are measuring to. This measurement is made from the origin of the coordinate system to a point

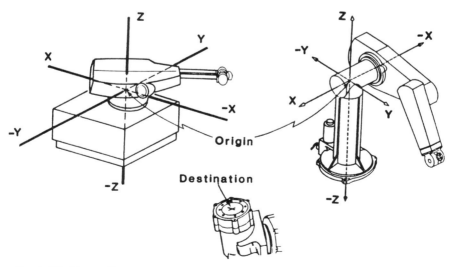

Fig. 4. World coordinate system of a robot using Cartesian coordinates. (*Source:* Ref. 2)

264 Robot Technology Fundamentals

that is exactly in the center of the circle on which the tool (end effector) is to be mounted. This system moves with the tool and is aptly called the *tool coordinate system*. In the tool coordinate system, the X and Y lines lie at right angles flat on the tool mounting surface. The Z line is the same as the axis of rotation for the joint; that is, it points directly through the tool in one direction and through the wrist in the other direction. This system is *not* tied to the world. Instead, it stays in position on the tool mounting surface and moves wherever the tool moves. While the origin of this system is thus allowed to move around, the destination (where it measures to) is left to the discretion of the user. Sometimes the tool coordinate system is actually used to measure where the tip of the tool lies relative to where it is mounted; sometimes it is used to measure where one position in space lies relative to some other point in space. (See Fig. 5.)

Cylindrical Coordinates. Robots incorporating cylindrical coordinates have a horizontal shaft that goes in and out and rides up and down on a vertical shaft, which rotates about the base. (See Fig. 6.) Additional rotary axes are sometimes used to allow for end-effector orientation. Cylindrical-coordinate robots are often well suited where tasks to be performed or machines to be serviced are located radially from the robot and where no obstructions are present. A robot that incorporates cylindrical coordinate has a working area or envelope that is a portion of a cylinder.

Spherical (Polar) Coordinates. Robots using this coordinate system may be likened to a tank turret; that is, they comprise a rotary base, an elevation point, and a telescoping extend-and-reach boom axis. (See Fig. 7.) Up to three rotary

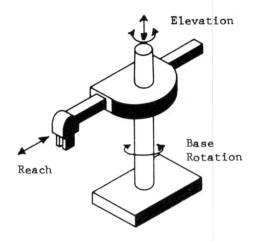

Fig. 6. A manipulator incorporating cylindrical coordinates.

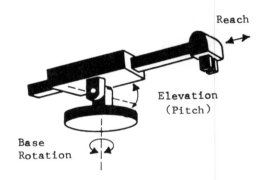

Fig. 7. A spherical-coordinate manipulator, the operation of which is comparable to a tank turret.

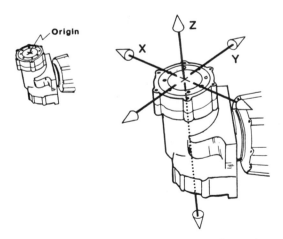

Fig. 5. Tool coordinate system of a robot using Cartesian coordinates. (*Source*: Ref. 2)

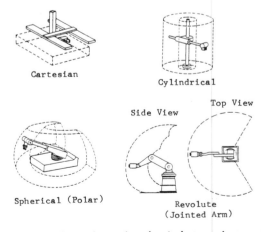

Fig. 8. Work envelope of a robot is that area in space which the robot can touch with the mounting plate on the end of its arm.

wrist axes (pitch, yaw, and roll) may be used to control the orientation of the end effector. The arm moves in and out and is raised and lowered through an arc while rotating about the base. The end effector moves in a volume of space that is a portion of a sphere.

Work Envelope. The work envelope of a robot is that area *in space* that the robot can touch with the mounting plate on the end of its arm. Actually, the envelope will be somewhat larger, depending upon the dimension of the end effector that is fastened to the mounting plate. As previously mentioned, the three-dimensional shape of the work envelope is related to the coordinate system used. (See Fig. 8.)

ROBOT CONTROL SYSTEMS

Most authorities classify robots as (1) Non-servo-controlled, or (2) servo-controlled. For this immediate discussion, the type of power used (Electric, pneumatic, or hydraulic) is not important. Some of the earlier robot control schemes are inlcuded here.

Non-servo-Controlled Robots

In a non-servo-controlled robot, the directional controls are fully off or fully on, causing essentially constant speed movement along an axis in one direction or in the reverse direction. Some form of limit switch is used to stop the movement at the desired point. Non-servo-controlled robots are the least complex of robots. They move in an open-loop fashion between two exact end points on each axis or along predetermined paths in accordance with fixed sequences. Such robotic systems can operate over an infinite number of points enclosed by their operational envelope. Non-servoed robotic units are given start and end points on each axis that must be passed. There is little or no control of end effectors between these points. Technically, controlled trajectory is possible on a non-servo-controlled system only if the unit is given the coordinates of all points lying between the start and end parameters. This specific type of programming will allow the robotic system to perform motions, such as straight lines and circles.

Non-servo-controlled robots are sometimes called *limited-sequence* robots. The number of limb articulations is usually few. Because of their control characteristics, the non-servo-controlled robot is sometimes called a *"bang-bang"* device or *pick-and-place* unit (commonly used for picking up an object, transporting the object to a predetermined location, and placing the object in that location). Robotic systems of this type are capable of high speed and usually have good repeatability. Other characteristics of non-servo-controlled robots are compared with those of servo-controlled robots in Table 1. A block schematic diagram of a representative limited-sequence robot is shown in Fig. 9. Improvement through the use of an analog servo system is shown in Fig. 10.

Servo-Controlled Robots

This type of robot will incorporate one or more servomechanisms that enable the arm and gripper to alter direction in midair, without having to trip a mechanical switch. Servo-controlled robots generally have larger program and memory capacity than do their non-servo counterparts.

TABLE 1. Comparison of Non-Servo and Servo-Controlled Robots.

Characteristic	Non-servo Robot	Servo-Controlled Robot
Flexibility	Limited in terms of program capacity and positioning capability. Arms can travel at only one speed and can stop only at end points of its axes.	Maximum flexibility provided by ability to program the axes of manipulator to any position within limits of its travel. Can vary speed at any point within envelope. Ability to move heavy loads in a controlled fashion.
Speed	Relatively high.	Relatively slow.
Repeatability	Good—to within 0.25 mm.	To within ± 1.5 mm. End-of-arm positioning accuracy of 1.5 mm.
Cost	Comparatively low.	Comparatively high.
Complexity	Simple operation, programming, and maintenance.	Permits storage and execution of more than one program, with random selection of programs from memory via externally generated signals. Subroutining and branching capabilities may be available, permitting robot to take alternative actions within a program when commanded.

(*Data Source:* Ref. 3)

266 Robot Technology Fundamentals

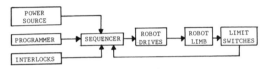

Fig. 9. Highly schematic system configuration of an early and some contemporary *limited-sequence* robots. Depending on the type of robot drive used, the power source may be electric, hydraulic, or pneumatic. The program sequence may simply be established by arranging plug-in jumper leads in an appropriate peg-board pattern. Each robot limb (arm, wrist, and hand mechanisms) movement is set to start and stop by means of adjustable end-stops. Limit switches at each of the end-stops operate to input the sequence that particular movements have been completed. The sequencer, controlled by signals from the limit switches, rotates step by step to furnish power to each robot drive as needed. Interlock switches on the operated machine or other equipment ensure that the robot is kept in step with the manufacturing cycle and prevent the robot from causing accidents or damage by collision.

In a system of this type, upon start of program execution, the controller addresses the memory location of the first command position and also reads the actual position of the various axes as measured by a position feedback system. These two sets of data are compared, and their differences (error signals) are amplified and transmitted as command signals to servo valves for the actuator of each axis. Thus, a servo-controlled robot is a closed-loop system.

This closed loop refers to the robot system per se and does not embrace the total machine system that is served by the robot. The latter would require feedback of measurements made on the product of the handling operation and adjustments of the total system (machine plus robot) required to ensure desired control of the total system. This requires additional inspection, gaging, and other hardware and, in more sophisticated instances, may involve machine vision.

The servo valves, operating at constant pressure, control flow to the manipulator's actuators. As the actuators move the manipulator's axes, feedback devices, such as encoders, potentiometers, resolvers, and tachometers, send position (and in some cases, velocity) data back to the controller. These feedback signals are compared with the desired position data, and new error signals are generated, amplified, and sent as command signals to the servo valves. This process continues until the error signals are effectively reduced to zero, whereupon the servo valves reach *null*, flow to the actuators is blocked, and the axes come to rest at the desired position. The controller then addresses the next memory location and responds to the data stored there. This may be another positioning sequence for the manipulator or a signal to an external device.

Normally, the memory capacity of a servo-controlled robot will be sufficiently large to store up to 4000 points or more in space. Specific programs select and sequence activity points for a given operating scheme. Programs can be varied to maintain the scheme while changing the activity points. Both continuous-path and point-to-point capabilities are possible, and, in this regard, robot system control is reminiscent of numerical control of machine tools. Accuracy can be varied, if desired, by changing the magnitude of the error signal, which is considered *zero*. This can be useful in "rounding the corners" of high-speed continuous motions. Programming, described in more detail a bit later, is accomplished by manually initiating signals to the servo valves to move the various axes into a desired position and then recording the output of the feedback devices into the memory of the controller. This process is repeated for the entire sequence of desired positions in space. A servo-controlled robot will fall into one of two categories: (1) point-to-point, and (2) continuous path.

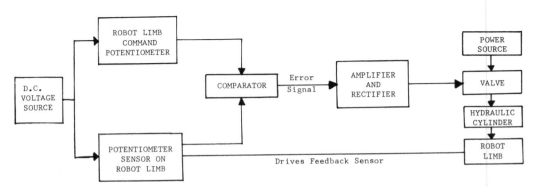

Fig. 10. Analog servo system where potentiometer sensors are fitted onto the robot limbs. These signals go to a comparator which produces an error signal that is directly proportional to limb displacement from desired position.

Point-to-Point Robots. With this type of robot, there are two main commands given: (1) the attitude of all limbs at the start of the move, and (2) the new attitude of those limbs when a particular move has been completed. While the robot makes the move as fast as possible and while it moves all limbs simultaneously to fulfill a given command, there is no definition of the paths that the robot limbs will trace. Thus, the term point-to-point control. In programming a robot of this type, however, the system designer must take into consideration certain *intervening points* between start and destination. The robot, for example, may have to clear an object that may fall in its "direct line" path, or it may be desired for the robot arm to approach its destination at the best angle (in the instance of picking up a pallet). Point-to-point robots can do any job performed by a limited-sequence robot previously described. With sufficient memory capacity, these robots can handle such jobs as palletizing, stacking, and spot welding, among others.

Continuous-Path Robots. There are applications where it is required to control not only the start and finish points of each robotized steps but also the path traced by the robot hand as it travels between these two extremes. Seam welding is an example of where a robot wields a welding gun and moves it along some complex contour at the correct speed to produce a strong and neat weld. Theoretically, the continuous-path robot is an extension of the point-to-point concept because the curved path is made up of many straight-line segments. This requirement, of course, calls for a very substantial memory as in the case of contour numerical control for machine tools.

PROGRAMMING ROBOTS

With the availability of sophisticated electronic hardware and system software developed during the past decade, programming represents one of the most advanced areas of robot technology, particularly as compared with earlier programming methodologies.

Development of Programming Techniques

Of historical interest and also because some of the earlier programming techniques are still in use and may persist for a while, it can be of value to trace the steps taken over the years to program robots. Early in the development of automated systems, at a time when robots and other automation techniques were largely associated with replicating the skills of human operators, recording the detailed manipulations of the operator and of later retrieving this information for use in guiding a mechanical device was the main source of programming information. Constrained by the available state-of-the-art equipment and methodologies then available, early systems posed difficult problems.

Playback Concept. One of the early techniques and one that is still used in connection with some robots is the so-called "playback" concept. In this method, the robot is programmed through a procedure known as "teaching." When one considers the numerous variables and complexities that can be encountered in applying robots, the need for a short cut to programming becomes evident. For example, in any reasonable volume of factory space, there are literally many thousands (depending upon the resolution needed) of points that may become part of a robot program. For example, if cartesian (X,Y,Z) coordinates are used, it immediately becomes apparent that, if each point is to be separately identified, this will require very large amounts of data to be stored. Furthermore, in the planning of a robotic system, the designer, without a short cut, must visualize just how the system will operate in three-dimensional space and express design objectives in terms of very long lists of coordinate positions.

In the early days of robot development, the designer did not have computer graphics and memory and computing systems in the form of minicomputers and microprocessors to turn to. But, as a result, early designers developed a clever and innovative "teach-and-playback" technique for robot programming.

In the *teach mode*, the robot is directed through various movements in sequence while the necessary information is recorded in memory. In the *playback mode*, the robot repeats, as desired, the sequence of movements taught by the operator and retrieved from memory. Incidentally, control of the motion may be point-to-point or continuous-path.

Of historical interest, diagrams illustrating earlier versions of the teach-and-playback method are shown in Fig. 11. These systems were used in the early 1970s.

Today the playback systems in use are the same in principle but utilize advanced digital hardware in place of earlier electronic and mechanical parts.

Levels of Robot Programming

Some authorities recognize three types or levels of robot programming and controlling systems:

Level 1—Lead-by-hand teaching (manual) and front-panel programming. Applications include some spot welding and pick-and-place tasks.

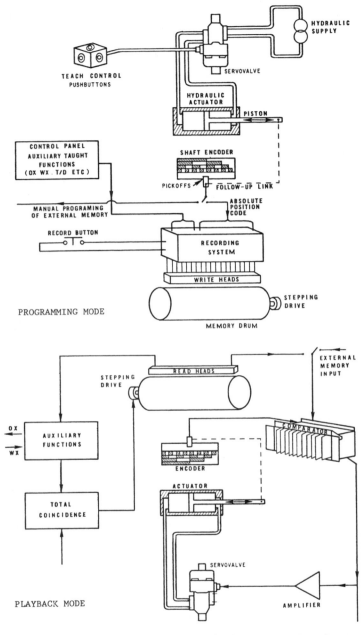

Fig. 11. Early (circa 1970) programming procedure for a playback point-to-point robot. In the *programming mode*, teach buttons control hydraulic power to an actuator for motion of arm-hand assembly to the desired position. A shaft encoder continuously monitors position. The record button is depressed to impress on the memory drum the position data for all five axes of motion. A stepping drive then indexes the memory drum one step. The process is repeated to record all steps in the program. An end-of-the-program signal is recorded for the automatic drum to return to the first step.

In the lower view, the *playback mode* is illustrated. Data for the desired position of each axis of motion, as recorded on the memory drum, are transferred by read heads to the comparator. Actual position, as shown by the encoder, is compared with the desired (memory) position. A difference in information activates the servomechanism to drive the actuator until the positions coincide. When all five axes of motion are at their memorized positions and compliance with other controls and auxiliary functions is obtained, the memory drum is indexed to the next step. (*An early system designed by Unimation.*)

Level 2—Programs are written in simple robot-programming languages, that is, techniques that assist programmers in entering motion, branching, coordinate-transformation, and signal instructions. Such systems may also provide a number of the lead-by-hand and control-panel operations that Level-1 systems provide. These programming capabilities permit the running of *user programs* that are more complex than Level 1 systems. Representative applications are found in some palletizing and arc welding uses.

Level 3—These are the most modern and growing methods of robot programming. The programming languages incorporate extended capabilities, including structured constructs, full arithmetic functions, external robot-path modifications, and supervisory computer-communications support. Since these systems support the functions and features also found on the two lower levels, Level-3 systems can handle Level-2 applications as well as modifying the robot arm's path, based upon data transmitted from external sensing devices, including machine vision.

Example of an Advanced Level-2 Programming System. The VAL-II programming system, developed by *Unimation (Westinghouse)*, provides the functionality of some Level-3 systems. As shown by Fig. 12, the basic hardware comprises a standard 19-inch Q-bus backplane, a servo control subsystem, a trajectory-human-interface subsystem, and a power-distribution system. VAL-II, which supplements an earlier VAL-I system, has 64-K bytes of C-MOS memory with battery backup (about twice that of the earlier VAL-I system). Another 64-K bytes of C-MOS memory contain the operating system. The system also includes a double-sized/double-density disk unit that stores up to 10,000 program steps per diskette, or about 1 megabyte of memory. VAL-II also makes eight serial ports available to users.

The VAL-II operating system uses a time-sliced control scheme. During operation about half of each major clock cycle is devoted to arm-trajectory planning and computation. The remaining portion is divided into time slots for the CRT interface, the supervisory computer interface, external path modification, monitor-command execution, main-program step execution, and process-control program step execution. Each task is scheduled to run for a given amount of time during each major clock cycle to ensure that the system can respond to time-critical events as quickly as possible. Any event will be received and acted upon within 28 milliseconds of occurrence.

Path Modification is accomplished through the use of an interface, named *Alter*. An RS-232-C interface is used. Robot paths can be modified in either of two modes (world or tool). As mentioned briefly before in this article (see Figs. 4 and 5), the world-coordinate mode refers to the fixed X,Y,Z directions, which are predetermined for each robot. The tool coordinate refers to the direction in which a robot is pointing at any given instant. To modify robot motion in real time, users connect an external device to the hardware port of *Alter*, which every 28 milliseconds transmits data consisting of cartesian positional offsets in the X,Y,Z directions and rotational changes about the three axes. *Alter* also makes it easier to interface smart sensors to VAL-II systems. A vision system can correct the position of a robot by using data received from the machine vision system's fields of view. VAL-II also can be interfaced to a supervisory com-

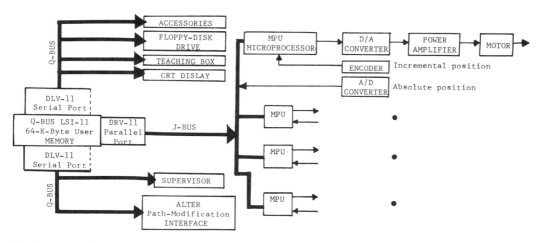

Fig. 12. VAL-II robot controlling system. A Q-bus (*Digital Equipment Corporation*) is used to connect the LSI-11 to other components of the robot system. A J-bus connects the 64K byte user memory to the joint processor-control system. (*Unimation-Westinghouse.*)

TABLE 2. Program for Drawing a Circle in the X-Y Plane and Modifying It in the Z Plane. (VAL-II Program)

```
• PROGRAM circle
 1    PROMPT "input radius in mm," rad
 2    TYPE "Move to center of circle"
 3    TYPE "Press COMP button when done."
 4    DETACH; Allow user to use Teach Pendant
 5        ; Wait for COMP button on Teach Pendant
 6    DO
 7    UNTIL PENDANT (2) BAND 20
 8    ATTACH ; Regain control of arm
 9    DECOMPOSE c[] = HERE ; get XYZDAT data for center
10    TYPE /B, "Moving in 5 seconds" ; Beep terminal
11    DELAY 5
12    PCEXECUTE pc. alt,-1.0 ; Start ALTERing program
13  Set internal ALTER, WORLD mode, cumulative
14    ALTER (-1,19)
15    WHILE SIG (1032) DO ; Continuously make circles
16        ; until signal stops it
17        FOR ang = 0 TO 360 STEP 5
18            x = rad*COS(ang) + c [1]
19            y = rad*SIN(ang) + c [2]
20            MOVES TRANS(x,y,c [3], c [4], c [5], c [6])
21        END
22    END
23    PCEND ; Finish up ALTERing program
• END

• PROGRAM PC . ALT
 1   ; This program will ALTER the z component of the
 2   ;   circle, dependent upon inputs from the external
 3   ;   binary signal lines 16-21
 4   ;
 5   ; E.G. If the binary signals are set up like:
 6   ;           BITS  1021 1020 1019 1018 1017 1016
 7   ;                  0    1    0    1    1    0
 8   ;
 9   ; Then this corresponds to a 26(8) = 16(H) = 22.
10   ;  correction in the z direction. Because
11   ;  this value will be divided by two (see ALTOUT),
12   ;  the correction would be 11 mm.
13   ;
14         ALTOUT 0,0,0,BITS(1016,6)/2,0,0,0
```

Note: VAL-I and VAL-II Programs were developed by *Unimation* (*Westinghouse*).

puter system through Digital Data Communications Message Protocol (DDCMP), used by Digital Equipment Corp. in its network communications. The physical connection is a standard RS-232-C serial link running at 9,600 bits a second.

Process Control Program, which is a background task that contains no arm-motion instructions, also can be run on the VAL-II system concurrently with a motion-control program. The process-control task also can be used to modify the robot path in real time by using *Alter*. This path modification can be calculated within the VAL-II system or based on various external inputs. See Table 2.

Structured Constructs, such as Case and If-Then-Else statements, can be implemented with VAL-II. These are valuable programming tools for complex robot programs. For these constructs, the system automatically generates intended decision levels whenever program steps are displayed on the CRT. See Table 3. The constructs provide the VAL-II programming language with a degree of power that compares with Pascal or PL/I.

VAL-II also has **procedural motion capability**. This allows users to define robot motion with mathematical formulas.

Future expansions of VAL-II include the ability to pass parameters to subroutines, allowing users to work with languages more familiar to them. Also, the ability to interface and move robots in tandem with welding positioners and gantry systems will be enhanced. Welding positioners, which hold the piece the robot is welding, are used when the robot's range does not extend sufficiently far to reach the whole area to be welded.

More detail on the VAL-II system can be found in Ref. 4.

EXAMPLE of a Microprocessor-Based Operator-Friendly Programming System:

In the T^3 ACRAMATIC (*Cincinnati Milacron*) system, a lightweight hand-held teach pendant is used to lead the robot through its required moves. A *controlled path motion* feature of the controller automatically coordinates all six axes to move the robot from one point to the next in world coordinates at the programmed velocity. An average of 3000 points can be programmed and stored within the control memory. (See Fig. 13.)

The straightforward menu-driven keyword approach to programming provides a simple interface with the robot. Powerful keyboard commands allow creation and editing of application data. While teaching, the status of the robot's program and operating statistics, I/O, or available memory, may be examined. The communications interface can provide the building block for an FMS (flexible manufacturing) system. The interface allows communication with host computers as well as with intelligent sensors.

Keyboard commands are given in Table 4. Hardware, operating system, and control specifications are summarized in Table 5.

EXAMPLE of an Off-Line Personal Computer Workstation Programming System:

Large manufacturers and users of robots are constantly aware of the fact that a day of lost production of an assembly line can cost the automaker the equivalent of $15 million in sales

TABLE 3. Structured VAL-II Program.

```
• PROGRAM robot
 1  ; Program to execute one piece of code for odd-numbered
 2  ;     rows, and a different section of code for
 3  ;     even numbered rows until all rows are done
 4         10 row = 1 ; Start at row #1
 5         DO
 6            IF row MOD 2==0 THEN ; Check for an even row
 7  ; ODD row
 8               FOR part = 1 TO col, odd, end
 9                  MOVE odd [row*col . odd.end + part]
10               END
11            ELSE
12  ; EVEN row
13               FOR part = 1 TO col.even.end
14                  MOVE even [row*col.even.end + part]
15               END
16            END
17         UNTIL row == row.end + 1
• END
```

Note: VAL-I and VAL-II Programs were developed by *Unimation* (*Westinghouse*).

Fig. 13. Checking performance of robot program against control system shown at left. While particular robot shown is fitted with arc-welding hardware, this style robot has numerous other industrial uses. (*Cincinnati Milacron*.)

at the showroom level. Thus, automakers are concerned with robot downtime attributed to the normally slow process of entering program logic statements for machine, peripheral and sensor interfaces, conditional and unconditional program branching, looping, register manipulation, digital I/O, indirect addressing, timers, macro instructions, and other commands. This input is essentially the same for all robots performing a similar task and can represent up to 80% of total programming time requirements. This problem can largely be resolved in installing a system that permits the logic portion of a program to be generated *off-line* before robots are installed.

The approach taken by *GMF Robotics Corp.* is reflected in their *SmartWare*™ Personal Computer Workstation system, shown in Fig. 14. The system consists of four modules, which are supplied in either of two basic combinations to fit a user's needs: (1) The *SmartWare* PC Workstation is a turnkey system, including a 32-bit CPU, CRT, Keyboard, dual disk drives and printer, plus an Edit/Store software package and a bubble cassette interface unit. Or (2), the bubble cassette interface unit and the Store Software Package may be procured for use with a PDP-11/RSX-11M and VAS/VMS computers. The latter option provides upload/download, storing, copying and printing capabilities as well as the ability to view programs on remote CRT screen and change

TABLE 4. Keyboard commands used in T^3 ACRAMATIC Version 4 Industrial Robot Programming and Control System.

DELAY causes the robot to pause in cycle for a specified time period (up to 30 seconds).

DISABLE cancels the effects of previously executed "Perform-On-Interrupt" or "Event" statements.

EXECUTE macro type facility, which allows user-defined functions consisting of groups of standard functions.

FLAG (optional) records and monitors events that occur during a user task; can be used to condition functions, such as "Perform," "Execute," etc.

INDEX (optional) allows multiple entry and exit points from any sequence except the mainline; useful in palletizing.

MESSAGE (optional) prints messages on the *Teach Station* during *Auto* mode; can optionally be spoken.

NOP—*No Operation*—provided for use where only robot location is important and no function is to be executed.

OUTPUT controls a single output contact or several contacts (optional); can occur concurrently with robot motion.

PERFORM permits entry into specified sequences (subroutines); may be conditioned; interrupt driven.

POSITION (optional) ability to capture the current Tool Center Point location for later use; used with palletizing applications to reduce cycle time.

PWP—*Programmable Workpiece Positioner* (optional) provides point-to-point control of auxiliary positioning devices (up to 4 axes/2 positioners); can move simultaneously with robot; both absolute and incremental commands possible.

REMOVE (optional) provides ability to communicate with a host computer using DDCMP protocol over RS232C, RS422, or 20 mA link at speeds up to 19.2 k baud. Capabilities include: up/down loading of both operating system and user data, error reporting, cycle control, dynamic program modification while in-cycle.

SEAM TRACKER (optional) provides the robot with the ability to "track" a weld seam using "Through-the-Arc" sensing.

SEARCH (optional) dynamically relocates a taught point based upon changes in the physical environment (such as changes in stack height in a palletizing application).

VARIABLE provides ability to count occurrences of events, read numeric values from input contacts, and define numeric values via keyboard entry; addition, subtraction, multiplication, and division are possible; can be used to condition program execution.

VELOCITY (optional) allows velocity table entries to be altered through program control; useful when more than 15 velocities are required.

VISION (optional) provides ability to identify objects and account for *X, Y,* and Roll positional differences from taught location.

WAIT stops robot motion and waits for a condition to be satisfied; may be timed if desired.

WEAVE (optional) causes the robot to superimpose a planar weave pattern over its programmed straight line path; can also be used to produce an oscillation motion when the arm is "stationary" and the part is moved by a positioner.

Note: T^3 ACRAMATIC was developed by *Cincinnati Milacron, Industrial Robot Division.*

TABLE 5. Specifications of T³ ACRAMATIC Industrial Robot Control.

Hardware
Distributed microprocessor architecture
Operator panel with membrane keys, LED indicators, key start switch
Teach pendant with full programming capability and message scrolling
Portable Teach Station with 24 x 80 display and keyboard (optional)
Hardcopy print out (optional)
Cartridge tape loader with 9600 baud data transfer rate (required option)
Self diagnostics with error display
Microprocessor controlled servos with absolute position measurement
Serial communications RS232C, RS422, 20mA current loop (optional)
Memory battery backup, 5 days
Special Interfaces for Voice, Vision, and Seam Tracking (optional)
Dedicated system input/output contacts 6/10
User programmable input contacts/Maximum optional 16/32
User programmable output contacts/Maximum optional 16/32
Analog output channels (12 bit, ± 9 V) (optional) 8
Servoed axes maximum 10

Operating System
Programmed in "world" coordinates (X, Y, Z, D, E, R)
Controlled path motion, straight line path at TCP*
Velocity and ACC of TCP* controlled along the path
Teach mode coordinate systems: Rectangular, Cylindrical, Hand
Two programmable pendant keys for tooling operation in Teach mode
Teach mode password
Auto/Teach mode transition
Event servicing (Stop, Restart, Hold, Hold Clear, Error, Power Up)
Restart cycle after power failure without realigning
Programmable "Home" position
Velocity scaling factor (0.1 to 9.9) and clipping
Inch/Metric unit selection
Interrupt status assignable to any user input
Relocatable and Absolute Sequences
Sequences (number of sub-programs) 256
Memory area for data/average number of data points 48k bytes/3000
Variables (16 bit integer) 64
Flags (optional) 64
Velocity table entries 15
Tool dimension table entries, (3 dimensional) 16
Analog output schedule entries, (2 values/schedule) (with option) 32
Weave pattern entries, (5 parameters/weave pattern) (with option) 32
Trigger table entries, (distance or time) (with option) 16

Control
Control size, width x depth x height, in/mm .... 34.3x34x60/(890x865x1525)
Weight (including drive amplifiers) .... 1470 lb (670 kg)
Ambient temperature .... 40 to 105°F (5 to 40°C)
  with air conditioner option .... 40 to 120°F (5 to 50°C)
Power requirement .... 460 volt, 3φ, 60 Hz**

*TCP represents Tool Center Point   **Other voltages and 50 Hz available

Note: *Milacron*, T³, and *ACRAMATIC* are trademarks of *Cincinnati Milacron Industries, Inc.*

program names. At the workstation the operator can create and modify programs off-line, even while the robot is working in the plant (or before installation). The user can edit stored programs and enter, store, and display programming or editing comments, notes, and reminders.

The *SmartWare* program is user friendly. Programmers can create, edit, and read robot programs in a descriptive, English-like language or work with the standard manufacturer's syntax, which is derived from the standard numerical control (NC) language. The bilingual capability of the *SmartWare* system provides instantaneous translations from one of these languages to the other on the CRT screen or in hard-copy form.

There are three options for entering programs: (1) typing the commands in either of the two previously mentioned languages; (2) selecting

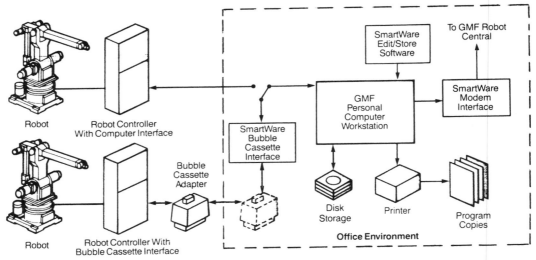

Fig. 14. Off-line personal computer workstation programming system. (*SmartWare™ System, GMF Robotics Corporation.*)

commands that are "scrolled" on the CRT; or (3) picking commands from a menu. The inexperienced programmer may prefer a menu—shown in reverse video at the bottom of the CRT screen—which corresponds with the keyboard and guides the sequence of entries. Using the scrolling method is also simple. By rotating a dial on the keyboard, a programmer can scan through a list of possible commands and, upon finding the next desired step, can punch one key to enter it into memory. The experienced user can bypass these two methods and type commands in at regular typing speed.

Another user-friendly feature is the "comment" memory, which the programmer files while constructing the logic. Notes so entered can make it easier to program the robot path later in the shop. This comment area also can be used to annotate the robot program listing.

Off-line programming may not be needed for single-robot installations that rarely require program changes, but it is an invaluable aid for multiple-robot operations and undoubtedly essential for FMS (flexible manufacturing systems).

EXAMPLE of Software-Based Motion Control System:

Consider the robot (*Model 660—Intelledex Incorporated*) designed for light industrial assembly and shown in Fig. 15. This is a six-axis robot, the

Fig. 15. A six-axis robot with a human-like (anthropomorphic) joint configuration, giving it exceptional dexterity and allowing it to approach any point in the work envelope from any angle. This provides easy adaptability to existing workstations. (*Model 660, Intelledex™.*)

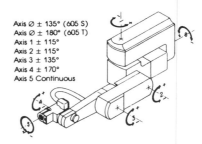

Fig. 16. Robot arm with six axes. Each axis has an angular coordinate or motor angle. Software is used to convert the motor angles to world coordinates and tool coordinates. (*Intelledex Corporation.*)

motion control of which uses fifteen microprocessors, six optical encoders, and six stepper motors. The robot design has been optimized for speed, repeatability, accuracy, flexibility, and low cost. A diagram of the axes system is given in Fig. 16.

Six degrees of freedom are required to minimize the demands on the construction of the robot workstation. In the interest of low cost and speed, a distributed processing system was selected. Stepper motors were selected because of relatively low cost and ease of control. Problems of accuracy and motor control are solved by software. (See Fig. 17.)

Some of the tasks assigned to software include calibrating the arm, control of end-effectors and safety devices, and image processing analysis by an integrated vision system. The software is also responsible for motor control functions and software control of motion. Although the robot is built with as much mechanical accuracy as possible considering cost restraints, the arm-parameter calibration, for example, can make up for certain manufacturing tolerances. Nearly twenty sources of mechanical error have been identified. These errors, however, are constant for a given robot but may vary from robot to robot. Once these constants are determined for a specific robot, they can be used by the software in performing calculations that accurately position the robot.

Software also interacts with safety devices, such as light curtains, safety stop switches, or pressure mats. These can be plugged into any of twelve safety ports, four of which are nonmaskable. The maskable ports allow software to ignore a safety switch shutdown signal.

Five requirements of the motor control system include: (1) microstepping of the stepper motors, (2) high-speed switching of the motor coil currents, maximizing power during high-speed moves, (3) motor synchronization, (4) backlash control, and (5) stall monitoring.

As shown by Fig. 18, stepper motors generally move in "full" or "half" steps. In the diagram it will be noted that the motor is shown at a position between coils 2 and 3. This position would be maintained as long as current is flowing through the coils. Half-steps are achieved in the stepping motors by phase changes of the coils. To obtain the required accuracy, microsteps are achieved by reducing the current slightly in a coil, resulting in sixteen positions for each

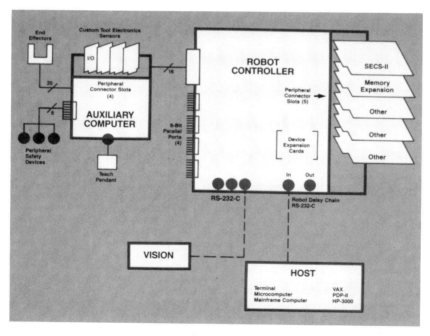

Fig. 17. Robot controller for *Intelledex™ Assembly I Plus* robots, such as the *Model 660* previously illustrated. The controller utilizes an *Intel 8086/8087* microprocessor pair and a *Motorola 68121* microprocessor and accepts plug-in hardware for *SECS-II*, and other communication interfaces, including MAP. A special auxiliary computer uses an *Intel 8088* microprocessor in multitasking fashion to manage information received from the end-effector sensors parts presentation mechanisms and optional safety devices, thereby allowing the CPU (central processing unit) to attend to system needs. Programs can be developed off-line on a personal computer and then loaded into the memory of the robot controller, eliminating much of the requirement to shut down the production line to teach points to the robot.

276 Robot Technology Fundamentals

Diagram of Coils in a Stepping Motor

```
      3
      ↑
0 ──┼── 2
      1
```

PHASE CHANGES FOR HALF-STEPPING

3	2	1	0	Hex Equivalent
1	1	0	0	C
0	1	0	0	4
0	1	1	0	6
0	0	1	0	2
0	0	1	1	3
0	0	0	1	1
1	0	0	1	9
1	0	0	0	8

Fig. 18. Half-steps are achieved in stepping motors by phase changes of the coils.

step. It is possible to obtain sixteen discrete steps by controlling the balance between the various coil currents. A pair of microprocessors is used to control the current flow to each of the six motors.

To obtain the high speeds required, the stepper motors must be driven at a rate of 5200 full steps per second. This reduces to the fact that the coil currents must be changed at a rate of 10,400 times per second, or once every 96 microseconds. Use of the microprocessors greatly simplifies the task as compared with traditional electronic hardware. Also, by controlling coil voltage (as opposed to current), these same two microprocessors are used to accomplish the coil phase changes. See also the article on stepper motors in *Section 4* of this *Handbook*.

Backlash is the amount that a motor moves when changing direction before an arm is actually moved. Antibacklash gears help to minimize this error, but that remaining can be further reduced by use of a feedback system based upon precisely made optical encoders and sensors. This is another task that can be managed by microprocessors. The combination of motor driving and encoder monitoring allows for creation of a safety feature. Should the arm hit an object, the motors will stop, but the encoders register no movement, thus indicating a stalled condition.

With further reference back to Fig. 16, corresponding to each axis there is an angular coordinate (*motor angle*), which describes the rotation of the motor shaft. Six degrees of freedom enable arm position to be defined by six variables in real space.

Transformation refers to the mathematical conversion of motor angles to world coordinates and tool coordinates, and vice versa. This is the main task of software. The need for rapid calculations is met by combining an *Intel 8087* math coprocessor with the robot's main controlling processor, an *Intel 8086*.

A movement command can be issued from a host computer in direct mode (the movement is executed immediately), from a teach pendant (directed by a group of buttons), or from a running ROBOT BASIC program. (See Fig. 19.) The movement command can be in world coordinates or motor angles. The number of steps that each motor has to move (delta steps) is calculated. Maximum speeds for the various motors are then adjusted in a synchronization code so the motors are timed to complete their moves at the same time. That is equivalent to slowing down the motors that have less distance to travel. The information, computed by the *8086/8087*, is then transmitted to six pairs of Zilog (Z8) motor controllers, each of which controls one of the six motors. For each motor there is a master Z8 and a slave Z8. The master Z8 receives information from the *8086*, defining a desired move for that motor, and also reads the optical encoders. It directs the slave Z8 to issue the motor pulses, which move the motor the calculated number of delta steps. Because of backlash in the system, the arm may not be exactly at the intended position at the end of the move. The actual position of the arm is known by the master Z8 since it is monitoring the encoder. Therefore, the master can order the slave to make a move attempting to correct for the position error. This procedure is repeated until the position is obtained that is within the desired degree of accuracy. Higher accuracy requires more iterations and thus time required. High precision moves can require as much as 0.5 second of extra time.

The specialized modes of motion used are "straight-line," "continuous-path," and "move-through." How fast a stepper motor moves is based upon how long a delay exists between coil changes. A stepper motor will not respond, but will in fact stall, if an attempt is made to change the velocity too rapidly. Therefore, a typical move has three phases: (1) acceleration, (2) deceleration, and (3) maximum velocity (called maximum slew).

The typical move of a robot arm with rotary joints is in an arc, whereas many applications require a straight-line motion. To move in a straight line, the path between two given points is broken down into a number of smaller paths. Thus, instead of a big arc, the path is broken down into a number of smaller arcs that approximate a straight line. The finer the breakdown into smaller arcs, the smoother the path. Thus, to move in a straight-line path between two points, a large number of points must be calculated between the end points.

If one uses the normal motor ramp for a straight move, excessive vibration is produced from the abrupt and rapid ramping, both up and down, of the motors. This can be avoided through effective application of software. During the time that a move is occurring between two points of a straight-line move, the next point is calculated and transmitted to the motor. The move then continues without the necessity of the motors

Robot Technology Fundamentals 277

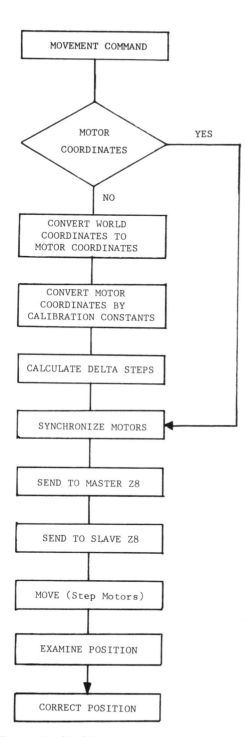

Fig. 19. Simplified flowchart illustrating steps taken when a movement command is given to any of the motors.

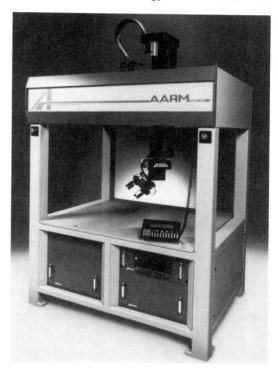

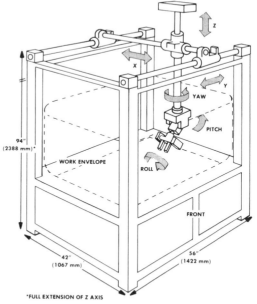

Fig. 20. Intelligent, high-precision parts-assembly robot that incorporates both optical (vision) and tactile (force) sensors for feedback. Modular workstation concept of overhead gantry, as shown at right, facilitates integration into assembly flow. (*Adaptive Intelligence Corporation.*)

ramping down. Speed of movement, of course, is limited by computation time.

More detail can be obtained from the Harms article (Ref. 5).

EXAMPLE of Software for Intelligent, High-Precision Parts Assembly Robot

Consider the robot illustrated in Fig. 20. This is an electric-drive, overhead gantry robot that incorporates tactile sensor technology. As designed, the machine is a completely self-contained workstation that allows either stand-alone or integrated assembly line configuration. It has a high level of dexterity of the kind required for the assembly of disk drives, printed circuit boards, wire harnesses, and telecommunication and electromechanical devices. Automatic laser inspection will be added to the system at a later date.

As shown by the close-up of Fig. 21, the intelligent gripper ("*Adaptive Touch*"™) incorporates both optical sensing and a strain gage design that allows force-sensing thresholds to be programmed in three orthogonal axes. Parts that are out of tolerance or that are not properly oriented are detected quickly, enabling the robot to respond as required. It will take intelligent action based on the specific "force signatures" programmed into the gripper for recognizing and accommodating specific assembly tasks. Design of the machine (by *Adaptive Intelligence Corporation*) allows expandable modular configurations of three, four, and six axes. These axes handle payloads of 17, 15, and 9 pounds (7.6, 6.7, and 4.0 kilograms), respectively, at a maximum speed of 40 inches/second (1016 mm/s). The six-axis configuration consists of the three linear axes (X, Y, Z) and three rotary axes (yaw, pitch, and roll). All are driven by DC servomotors. Accuracy and repeatability are ±0.002 and ±0.001 inch (0.05 and 0.02 mm), respectively. Resolution is specified as 0.0005 inch (0.013 mm).

A teach pendant is provided for on-line programming and reprogramming to accommodate production line changes. A hand-held tool allows production personnel to control all robot movements, time delays, high-level commands for self-calibration, and to call up complex subroutines.

Supporting the teach pendant, including on-line and off-line programming through an IBM PC compatible computer, is the firm's Adaptive Assembly Machine Programming Language (AAMPL)™. Programs can be written and tested interactively on-line or written off-line and downloaded from disk for implementation. The system's architecture is shown in Fig. 22 and consists of: (1) a teach pendant, (2) the robot programmer, (3) the robot controller, and (4) the robot power chassis.

The AAMPL program allows for three possible situations as delineated in Table 6. A sample robot program is shown in Table 7. A summary of robot and I/O control statements is given in Tables 8 and 9, respectively.

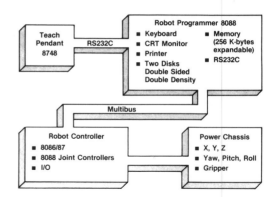

Fig. 22. Architecture of AAMPL system consists of four separate subsystems. (*Adaptive Intelligence Corporation*.)

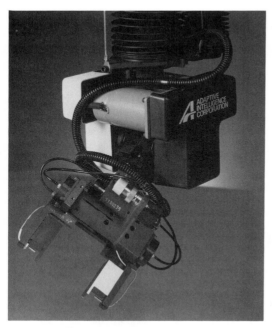

Fig. 21. Close-up of intelligent gripper that incorporates both optical sensing and strain gage sensors. (*Adaptive Touch*™; *Adaptive Intelligence Corporation*.)

TABLE 6. Accommodation of AAMPL™ Programming Language to Three Major Installation Situations.

1. Within Its Own Work Space

The robot's multi-tasking software enables the control of all activities within the work envelope, including:
   (a) Gripper arm movement to a given site with a specified speed, acceleration, clamp force, and wait on arrival.
   (b) Use of force feedback and/or optical sensing to test for particular conditions, or to determine precise location of an object whose position may vary.
   (c) Exception recognition by monitoring how the arm travels and what patterns of forces it encounters and automatically comparing actual values with anticipated values.
   (d) Exception handling by means of responses prescribed by the user.
   (e) Logging of production statistics and exception data.
   (f) Switching from one batch-oriented procedure to another simply by changing the tooling plate and the assembly program and executing a self-calibration routine.

2. Within Its Vicinity

Like the robot controller, the integrated I/O controller has multi-tasking software and a dedicated command set. With input from 64 sources and output to 32 devices, it can:
   (a) Program the AARM's interactions with conveyors, auto-guided vehicles, and other ancillary parts-feeding equipment.
   (b) Schedule multiple events concurrently and time I/O to be synchronous or asynchronous with the motions of the gripper.
   (c) Operate two or more robots as a cluster, passing information, parts, and tools between them.

3. Within A Factory-Wide System

The full upward compatibility of the system permits:
   (a) Integration of the robot into an existing manufacturing structure. Software is available to interface the AARM to a master factory computer or CAD/CAM system. Instructions can be downloaded to the robot, and data compiled by the robot can be uploaded.
   (b) The AARM can be incorporated into a true manufacturing network. Full compatibility (hardware and software) with standards set by MAP (Manufacturing Automation Protocol).

Note: AAMPL™ Programming Language was developed by *Adaptive Intelligence Corporation*.

TABLE 7. Representative Robot Program (AAMPL™).

```
COMMENT: FIRST PICK UP THE SPECIAL RING
         ASSEMBLY TOOL
  MOVE TO OVERTOOL
  MOVE TO TOUCHING, ACC=10, SPD=10
  MOVE TO LOCKON, ACC=5, SPD=5
  MOVE TO WITHDRAW, ACC=25, SPD=25
COMMENT: NEXT GO TO GET THE RING
  MOVE TO OVERRING
GETRING:
  MOVE TO TOPRING
COMMENT: THE GRIPPER WILL SOON BE HOLDING
         THE RING...PROGRAM WILL BRANCH WHEN
         GRIPPING FORCE IS 0.5 POUNDS INDICATING
         RING IS PRESENT
  MONITOR ENABLE, USING PINCH 0.5 TO HAVRING
  MOVE TO GRABRING, SPD=25
COMMENT: WE GET HERE IF THERE IS NO RING
  MOVE TO OVERRING
  DISPLAY WE HAVE GONE TO PICK UP THE RING
    BUT IT IS NOT THERE
  GO TO GETRING
HAVRING
  MONITOR DISABLE USING PINCH
  MOVE TO LIFTRING
COMMENT: NOW MOVE TO POSITION OVER THE POST
  MOVE TO OVERPOST
  MOVE TO POSTBASE, SPD=15
COMMENT: GRIPPER OPENS TO DROP THE RING,
         THEN LIFTS THE TOOL TO CLEAR THE POST
  MOVE TO RELEASE
  MOVE TO OVERPOST
COMMENT: NOW RETURN THE TOOL TO ITS PARK
  MOVE TO WITHDRAW
  MOVE TO LOCKON, ACC=10, SPD=10
  MOVE TO TOUCHING, ACC=5, SPD=5
  MOVE TO OVERTOOL, ACC=25, SPD=25
```

TABLE 8. Robot Control with AAMPL™ Programming Language.

DATA TYPES:	Site	A point in space defined by coordinates for X, Y, Z, Yaw, Pitch, Roll and width of gripper opening.
	Counter	An integer used to count parts, errors, or repetitions of a procedure.
COMMANDS:		
Motion Control	Move	Moves the arm to a predefined site. You can specify acceleration, speed, gripper clamp force, and time to pause after moving, or select default values.
	Stop	Stops the current motion of the gripper.
	Home	Places the gripper at a predefined "home" position in the work envelope.
Site Modification	Origin	Used to execute a set of moves relative to a chosen location.
	Inc/Dec	Changes the location of a site along an axis, e.g., decrementing a Z value to get to the next lower part in a vertical pallet. Also used to increment or decrement counters.
	Store	Moves site data from one location to another, or from the current AARM position to another site. Also used to set a counter value.
Robot Program Control	Repeat	Repeats a series of AAMPL statements. A counter specifies the desired number of repetitions.
	If	Compares a counter value to a constant and redirects program flow depending on the result.
	Label	Marks a spot in the program to which control can be transferred from commands like Go To.
	Go To	Transfers program control to the specified label.
	Perform	Executes a sequence of program steps, then returns program control to the statement following the Perform command.
	Monitor	Looks for input from the gripper sensors (a light beam between the fingers, or Pinch, Tip or Side sensing with the fingers), or from up to 64 external intelligences. When input occurs, the motion of the gripper stops and program control resumes at the specified label.
	Call	Executes an AAMPL sub-program stored on disk.
Operator Interface	Display	Displays a message for the operator, who must press CLEAR key to resume program execution.
	Comment	Annotates a program listing with information about the intent of the instructions.
	Log	Records date and time along with a label in a disk file. Reports exception and production data.
SUB-ROUTINES:	Findpost	With the gripper fingers positioned on either side of a post, locates the exact center of the post. This position can be stored as a site.
	Findtop	With the gripper fingers positioned over an object, locates the top of the object.
	Grasp	With the gripper fingers on either side of the object but not necessarily centered on it, closes the gripper and moves it so as to grasp the object.

In the short AAMPL program shown above, the gripper picks up a special assembly tool from its park and then uses the tool to pick up a ring and place the ring on a post. By monitoring the gripper's pinch strain gage, the AARM can detect the absence of the ring. The program defines the desired response to this exception—i.e., to alert the operator through a message on the teach pendant display. The final step is replacing the tool into its park. Three operations are performed here, with the first repeated in reverse order as the final step. Each has four components: *over, down, clamp* (or release), and *up*.

TABLE 9. I/O Control with AAMPL™ Programming Language.

DATA TYPE:	Channel	Collection of bits comprising input from a sensor or output control for an external device.
COMMANDS:		
I/O Activity	Input	Reads the information from an input sensor and stores the value into a counter.
	Output	Sends a value to an output channel.
Task Control	Task	Defines an independent set of I/O statements which can execute concurrently with AARM motion and with other I/O tasks.
	Start	Starts an I/O task.
	Stop	Stops an I/O task.
Time Control	Wait	Suspends task execution until a specified input channel reaches some value.
	Delay	Suspends task execution for a specified time period.
I/O Program Control	Loop	Restarts a task a specified number of times.
	Compare	Starts one of three other tasks depending on whether a given input channel is less than, equal to, or greater than some value.
Operator Interface	Notify	Sends an event notification to the robot programmer. A Monitor statement can watch for the event and redirect program flow when it occurs.

LOAD CAPACITY AND POWER REQUIREMENTS OF ROBOTS

A 1985 survey of manufacturers of robots in the United States, Europe, and Japan, based upon nearly 140 suppliers and not including a number of small and very specialized robot manufacturers, indicated that models available (*not* in terms of total units sold) were designed to handle loads ranging from about one pound (0.5 kilogram) upward to about 2300 pounds (1043 kilograms). Most likely these figures do not bracket *all* robots ever made, either in terms of very heavy or very light loads. In terms of models available, the breakdown was approximately that shown in Table 10. Applications listed in the survey for robots included die casting, forging, plastic molding, machine tools, investment castings, a general and miscellaneous category, spray painting, welding, and machining. It is well established, of course, that the use of robots in light manufacturing and inspecting operations, such as are found in the electronics industry, has increased markedly during the past few years. The surprisingly uniform distribution of robot models by carrying capacity gives some insight to the large spectrum of power requirements facing the designers of robots unless, of course, they are specializing in just a few select applications.

A separate survey (1984) indicated that, in terms of total robots made (*not* models available) electric drives account for about one-half of the robot drives used; hydraulic drives, about one-third of the total; and pneumatic drives, about one-sixth of the total. Some authorities believe that these ratios will hold rather steady; others profess a solid trend toward electric drives.

Many electric robots utilize DC stepping motors. Hydraulic robots usually employ hydraulic servo valves and analog resolvers for control and feedback. Digital encoders and well-designed feedback control systems can provide hydraulically actuated robots with an accuracy and repeatability generally associated with electrically driven robots. Pneumatically driven robots normally are found in light-service, limited-sequence, and pick-and-place applications.

Electric Drives for Robots

Electric motors provide the greatest variety of choices for powering manipulators, especially in the low- and moderate-load ranges, and for low-speed high-load operations. They are relatively easy to control, and a number of control techniques can be used. Electric motors are not so responsive as hydraulic systems and are considerably stiffer than pneumatic systems, unless the latter are operated under very high pressure.

Motors generally operate at speeds that far exceed those desirable for manipulator joints. Thus, speed reducers are required. Although speed reducers have the advantage of amplifying available torque and in preventing or inhibiting back driving, they are usually a major source of inefficiency and error. In demanding applications, the most important factors in choosing a motor include the power-to-weight ratio and torque-speed characteristics. The ability to accelerate and decelerate the working load quickly is a very desirable attribute. Also required is the ability to operate at variable speeds.

The development of microstepping techniques, pioneered by the office automation and instrument industries, contributed to the knowledge of and effective use and control of electrical stepping motors. For example, microstepping had led to the use of stepping motors in applications requiring incremental rotary motion of only a fraction of the particular motor's primary step angle. Hybrid step motors combine the fast response of variable reluctance motors with the detent torque of permanent-magnet step motors. Step motors are easily used in closed-loop servo systems and may also be operated in a synchronous mode at their slew speed. See article on stepper motors in *Section 4* of this *Handbook*.

Because electric robots do not require a hydraulic power unit, they conserve floor space and decrease factory noise. In an electric manipulator the motors generally connect to the joints through some kind of mechanical coupling, such as a leadscrew, pulley block, spur gears, or harmonic drive. This is because electric motors generally produce much less force or torque than a hydraulic actuator of the same size, so they require a mechanical impedance matcher between them and the joint if they are to overcome the loads that are encountered in a typical manipulator. A hydraulic actuator can usually drive a joint directly.

Permanent magnet DC motors have proved a good choice for medium- and small-size manipulators in the past. They are generally more efficient, less expensive, lighter, and smaller than wound-field motors. These servoed motors have been and continue to be a good choice for a number of applications. The brushless, electronically commutated versions have very long lives.

Printed-circuit motors have high torque relative to their rotor inertia and, therefore, fast response times. These motors are capable of driv-

TABLE 10. Load-Carrying Capacity of Robots

Load Capacity		Percent of Robot Models Offered
Pounds	Kilograms	
300 to 2300	136 to 1043	11%
100 to 299	45 to 136	16
50 to 99	23 to 45	15
20 to 49	9 to 22	21
10 to 19	5 to 9	17
Less than 10	Less than 5	20

ing at low speeds without the need for speed reducers.

Many motors manufacturers provide speed reducers, but most of them are of limited commercial quality, which may introduce backlash and therefore result in inferior performance. During the past decade, a number of motor manufacturers have tackled the design of motors strictly for robot applications, and where exceptional performance is required, there are several sources of customized robot motor systems. For the robot designer, considerable useful information is contained in Ref. 6.

Hydraulic Actuators for Robots

Hydraulic actuators are either (1) linear piston actuators, or (2) a rotary vane configuration. If the vane type is used as a direct drive, the range of joint rotation is limited to less than 360° because of the internal stops on double-acting vane actuators. Hydraulic actuators provide a large amount of power for a given actuator (of course, not including the weight of valves, pumps, piping, etc.). The high power-to-weight ratio makes the hydraulic actuator an attractive choice for moving moderate-to-high loads at reasonable speeds. Hydraulic systems are relatively easy to control because the low compressibility of hydraulic fluids results in the systems being very *stiff*. Hydraulic systems are characterized by fast response time, high natural frequencies, and low noise levels.

Where exceptional dynamic performance is required, the choice lies between making the piston actuators greatly oversized or going to hydraulic motors. Motors usually provide a more efficient way of using energy to achieve a better performance, but they are more expensive. Not only is the cost of hydraulic motors considerably more than that of actuators, they also require auxiliary devices, such as gearing or ball screws, to complete the system. Whichever system is selected, an electrohydraulic servovalve will be required. A common characteristic of hydraulic robots is oscillation or bounce in moving and decelerating to a point. By programming a delay, to allow for settling before a tool function, the difficulties caused by oscillation essentially can be eliminated.

A major disadvantage of hydraulic systems is their requirement for an energy storage system, including pumps and accumulators. Hydraulic systems also are susceptible to leakage, which may reduce efficiency or require frequent cleaning and maintenance. The working fluid must always be kept clean and filtered free of particles. Also, air entrapment and cavitation effects can sometimes cause difficulties. Another major disadvantage for the designer is the bulkiness of hydraulic systems and the requirement for accommodating the relatively stiff lines that circulate the fluid from the accumulator to the actuators. To make the control systems as stiff as possible, the servovalves, which control the fluid flow to the actuators, must be placed close to each actuator, and this further adds to the bulkiness of the design.

Several years ago a number of machine tool builders switched from hydraulic to electric machine drives. The principal reason then was reliability and leakage problems associated with hydraulics. During the intervening years, however, improvements in hydraulic systems have included better filters and seals so that today a hydraulic servo system no longer can be regarded as second in reliability to a DC servomotor system. Oil leaks in hydraulic drives generally develop gradually and progressively. Leaks are noticeable and frequently can be repaired when convenient, and they rarely cause unscheduled downtime. The danger of fire has been greatly reduced through the use of improved hydraulic fluids (phosphate ester and water-glycol types).

In paint spraying and other applications where the environment may present an explosion hazard, the robot either must be explosion proof or intrinsically safe. In such cases the hydraulically driven robot has obvious advantages over its electric counterpart.

Hydraulic power lends itself to some robot applications because energy can be easily stored in an accumulator and released when a burst of robot activity is called for[1]. Because there are no convenient means to store electrical energy, designers of electrically driven robots tend to underpower the drives. To obtain the necessary dynamic performance, they often use gear ratios that are too high. This results in a *snappy* robot for small moves and an unacceptably slow machine for large transfer moves.

Pneumatic Drives for Robots

Pneumatic drives are generally found in relatively low-cost manipulators with low load-carrying capacity. When used with non-servo controllers, they usually require mechanical stops to ensure accurate positioning. Pneumatic drives have been used for many years for powering simple point-to-point motions. Most often used configurations are a linear single or a double-acting piston actuator (for single acting, a spring powers the return motion). Rotary actuators also are used. In converting linear actuation to rotary motion, a drive pulley connected to the actuator by a cable (with two or more wraps around the pulley) may be used, thus avoiding the nonlinearities of joint motion inherent in linkwork conversion of linear to rotary motion.

An advantage of the pneumatic actuator is its inherently light weight, particularly when operating pressures are moderate. This advantage coupled with readily available compressed air supplies makes pneumatics a good choice for

---

[1] For example, some contemporary robots will require as much as 60 gallons per minute of fluid, which requires a 17-gpm pump operating part time for loading an accumulator.

moderate-to-low-load applications that do not require great precision. Because of their light weight, pneumatics is often used to power end-effectors even when other power sources are used for the manipulator's joints.

The principal disadvantages of pneumatic actuators include their inherent low efficiencies, especially at reduced loads, their low *stiffness* (even at the high end of practical operating pressure), and problems of controlling them with high accuracy.

Fluid power for robots (hydraulic and pneumatic) is reviewed briefly in Ref. 7.

Electromechanical Drive Hardware

Electromechanical systems are used in about 20% of the robots available today. Typical forms are servomotors, stepping motors, pulse motors, linear solenoid and rotational solenoids, and a variety of synchronous and timing belt drives. In the design of their Model M-100 robot, *Citizens Robotics Corporation* sought means to reduce costs and overcome weight problems. The robot per se weighs just 125 pounds (57 kilograms), but can lift a load of 25 pounds (11 kilograms), giving it a lift ratio of 5:1. Chain drives were found to be too heavy and ball-tendon member drives too restrictive. The robot is an articulating design of jointed spherical members. It uses six DC motors and 15 synchronous belts for power transmission. The robot arms are counterbalanced by the motors to achieve dynamic and static balance for all movements. It is claimed that this arrangement reduces the power needed to lift the maximum load to about 300 watts.

The designers used 5 mm pitch, 40-inch drive belts. At each axis of movement, the belts come off gear heads on the motors, go to a large timing belt sprocket to effect a gross torque reduction, and then distribute the power on an axial basis. Reduction ratios are variable and dependent on the individual axis and the velocity and amount of torque needed. The belts were selected because they have a good torque-to-weight ratio, are relatively inexpensive, require negligible maintenance, and have a high degree of dimensional accuracy from one belt to the next.

DYNAMIC PROPERTIES OF ROBOTS

Included among the important dynamic properties of robots are: (1) stability, (2) resolution, (3) repeatability, and (4) compliance. Considering these factors, the design of a robot is innately complex because of the manner in which these properties interrelate. Complex interrelationships also make optimization difficult. These complexities become evident from a study of Figs. 23 through 26 and their respective captions, which describe a specific example.

Stability

Stability is associated with the oscillations in the motion of the tool. The fewer the oscillations present, obviously the more stable is the operation of the robot. Negative aspects of oscillations include: (1) Additional wear is imposed on the mechanical, hydraulic, and other parts of the robot arm, (2) the tool will follow different paths in space during successive repetitions of the same movement, thus requiring more distance between the intended trajectory and surrounding objects, (3) the time required for the tool to stop at a precise position will be increased, and (4) the tool may overshoot the intended stopping position, possibly causing a collision with some object in the system.

Oscillations may be damped or undamped. Damped (transient) oscillations will degrade and cease with time. Undamped oscillations may persist or may grow in magnitude (runaway oscillation) and are the most serious because of the

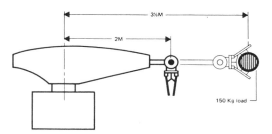

Fig. 23. Diagram illustrating robot arm performance. Consider a robot arm that has a retracted hand position of two meters and an extended hand position of three and one-half meters. Consider also that this arm might carry a load of 150 kilograms, and that the arm should go from position to position, with or without load, at any extension and without overshoot. For the configuration shown, the variation in moment of inertia is from 70 KgMsec^2 when tucked in and unloaded to 230 KgMsec^2 when fully extended and loaded. To achieve a critically damped servo with position repeatability of 0.5 mm under all operating conditions is difficult. Note that 0.5 mm resolution for an arm with 300 degrees of rotation requires position encoding to an accuracy of 1 part in 33,000 or 2^{15}. The foregoing deals only with a major robot arm articulation. In a full arm, the interactions among the various articulations complicate both dynamic performance and accuracy. For example, a robot arm designed to achieve an individual articulation natural frequency of 50 Hz degenerates to an overall 17 Hz in a six-articulation arm. (*Unimation-Westinghouse.*)

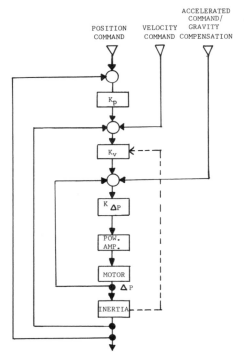

Fig. 24. Block diagram for functionally describing key elements of a single-articulation servo system, including velocity and acceleration feedback and inter-articulation bias signals. In estimating the time to complete a task (without actually simulating the entire process), the interface with the workplace complicates the process. Paths to avoid obstacles add program steps. Some steps must be very precise, calling for closing out to zero error before the program advances. Other steps may be the corners in a motion path which can be passed through 'on the fly' so to speak. The use of interlock switches may introduce transport lags. Simple programs often permit using a rule of thumb.

For a 2000 Series *Unimate* (*Unimation-Westinghouse*), if one allows 0.8 sec for each motion taught, short steps as well as long, a time for program completion can be estimated quite closely. However, if a program is complex, as in spot welding a car body, there are too many variables to permit use of such methods. For example, *Unimation-Westinghouse* has developed an 18-page treatise to aid in forecasting program time for a multiplicity of robots used in spot welding an auto body. Factors considered include weld gun inertia, weld gun operating time, metal thickness, proximity of spots one to another, among other factors. (*Unimation-Westinghouse*.)

potential damage they may cause to the surroundings.

Variations of inertial and gravitational loads on the individual joint servos (as the arm's posture changes) make the operation of the robot prone to oscillation. Furthermore, the servos must operate over a wide dynamic range of position (and in some cases, velocity) error. The servos must operate reliably in all situations despite the limits on velocity and acceleration imposed by the actuators used. Certain exceptional conditions can be extremely unstabilizing to a joint servo system, as, for example, when the load accidentally slips out of the end-effector. This causes a step change in the gravity loading on one or more joints and can cause a poorly designed arm to go into oscillation. Motion of a joint can also exert various combinations of inertial, centrifugal, and Coriolis forces on the other joints. The reactions of the other joints to these forces can exert forces on the original joint, and this is another potential source of oscillation. Finally, two manipulators working in close proximity can excite oscillations in each other. This can be through a mechanical coupling, such as a common mounting, or through a workpiece held simultaneously by two machines.

Solutions to Oscillation Problems. Two solutions are suggested in Ref. 8. (1) In one method the joint servos operate continuously. Some sophisticated servo designs (as the result of experience from numerical control of machine tools) prevent oscillations from starting regardless of the load carried. (2) In another approach, the robot controller locks each joint independently the first time it reaches its set point. Special circuitry also decelerates the joint after it comes within a prescribed distance of that position. The joint in this robot may lock in any order. When

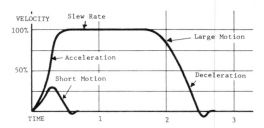

Fig. 25. It is common for robots to be offered with abbreviated specifications that list the slew rates of each articulation and the repeatability of each articulation. What is really needed is block point of time to go from position to position and net accuracy of all articulations in consort. Shown here are two typical velocity traces for a short arm motion and a large motion. It is evident that slew rate is no measure of elapsed time in making a motion, particularly a short motion in which slew rate may not be attained at all. (*Unimation-Westinghouse*.)

Robot Technology Fundamentals

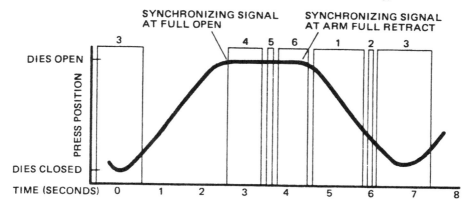

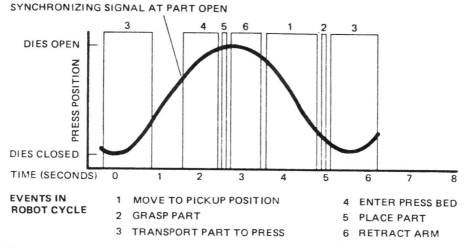

Fig. 26. For some operations, program time is critical, such as when a robot is serving heavy, expensive capital equipment. If the production rate is paced by the robot rather than the equipment, the project would not seem viable because of loss in throughput. Optimizing such a program may involve a range of techniques. A typical application might be press-to-press transfer of sheetmetal parts.

A line of presses runs at a gross production rate of up to 700 parts per hour. At this rate, a robot must make a complete transfer and return for the next pickup in 5.16 sec. With presses on center-to-center distances of six meters, this is a demanding transfer speed. To meet this rate, a robot was modified by increasing the capacity of both hydraulic supply and servo valves. Acceleration and deceleration times were reduced at some sacrifice in damping and accuracy. This was compensated by providing die nests with leads or strike bars. Finally, interlocks were refined so that the robot could make approaches and departures during the rise and fall of the moving platens of the press. The curves given here show how the time can be shortened by tight interlocks that do not wait for press cycle completion. For safety, this approach cannot be used with human operators. (*Unimation-Westinghouse.*)

the joints are all locked (*total coincidence*), the arm is stationary and it can then begin to move to the next position. If the position is held for more than a few seconds, the tool slowly creeps away from its programmed position (as oil leaks out of actuator cylinders). When the position error accumulates sufficiently, the joint servos are allowed to operate again to return the tool to the original position. Technically, this is a form of instability inasmuch as the tool position can vary periodically (although the period may be on the order of 30 or 60 seconds). However, it is part of the machine's normal operation and causes no problems.

Resolution

Resolution is a function of the design of the robot control system and specifies the smallest increment of motion by which the system can divide the working space. This may be a function of the smallest increment in position the control can command, or it may be the smallest incremental change in position that the control measurement system can distinguish. Spatial resolution is the control resolution combined with mechanical inaccuracy. In order to determine spatial resolution, the range of each joint on the manipulator is divided by the number of control increments.

Differences Between Sliding and Rotary Joints. Two manipulator positions that differ by only one increment of a single joint are called *adjacent*. A unit change in the position of a *sliding joint* will move the tool tip the same distance, regardless of where it is in the workspace. A manipulator with an *XYZ* geometry thus has essentially a constant spatial resolution throughout its work volume. This consideration could be important if the arm is to be trained to perform a precise manipulation in one location of its workspace and then is to repeat it elsewhere in the workspace.

In contrast, a unit change in the position of a *rotary joint* will move the tool tip through a distance that is proportional to the perpendicular distance from the joint axis to the tool tip. There is an angular-position error on the final tool tip position that depends on how far the boom is extended. The farther the boom is extended, the larger the distance that the tool tip will move when the rotary joint moves to an adjacent position.

Repeatability

Repeatability is the ability of the robot to reposition itself to a position to which it was previously commanded or trained. Repeatability is affected by resolution and component inaccuracy. Both short- and long-term repeatability exist. Long-term repeatability is of concern for robot applications requiring the same identical task to be performed over several months. Over a long time period, the effect of component wear and aging on repeatability must be considered. For some applications where the robot is frequently reprogrammed for new tasks, only short-term repeatability is important. Short-term repeatability is influenced most by temperature changes within the control and the environment, as well as by transient conditions between shutdown and startup of the system. Factors that influence both short-term and long-term repeatability are referred to as *drift*.

Obtaining good repeatability is more difficult in a computer-controlled manipulator that records tool positions rather than joint positions because three additional data processing steps are required: (1) converting the several joint positions to a tool position and storing it (sometimes called *back solution*); (2) transforming a tool position in some useful way, such as by translating, rotating, or scaling it; and (3) converting the transformed tool position back to a set of joint positions (sometimes called the *arm solution*). The kinematic equations used in the arm solution and back solution must accurately reflect the design of the manipulator. The accuracy of these computations depends on the accuracy with which the following values (*joint parameters*) are known: (1) joint extensions and rotations, (2) link lengths, (3) offset distances between successive joint axes, and (4) angles between successive joint axes.

The repeatability of a manipulator system is measured in the following way: The manipulator is moved by its control system from a reference position to a specified second position with a known nominal location of the center of the end-effector. The distance of the actual position of the center of the end-effector from its nominal position is measured and recorded. This value is denoted as δ. (See Fig. 27.) The experiment is repeated a number of times and the largest value of (δ_{max}) is the manufacturer's value of repeatability, which is usually given as $\pm\delta_{max}$. Obviously, the results of such a test can be used only as a rough guide as to the accuracy one can expect to obtain. One frequently will be delivered a different machine, i.e., not the same manipulator that was tested. Also, manufacturers rely on a nominal-size model and do not have the dimensions of the actual manipulator in its own controller. Thus, the actual travel, from reference to final position, may be different from that of the manufacturer's test. Furthermore, other important factors include ambient conditions, rigidity of the structure, and the tendency for the precision of the drive trains and transmissions to differ somewhat from one production model to the next. A manufacturer can improve the repeatability simply by decreasing the load rating below its true value (Ref. 6).

Accuracy can range from several hundredths of an inch for a simple robot to several thousandths of an inch for a robot doing precision assembly or handling small parts. In the case of

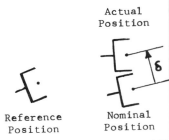

Fig. 27. Determination of repeatability.

a robot used in testing printed-circuit boards, the end effector must get close enough to the board to pick it up and close enough to the fixture on the tester to place the board on it with sufficient alignment. This could call for an accuracy of a few thousandths of an inch. A less accurate robot can perform adequately but requires guides to direct the board onto the end effector. Adding these extra features to a less accurate robot is usually more costly than purchasing a more expensive, more precise robot.

A random scan of manufacturers' literature giving their claims of repeatability pertaining to widely procured contemporary robots yields the data given in Table 11.

Compliance

The compliance of a manipulator is indicated by its displacement relative to a fixed frame in response to a force (torque) exerted on it. The force may be a reaction force (torque) that arises when the manipulator pushes (twists) the tool against an object, or it may be the result of the object pushing (twisting) the tool. High compliance means the tool moves a lot in response to a small force, and the manipulator is then said to be spongy or springy. If it moves very little, the compliance is low and the manipulator is said to be stiff (Ref. 8).

Compliance is a complicated quantity to measure properly. Ideally, one would find the relationship between disturbances and displacements to be *linear* (displacement or rotation proportional to force or torque); *isotropic* (independent of the direction of the applied force); and *diagonalized* (displacement or rotation occurring only in the same direction as the force or torque); constant with time; and independent of tool position, orientation, and velocity.

In practice, a manipulator's compliance turns out to be none of these. It is a nonlinear, anisotropic tensor quantity that varies with time and with the manipulator's posture and motion. It is a tensor because a force in one direction can result in displacements in other directions and even rotations. A torque can result in rotation about any axis and displacement in any direction. A six-by-six matrix is a convenient representation for a compliance tensor. Time can affect compliance through changes in temperature and hence viscosity, of hydraulic fluid, for example. Furthermore, the compliance will often be found to be a function of the frequency of the applied force or torque. A manipulator may, for example, be very compliant at frequencies around 2 Hz but very stiff in response to slower disturbances.

Compliance may exhibit hysteresis. For example, the servos in one design of hydraulic manipulator turn off when the arm stops moving. In this condition all the servo valves are closed, and the compliance has a value that is determined by the volume of incompressible hydraulic fluid trapped in the hydraulic hoses and the elasticity of the hoses. However, if an outside force on the tool should move any of the joints more than a given distance from the position at which they are supposed to remain, then the servos on all joints will turn on again. The compliance then changes to a completely different value (presumably stiffer in some sense).

Both electric and hydraulic manipulators have complicated compliance properties. In an electric manipulator the motors generally connect to the joints through a mechanical coupling. The sticking and sliding friction in such a coupling and in the motor itself can cause strange effects on the compliance measured at the tool tip. In particular, some of these couplings are not very *back drivable*. For example, if one pushes on the nut of a leadscrew (back-drive), the leadscrew will not turn (unless the screw's pitch is very coarse and ball bearings are used between the threads to reduce friction). But one can turn the screw easily, and the nut will move.

Most manipulators are operated open loop in the sense that they go blindly to a given point in space without regard to the actual position of the object in the environment or to any reaction forces (feedback) that those objects exert on the arm (or tool). In this case, less compliance than that of surrounding objects is advantageous because it means contact with objects would cause high-frequency oscillations that can be filtered out without degrading overall response. Such filtering actually requires no special effort since the combination of servo valves and actuators commonly used have relatively low bandwidths (1 to 2 Hz).

Tactile sensors that measure forces and moments exerted on the tool can allow the manipulator to track or locate objects. Even in such cases, however, oscillations may arise in the force-feedback control loop if the compliance at the point of sensing is too low (stiff). Examination of a particular servo design is required to reliably predict whether it will provide the kind of compliance needed for a specific task. There is no substitute for an actual test with the real tool on the manipulator.

TABLE 11. Manufacturers' Claimed Repeatability of Randomly Selected Contemporary Robots

LOAD CAPACITY		CLAIMED REPEATABILITY (±)	
Pounds	Kilograms	Inches	Millimeters
5	2.2	0.004	0.1
14	6	0.004	0.1
22	10	0.008	0.2
35	16	0.001	0.03
66	30	0.002	0.05
110	50	0.020	0.5
132	60	0.020	0.5
150	68	0.010	0.25
176	80	0.020	0.5
200	90	0.010	0.25
264	120	0.04	1.0

END-EFFECTORS

The end-effector is the device fastened to the free end of a manipulator. It provides the means for the manipulator to interact with its surroundings. An end-effector's usual function is first to grasp an object or a tool and then hold it while the manipulator moves—thereby also moving the object—and finally to release the object. End-effectors are also called *hands* or *grippers* (Ref. 6).

Manufacturers usually offer a number of standard grippers, often in the form of clamping devices with a choice of a variety of clamping "fingers" and vacuum devices with a variety of suction cups. Many dozens of types of grippers have been designed, but even then, some applications require customized end-effector design and engineering. Normally, the cost of the end-effector is not included in the price of the robot. In some applications, the end-effector can represent a considerable expense.

Four main methods are used for holding parts or tools in an end-effector: (1) mechanical clamping, (2) vacuum suction, (3) magnetic attraction, and (4) plug-in or detent fittings. Scooping and ladling and sticky fingers using adhesives are among numerous specialized designs.

Mechanical Clamping

Two principal types of mechanical grippers are shown in Fig. 28: (1) *rotating fingers*, and (2) *translating fingers*. As shown in Fig. 29, a three-fingered design is available for spherical and

Fig. 29. Three-fingered gripper. (*Data Source*: Ref. 6)

cylindrical shapes. In these designs, the fingers can be straight, shaped for a specific object, or have a flexible padded surface to accommodate a number of different shapes. The grippers hold by either grasping from the outside and pushing inward onto the object, or by reaching inside a cavity in the object and pushing outward. A complex set of grippers can combine various types of grips on different surfaces of the fingers and can accommodate several independently acting sets of fingers. A variety of grips is sometimes required, at different parts of a cycle, in handling the same part.

Mechanical clamping also can be accomplished with expanding bladder-type grippers. The flexible bellows or bladders are pressurized and expanded after the gripper has reached a required position. (See Fig. 30.)

Some standard and special hands and grippers offered with *Unimation* (*Westinghouse*) robots are illustrated in Fig. 31. The grippers shown are by no means representative of all designs that are available or possible.

Design Guidelines. For determining how mechanical grippers should grasp and the force required, guidelines include:

1. The surface which the robot's hand is to grasp must be reachable. *Example:* It should not be hidden in a chuck.
2. The tolerance of the surface to be grasped influences the accuracy of a part. *Example:* If

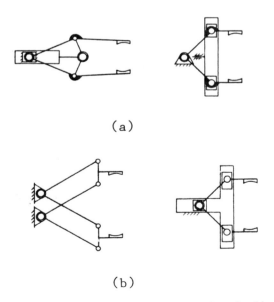

Fig. 28. Mechanical grippers: (a) Equipped with rotating finters; (b) translating fingers. (*Data Source*: Ref. 6)

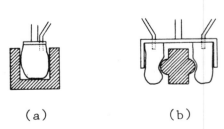

Fig. 30. Inflatable grippers: (a) Internal contact, (b) external contact of object. (*Data Source*: Ref. 6)

Robot Technology Fundamentals 289

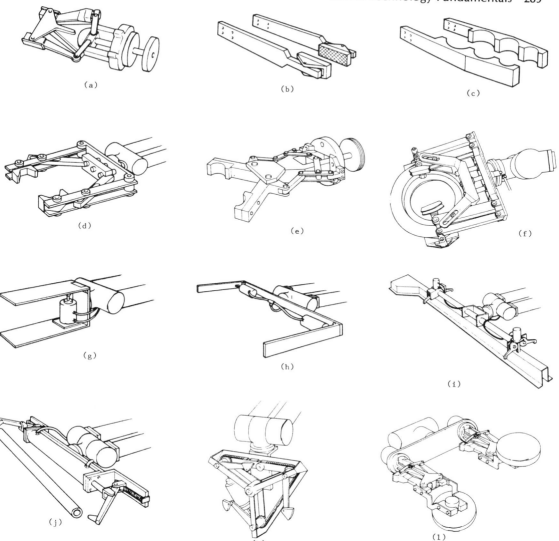

Fig. 31. Some standard and special mechanical hands and grippers: (a) **Standard hand**. Inexpensive, all-purpose hand. Will accept a wide variety of custom fingers, tailored to the parts to be manipulated or moved. Parts should be of moderate weight. Simple linkages provide both finger action and force multiplication needed to grip object sufficiently lightly. At the completion of finger closure, the fingers exert their maximum clamping force on the part. (b) **Fingers self-aligning**. Self-aligning pads for fingers are valuable for assuring a secure grip on a flat-sided part. 'Cocking' of the part is highly unlikely when these pads are used. (c) **Fingers for grasping different size parts**. A particular finger design need not be restricted to parts within a limited range of sizes. Perhaps the fingers can be equipped with extended pads having several cavities for parts of differing sizes and shapes, or for parts that change shape duringf processing. Then, the industrial robot is pre-programmed to position the hand so that the proper cavity will match the location of the part. (d) **Cam-operated hand**. Heavy weights or bulky objects can be easily handled by cam-operated hand. More expensive than the standard hand, the cam-operated hand is designed to hold the part so that its center of gravity (CG) is kept very close to the 'wrist' of the hand. The short distance between the CG and wrist minimizes the twisting tendency of a heavy or bulky object. To achieve this 'close coupling' of hand and part, there is a sacrificie, i.e., a specific design can accommodate only a very narrow range of object sizes. (e) **Wide-opening hand**. When the part to be picked up is not always to be found in a constant orientation or at the same site, a wide-opening hand may be recommended. As it closes, this hand will sweep the inexactly located part into its grasp. If the part is always

the machined portion of a cast part is to be inserted into a chuck (where the robot must grasp the cast surface), the opening in the chuck must be larger than the eccentricity between the cast and the machined surfaces.
3. The hand and fingers must be able to accommodate the change in dimension of a part that may occur between the part loading and the part unloading operations.
4. Grasping should not distort or scratch delicate surfaces.
5. In grasping a part of two different dimensions, grasping the part along the larger dimension will ensure better control in positioning the part.
6. Fingers should have either resilient pads or self-aligning jaws that will conform to the part to be picked up. Self-aligning jaws ensure that each jaw contacts the parts on two spots. If each jaw contacted the part on only one spot, the part could pivot between the jaws.

Firmness of Grasp. How hard the robot must grasp the part depends on the wieght of the part, the friction between the part and the fingers (or vacuum cups or magnets), how fast the robot is to move, and the relation between the direction of movement to the fingers' position on the part. The worst case occurs when the acceleration forces are parallel to the contact surface of the fingers, in which case friction alone has to hold the part.

A robot at normal full speed may, during acceleration and deceleration, exert forces on a part of about 2g. The following relationships are germane:

1. A part transferred by a robot in the horizontal plane will exert a force on the hand tooling of twice the weight of the part.
2. If the part is lifted, it will exert a force three times the weight of the part: 1g due to gravity and 2g due to upward acceleration made by the robot.

An example showing calculation of grasping force is given in Fig. 32. If the center of gravity of the part is outside the line between the two jaws, a moment caused by acceleration forces will tend to pivot the part. To prevent pivoting the product of the clamping force, the spread between contact points and friction must be greater than the moment.

Vacuum Systems

As shown by Fig. 33, there are several types of vacuum cups used for grasping and holding parts to be handled by a robot. The holding force of a vacuum cup is the effective area multiplied by the

---

(*continued from previous page*)
precisely positioned for pickup, the wide-opening hand can shorten the time needed to reach for the part. The hand can travel the shortest path to the part and skip the extra step of making its final approach to the part from one specific direction. The hand develops low force when open and maximum force when closed. It is for parts of moderate weight. (f) **Cam-operated hand with inside and outside jaws.** Assume that a part is reoriented between the time when the part is placed in a machine and when it is removed. This special hand is one of those which will deal with the problem. When the part is oriented as shown, the hand can grasp it on the outside dimension (OD) by using the outer self-aligning pads. If the part is turned over, the inner pads will grasp the inside dimension (ID). A similar principle applies when the grasped surface of a part is changed significantly between the time when it is placed in a machine and when it is removed. A special hand can be designed to deal with most changes in ID, OD, or other dimension. (g) **Special hand with one movable jaw.** A hand with single-acting jaw should be considered when there is any access underneath a part, as when it is on a rack. In such cases, this hand will scoop up a part quickly. Simplicity of the design makes it one of the most economical hands for many uses. (h) **Special hand for cartons.** The dual-jaw hand will open wide to grasp inexactly located objects of light weight. Lifting and placement of cardboard cartons is an example. Actuators and jaws can be remounted in any of several positions on the fixed back plate, making it practical for the same dual-jaw to move large cartons as well as smaller cartons. (i) **Special hand with modular gripper.** This special hand, with a pair of pneumatic actuators, is suitable for parts of light weight. Lifting capacity is dependent upon friction developed by the fingers, but heavier parts can be handled if the fingers can secure a more positive handle on the object—as under a flange or lip. (j) **Special hand for glass tubes.** Secure grasping of relatively short tubes is easy with this hand. Pickup can be effective even when the tube length varies some. The fingers of the hand close in two stages: (1) They travel through an arc until they are vertical, after which (2) the actuator draws them together axially. Linear travel in the second stage of closure is selected to accommodate the range of tube lengths to be handled. (k) **Special hand chuck type.** It is practical to handle drums and similar large cylindrical pieces with a relatively simple mechanism consisting of three fingers and a single actuator. The actuator drives all three fingers simultaneously by means of a chain and sprockets. The fingers expand against the inside diameter of the drum. One hand of this type will pick up drums of various diameters. (l) **Double hand.** A double hand with double actuators may be used for removing a finished part from a machine and replacing it with an unfinished part. The double hand will pick a part out of the chuck of a macine, swivel, and place a new part back into the chuck. Thus, a robot with this hand does not have to expend time putting one part down before it manipulates another. The hand seldom makes a trip while empty. Parts should be of moderate weight. (*Unimation-Westinghouse.*)

Robot Technology Fundamentals 291

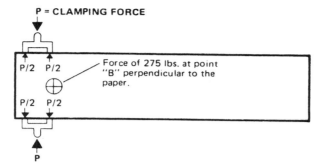

clamping force x friction coefficient (μ)
$\geqslant$ force at point "B"
2 (P/2 x μ) = B
2 (P/2 x 0.15) = 275
P = 1830 lbs
with a safety factor of 2, the required
clamping force is 3660 lbs.

This is a very large clamping force for a 25 lb part, which indicates that the design is not efficient. The force can be reduced in two ways: first, try to grasp the part closer to CG or make the pads longer (6" instead of 3" will reduce the clamping force from 3660 to 1340 lbs).

The situation may be represented thus:

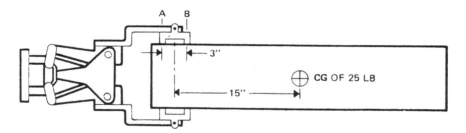

This simplifies into the following force diagram:

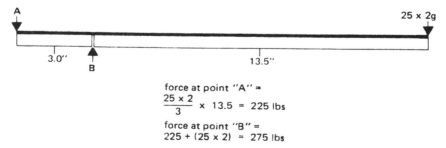

force at point "A" =
$\dfrac{25 \times 2}{3} \times 13.5 = 225$ lbs

force at point "B" =
225 + (25 x 2) = 275 lbs

Fig. 32. Example showing calculation of robot grasping force. A part weighs 25 pounds and has a center of gravity 15 inches off the point where it is grasped. The friction coefficient is 0.15 and the spread between the contact point on each jaw pad is 3 inches. How hard must the robot grasp to have a safety factor of two on its hold of the part? It is necessary to design for the highest force which will occur at the point where slippage between fingers and the part will first arise if the clamping force is not high enough. (*Unimation-Westinghouse.*)

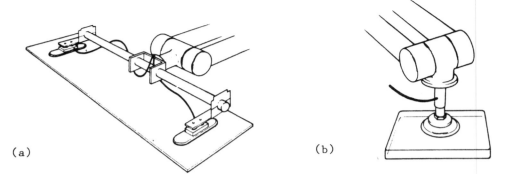

(a)   (b)

Fig. 33. Examples of use of vacuum pickup systems (a) Vacuum cup hand—this device has the advantages of the magnetic pickup and is much less susceptible to workpiece side slip. For light-to-moderate-weight glass, plastic, ferrous, and nonferrous parts, the vacuum pickup frequently makes an excellent choice. (b) Simple vacuum cap hand—fragile parts, such as cathode ray tube face plates (as shown), are handled easily by a simple pickup of this kind. The reliability is usually better than for magnetic devices. Well-designed telescoping vacuum lines for long-reach arms are obtainable. (*Unimation-Westinghouse*.)

differential pressure between the outside and the inside of the cup. To obtain the best utilization of a cup, the largest possible vacuum or pressure differential should be used. The vacuum will not form until the cup has sealed on the part. Thus, to get speed out of a vacuum system, it is advantageous to mount the cups on spring-loaded stems and have the robot programmed so that the cup touches the part an instant before the arm reaches its final pickup position. This will eliminate a large portion of the deceleration time from the cycle. Spring loading will also compensate for any variation in the height or level of the part. Cups can also be put on ball joints for accommodating variations between parts to be handled, such as distorted sheet stock.

For creating a vacuum, a vacuum pump or venturi may be used. The vacuum pump will be a piston or vane-type driven by an electric motor. Relative advantages and limitations of the two methods include:

	Advantages	*Disadvantages*
Vacuum pump	Can create a high vacuum	High initial cost
	Low operating cost	System more complex than venturi
	Relatively quiet	
Venturi	Low initial cost	Noisy
	Less complex than pump	High operating cost
	High reliability	

To release a part rapidly, a blow-off system is required.

Magnetic Pickups

Where parts to be handled have an iron content, magnetic handling should be included among the candidate pickup systems. Magnets are either: (1) permanent magnets, or (2) electromagnets.

Permanent Magnets. These do not require a power source for operation, which makes them well adapted for hazardous atmosphere. However, they do require a means of separating material from the magnet—generally a stripper device is used—or where the part is in position and in some way firmly secured, the magnet simply can be pulled away from the part by applying necessary force. The permanent magnet can be designed to produce extremely shallow magnetic penetration, useful when it is required to remove single thin ferrous metal sheets from a stack. Single sheets as thin as 0.031 inch (0.08 cm) can be lifted. Permanent magnets are also used as sheet "fanners" or separators. Magnetic induction of the sheets with like polarity causes them to repel each other and to tend to rise in midair. Most permanent magnets are fully effective up to material temperature of 200° F (93° C), although they have been used in specially designed systems up to 900° F (482° C).

Electromagnets. These are well suited for remote control as well as for moderately high-speed pickup and release of parts. A source of dc power is required. To assist in achieving fast release of parts, a so-called drop controller may be incorporated in the circuit. Basically, this is a multi-function switch through which power is supplied to the magnet and, as it interrupts the power supply, it reverses the polarity and supplies power at a reduced voltage for a short duration before completely disconnecting the magnet from the line. This reverse polarity tends to cancel any residual magnetism in the part, thus causing it to release almost instantaneously. Electromagnets of standard design will handle

Robot Technology Fundamentals

materials at temperatures up to 140°F (60°C), but modified designs will accommodate temperatures up to 300°F (150°C). A representative electromagnetic pickup is shown in Fig. 34.

Fastening Tools to Robot Wrists

Apart from their mounting characteristics, tools fastened to robot wrists will generally have the same capabilities as if they were manipulated manually. Where a robot has two or more tools to manipulate, quick-disconnect provisions are in order. Representative tool-fastening methods are shown in Fig. 35.

Standardized End-Effectors Available from Stock

A surprisingly large number and wide variety of end-effectors, grippers, and associated robot hardware are available from stock. A very abridged representation of the devices available from stock is given in Table 12.

WORKPLACE CONFIGURATION AND ENVIRONMENTAL FACTORS

Robots encounter the same environmental hazards as other industrial instruments and controls that must be installed on the factory floor. Among the most important of possible adverse conditions are (1) high temperature, (2) dusty, dirty, corrosive, and sometimes potentially explosive and fire-prone atmosphere, (3) shock and vibration, and (4) electromagnetic interference. In turn, of course, robots contribute their share to some of these and other environmental problems, not the least of which is noisy operation. Furthermore, safe shutdown of robots (good failsafe protection) is required because a meandering, heavily loaded robot can cause considerable damage to surrounding equipment within a fraction of a second, and unless supervisory personnel are amply protected, injuries can occur. Such situations are far from commonplace because of the excellent engineering that goes into robot installations. However, as robots become more and more popular and are used increasingly by less competent people in medium- and small-size facilities, manufacturers of "stock" robots must be vigilant in their preparation of appropriate instructions and warning information.

In many applications today it is acceptable design practice to package the robot as a self-contained entity, but there is an advantage to mounting the electronics separately. In extreme shock conditions, it may be desirable to mount the control console on a shock-absorbing pad in a remote location for protection against hostile atmospheres. If the power supply of a robot can be separated from the robot's arm, then it is possible to introduce only the arm into an explosive atmosphere, such as a paint room, or some potentially dangerous area as may be found in a nuclear facility. Following the design practices for intrinsic safety is always a worthwhile consideration. Particularly for protection against abrasive dust, the robot joints may be booted. Nonflammable fluids for lubrication and hydraulics represent good robot design practice. Where atmospheric air is particularly dirty, cooling air should be well filtered and should enter enclosures to provide positive internal pressure. Robot logic design should be well protected from power line spikes and noise pickup that may enter through any of the robot's communication links with surrounding equipment.

The Robot Setting

There are four basic situations pertaining to the flow of work and the floor location of the robot: (1) Work may be arranged *around the robot*, (2) work may be *brought to the robot*, (3) work may *travel past the robot*, and (4) the *robot travels to the work*. Arrangement (3), of course, is a variant of (2).

Work is Around the Robot. In the early installations of robots, the work was arranged around the robot. This arrangement caused the least commitment and the least plant disruption. The system is still used and frequently found in operations, such as forging and trimming, press-to-press transfer, plastic molding and packaging,

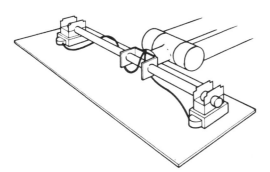

Fig. 34. Electromagnetic pickup for use with flat surfaces. These pickups are applicable to the handling of ferrous sheets or plates and will handle objects of several sizes. Weight of the part should be moderate so that side slippage can be avoided. Positioning for pickup does not have to be precise. Grasping is essentially instantaneous. (*Unimation-Westinghouse*.)

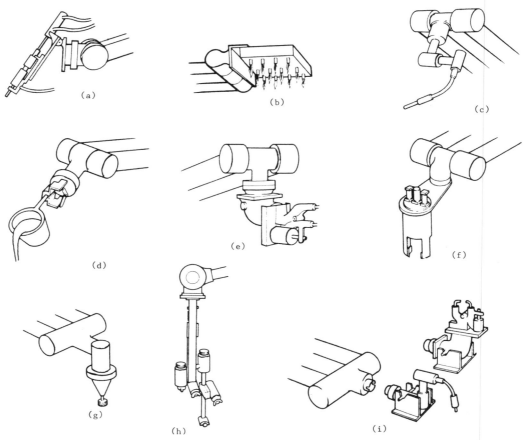

Fig. 35. Methods for mounting tools to wrist of a robot: (a) **Stud welding head**. Equipping a robot with a stud welding head is practical. Studs are fed to the head from a tubular feeder suspended from overhead. (b) **Heating torch**. This may be used to bake out foundry molds by playing the torch over the surface, allowing the flame to linger where most heat input is needed. Fuel is saved because heat is applied directly. Also, this method is faster than where molds are conveyed through a gas-fired oven. (c) **Inert gas arc welding torch**. Welds can be single- or multiple-pass. The most effective use is for running simple-curved or compound-curved joints, as well as running multiple short welds at different angles and on various planes. (d) **Ladle**. Ladling hot materials, such as molten metal, is a hot and hazardous job for which robots are well-suited. In piston casting, permanent mold die casting, and related applications, the robot can be programmed to scoop up and transfer the molten metal from the pot to the mold, and then do the pouring. In cases where dross will form, dipping techniques will often keep the dross out of the mold. (e) **Spotwelding gun**. A general-purpose industrial robot can maneuver and operate a spotwelding gun to place a series of spot welds on flat, simple-durved or compound-durved surfaces. (f) **Pneumatic nut-runners, drills, and impact wrenches**. General-purpose robots are well-suited for performing these kinds of tasks, particularly when in hazardous environments. Drilling and countersinking with the aid of a positioning guide is practical. Mechanical guides increase locating accuracy and help shorten positioning time. (g) **Routers, sanders, and grinders**. A routing head, grinder, belt or disc sander can be mounted on the wrist of the robot. Thus equipped, the robot can rout workpiece edges, remove flash from plastic parts, and do rough snagging of castings. (h) **Spray gun**. Ability of robots to do multipass spraying with controlled velocity fits them for automated application of primers, paints, and ceramic or glass frits, as well as application of masking agents used before plating. For short- or medium-length production runs, the robot will often be a better choice than a special-purpose setup requiring a lengthy changeover procedure for each different part. The robot can spray parts with compound curvatures and multiple surfaces. (i) **Tool changing**. A single industrial robot can handle several tools sequentially, with an automatic tool-changing operation programmed into the robot's memory. The tools can be of different types or sizes, permitting multiple operations on the same workpiece. To remove a tool, the robot lowers the tool into a cradle that retains the snap-in tool as the robot pulls it away. The process is reversed to pick up another tool. (*Unimation-Westinghouse*.)

TABLE 12. Representative End-Effectors and Grippers Available from Stock

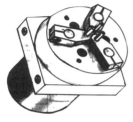

Rotary-Action, Three-Finger Gripper. Air-vane motor with single-gear drive simultaneously activates three fingers in a centralizing motion. Two models. Adjustable stroke for different bores: 5–45; 8–70mm. Wt: 700–1950g (25–69oz). Closing force: 80–250N (18–56 lbf). Dimens.: 86×49×49—185×88×88mm. Well suited to cylindrical shapes.

Rotary-Action Internal Gripper. Bellows opens; spring closes. Three models. Stroke: 1mm. Wt: 60; 140; 250g (2.2; 5; 8.8 oz). Opening force: 20; 25; 30N (4.5; 5.6; 6.7 lbf). Dimens. (closed): 60×20ϕ; 65×30ϕ; 65×50ϕ. Well suited for irregular shapes. Patent applied for.

Rotary-Action External Gripper. Spring and expanding bellows drive. One model. Stroke: 1mm. Wt: 400g (14 oz). Closing force: 40N (9 lbf). Dimens. (closed): 69×45ϕ mm. Effectively adapts to irregular shapes that normally require custom design. A two-way solenoid is required to open/close gripper.

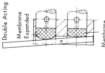

Rotary-Action Membrane Drive. Fingers are opened/closed by expansion/contraction of air-powered membranes. A three-way solenoid valve required. Fingers can be extended/shortened. Three models. Stroke: 5, 5, 10mm. Opening/closing force: 15N (3.37 lbf) at 20mm; 30N (6.74 lbf) at 40mm; 30N (6.74 lbf) at 40mm. Wt: 100; 250; 300g (3.53; 8.82; 10.58 oz). Can grip objects externally or internally. Dimens. (closed): 115×40×25mm.

296 Robot Technology Fundamentals

TABLE 12. *continued*

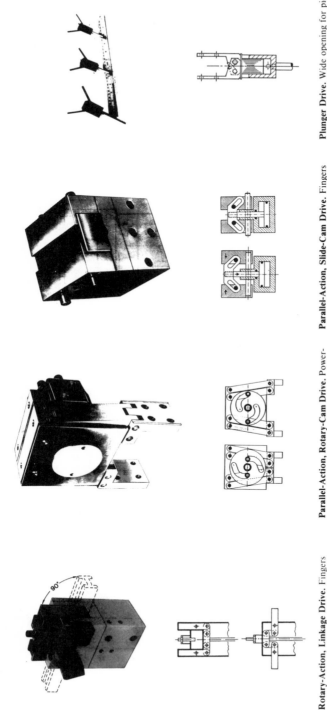

Rotary-Action, Linkage Drive. Fingers open/close in precise symmetric motion by means of a linkage driven by an air-powered cylinder. Self-locking mechanism remains closed with air loss. Designed for easy modification; screw adjustable stroke (90°); counterbored for quick, precise alignment. A three-way solenoid required to control opening. Wt: 1700g (60 oz). As low as 20g (.71 oz) and other lighter weights available. Dimens. (closed): 50×28× 20mm. Other dimens. available. Closing force: 25N at 40mm to 140N at 80mm. depending on model.

Parallel-Action, Rotary-Cam Drive. Powerful with long stroke and low weight. Grips objects externally or internally. Fingers move in precise symmetric parallel action. A three-way solenoid controls opening/closing. Fail safe on loss of air. Stroke: 60mm. Wt: 3100g (6.83 lb) Closing/opening force: 400N (89.92 lbf). Length: 152mm.

Parallel-Action, Slide-Cam Drive. Fingers move in precise symmetric parallel action driven by air-actuated slide mechanism. A 3-way solenoid controls opening/closing. Fingers designed to easily accommodate extensions and adaptors. Powerful with long stroke and light weight. Grips object internally/externally. Stroke: 10mm. Wt: 250g (8, 82 oz). Closing/opening force: 200N (80.93 lbf). Length: 50mm. Lower-weight/force models available. Counter-bored for quick, precise alignment.

Plunger Drive. Wide opening for pickup of objects not positioned accurately. Self-entering; both jaws are activated by same plunger. Separate torsion spring on each jaw opens gripper and restores plunger after air pressure is relieved. A two-way solenoid valve controls opening/closing. Designed to accommodate extensions and adaptors. Three models. Jaw motion: 30, 35, 40 mm. Grip at 5.5 bar: 25N at 20mm; 32N at 30mm; 41N at 40mm. Wt: 42; 100; 157g (1.5; 3.5; 5.5 oz). Max op air pressure: 10 bar. Min op air pressure: 2; 2.5; 2.5 bar). Similar plunger drives available with up to three-axis of electrical actuated rotation.

Note: Information for this table provided by Stock Drive Products, Division of Designatronics, Inc.. New Hyde Park, New York.

Robot Technology Fundamentals

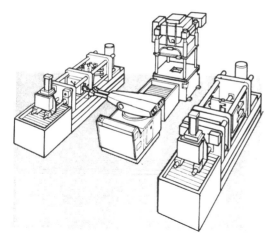

Fig. 36. Die casting installation to unload, quench and dispose of part. In this installation, quite exemplary of the earlier robot installations, the work is arranged around the robot. (*Unimation-Westinghouse.*)

Fig. 37. Automatic welding line where work (unitized car body parts, such as side aperture panels, roof panels, flat floor pan, and bolt-on front fenders) is brought to the computer-controlled robots by way of a conveyor line. (*Chrysler Corporation, Windsor, Ontario Assembly Plant.*)

and investment casting. (See die casting operation of Fig. 36.)

Work Is Brought to the Robot. Flexibility has been enhanced through the use of computer control. For example, the robot can be made to track a workpiece that is being carried on a conveyor, the robot performing its task as the work passes by. The versatility of such a system

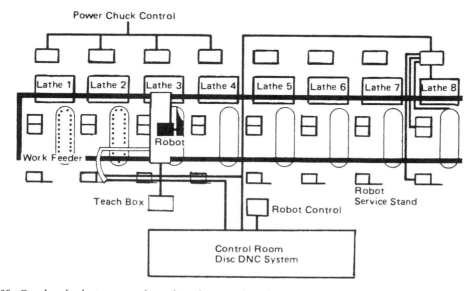

Fig. 38. Overhead robot system where the robot travels to the work. In the system shown, an overhead robot system allows one robot to serve eight numerically-controlled lathes. (*Unimation-Westinghouse.*)

can be extended to handle variations in the conveyor speed. (See automatic welding system of Fig. 37.) In this installation, the robots are mounted in a fixed position relative to the line (conveyor). The configuration is sometimes called "stationary-base line tracking." In a contrasting situation the robot(s) may be mounted on some form of transport system (e.g., rail-and-carriage), which moves parallel to the line at line speed. This configuration is known as "moving-base-line tracking." In addition to the cost of the transport system, there can be interference problems between adjacent stations. The transport system also requires a powerful drive system so the robot can be returned to its starting point from the other end of its tracking range in the fastest possible time.

Robot Travels to the Work. When machining cycles are quite long, a robot can be mounted on a track to enable it to travel among more machines than can conveniently be grouped around a stationary robot. This type of configuration is shown in Fig. 38.

Work Cells. A robotic work cell may be defined as a cluster of one or more robots and several machine tools or transfer lines that are interconnected in such a way that they work together in unison. All of the necessary accessory equipment is embraced within the work cell and, together, establishes a particular work environment. Criteria for establishing and utilizing a robotic work cell (Ref. 9) include:

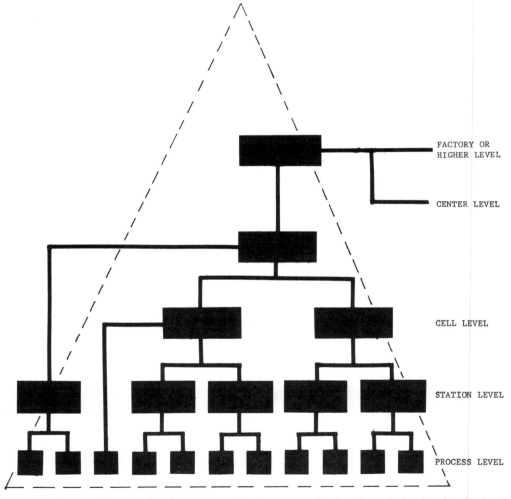

Fig. 39. Pseudopyrimidal hierarchy where communications are predominantly vertical rather than horizontal. (*Source:* Ref. 9)

1. Must be capable of performing required machining functions on a limited number of parts of sizes within predetermined limits.
2. One worker must be capable of operating the work cell with minimal skill requirements.
3. Selection of operator personnel from the shop floor must be possible.
4. Manual parts fabrication in the work cell must be feasible in case of system failure.
5. Compliant end-effectors must be used because of inherent inaccuracies of available commercial robots.
6. Part programming must be done on line.
7. Safety sensors must be installed for protection of personnel, equipment, and work in process.
8. Consistent quality must be maintained on parts.
9. Productivity must be increased to make system economical.
10. Early implementation must be made for quickest payoff.

As reported by the Comptroller General's report to Congress in 1975, concerning the state of productivity in the United States, it was of interest to learn that (1) of the time consumed in manufacturing a part, only 5% is used on the machine itself; and (2) of the time consumed on a machine tool, only 1.5% is used in making chips. Thus, 95% of the time is used for handling, record keeping, and similar nonproductive activities. These observations lead to three criteria of profitability from work cells: (1) Work cells should not be isolated from the remainder of the plant; (2) the product should be made with automation in mind; and (3) the work cell should be designed for maximum product-making flexibility. Furthermore, studies have shown maximum profitability is obtained when work cells can communicate with the remainder of the plant. The ICAM project developed a pyramidal hierarchy where communications are exclusively vertical (see Fig. 39) rather than horizontal. Also refer to *Section 5* of this *Handbook*.

OVERVIEW OF SOME CONTEMPORARY ROBOTS AND APPLICATIONS

Thus far in this article, a number of specific robots have been illustrated and described in connection with specific design principles. There are several thousand engineers currently engaged in designing, procuring, applying, and installing robots, and there are many more thousands of engineers and technical management personnel who are just becoming seriously interested in robot technology. A recent survey indicated a thirst for information pertaining to robot hardware (and software) available and working today—that is, systems that portray the current state-of-the art. The next several pages of this article are devoted to descriptions of specific robots in the marketplace as of the last half of the 1980s. Listing of current specifications can be helpful to the engineer who is charged with working up the detailed needs of a given application. Most of the robots and their associated systems are in wide use and represent intense experience over the past decade. The editor's selection, of course, carries no specific endorsement and, regrettably, space available here does not permit including the products of many other suppliers.

The general configuration and operating parameters of the *Milacron T³™* series of industrial robots are given in Fig. 40. Four of the T³ Series of robots are illustrated, along with a summary of specifications for each. See Figs. 41 through 44 and Tables 13 through 16.

The *A Series* of electromechanically operated industrial robots, designed primarily for assembly operations and manufactured by *GMF Robotics Corporation*, is shown in Fig. 45, and their specifications are summarized in Table 17. The *M Series* of robots, designed mainly for machine loading operations, is illustrated in Figs. 46 and 47 and summarized in Tables 18 and 19. The GMF *S Series* of industrial robots is designed for special-purpose applications, including welding, machine loading, flame cutting, materials handling, inspection, finishing, and sealing operations. (See Fig. 48 and Table 20.) The GMF *P Series* of robots is used primarily in painting and coating operations. (See Fig. 49 and Table 21).

For use with its robots, GMF engineers have designed a number of standard robot hands. Before investing a lot of hand engineering, robot system designers should ascertain what "stock" hands are available from various robot manufacturers. Several GMF standard hands are shown in Fig. 50.

The *Hitachi HPR1011* process robot, shown in Fig. 51, has a carrying capacity of 22 pounds (10 kilograms) and is used for a variety of tasks, including arc welding, sealant application, adhesive application, and materials handling. The robot can be mounted upside-down.

A low-cost, four-axis high-performance robot for use in electromechanical and electronic assembly operations is shown in Fig. 52. This *Hitachi A4010H* has a capacity of 2.2 pounds (1 kilogram).

A four-axis, DC servo controlled assembly robot with a capacity of 22 pounds (10 kilograms) is shown in Fig. 53. This Hitachi A4100L robot is used in electronic assembly, sealant, and material handling of small components.

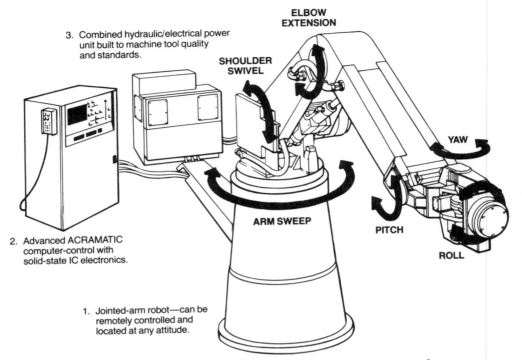

Fig. 40. The general configuration and operating parameters of the Cincinnati *Milacron* $T^{3™}$ series of industrial robots.

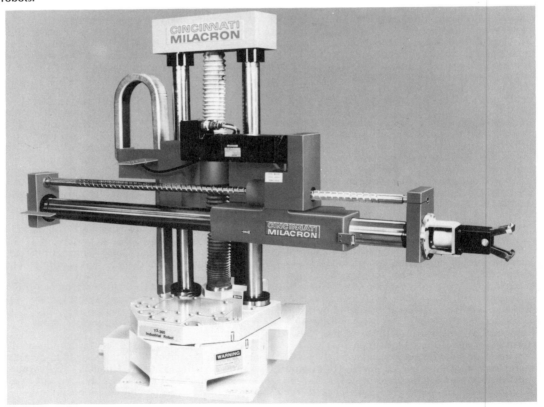

Fig. 41. The Cincinnati *Milacron* T^3 363™ industrial robot designed primarily for materials handling operations.

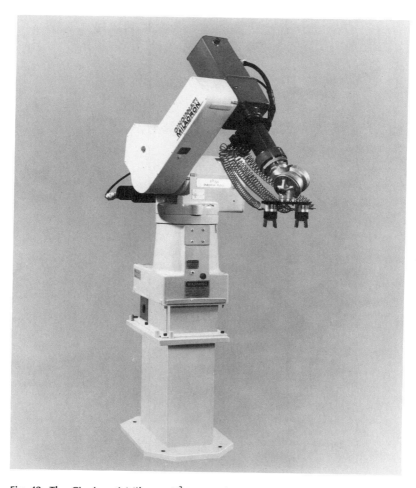

Fig. 42. The Cincinnati *Milacron T³ 726*™ industrial robot designed primarily for part handling, assembly, and welding applications.

Fig. 43. The Cincinnati *Milacron* T^3 776™ industrial robot designed primarily for heavy-payload process applications.

Fig. 44. The Cincinnati *Milacron T^3 800*™ gantry-type industrial robot designed primarily for large work volume applications. Robots in this series are particularly applicable where there is a large work volume, or where space is a premium. Their modular design allows up to 40 feet (12 meters) of *X*-axis travel and 200 inches (5 meters) *Y*-axis travel. Side-by-side gantries with a common center rail are available. Gantry robots are used in spot welding, arc welding, light machining, material handling, machine loading, and other general-purpose robotic applications.

TABLE 13. General Specifications of the *Milacron T³ 363*™ Industrial Robot.

Load Capacity
Load 8 in (200 mm) from tool mounting plate .... 110 lb (50 kg)
  with Pitch or Yaw axis option .... 75 lb (35 kg)

Number of axes, configuration, servo system, control type
Number of servoed axes 3
  with Pitch or Yaw axis option 4
Configuration .... Cylindrical
Drive system .... DC Motor
Power amplifiers .... PWM
Control type .... Sequential Point-to-Point
  with control option .... Simultaneous Point-to-Point

Positioning repeatability
Repeatability, in cycle .... ±0.020 in (±0.5 mm)

Range of motion, velocity	Range	Velocity
Base, θ	300 deg	90 deg/sec
Vertical, Z*	24 in (610 mm)	20 ips (0.5 m/sec)
Horizontal, R	26 in (666 mm)	40 ips (1.0 m/sec)
optional extended, R	45 in (1146 mm)	40 ips (1.0 m/sec)
Pitch, D, option	240 deg	90 deg/sec
Yaw E, option	240 deg	90 deg/sec

Reach	3 Axis	4 Axis
Horizontal	57 in (1450 mm)	70 in (1785 mm)
with extended reach option	76 in (1930 mm)	89 in (2265 mm)

Memory capacity, I/O contacts
Data area size/Average number of data points 8 kb/500
  with memory expansion option 24 kb/1500
Number of input contacts/Maximum optional 8/24
Number of output contacts/Maximum optional 8/24

Floor space and approximate net weight
Robot .... 4.5 ft² (0.4 m²)/1550 lb (705 kg)
Robot control .... 7.6 ft² (0.70 m²)/950 lb (430 kg)

Ambient temperature .... 40° to 105°F (5° to 40°C)
  with air conditioner option .... 40° to 120°F (5° to 50°C)

Power requirements .... 230/460 volts, 1 φ, 60 Hz**
Power rating/Power required for typical cycle 3.7 kVA/1.0kW
  with air conditioner option .... 5.7 kVA/***
  with simultaneous control option .... 7.5 kVA/***
  with air conditioner and simultaneous control options .... 9.5 kVA/***

*Consult factory for greater vertical range
**Other voltages and 50 Hz available
***Consult factory.

TABLE 14. General Specifications of the *Milacron T³ 726*™ Industrial Robot.

Load Capacity
Load 5 in (125 mm) out and 2 in (50 mm) offset
   from tool mounting plate .... 14 lb (6 kg)

Number of axes, configuration, servo system, control type
Number of servoed axes 6
Configuration .... Articulated
Drive system .... DC Motor
Position feedback .... Brushless resolver
Control type .... Controlled path at Tool Center Point

Positioning repeatability
Repeatability to any previously taught point .... ±0.004 in (±0.10 mm)

Range of motion, velocity
Base sweep .... 285 deg
Horizontal reach to tool mounting plate .... 41 in (1040 mm)
Vertical reach to tool mounting plate .... 72 in (1830 mm)
Pitch and Yaw relative to forearm .... 238 deg
Roll relative to wrist .... 900 deg
Nominal velocity at Tool Center Point (TCP) .... 40 ips (1000 mm/sec)
Base slew rate .... 250 deg/sec

Memory capacity, I/O contacts
Data area size/Average number of data points .... 48 k byte/3000
Number of input contacts/Maximum optional 16/32
Number of output contacts/Maximum optional 16/32

Floor space and approximate net weight
Robot .... 2.1 ft² (0.2 m²)/1,050 lb (480 kg)
Robot control .... 8.1 ft² (0.75 m²)/1,470 lb (670 kg)

Ambient temperature .... 40 to 105°F (5 to 40°C)
   with air conditioner option .... 40 to 120°F (5 to 50°C)

Power requirements .... 460 volts, 3φ, 60 Hz*
   Power rating/Power required for typical cycle 10kVA/2kW

*Other voltages and 50 Hz available

TABLE 15. General Specifications of the *Milacron T³ 776*™ Industrial Robot.

Load Capacity
Load 10 in (250 mm) out and 5 in (125 mm) offset
   from tool mounting plate .... 150 lb (68 kg)

Number of axes, configuration, servo system, control type
Number of servoed axes 6
Configuration .... Articulated
Drive system .... DC Motor
Position feedback .... Brushless resolver
Control type .... Controlled path at Tool Center Point

Positioning repeatability
Repeatability, in cycle .... ±0.010 in (±0.25 mm)

Range of motion, velocity
Base sweep .... 270 deg
Horizontal reach to tool mounting plate .... 101 in (2565 mm)
Vertical reach to tool mounting plate .... 139 in (3530 mm)
Pitch and Yaw relative to forearm .... 238 deg
Roll relative to wrist .... 900 deg
Nominal velocity at Tool Center Point (TCP) .... 25 ips (635 mm/sec)
Base slew rate .... 95 deg/sec

Memory capacity, I/O contacts
Data area size/Average number of data points .... 48 k byte/3000
Number of input contacts/Maximum optional 16/32
Number of output contacts/Maximum optional 16/32

Floor space and approximate net weight
Robot .... 15ft² (1.4 m²)/5,250 lb (2385 kg)
Robot control .... 8.1ft² (0.75 m²)/1,470 lb (670 kg)

Ambient temperature .... 40 to 105°F (5 to 40°C)
   with air conditioner option .... 40 to 120°F (5 to 50°C)

Power requirements .... 460 volt, 3ɸ, 60 Hz*
Power rating/Power required for typical cycle 23kVA/4kW

*Other voltages and 50 Hz available

TABLE 16. General Specifications of the *Milacron T³ 800*™ Gantry Series of Industrial Robots.

Load Capacity
Load 10 in (250 mm) out and 5 in (125 mm) offset
from tool mounting plate .... 200 lb (90 kg)*
Number of axes, configuration, servo system, control type
Number of servoed axes 6
Configuration .... Gantry
Drive system .... DC Motor
Position feedback/type .... Brushless resolver/Absolute
Control type .... Controlled path at Tool Center Point
Repeatability
Repeatability to any previously taught point .... ±0.010 in (±0.25 mm)

Range of motion	T³886**	T³896**
X Axis	80/120/160 in (2.0/3.0/4.0 m)	192/288/384/480 in (4.8/7.2/9.6/12.0 m)
Y Axis	36/56/76 in (0.9/1.4/1.9 m)	70/128/200 in (1.8/3.3/5.0 m)
Z Axis	40 in (1.0 m)	40/60 in (1.0/1.5 m)

Pitch and Yaw relative to forearm .... 238 deg
Roll relative to wrist .... 900 deg
Nominal velocity at Tool Center Point (TCP) .... 40 ips (1.0 m/sec)
Memory capacity, I/O contacts
Data area size/Average number of data points .... 48 k bytes/3000
Number of input contacts/Maximum optional 16/32
Number of output contacts/Maximum optional 16/32
Ambient temperature .... 40 to 105°F (5 to 40° C)
    with air conditioner option .... 40 to 120°F (5 to 50°C)
Power requirements .... 460 volt, 3ϕ, 60 Hz***

*Consult factory for larger load capacity
**Height under side rails
***50 Hz and other voltages available

(b)

Fig. 45. GMF *A-Series* of electromechanically-operated industrial robots, designed primarily for assembly operations. (a) The *A-100* is a 3-axis cylindrical coordinate robot. One or two additional axes of motion are added by installing one of six different wrist options. All five axis motions are controlled simultaneously. A variety of standard and special hand options are available. (b) The *A-200* robot is a 2-axis cylindrical coordinate robot. One or two additional axes of motion are added by installing one of six different wrist options. Detailed specifications are given in Table 17. (*GMF Robotics Corporations.*)

TABLE 17. Series of Robots Designed Primarily for Assembly Operations. (Also used for inspection, machine loading, parts transfer, sealing, deburring, and materials-handling operations.) *GMF Robotics Corp.*

"A" Series

2-4 Axes	2-4 Axes	3-5 Axes	3-5 Axes
2-4 Simultaneous Axes	3 Simultaneous Axes	3-5 Simultaneous Axes	3-5 Simultaneous Axes
Axes Travels			
θ = 290° R = 11.8" (300mm) Z = 4.7" (120mm) α = 300°	θ = 300° R = 11.8" (300mm) Z = 5.9" (150mm) α = 300° (roll)	θ = 300° Z = 11.8" (300mm) R = 11.8" (300mm) α = 360° (roll) β = 190° (pitch)	θ = 300° Z = 19.7" (500mm) R = 19.7" (500mm) α = 360° (roll) β = 190° (pitch)
Axes Drive Method			
AC servo	α, θ, R—AC servo Z pneumatic	AC servo	AC servo
Position Repeatability			
±0.001" (±0.03mm)	±0.002" (±0.05mm)	±0.002" (±0.05mm)	±0.002" (±0.05mm)
Maximum Operating Load at Wrist-without hand.			
Without α, Z: 35.2 lbs. (16kg) With α: 26.4 lbs. (12kg) With Z: 22.0 lbs. (10kg) With α, Z: 13.2 lbs. (6kg)	Without α, Z: 22.0 lbs. (10kg) With α: 11.0 lbs. (5kg) With Z: 15.4 lbs. (7kg) With α, Z: 6.6 lbs. (3kg)	Without α, β: 22.0 lbs. (10kg) With α, β: 6.6 lbs. (3kg)	Without α, β: 66.0 lbs. (30kg) With α, β: 19.8 lbs. (9kg)

Fig. 46. GMF *M-Series* (M-100) industrial robot. This is a 3-axis cylindrical coordinate, electromechanically-operated robot. One or two additional axes of motion are added by installing one of five different wrist operations. All five axis motions are controlled simultaneously. A variety of standard and special hand options are available. Detailed specifications are given in Table 18. (*GMF Robotics Corporations.*)

Fig. 47. GMF *M-Series* (M-300) industrial robot. This is a 3-axis cylindrical coordinate, electromechanically-operated robot. One or two additional axes of motion are added by installing one of five different wrist operations. All five axis motions are controlled simultaneously. A variety of standard and special hand options are available. Detailed specifications are given in Table 19. (*GMF Robotics Corporation.*)

TABLE 18. Series of Robots Designed Primarily for Machine Loading. (Also used for parts transfer, materials handling, press loading, die casting, finishing, and inspection operations.) *GMF Robotics Corp.*

"M" Series

5 Axes	4 Axes	3-5 Axes
1 Simultaneous Axes	1 Simultaneous Axes	1 Simultaneous Axes
Axes Travels		

A = 60°
X = 5.9" (150mm)
α = 270°
Z = 11.8" (300mm)
C = 20 pallet parts feeder standard
	Z = 5.9" (150mm)
	X = 5.9" (150mm)
	A = 90°
	B = 180°
	C = 120°
	α = 270°, 4 positions, (−90°, 0°, +90°, +180°)
		θ = 300°
		Z = 19.7" (500mm)
		R = 19.7" (500mm)
		    31.5" (800mm)
		    43.3" (1100mm)
		α = 270° (roll)
		β = ±5° on/off control (yaw)

Axes Drive Method		
α, A, X pneumatic C, Z motorized	A, B, C, X, Z–DC servo α pneumatic	θ, Z, R—AC servo α, β pneumatic
Position Repeatability		
α, A, X = ±0.012" (±0.3mm) Z, C = ±0.039" (±1.0mm)	±0.020" (±0.5mm)	±0.039" (±1.0mm)
Maximum Operating Load at Wrist—without hand		
44 lbs. (20kg)	44 lbs. (20kg)	Without α, β: 103 lbs. (47kg) With α, β: 68 lbs. (31kg)
Maximum Graspable Load with standard single-part hand		
—	22 lbs. (10kg) 5.9" (150mm) max. part dia.	44 lbs. (20kg) 9.8" (250mm) max. part dia.
Maximum Graspable Load with standard double-part hand		
11 lbs. (5kg) each 3.9" (100mm) max. part dia.	11 lbs. (5kg) each 5.9" (150mm) max. part dia.	22 lbs. (10kg) each 5.9" (150mm) max. part dia.

TABLE 19. Series of Robots Designed Primarily for Machine Loading. (Also used for parts transfer, materials handling, press loading, die casting, finishing, and inspection operations.) *GMF Robotics Corp.*

"M" Series

3-5 Axes	4 Axes	5 Axes
3-5 Simultaneous Axes	1 Simultaneous Axes	5 Simultaneous Axes
Axes Travels		
$\theta = 300°$ $Z = 21.6"$ (550mm) $\quad\;\; 51.2"$ (1300mm) $R = 19.7"$ (500mm) $\quad\;\; 31.5"$ (800mm) $\quad\;\; 43.3"$ (1100mm) $\alpha = 300°$ (roll) $\beta = \;\;0°-190°$ servo (yaw)	$Y = 11.8"$ (300mm) $Z = 19.7"$ (500mm) $C = 180°$ $\alpha = 300°$ (roll)	$\theta = 300°$ $Z = 47"$ (1200mm) $R = 47"$ (1200mm) $\alpha = 300°$ (roll) $\beta = 190°$ (pitch)
Axes Drive Method		
AC servo	DC servo	DC servo
Position Repeatability		
$\pm 0.039"$ (± 1.0mm)	$\pm 0.039"$ (± 1.0mm)	$\pm 0.039"$ (± 1.0mm)
Maximum Operating Load at Wrist—without hand		
Without α, β: 103 lbs. (47kg) With α: $\quad\quad\;$ 68 lbs. (31kg) With α, β: $\;\;\;$ 44 lbs. (20kg)	With α: 132 lbs. (60kg)	With α, β: 176 lbs. (80kg)
Maximum Graspable Load with standard single-part hand		
44 lbs. (20kg) 9.8" (250mm) max. part dia.	—	110 lbs. (50kg) 11.8" (300mm) max. part dia.
Maximum Graspable Load with standard double-part hand		
22 lbs. (10kg) each 5.9" (150mm) max. part dia.	44 lbs. (20kg) each 9.8" (250mm) max. part dia.	—

Fig. 48. The *S-300* robot is a 6-axis articulated arm, electric servo driven robot. All six axis motions are controlled simultaneously. This robot finds use in spot welding, machine loading/unloading, sealing, material handling, arc welding, parts-transfer, and water jet cutting operations. (Detailed specifications are given in Table 20. (*GMF Robotics Corporation*.)

Fig. 49. GMF *P-Series* (P-150) industrial robot designed specifically for painting operations. This is a 6-axis (7th axis optional) articulated arm, electric servo driven robot. All axis motions are controlled simultaneously. A variety of applicators are available. The robot finds use in automotive and non-automotive painting, application of underbody deadeners, sealing operations, water jet booth cleaning and can be installed for moving-line or stationary applications. Detailed specifications are given in Table 21. (*GMF Robotics Corporation*.)

TABLE 20. Series of Robots Designed for Special-Purpose Applications. (These include welding, machine loading, flame cutting, materials handling, inspection, finishing, and sealing.) *GMF Robotics Corp.*

"S" Series

5 Axes	6 Axes	6 Axes	6 Axes
5 Simultaneous Axes	6 Simultaneous Axes	6 Simultaneous Axes	6 Simultaneous Axes
Axes Travels			
Y = 39" (1000mm) 79" (2000mm) W = 75° U = 85° α = 360° (roll) β = 210° (pitch)	θ = 300° W = 90° U = 105° γ = 360° (roll) β = 210° (pitch) α = 360° (roll) Available on Y axis servo slide.	θ = 300° W = 90° U = 105° γ = 360° (roll) β = 210° (pitch) α = 360° (roll) Available on Y axis servo slide.	Y = 79" (2000mm) 118" (3000mm) W = 90° U = 105° γ = 360° (roll) β = 210° (pitch) α = 360° (roll)
Axes Drive Method			
DC servo	AC servo	AC servo	AC servo
Position Repeatability			
±0.008" (±0.2mm)	±0.02" (±0.5mm)	±0.02" (±0.5mm)	±0.02" (±0.5mm)
Maximum Operating Load at Wrist-without hand.			
17 lbs. (8kg)	132 lbs. (60kg)	132 lbs. (60kg)	132 lbs. (60kg)

TABLE 21. Robots Designed for Painting Operations. (At left is robot for numerically controlled (NC) tracking; at right is robot that is pedestal mounted for NC tracking.) *GMF Robotics Corp.*

"P" Series

7 Axes	6 Axes
7 Simultaneous Axes	6 Simultaneous Axes
Axes Travels	
X = 16 ft. (4.9m) W = 160° (waist rotation) S = 128° (shoulder rotation) E = 140° (elbow rotation) R = 1080° (wrist rotation) P = 180° (wrist yaw) F = 1080° (fan rotation)	W = 160° (waist rotation) S = 128° (shoulder rotation) E = 140° (elbow rotation) R = 1080° (wrist rotation) P = 180° (wrist yaw) F = 1080° (fan rotation)
Axes Drive Method	
Hydraulic servo	Hydraulic servo
Position Repeatability	
±0.25" (±6.4mm)	±0.25" (±6.4mm)

Robot Technology Fundamentals 315

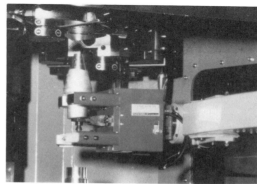

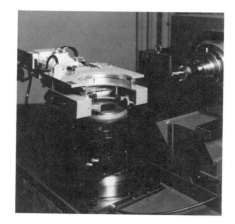

Fig. 50. Some of the large variety of standardized hands available from one robot manufacturer. (*GMF Robotics Corporation.*)

Fig. 51. Versatile robot, including ability to be mounted in an upside-down position, is used in arc welding, sealant, adhesive, and material handling applications. Capacity is 22 pounds (10 kilograms). Control is by a 16-bit (8088) microprocessor-based unit, with an integral 9-inch CRT. Standard systems feature software for palletizing routines, management data, self-diagnostics, branching, and linear and circular interpolation with weaving. A large nonvolatile one-megabit bubble memory is provided with an additional megabit of expansion available. Software changes can be upgraded in the field and a full RS232C communications package is available. (Hitachi America, Ltd.)

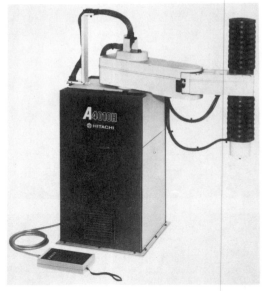

Fig. 52. Low-cost, four-axis, high-performance robot with a capacity of 2.2 pounds (1 kilogram). This point-to-point controller is self-contained in the compact robot body. The robot is taught by the direct-teach method. Maximum operational speed is 47 inches/sec (1200 mm/sec); the maximum vertical stroke is 6 inches (150 mm) at a speed of 10 inches/sec (250 mm/sec). Resident memory capacity is 200 steps with RS232C up-and-down load function as standard. Power consumption is low. This light-weight robot finds uses in electromechanical and electronic assembly operations. (Model A4010H, Hitachi America, Ltd.)

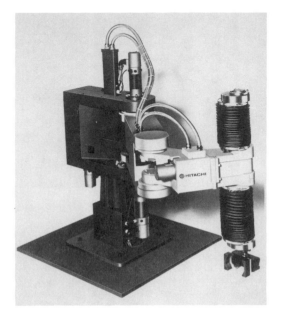

Fig. 53. A four-axis, DC servo controlled robot with a capacity of 22 pounds (10 kilograms). The fully programmable Z axis motion has a 12-inch (300 mm) stroke at a speed of 10 inches/sec (250 mm/sec). Control is provided by a 16-bit (8088) microprocessor-based unit with an interal 9-inch CRT. Standard system software features include palletizing routines, management data, self-diagnostics, branching, coordinate transformation, and linear and circular interpolation. A large nonvolatile one megabit bubble memory is provided with an additional megabit of expansion available. Software changes can be upgraded in the field and a full RS232C communications package is available. Typical applications include electronic assembly, sealant, and material handling of small components. (Model A4100L, Hitachi America, Ltd.)

Robot Technology Fundamentals 317

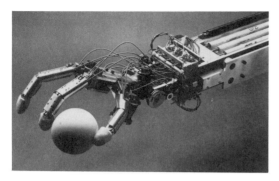

Fig. 54. Multi-finger robot hand. The hand has fourteen joints which are driven by special 'shape memory' alloy actuators. The fingers are dexterous and gentle and are well-suited to a number of assembly and maintenance operations.

An interesting, multi-finger robot hand developed by Hitachi is shown in Fig. 54.

A new concept in remote robot arms for use in certain manufacturing operations incorporates air-compressed rubber actuators. The actuators, which act very similarly to human muscles, expand and contract with compressed air pressure and dictate the movement of the arm with the help of wire rope and pulleys. The arm also has rubber hands, which are capable of handling objects up to 4.5 pounds (2 kilograms). If the

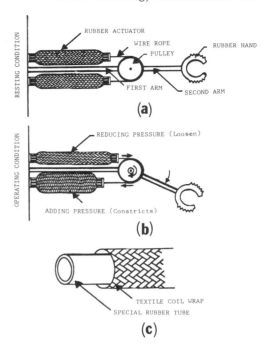

Fig. 56. Details of robot with rubber muscles: (a) Resting position of robot, (b) example of one possible movement, and (c) tubular construction. (*Developed jointly by Bridgestone Corp. and Hitachi.*)

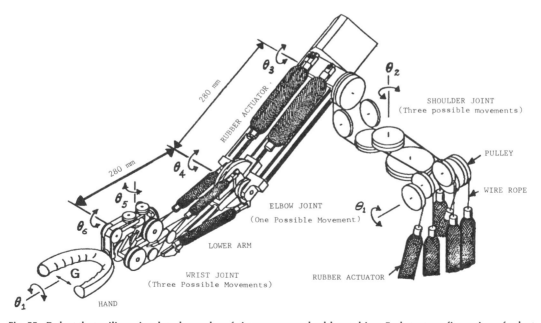

Fig. 55. Robot that utilizes simulated muscles of air-compressed rubber tubing. Early test configuration of robot is shown. (*Developed jointly by Bridgestone Corp. and Hitachi.*)

"forearm" is kept straight, the capacity from the "shoulder" increases to 13.2 pounds (6 kilograms). The new light-weight robot has seven degrees of freedom of motion. The seven possible places for movement are in the "joints" of the arm; three in the "shoulder" and three in the "wrist" and one in the "elbow." The hand is shaped like a horseshoe and can open and close with the aid of a 16-bit microcircuit that controls the movement program. (See Figs. 55 and 56.)

An advantage of the design is that it can be used where space is limited—it can work independently of its compressor location. The soft rubber grip permits handling of fragile items. The actuators consist of a rubber tube covered with braided fiber. Both ends are secured with a flange, which opens to accept and release air pressure. Round joints are used for movements up to 90°, giving them much flexibility in many positions.

EVALUATING POTENTIAL OF ROBOTS FOR A GIVEN APPLICATION

At least one robot supplier[1] has developed a plan for evaluating the potential of robots for a selected use. This may take the form of a checklist as given below.

1. **Analysis of the Application**
   Long- and short-term objectives.
   Manufacturing processes involved.
   Space availability.
   Budget.
   Initial development of automation recommendations.
   System objectives.
   Functional performance specifications.
   Equipment recommendations.
2. **Feasibility Study**
   How will a more automated system affect related operations in the plant?
   Material handling methods.
   Potential process changes.
   Commercial equipment available.
   Updating existing equipment (partial automation) for maximum cost effectiveness.
   Preparation of specifications.
   CAD layouts of floor space and cell configuration.
   CAD cell simulation.
3. **System Proposal**
   Functional specifications.
   System operation.
   Robot types.
   Tooling.
   Fixturing.
   Peripheral equipment.
   Electronic communications requirements.
4. **System Design**
   Microprocessor controls.
   Software.
   Multiple levels of control.
   Management data.
   Preplanned expandability.
5. **Construction Phase**
   Note: It normally is a good procedure for the system to be set up and thoroughly tested at the supplier's facility. This will minimize the interruption of current production procedures.
6. **Installation Phase**
   Note: It is good practice for the supplier to supervise the step-by-step installation of the system.
7. **Training and Documentation**
   Note: Hands-on robot training should be provided by supplier for all persons who will interface with the new automated system. Training can take place at the supplier's facility or at the user's plant. The supplier should provide design drawings and documentation for system control, operation, and maintenance.

---

[1] Cincinnati *Milacron*™.

ADVANCED ROBOTIC SYSTEMS/EXPANDED USES

Commercially available "blind" robots of the mid-1980s are adequate for a high percentage of the tasks required where parts positions are changed relatively infrequently and where extensive software packages have already recorded positions of all parts. But a lot of industry still destined to feel the impact of robotics is not stable and well ordered. To expand their usefulness, robots must be *automatically adaptive*, capable of recognizing, reorienting, and then manipulating disordered parts. For many assembly operations and installation procedures, this adaptive ability will be essential. The keys to automatic adaptability are (1) equipping robots with sensors that enable them to adjust to their working environment and to make the last few millisecond decisions "locally" and without having to have countless details programmed into its

Fig. 57. Example of bin-picking or occlusion problem where parts are randomly jumbled in a bin prior to being drawn out for assembly. As of 1986, the problem is verging on partial solution through application of complex machine vision system. See "Machine Vision" in Section 2 of this Handbook.

memory, and (2) using some of the techniques previously developed and others currently being researched that come under the general umbrella of artificial intelligence (AI).

For decades fixed or hard automation systems have taken advantage of very effective sensors, but they are not considered highly sophisticated. These, of course, would include a variety of position-sensing switches, such as electromechanical switches (both large and small, the latter sometimes called miniature or *micro* (TM Honeywell) switches; pressure and vacuum switches; photoelectric devices (visible portion of the electromagnetic spectrum); and infrared radiation detectors, among many others. More recently, new configurations of touch or tactile sensors have been developed and have been used in a relatively few test robots.

Giving robots an ability to "see" (machine vision) is and will continue to be one of the most active research and development areas directed toward the design of "smart" robots. A number of first-generation (and some second-generation) artificial vision systems have been put to work on the factory floor.

Other targets of robot R&D include: (1) increasing robot speed, (2) coordination of multiple arms, (3) robots with mobility (locomotion), including the creation of different robot configurations, such as legged robots, (5) voice communication to and from robots, (6) improved integration of robotic systems into the manufacturing command and control system in accordance with the objectives of CIM (computer integrated manufacturing)—this including the development and utilization of standard communication protocols, such as MAP, (7) introduction of more self-diagnostics (fault tracing methodologies) (8) improved inherent safety, (9) creating general-purpose hands for increasing applicational flexibility—this involving studies of robot geometry, kinematics, and dynamics to achieve high-level dexterity, (10) further refinement of robot simulation techniques, (11) continuing efforts to improve robot programming—with emphasis, for example, on natural language processing as well as the use of "expert systems" involving both factual and heuristic knowledge—among many other aspects of current R&D, not the least of which is to improve robot cost effectiveness.

Just as a way to put matters in perspective as of the late 1980s, the so-called occlusion problem or bin-picking problem should be mentioned. A human with very low intelligence can look into a basket of parts and pick them out one at a time and orient them for placement in a secondary operation. But, thus far this problem has eluded research in artificial intelligence. See Fig. 57 and Ref. 10. Also see "Machine Vision—State-of-the-Art Systems" in *Handbook Section 2* for partial solution.

REFERENCES

1. Tver, D. F., and R. W. Bolz: "Encyclopedic Dictionary of Industrial Technology," Chapman & Hall, New York, 1984.
2. Trickey, A. F.: "Industrial Robots and Cartesian Coordinates," *Pubn. 397K1*, Unimation (Westinghouse), Danbury, Connecticut, January 1983.
3. Engelberger, J. F.: "Robotics in Practice," Kogan Page Limited, London, 1981.
4. Larson, T. M., and A. Coppola: "Flexible Language and Control System Eases Robot Programming," *Electronics*, 156-159 (June 14, 1984).
5. Harms, D.: "Robot Motion Control: A Software Problem," *Cont. Eng.*, 314-324 (September 1984).
6. Roth, B.: "Introduction to Robots," in "Design and Application of Small Standardized Components," *Data Book 757*, published by Stock Drive Products (New Hyde Park, New York) and distributed by Educational Products, Mineola, New York, 1983.
7. Bailey, S. J.: "Fluid Power Control Provides a Wide Torque/Force Choice," *Cont. Eng.*, 69-72 (March 1982).
8. Hunt, V. D.: "Smart Robots," Chapman and Hall, New York, 1985.
9. USAF: "ICAM Robotics System for Aerospace Batch Manufacturing—Task A," Materials Laboratory, U. S. Air Force, Wright Aeronautical Laboratories, Air Force Systems Command, Wright-Patterson Air Force Base, Ohio, 1984.
10. Horn, B. K. P., and K. Ikeuchi: "The Mechanical Manipulation of Randomly Oriented Parts," *Scientific American*, 100-111 (August 1984).

ADDITIONAL READING

Aleksander, I.: "Artificial Vision for Robots," Chapman & Hall, New York, 1983.

Aleksander, I.: "Computing Techniques for Robots," Chapman & Hall, New York, 1985.

Bailey, S. J.: "Fluid Power Control Provides A Wide Torque/Force Choice," *Cont. Eng.*, 69–72 (March 1982).

Bolz, R. W.: "Manufacturing Automation Management," Chapman & Hall, New York, 1985.

Carter, W. C.: "Modular Multiprocessor Design Meets Complex Demands of Robot Control," 73–77, *Cont. Eng.*, March, 1983.

Gallagher, M.: "Low Cost Networking for Islands of Automation," *Cont. Eng.*, 56–59 (October 1985—2nd Edition)

Gardner, L. B., Ed.: "Automated Manufacturing," *ASTM Spec. Tech. Pubn. 862*, American Society for Testing and Materials, Philadelphia, Pennsylvania, 1985.

Horn, B. K. P., and K. Ikeuchi: "The Mechanical Manipulation of Randomly Oriented Parts," 100–111, *Sci. Amer.*, August 1984.

Hunt, V. D.: "Artificial Intelligence and Expert Systems Sourcebook," Chapman & Hall, New York, 1986.

Kinsey, G.: "Matching the Robot to the Job," *ElectronicWeek*, 125–128 (September 3, 1984).

Laduzinsky, A. J.: "Factory Automation Control Engineers Take It One Step at a Time," 81–84, *Cont. Eng.*, May 1985.

Laduzinsky, A. J.: "Factory Automation/U.S.A.: Just the Beginning," 88–89, *Cont. Eng.*, June, 1985.

Laduzinsky, A. J.: "Factory Automation/U.S.A. and the Land of Oz," 98–103, *Cont. Eng., June, 1985.*

Morris, H. M.: "Where Do Robots Fit in Industrial Control?" 58–64, *Cont. Eng.*, February, 1982.

Morris, H. M.: "Adding Sensory Inputs to Robotic Systems Increases Manufacturing Flexibility," 65–68, *Cont. Eng.*, March 1983.

Morris, H. M.: "Profitable Robotic Work Cells Result from Interconnecting the Islands of Automation," *Cont. Eng.*, 81–84 (May 1985).

Todd, D. J.: "Walking Machines—An Introduction to Legged Robots," Chapman & Hall, New York, 1985.

von Alten, J. W.: "Electronic Factories Gearing up for Industrial Robots," 40–47, *Electronic Products*, May 15, 1985.

A Robot Dynamics Simulator

By Charles P. Neuman[1]

Robotic manipulators are complex (coupled and nonlinear) multi-variable mechanical systems that are designed to perform specific tasks. The versatile robot *arm* dynamic *s*imulation *t*ool VAST (Ref. 1) has been created to provide a user-friendly working environment in which to simulate and interpret the physical characteristics of robot and actuator dynamics and design and evaluate feedback controllers for robotic manipulators. The development and features of VAST are highlighted in this article. The simulator structure is flexible, versatile, and amenable to further development. The simulator has been designed to become the foundation of a robot-oriented CAD (computer-aided design) system.

An industrial robot is a general-purpose manipulator that consists of a chain of rigid bodies (*links*) connected serially by *joints* (Ref. 2). One end of the chain (the *base*) is attached to a support, while the other end (the *end-effector*) is free to move and manipulate objects. The number of joints is the number of degrees-of-freedom (DOF) of the robot. An industrial robot typically has six DOF.

Robot dynamics (including the dynamic interactions of the inertial, centrifugal, Coriolis, and gravitational forces/torques) are highly coupled and nonlinear. With demands for enhanced productivity through higher speeds and improved precision, robot control system design requires a physical understanding of manipulator dynamics. The objective of this article is to describe the versatile simulator that enables the engineer to gain a physical understanding of robot dynamics and interpret the relative significance of the aforementioned dynamic force/torque components. Such a simulator provides the software to design prototype arms, select actuators, and test control algorithms for existing and novel robots.

This article is organized as follows. Under "Robot Model," we review the concepts of robot and actuator dynamics. The closed-form Lagrange-Euler formulation (Ref. 2) is applied to model the robot dynamics, and the robot and actuator dynamics are combined to produce the complete dynamic model. Under "Simulator," we describe the simulator, highlight its features, and outline its applications. A command interpreter provides user-friendly interaction during input and output, and the database is easily alterable by the user. Since VAST is conceived to become a versatile tool for the design and evaluation of robot control algorithms, we outline (under "Simulation of Robot Control Algorithms") the implementation of robot control algorithms in the simulator. Finally, under "Conclusions and Extensions," we formulate directions for future software development.

ROBOT MODEL

The heart of a robot simulator lies in the dynamic robot model. Under "Robot Dynamics" and "Complete Dynamic Model," we review robot and actuator dynamics for dynamic analysis and control system design. We then combine (under "Complete Dynamic Model," the robot and actuator dynamics and describe the computer implementation of the complete dynamic robot model.

Robot Dynamics

Robot dynamics are modeled by the mathematical formulation of the dynamic equations-of-motion for a robot. These equations-of-motion lay the foundation for the dynamic computer

[1] Professor, Department of Electrical and Computer Engineering and the Robotics Institute, Carnegie Mellon University, Pittsburgh, Pennsylvania 15213. The author expresses his appreciation to Michael S. Pfeifer for implementing VAST and to Professor Arthur C. Sanderson for making computational facilities available.

simulation of a robot and the design of control systems.

The dynamic equations-of-motion describe the forces and torques applied to each link as a function of the joint positions, velocities, and accelerations. Dynamic robot models are formulated by theoretical approaches and computational algorithms (Refs. 2, 3). Foremost among these approaches are the closed-form Lagrange-Euler formulation (Refs. 9, 22) and the recursive Newton-Euler formulation (Refs. 4, 5, 6). Both closed-form and recursive formulations are essential ingredients of the control engineering repertoire. For robots with more than three DOF, the (general-purpose) recursive Newton-Euler formulation is more computationally efficient than the closed-form Lagrange-Euler formulation (Ref. 3). The computationally efficient, *recursive* Newton-Euler formulation (Ref. 7) sacrifices the compact structure of the closed-form formulation and is more amenable to simulation and real-time control applications (Ref. 2). The closed-form structure of the Lagrange-Euler formulation, however, allows easier access to the dynamic force/torque components than does the recursive structure of the Newton-Euler formulation. The Lagrange-Euler formulation thus provides more physical insight into robot dynamics (Ref. 2). Since the goal of our software development has been to develop a dynamic simulator and enable the engineer to obtain a physical understanding of the dynamic behavior of robots, we have implemented the Lagrange-Euler formulation of robot dynamics.

Upon neglecting mechanical dissipation, the dynamic equations-of-motion for an N DOF robot with rigid links (Ref. 2) are:

$$D(q)\ddot{q} + C(q, \dot{q}) + G(q) = F(t) \quad (1)$$

or

$$\sum_{j=1}^{N} d_{ij} \ddot{q}_j + \sum_{j=1}^{N} \sum_{k=1}^{N} \dot{q}_j c_{jk}(i) \dot{q}_k + G_i = F_i(t) \quad (2)$$

for $i = 1, 2, ..., N$.

Robot dynamics are characterized by the system of N coupled, nonlinear, second-order differential equations in (1), whose parameters are functions of the instantaneous joint positions and velocities. The parameters of the dynamic robot model are defined in Table 1 (Ref. 8). We use SI units throughout our software and follow the standard robotics nomenclature in Refs. 9 through 12.

Actuator Dynamics

Actuator dynamics relate the applied joint forces/torques $F(t)$ in (1) to the applied actuator inputs. Robot actuators are electric (for relatively small loads) or hydraulic (for relatively large loads) (Refs. 9, 13). For control engineering applications, each actuator in VAST is modeled by

TABLE 1. Lagrange-Euler Robot Dynamics.

	Joint Coordinates
$q_i(\dot{q}_i, \ddot{q}_i)$	Joint coordinate (velocity, acceleration) of joint i: q_i is an angular displacement if joint i is revolute; and q_i is a linear displacement if joint i is prismatic.
q ($\dot{q}, \ddot{q}$)	Joint coordinate (velocity, acceleration) N-vector
	Dynamic Robot Model Parameters
$D(q)$	Inertial coefficient ($N \times N$) matrix, with elements d_{ij}:
	d_{ii} is the self-inertia of joint i.
	d_{ij} is the mutual-inertia between joints i and j.
$C(q,\dot{q})$	Centrifugal and Coriolis force/torque N-vector with elements $C_i = \dot{q}^T C(i) \dot{q}$
$C(i)$	Coupling ($N \times N$) matrix for joint i, with elements $c_{jk}(i)$:
	$c_{jj}(i)$ is the coefficient of the centrifugal force/torque joint i because of the velocity of joint j.
	$c_{jk}(i)$ is the coefficient of the Coriolis force/torque at joint i because of the velocities of joints j and k.
$G(q)$	Gravitational force/torque N-vector, with elements G_i
	Input
$F(t)$	Applied joint force/torque N-vector, with elements $F_i(t)$
	Complete Dynamic Model Parameters
$J(q)$	Total inertial coefficient ($N \times N$) matrix, with elements:
	$J_{ij}(q) = \begin{cases} d_{ii}(q) + J_{Ai} N_{Ai}^2 & \text{for } i = j \\ d_{ij}(q) & \text{for } i \neq j \end{cases}$
F	Diagonal frictional ($N \times N$) matrix, with elements:
	$F_{ii} = F_{Ai} N_{Ai}^2$
K	Diagonal gain ($N \times N$) matrix, with elements:
	$K_{ii} = K_{Ai} N_{Ai}^2$

a second-order (type 1) system[2] with inertia J_A, damping coefficient F_A, gear ratio N_A, and gain K_A. The dynamic model of the i-th actuator is thus.

$$J_A\ddot{q} + F_A\dot{q} + F(t)/N_A^2 = K_A u(t) \quad (3)$$

where q is the position of joint i, $F(t)$ is the force/torque applied to joint i, and $u(t)$ is the actuator input to joint i. Armature-controlled DC motors with negligible armature inductance (Ref. 14) are currently implemented in VAST.

Complete Dynamic Model

The complete dynamic model is obtained by combining the robot dynamic model in (1) with the actuator dynamic model in (3). The following simplifying assumptions are made to formulate the model (Ref. 15):

Rigid links and actuators shafts.
Ideal gearing (no friction and backlash).
Negligible mechanical friction.
Absence of actuator saturation limits.
Negligible noise in the joint position transducers and actuator inputs.

The resulting 2Nth-order complete dynamic model is

$$J(q)\ddot{q} + F\dot{q} + C(q, \dot{q}) + G(q) = K u(t) \quad (4)$$

or

$$\sum_{j=1}^{N} J_{ij}\ddot{q}_j + F_{ii}\dot{q}_i + \sum_{j=1}^{N}\sum_{k=1}^{N} \dot{q}_j c_{jk}(i)\dot{q}_k + G_i \quad (5)$$
$$= K_{ii} u_i(t) \quad \text{for } i = 1,2,\ldots,N.$$

The parameters introduced in (4) and (5) are defined in Table 1.

The complete dynamic model in (4) and (5) is *linear* in the joint accelerations $\ddot{q}$. The inertial coefficient matrix $D(q)$ is positive definite (Ref. 16), and $\mathbf{J}(q)$ is the sum of $\mathbf{D}(q)$ and the positive definite diagonal actuator inertial matrix $[J_{Ai}N_{ai}^2]$. Since $\mathbf{J}(q)$ is positive definite, this system of N *l*inear *a*lgebraic *e*quations (LAE) can be solved for the joint accelerations by such computational algorithms as Gauss elimination and Cholesky decomposition (Ref. 17). The resulting joint accelerations are directly amenable to numerical integration and become the foundation of the robot simulator highlighted later.

SIMULATOR

In the following paragraphs, we highlight the development and computer implementation of VAST (Ref. 1). We present the structure of VAST and describe its flexible operation and simulation modes. We then delineate engineering applications of VAST.

Structure

The user-friendly and versatile software VAST can simulate any N DOF robot containing both revolute and prismatic joints. The number of DOF of the robot arm is limited only by the computational time and memory of the computer. VAST is written in the C programming language on a multi-user VAX 11/750 minicomputer, under the UNIX (TM of Bell Laboratories) operating system.

The *modular* structure of VAST, depicted by Fig. 1, is flexible, versatile and amenable to future evolution., Modules that perform the *same* function can be interchanged easily, without affecting the remainder of the simulator. Additional capabilities can therefore be directly incorporated into the modular structure of VAST. This prototype structure is a feature of our software development.

VAST consists of three components: (1) the *command interpreter*, (2) the *database modules* (in Table 2), and (3) the *simulator modules* (in Table 3). The *c*ommand *i*nterpreter (CI) (ref. 18) is the interface between the user, the database, and the simulator. The CI initializes the database parameters, runs the simulator, and displays the simulation results. Features of the CI include:

User-friendly input and output interaction.
A complete library of help functions.
Capability to alter easily the database parameters.

The function of each database software module (in Table 2) is to:

Compute the parameters for the simulator.
Input and output the module parameters through the CI.

The computational function and current VAST implementation of each database module are described in Table 2.

The simulator modules (in Table 3) generate the transient response (joint positions, velocities, and accelerations) of the manipulator. The dynamic equations in (4) and (5) are solved for the joint accelerations and integrated numerically

---
[2]Higher-order actuator models increase the order of the system.

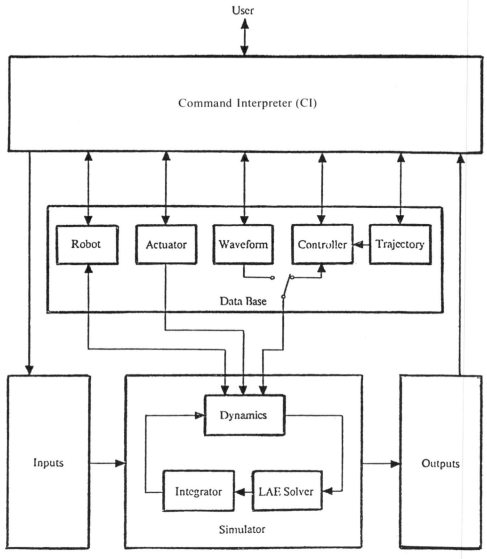

Fig. 1. Structure of VAST.

to generate the joint positions and velocities. The computational function and current VAST implementation of each simulator module are described in Table 3.

VAST Operation

The operation of VAST involves three phases: (1) parameter initialization, (2) dynamic simulation, and (3) output display. The inputs and outputs are listed in Table 4. Parameter initialization enables the user to initialize the database parameters, i.e., the parameters in the robot, actuator, waveform, controller, and trajectory database modules (in Fig. 1). Upon startup of VAST, these parameters are initialized to their default values. The user has the option of retaining these default values, reading a set of parameter values from a previously stored file, or altering a user-specified set of values. Any changes can be saved by writing the database to an off-line file.

The dynamic simulation phase generates the robot motion. The program prompts the user for the simulation title, total simulation time, sam-

TABLE 2. Database Modules

Module	Function
Robot	Compute the inertial matrix, centrifugal and Coriolis vector, and gravitational vector in the dynamic robot model (in Table 1).
	VAST implementation: 3 DOF articulated robot.
Actuator	Compute the actuator inertia, damping coefficient, gear ratio, and gain of the actuator model in (3).
	VAST implementation: Armature-controlled DC motor with negligible armature inductance (Ref. 14).
Waveform	Sample (at each integration step) analog input waveforms for the joint actuators.
	VAST implementation: Step, ramp, pulse and sinusoidal input waveforms.
Controller	Compute (at each sampling instant) the control signals for the joint actuators using data from the trajectory and integrator (Table 3) modules.
	VAST implementation: Fixed controller (Refs. 9, 19) computed torque controller (Refs. 20, 21) and reduced computed torque controller (Ref. 22).
Trajectory	Compute (at each sampling instant) the desired joint positions, velocities, and acclerations for the controller module.
	VAST implementation: Constant and set point; and linear, cubic, and quintic joint positions (Ref. 10).

pling period, and simulation mode (described shortly). The simulation then begins, and the user can observe (at each sampling instant) the joint positions, velocities, accelerations, and actuator inputs. At each sampling instant, the program stores *on-line* the joint positions, velocities, accelerations, actuator inputs, the force/torque components in (5), and, if the controller mode is selected, the desired joint positions and velocities, and the joint position and velocity errors.

Outputs from the dynamic simulation phase can be stored *off-line* in the output display phase. The user has the option to write or plot the robot and actuator variables, force/torque components

TABLE 3. Simulator Modules.

Module	Function
Dynamics	Formulate the complete dynamic model in (4) using data from the robot, actuator, controller, and waveform modules.
LAE Solver	Solve the complete dynamic model (generated by the Dynamics Module) for the joint accelerations.
	VAST implementation: Gauss elimination with partial pivoting (Ref. 17).
Integrator	Integrate numerically the complete dynamic model.
	VAST implementation: Fourth-order Runge-Kutta algorithm (Ref. 23) with adaptive step-size (Ref. 24).

TABLE 4. Simulator Inputs and Outputs

Inputs
— Simulation title.
— Total simulation time.
— Sampling period (time between synchronously sampled outputs)
— Simulation (waveform or controller) mode.

Outputs
— Robot and actuator variables: Position, velocity, acceleration, and actuator input for each joint.
— Force/torque components: Total force/torque, self- and mutual-inertial force/torque, centrifugal force/torque, Coriolis force/torque, gravitational force/torque, actuator inertial force/torque, and actuator damping force/torque for each joint.
— Controller variables: Desired position, desired velocity, position error, and velocity error for each joint (if the controller mode is selected).

in (5), or controller variables. The user must enter this phase before commencing the next simulation. When a new simulation begins, the results of the previous simulation are destroyed.

Simulation Modes

VAST accommodates two modes of dynamic robot simulation: (1) the (open-loop) *waveform* mode and (2) the (closed-loop) *controller* mode (see Fig. 1). In the waveform mode the robot is simulated by applying a user-specified *input* waveform to each joint actuator. In the controller mode the robot is simulated by connecting a user-specified controller to each joint actuator.

If the *waveform* mode is selected, the user must specify (in the parameter initialization phase):

Actuator input waveform for each joint. (The available waveforms in Table 2 are step, ramp, pulse, and sinusoidal.)

Parameters for each waveform. For the waveforms listed in Table 2, the parameters are as follows:

- Step: on-time and amplitude.
- Pulse: on-time, off-time, and amplitude.
- Ramp: on-time, off-time, and slope.
- Sinusoidal: on-time, amplitude, frequency, and phase.

Initial joint position, velocity and acceleration for each joint. The simulator calculates the initial voltage bias, which is required by each joint actuator to satisfy the prescribed initial joint positions, velocities, and accelerations. These biases are then added to the input waveforms to ensure continuity of the actuator inputs.

If the *controller* mode is selected, the user must specify (in the parameter initialization phase):

Controller. (The available controllers in Table 2 are fixed, computed torque, and reduced computed torque.)

Parameters for each controller. For the controllers listed in Table 2, the parameters are:

- Fixed controller: position gain and velocity gain.
- Computed torque controller: position gain and velocity gain.
- Reduced computed torque controller: position gain and velocity gain.

Desired trajectory for each joint. (The available joint position trajectories in Table 2 are constant and set point and linear, cubic, and quintic.)

Initial joint position, velocity, and acceleration for each joint.

Parameters for each trajectory. For the trajectories listed in Table 2, the parameters are:

- Constant: none.
- Set Point: final joint position.
- Linear: final joint position and initial and final path times.
- Cubic: final joint position and velocity and initial and final path times.
- Quintic: final joint position, velocity and acceleration and initial and final path times.

Engineering Applications

Our software tool VAST enables the user to:

Calculate the *forward* and *inverse* kinematic and dynamic solutions (Refs. 9, 10).
Determine the relative contributions of the inertial, centrifugal, Coriolis, gravitational, and actuator inertial and damping forces/torques to the dynamic behavior of the robot.
Evaluate the sensitivity of the manipulator dynamics with respect to the robot and actuator parameters (e.g., link mass and link length, and actuator inertia).
Design, implement and evaluate control algorithms.

Knowledge of the relative contributions of the force/torque components to the dynamic behavior is essential for control system analysis and design. For example, if certain force/torque components (such as mutual inertia and Coriolis forces/torques) are found to be relatively small, these components can be neglected to simplify the control algorithm. Sensitivity analysis facilitates the selection of design parameters for the robot and actuators. VAST may be used, for example, to determine counterbalance masses that lead to accurate trajectory tracking.

VAST enables the user to change the payload in the midst of the simulation. The user can thereby evaluate controller performance during payload variations. This option is especially useful for simulating a robot such as the *Unimation/Westinghouse Puma 600*, which operates in a flexible assembly environment and is required to handle a variety of payloads. The user also has the option to continue the simulation from the last stored sampling instant. This option enables the user to concatenate a family of trajectories in a single simulation run.

Simulation of Robot Control Algorithms

In this portion of the article, we outline the implementation of robot control algorithms in VAST. The user must supply the control algorithm and the input and output routines for the controller parameters (such as position and velocity gains). Under "Robot Control," we highlight robot control and outline *fixed* (Refs. 9, 19) and computed torque (Refs. 20, 21) control algorithms. Under "Controller Performance," we summarize the performance of these controllers in a case-study of an articulated 3 DOF robot.

Robot Control

Industrial robots are used for a variety of tasks, including spot welding, painting, assembly, and material handling (Refs. 13, 25). The current implementation of VAST focuses on point-to-point and continuous path control (Ref. 10).

A block diagram of a robot control system is shown in Fig. 2. The *inputs* to the robot (such as voltage or current inputs to actuators) are signals that can be varied dynamically. The *outputs* (such as joint positions or velocities) are signals which are directly measurable. The *actual trajectory* constitutes the controlled variables, which are often directly measured or estimated as outputs. The *disturbances* are features of the robot and its environment which are not known or included in the control design. The *desired trajectory* is the ideal trajectory of the robot. The *task description* (such as the path and the final joint positions) is the external description of the desired task. The objective of a robot control system is to ensure that the desired trajectory, generated by a *trajectory planner* (the trajectory module in VAST), is followed accurately in the presence of unpredictable disturbances arising from incomplete modeling of the robot, limitations of computational precision, actuator and transducer errors, and mechanical effects such as friction, backlash and vibration. The robot control problem is to design the *control algorithm* to compute the control signals from past and/or present errors. VAST provides an environment in which to simulate and evaluate robot control algorithms.

Fixed control (Refs 9, 19) is one of the most common approaches in industrial robots (Refs. 2, 13). The control algorithm *simplifies* design by neglecting the mutual inertia and centrifugal and

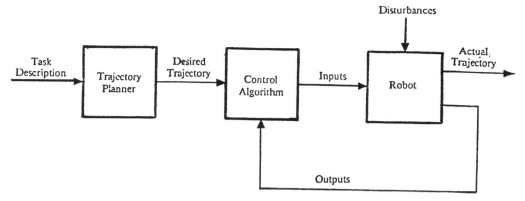

Fig. 2. Robot control system.

Coriolis forces/torques. The robot is modeled as a set of N uncoupled joints, each with a *constant* link inertia and *coordinate dependent* gravity loading. When the mutual inertia or centrifugal and Coriolis forces/torques become significant, or there is a mismatch between gravity loading and gravity compensation, the performance of the fixed control algorithm degrades. Position feedback decreases the position error and increases the speed of response; velocity feedback increases the damping; and feedforward gravity compensates for gravity loading. The engineer applies the concepts of classical control engineering to specify the position and velocity gains.

The complete dynamic model in (4) is coupled and nonlinear, and the parameters in Table 1 are explicit functions of the joint positions and velocities. A desired response cannot be achieved for the entire path by selecting *fixed* values of the feedback gains, since fixed values cannot apply to all configurations. Centrifugal and Coriolis forces/torques become significant when more than one joint is moving simultaneously at moderate and high velocities (Ref. 28).

The fixed control algorithm is designed to compensate only for the gravitational forces/torques and neglects the mutual inertia and centrifugal and Coriolis forces/torques. To achieve accurate trajectory tracking, the control algorithm must *account* for the coordinate dependencies in the complete dynamic model in (4) and *compensate* for deviations from the desired trajectory. The computed torque control algorithm is conceived to fulfill these requirements.

Compound torque (Refs. 20, 21) is an augmented feedforward nonlinear control algorithm. The underlying idea is to use a feedforward model of the robot to cancel the nonlinearities and reduce the system to a second-order *linear* model. The computed torque control signal is the sum of a *nominal* feedforward control signal (to cancel the nonlinearities) and an *augmented* feedback control signal (to specify the closed-loop response).

The feedforward control signal is sufficient to control the robot in the absence of modeling errors. Errors attributable to backlash, gear friction, and uncertainty in the robot and actuator parameters make deviation from the desired trajectory inevitable. The feedback control signal *augments* the feedforward control signal to compensate for deviations from the desired trajectory. The algorithm feeds back position and velocity and relies on the perfect modeling of the robot. When there are modeling errors, the performance of the computed torque control algorithm degrades.

To reduce the computational requirements, the computed torque control algorithm can be simplified by neglecting the mutual inertia and centrifugal and Coriolis forces/torques. This practical implementation is called the *reduced computed torque* algorithm (Refs. 9, 22).

Controller Performance

We have implemented the three control algorithms to drive 3 DOF articulated robots. (The parameter values are obtained from the *Unimation/Westinghouse Puma Series 500* and *600* robots.) The case studies are detailed, and graphical VAST output illustrating the tracking performance of the control algorithms (and the relative contributions of the significant force/torque components) are documented in the literature (Refs. 1, 26, 27).

To evaluate the point-to-point and continuous path performance of each control algorithm under payload variations, a point mass payload is attached to the third link. Each control algorithm is designed for maximum payload (the maximum payload for the Puma Series 500 and 600 robots is 2.2 kilograms [4.85 lb]) and is simulated under no payload (0 kg) and maximum payload conditions. In our case studies with geared robots, fixed control is attractive for point-to-point control, while reduced computed torque is adequate for continuous path control (Ref. 1). Gear-coupled actuators reduce the nonlinear character of robot dynamics.

CONCLUSIONS AND EXTENSIONS

VAST creates a user-friendly working environment in which to simulate robot dynamics and design and evaluate feedback controllers for robotic manipulators. The simulator structure is flexible, versatile, and amenable to future development. The software package contributes to both robot simulation development and control system design. The simulator enables the user to gain a physical understanding of robot dynamics and identify the relative contributions of the dynamic force components. Within the framework of VAST, we have incorporated fixed, computed torque, and reduced computed torque control algorithms for 3 DOF articulated robots (Refs. 1, 26, 27).

Future Development

Additional capabilities (including controllers, trajectories, and input waveforms) can be easily implemented in the flexible and modular dynamic robot simulator VAST. VAST may also be used to simulate the dynamics of a multitude of robot configurations. To simulate a new configuration, the user must formulate the dynamic equations-of-motion and update the input and output routines of the robot module (Table 1). To facilitate the formulation of the dynamic equations-of-motion, robot control engineers at Carnegie Mellon University have implemented the user-friendly *A*lgebraic *R*obot *D*ynamic *M*odeling Program (ARM), which generates *symbolically* the forward solution and complete closed-form and recursive dynamic robot models for control engineering applications (Ref. 12). A natural application of our software development is to interface ARM to VAST. The practical direction for software development is the implementation of a visualization module for the graphical display of dynamic robot motion. High-speed graphics packages will make such display of dynamic robot motion a reality. Progress in these directions will lead to a more versatile robot simulator program, provide an even greater understanding of robot dynamics, and culminate in practical robot control engineering analysis and design software.

REFERENCES

1. Pfeifer, M. S., and C. P. Neuman: "VAST: A Versatile Robot Arm Dynamic Simulation Tool," *Computers in Mechanical Engineering*, 3 (3) 57–64, November 1984.
2. Lee, C. S. G.: "Robot Arm Kinematics, Dynamics, and Control," *Computer* 15 (12) 62–80, December 1982.
3. Hollerbach, J. M.: "A Recursive Lagrangian Formulation of Manipulator Dynamics and a Comparative Study of Dynamic Formulation Complexity," *IEEE Trans. on Systems, Man, and Cybernetics SMC-10* (11) 730–736, November 1980.
4. Luh, J. Y. S., Walker, M. W., and R. P. C. Paul: "On-Line Computational Scheme for Mechanical Manipulators," *J. of Dynamic Systems, Measurement, and Control*, 102 (2) 69–76, June 1980.
5. Walker, M. W., and D. E. Orin: "Efficient Dynamic Computer Simulation of Robotic Mechanisms," *J. of Dynamic Systems, Measurement, and Control*, 104 (3) 205–211, September 1982.
6. Swartz, N. M.: "Arm Dynamics Simulation," *J. of Robotics Systems*, 1 (1) 83–100, Spring 1984.
7. Khosla, P. K., and C. P. Neuman: "Computational Requirements of Customized Newton-Euler Algorithms," *J. of Robotic Systems*, 2 (3) 309–327, 1985.
8. Neuman, C. P., and J. J. Murray: "Linearization and Sensitivity Functions of Dynamic Robot Models," *IEEE Trans. on Systems, Man, and Cybernetics*, 14 (6) 805–818, November/December 1984.
9. Paul, R. P. C.: "Robot Manipulators: Mathematics, Programming, and Control," MIT Press, Cambridge, Massachusetts, 1981.
10. Brady, M., et al. eds: "Robot Motion," MIT Press, Cambridge, Massachusetts, 1982.
11. Luh, J. Y. S.: "An Anatomy of Industrial Robots and Their Controls," *IEEE Trans. on Automatic Control* AC-28 (2) 133–153, February 1983.
12. Murray, J. J., and C. P. Neuman: "ARM: An Algebraic Robot Dynamic Modeling Program," in *Proc. of the First IEEE Int'l. Conf. on Robotics* (R.P.C. Paul, Ed), 103–114, Atlanta, Georgia, March 13–15, 1984.
13. Heginbotham, W. B., Ed.: "Industrial Robots of the World," *Industrial Robot* 10 (3) 214–217, December 1983.
14. Kuo, B. C.: "Automatic Control Systems," Prentice-Hall, Englewood Cliffs, New Jersey, 1982.
15. Stone, H. W., and C. P. Neuman: "Dynamic Modeling of a Three Degree-of-Freedom Robotic Manipulator," *IEEE Trans. on Systems, Man, and Cybernetics*, 14 (4) 643–654, July/August 1984.
16. Tourassis, V. D., and C. P. Neuman: "Properties and Structure of Dynamic Robot Models for Control Engineering Applications," *Mechanism and Machine Theory*, 20 (1) 27–40, 1985.
17. Rice, J. R.: "Numerical Methods, Software and Analysis," McGraw-Hill, New York, 1983.
18. "*UNIX* Programmer's Manual," 7th Ed. (CMU Abridged), Bell Telephone Laboratories, Inc., Murray Hill, New Jersey, 1981.
19. Luh, J. Y. S.: "Conventional Controller Design for Industrial Robots—A Tutorial," *IEEE Trans. On Systems, Man, and Cybernetics* SMC-13 (3) 298–316, May/June 1983.
20. Markiewicz, B. R.: "Analysis of the Computed Torque Drive Method and Comparison with Conventional Position Servo for a Computer-Controlled Manipulator," *Tech. Memorandum 33-601*, Jet Propulsion Laboratory, Pasadena, California, March 1973.
21. Luh, J. Y. S., Walker, M. W., and R. P. C. Paul: "Resolved-Acceleration Control of Mechanical Manipulators," *IEEE Trans. on Automatic Control* AC-25 (3) 468–474, June 1980.
22. Bejczy, A. K.: "Robot Arm Dynamics and Control," *Tech. Memorandum 33-669*, Jet Propulsion Laboratory, Pasadena, California, February 1974.
23. Carnahan, B., Luther, H. A., and J. O. Wilkes: "Applied Numerical Methods," Wiley, New York, 1969.
24. Collatz, L.: "The Numerical Treatment of Differential Equations," Springer-Verlag, Berlin, 1960.
25. Engelberger, J. F.: "Robotics in Practice," Amacon Div. of American Management, New York, 1980.

26. Tourassis, V. D., and C. P. Neuman: "Robust Nonlinear Feedback Control for Robotic Manipulators," *IEE Proc. D: Control Theory and Applications.* (Special issue on Robotics), **132** (4) 134–143, July 1985.
27. Tourassis, V. D., and C. P. Neuman: "Robust Feedback Control of an Articulated Robot: A Case-Study," *Proc. of 24th IEEE Conf. on Decision and Control,* 1505–1509, Fort Lauderdale, Florida, December 11–13, 1985.
28. Lewis, R. A.: "Autonomous Manipulation on a Robot: Summary of Manipulator Software Functions," *Tech. Memorandum 33-679*, Jet Propulsion Laboratory, Pasadena, California, March 1974.

Control of Actuators in Multilegged Robots

By Nader D. Ebrahimi[1]

The increased interest in robotics has led to the integration of robots in automated manufacturing systems. Since the robot's area of operation (*workspace*) is limited by its reach, modification of some working environments is required in order to put the workpiece within the robot's reach. One can also envision applications where considerable distance must be traversed by the robot to reach its workpiece. One example may be in crisis management, where the presence of human operators may not be possible. For instance, in the case of a nuclear reactor accident where the radiation level may exceed the safe level for human operators, a number of robots could be deployed to the accident site for assessment of damages and possible repairs.

Despite the possibility of wide-ranging applications for *legged robots*, during the last several years research has primarily focused on the control of fixed manipulator arms. Consequently, research in multilegged robots is still in the developmental stages. Although some basic studies have been reported, no legged robot is currently available on the commercial market.

In this article we will consider a legged structure that can be imagined as the lower torso of a robot. The objective is to find the configuration of the structure as it moves on a rough terrain as well as the corresponding force system to maintain its dynamic equilibrium. The linear programming approach used for this purpose bypasses the usual difficulties inherent in nonlinear methods.

BACKGROUND

The need for robot vehicles has been recognized for more than a decade, and one current solution involves legged locomotion. The vehicle is usually composed of a number of links joined together by a series of joints. The links and joints comprise the "legs." These legs support a frame that forms the cargo-carrying part of the vehicle, which may include a manipulator arm.

The motion of the vehicle is controlled by a set of actuators which, while maintaining the dynamic equilibrium, modify the joint angles continuously so that the legs will adapt to the terrain of travel.

In the United States a major center for the research on walking machines has been the Ohio State University (OSU). At OSU, under the direction of Prof. R. B. McGhee, researchers have designed what is known as the *OSU Hexapod Vehicle*. Other researchers throughout the United States and the Soviet Union have also contributed to the present state of knowledge on walking vehicles (Refs. 1 through 34). Research at OSU has centered on joint control. Although researchers there have solved some problems associated with legged vehicles, the problems of planning and navigation (obstacle avoidance) still remain.

STATEMENT OF THE PROBLEM

Let us consider a hypothetical structure as shown in Fig. 1. The structure is composed of several legs. Five legs are shown in Fig. 1. Each leg has two constant-length links as well as two joints. Each leg has an arbitrary (fixed) orientation indicated by an angle (ϕ), as shown in the plan view of Fig. 2. In addition to these legs, a system of actuating forces connect the different links. These forces, which can be compressive or tensile, will ensure the dynamic equilibrium of the system under load. Torque actuators acting on joints may also be used independently or in addition to the linear actuators to provide the required moments for dynamic equilibrium.

Kinematic Considerations

In order to determine the angular velocities as well as the angular accelerations of the links, we

---

[1] Department of Mechanical Engineering, The University of New Mexico, Albuquerque, New Mexico.

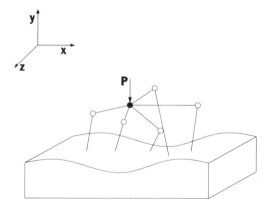

Fig. 1. A statically indeterminate structure.

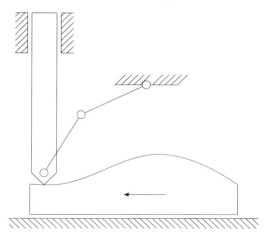

Fig. 3. Kinematic model.

can use a kinematic model to represent each leg. As shown in Fig. 3, a leg is modeled as a planar linkage. Each linkage is composed of a leg, a knife-edge follower, and a plate cam. The cam moves in the opposite direction of the actual movement of the structure and thus produces the same kinematic effect. Each of the individual linkages modeling a leg has one degree of freedom, which is defined by the motion of the structure.

This model assumes that the structure moves only along the x-direction with constant velocity, that the point of application of the external load travels at a constant level, and that the terrain of travel is known (*cam profile*).

Using the elementary methods of kinematics, we can perform a kinematic analysis of the model. For this purpose, we use the position of the cam to obtain the configuration of the leg at every instant. In addition, by using the velocity of the cam, we can also obtain angular velocities and angular accelerations of each link. Consequently, the positions, velocities, and accelerations of all the points on the links can readily be calculated. Using these accelerations, we can calculate the inertia forces (and torques) that are required in analyzing the equilibrium of the structure.

Equilibrium of the Structure

We can determine the amount of external forces (actuating forces) by considering the dynamic equilibrium of the structure. For this purpose, we can treat the inertia forces as applied forces and include them in the dynamic equilibrium equations of the system (*D'Alembert Principle*). This approach allows us to solve a set of algebraic equations rather than solving a set of differential equations.

For example, let us assume that a system of force actuators is connecting the middle point of every link to the middle point of all other links. In this case, we will have twenty actuators that can apply either tensile or compressive forces to the links they are connecting.

The system of algebraic equations to be solved is statically indeterminate—that is, the number of available equations is less than the number of unknowns. Since the equations to be solved are linear, we can use the linear programming approach to solve them.

In this approach, the solution vector (x) of the linear programming includes the actuating forces, joint reactions as well as joint moments, support (ground) reactions, and maximum vertical ground reaction.

The equality constraints of the linear programming are, then, equations of equilibrium. In addition to these equality constraints, we construct a number of inequality constraints in the linear

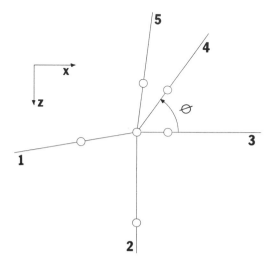

Fig. 2. Plan view of the structure.

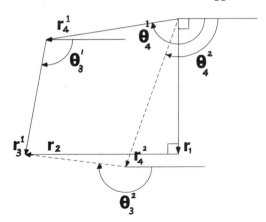

Fig. 4. Vector representation of links: (1) Acceptable solution, (2) unacceptable solution.

programming system. These inequality constraints are

$$R_{yi} - R_{y\,max} \leq 0; \; i = 1,5$$

where R_{yi} are the vertical ground reactions and $R_{y\,max}$ is the maximum of these reactions.

The objective function (performance index) of the linear programming system can be expressed symbolically as

$$J = \Sigma M + c_F \Sigma F + c_R R_{y\,max}$$

In this expression, M represents a joint moment, and F represents an actuating force. The cost coefficients for the actuating forces and maximum vertical ground reaction, c_F and c_R, determine the relative significance of minimizing these forces compared with minimizing the joint moments.

Including the maximum vertical ground reaction ($R_{y\,max}$) in the objective function has a particular implication: By minimizing the $R_{y\,max}$, we are, in effect, demanding a uniform distribution of these vertical ground reactions. In an ideal situation, all these vertical support reactions will be equal, which improves the stability condition of the vehicle.

Repeated solution of the linear programming system for consecutive positions of the structure, as it moves on a terrain, will give the required variation of the actuating forces as a function of time.

TABLE 1. Values of Some Input Parameters

leg (i)	φ(deg)	r_2	r_3	r_4
1	50.0	1.52	4.5	7.0
2	70.0	8.0	5.0	6.0
3	190.0	7.0	5.0	5.0
4	270.0	7.5	4.5	6.5
5	360.0	3.17	4.0	5.5

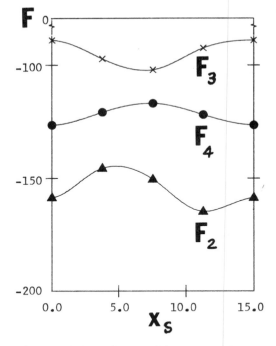

Fig. 5. Variations of some of the actuating forces.

By using this approach, we have separated an originally complex, nonlinear problem into two smaller, more manageable sub-problems. One of these problems is the linear programming part, which determines the linear system of forces and moments.

In solving the geometric equations, two solutions for each link configuration are possible, as illustrated in Fig. 4. In other words, the quadratic equations for the angles have two solutions. The physically acceptable solution is the one with the negative sign before the radical.

In the approach presented here, by separating the geometry from the force evaluation problem, we have not only simplified the solution by allowing the use of linear programming but also helped simplify the uniqueness problem of the geometric solution. In the original nonlinear problem, the

TABLE 2. Values of Actuating Forces.

Force #	Value	Force #	Value
1	−52.50	11	6.78
2	−152.42	12	0.00
3	−102.15	13	0.00
4	−116.87	14	0.00
5	−89.91	15	0.27
6	0.00	16	−15.02
7	−8.19	17	0.00
8	0.00	18	−0.71
9	0.00	19	−0.86
10	−1.20	20	−1.74

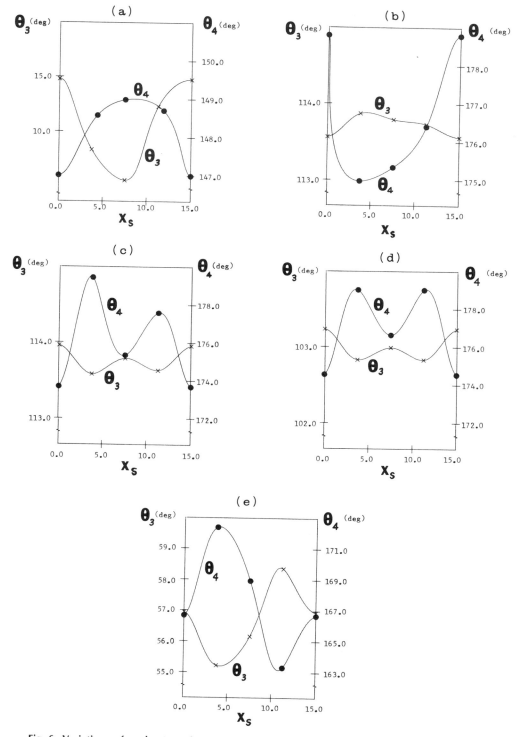

Fig. 6. Variations of angles θ_3 and θ_4: (a) Leg 1, (b) leg 2, (c) leg 3, (d) leg 4, and (e) leg 5.

unacceptable solution would have caused the appearance of local optimums whose recognition and avoidance might have been difficult during the optimization process.

Numerical Example. Let us consider an example for illustration. To calculate the mass and mass moment of inertia of the links from dimensions and material properties, we assume that these links have the shape of a connecting rod.

In this example, the point of application of the load P (platform) travels with a constant velocity (V) along the x-direction. The legs move on a terrain with a profile described by

$$y_s = A(1-\cos 2\pi \frac{x_s}{p})$$

The platform maintains a constant level (y_{so}) from the x_s axis. The goal is to find the optimal forces applied by the actuators.

We will use the values shown in Table 1 for the dimensions and for the angles ϕ (see Fig. 4). In addition, we will use the following values for the input parameters:

$P = 150.0$
$c_F = 0.1$
$c_R = 2.0$
$A = 0.5$
$p = 15.0$
$V = 5.0$
$y_{so} = 5.0$

The variations of some sample actuating forces over one motion cycle on the given terrain profile are shown in Fig. 5. In addition, Table 2 shows the values of all the actuating forces when the point of application of the external load (P) is at a 7.5 distance from the minimum point of the terain profile. Variations of angles θ_3 and θ_4 of the legs are shown in Fig. 6.

SUMMARY

The dynamic equilibrium of indeterminate structures can generally be represented by a non-linear system of equations. Applying a piece-wise linear method, instead of using a direct nonlinear one, greatly simplifies the problem. The method developed in this article is flexible and can be conveniently applied to similar systems. A general system of applied external loads can readily be accommodated in the analysis by appropriate formulation of the equations of equilibrium.

REFERENCES

1. Alekseeva, L. A., and Yu. F. Golubev: "Model of Dynamics of a Walking Apparatus," *Engineering Cybernetics*, **13**, (3) 55–63 (1975).
2. Bessonov, A. P., and N. V. Umnov: "The Analysis of Gaits in Six-Legged Vehicles According to Their Static Stability," *Proc. of the Symposium on Theory and Practice of Robots and Manipulators*, Int'l. Center for Mechanical Sciences, Udine, Italy (1973).
3. Bessonov, A. P., and N. W. Umnov: "Choice of Geometric Parameters of Walking Machine," in *On Theory and Practice of Robots and Manipulators*, 63–74, Polish Scientific Publishers, Warsaw (1976).
4. Frank, A. A.: "Automatic Control of Legged Locomotion Systems," *Ph.D. thesis*, University of Southern California, June 1968.
5. Frank, A. A., and R. B. McGhee: "Some Considerations Relating to the Design of Autopilots for Legged Vehicles," *J. of Terramechanics*, **6** (1) 23–35 (1969).
6. McGhee, R. B.: "Finite State Control of Quadruped Locomotion," *Simulation*, **9** 135–140 (1967).
7. McGhee, R. B.: "Some Finite State Aspects of Legged Locomotion," *Mathematical Biosciences*, **2** (1, 2) 67–84 (1968).
8. McGhee, R. B., and A. A. Frank: "On the Stability of Quadruped Creeping Gaits," *Mathematical Biosciences*, **3** (3, 4) 331–351 (1968).
9. McGhee, R. B., and A. L. Pai: "An Approach to Computer-Control for Legged Vehicles," *J. of Terramechanics*, **11** (2) (1974).
10. McGhee, R. B., and S. S. Sun: "On the Problem of Selecting a Gait for a Legged Vehicle," *Proc. of VI IFAC Symposium on Automatic Control in Space*, Tsakhkadzor, Armanian SSR, U.S.S.R. (1974).
11. McGhee, R. B.: "Robot Locomotion," *Proc. of Int'l. Conf. on Neural Control of Locomotion*, Philadelphia, Pennsylvania (1975).
12. McGhee, R. B., and D. E. Orin: "Mathematical Programming Approach to Control of Joint Positions and Torques in Legged Locomotion Systems," in *On Theory and Practice of Robots and Manipulators*, 231–239, Polish Scientific Publishers, 1976.
13. McGhee, R. B.: "Control of Legged Locomotion Systems," *Proc. 18th Joint Automatic Control Conf.*, 205–215, San Francisco, California (1977).
14. McGhee, R. B., and G. Iswandhi: "Adaptive Locomotion of a Multilegged Robot over Rough Terrain," *IEEE Trans. on Systems, Man, and Cybernetics*, **9** (4) 176–182 (1979).
15. Miller, J. W., and W. C. Baldwin: "Multi-legged Walker Final Report," Space General Corp., El Monte, California (1966).
16. Morrison, R. A.: "Iron Mule Train," *Proc. Off-road Mobility Research Symposium*, sponsored by The International Society for Terrain Vehicle Systems, Washington, D. C. (1968).
17. Mosher, R. S., and R. A. Liston: "The Development of a Quadruped Walking Machine," *ASME Trans. 34*, Paper No. 67 (1967).
18. Mosher, R. S.: "Test and Evaluation of a Versatile Walk-

ing Truck," *Proc. Off-road Mobility Research Symposium*, sponsored by The International Society for Terrain Vehicle Systems, Washington, D. C. (1968).
19. Mosher, R. S.: "Exploring the Potential of Quadruped," *SAE Paper No. 690191*, Intl. Automotive Engineering Conf., Detroit, Michigan, January 1969.
20. Muybridge, E.: "Animals in Motion," Dover Publishing, New York, 1957. (First published in 1899).
21. Okhotsimskiy, D. E., et al.: "Digital Computer Simulation of the Motion of a Stepping Vehicle," Izv. AN SSSR, *Engineering Cybernetics* (3) (1972).
22. Okhotsimskiy, D. E., and A. K. Platonov: "Control Algorithms for a Stepping Vehicle Capable of Overcoming Obstacles, "*Engineering Cybernetics* (5) (1973).
23. Okhotsimskiy, D. E., et al.: "Algorithm of Stabilization of Motion of Automatic Walking Apparatus," *Paper presented at the 6th IFAC Symposium*, Izd. Arm. NIINTI (preprint), Erevan, U.S.S.R. (1974).
24. Okhotsimskiy, D. E., et al.: "Control of Dynamic Model of Walking Apparatus," *Preprint No. 20, IPM AN SSR, Dep. VINITI, No. 908-74* (1974).
25. Okhotsimskiy, D. E., et al.: "Problems of Constructing and Modeling the Motion of an Operator-Controlled Walking Machine," *Preprint No. 125, Inst. of Applied Mathematics, Academy of Sciences of the U.S.S.R.*, Moscow (1974).
26. Okhotsimskiy, D. E., et al: "Control of an Integrated Locomotion Robot," *Engineering Cybernetics*, **12** (6) 43-47 (1974).
27. Orin, D. E.: "Interactive Control of a Six-Legged Vehicle with Optimization of Both Stability and Energy," *Ph.D. dissertation*, The Ohio State Univ., Columbus, Ohio, March 1976.
28. Orin, D. E., et al.: "Kinematic and Kinetic Analysis of Open-Chain Linkages Utilizing Newton-Euler Method," *Mathematical Biosciences*, **43** (1, 2) 107-130 (1979).
29. Sun, S. S.: "A Theoretical Study of Gaits for Legged Locomotion Systems," *Ph.D. dissertation*, The Ohio State University, Columbus, Ohio, March 1974.
30. Taguchi, K., Ikeda, K., and S. Matsumoto: "Four-Legged Walking Machine," *Proc. of the 2nd Intl CISM-IFToMM Symposium On Theory and Practice of Robots and Manipulators*, 172-181, Warsaw, Poland, September 14-17, 1976.
31. Tomovic, R.: "A General Theoretical Model of Creeping Displacement," *Cybernetica*, **IV** 98-107 (1961). (English translation)
32. Vasenin, V. A., et al.: "A Model of a Walking Vehicle and Its Control System," *Engineering Cybernetics*, **12** (6) (1974).
33. Ebrahimi, N. D.: "On Optimum Design of Stationary or Moving Legged Structures," *Ph.D. dissertation*, University of Wisconsin, Madison, Wisconsin, August 1983.
34. Ebrahimi, N. D., and A. A. Seireg: "Design and Control of Stationary or Moving Legged Structures," *Computers in Mechanical Engineering* (CIME), **3** (2) 62-69 (1984).

Servomotor and Servosystem Design Trends

By John Mazurkiewicz[1]

New products or an upgrading of existing products evolve as suppliers continually evaluate their equipment for improvements to maintain a competitive edge. Whether it be *X-Y* or point-to-point positioning or a constant or variable speed requirement, an electric motor provides precise motion control in a diverse group of products, ranging from simple conveyors to more complex machine tools and computer peripherals. The more complex systems utilize a four-quadrant servo drive system in conjunction with the servomotor. With emphasis on increased productivity and reliability, the technology in this area is being pushed at a good rate. This is leading to more effective use of the microprocessor in servo loop control. This article explores trends in both servomotors and servosystems and how these trends relate to applications when a manufacturing firm retrofits or upgrades its products.

INDUSTRIAL MOTORS—A PERSPECTIVE

The direct current (DC) motor was one of the first machines devised to convert electrical energy to mechanical power. Origin of the DC motor can be traced to machines conceived and tested by Michael Faraday, the experimenter who formulated the fundamental concepts of electromagnetism.

Since DC motor speeds can easily be varied, they are utilized for applications where speed control, servo control, and/or positioning tasks exist. These motors and their applications are discussed a bit later. Most small motors used in industry are alternating current (AC) motors. These motors are relatively constant speed devices and are applied where speed control is not required. The speed of an AC motor is determined by the frequency of the voltage applied (and the number of magnetic poles).

ALTERNATING CURRENT (AC) MOTORS

There are basically two types of AC motors: (1) induction, and (2) synchronous. If the *induction motor* is viewed as a type of transformer, with the stator as the primary and the rotor as the secondary, it becomes easier to understand its operation. The currents that flow in the stator induce currents in the rotor, and two magnetic fields are set up. These two magnetic fields *interact* to produce *motion*. The speed of the magnetic field going around the stator will determine the speed of the rotor. The rotor will attempt to follow the stator's magnetic field but will "slip" when a load is attached. Therefore, induction motors always rotate slower than the stator's rotating field. The *synchronous motor* is basically the same as the induction motor but with slightly different rotor construction. The rotor is either (1) self-excited (same as induction), or (2) directly excited to set up the rotor field. The salient poles (or teeth) construction prevents slippage of the rotor field with respect to the stator field. Thus, this type of motor always rotates at the same speed (in synchronization) as the stator field. A single phase AC motor is not self-starting. It employs a starting mechanism in order to start rotation—in the form of a start winding or a capacitor in a winding. Thus both motor types, induction and synchronous, utilize stators with rotating magnetic fields. They suffer from low starting torque, slow acceleration, and torque breakdown at overload.

---

[1] Manager, Application Engineering, Pacific Scientific, Rockford, Illinois.

DIRECT CURRENT (DC) MOTORS

In a DC motor the stator field can be set up by either permanent magnetics or a field winding. Thus, in contrast with the AC stator field, which is rotating, the stator field is stationary. The second field, the rotor field, is set up by passing current through a commutator and into the rotor assembly. The rotor field rotates in an effort to align itself with the stator field, but at the appropriate time (owing to the commutator) the rotor field is switched. Thus, by this method the rotor field never catches up to the stator field. Rotational speed depends on the strength of the rotor field, i.e., the more voltage on the rotor, the faster the rotor will turn. Thus, the DC motor is straightforward—it has predictable speed-torque characteristics and suffers none of the speed control problems associated with AC motors.

BRUSHLESS MOTORS

In recent years there has been a trend to favor *brushless* motors. The brushless motor technology emerged in the 1930s, along with vacuum tube power technology, more sophisticated control systems, and the industrial needs for velocity control and position control of basic motors. The transistor became an efficient power-handling device in the 1950s, when PWM (pulse width modulation) and PFM (pulse frequency modulation) techniques became practical. With subsequent developments of transistor circuits, analog operational amplifiers, low-cost logic components, memory arrays, and microprocessor chips, control systems became oriented toward the retention and processing of information and thus able to handle more complex calculations. In the

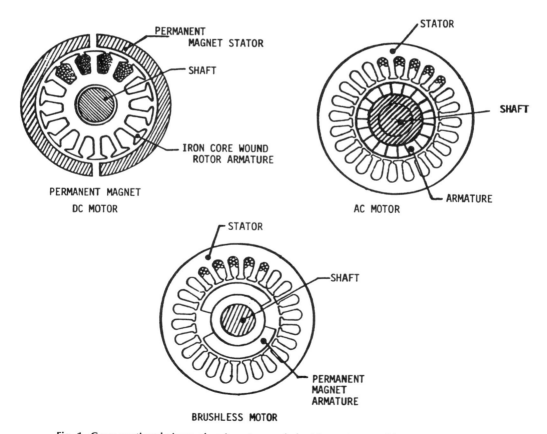

Fig. 1. Cross sectional views of various types of electric motors used in servosystems.

1970s the development of new magnetic materials provided the opportunity to explore and to design innovation in terms of pulse-modulated DC motors. It was not until this rapid expansion of modern semiconductors and new magnetic materials, along with the requirements for upgrading existing products for the automation of industry, that brushless motor development and use began in earnest.

Motivations for considering the brushless motor include the desire for improved productivity, improved product requirements in terms of higher speed, greater acceleration/deceleration, reduced maintenance, reduced size, improved power-to-weight ratio, and increased reliability.

Operating Principle of Brushless Motors

A cross-sectional view of a brushless motor is shown in Fig. 1. Brushless motors are similar to AC motors in that a moving magnet field causes rotor movement or rotation. Both motor types use stator windings and have no brushes. Brushless motors are also similar to permanent magnet (PM) DC motors since they have linear characteristics. Also, both motor types use permanent magnets to generate one field. The brushless motor is, in essence, a *hybrid*, which combines the best attributes of both the AC and DC motors.

The configuration of the brushless motor most commonly used in contemporary systems is shown in Fig. 2(a). In this motor, the rotor consists of permanent magnets and the stator consists of windings. These windings are termed "commutation" windings. By passing a current through a winding, a magnetic field is set up with which permanent magnets on the rotor interact. This results in rotation of the rotor. A representative family of modern brushless motors is shown in Fig. 2(b). Examples of brushless rotor assemblies are illustrated in Fig. 2(c).

Fig. 3 illustrates in simplified form how rotation occurs. With a current passing through *winding 1* (see Fig. 3[a]), a south pole is set up with which the permanent magnet will react and movement will begin. If, at the appropriate time, current is shut off in *winding 1* and turned on in *winding 2* (see Fig. 3[b]), then the rotor will continue to move. By continuation of this timing sequence, complete rotation will occur as the rotor repeatedly tries to catch up to the stator magnetic field. In this example, the operation is simplified for explanation by exciting only one winding at a time. In practical situations, two and sometimes three windings are energized at a time. This procedure permits the development of higher torques.

As indicated, if current is properly switched from winding to winding, the rotor will continue to rotate. Switching is accomplished in conjunction with a position sensor. Frequently, solid-state Hall-effect sensors are located on the shaft

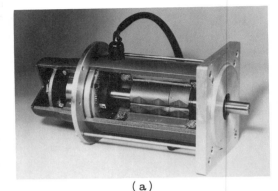

(a)

(b)

(c)

Fig. 2. Brushless motors: (a) Cutaway view; (b) representative family of contemporary designs; (c) brushless rotor assemblies. (*Pacific Scientific, Rockford, Illinois*)

assembly. These extremely rapid sensors note the position of the shaft and provide an output signal. This output signal informs the motor's electronics when to switch current from winding to winding.

In comparing two motors that develop the same torque, the brushless type has advantages. Figure 4 illustrates this point by showing a locus of safe operating areas. The maximum speed is the maximum recommended top speed of the motor, determined by (1) commutator bar-to-bar breakdown voltage in a brush type motor, and (2) by mechanical centrifuge conditions in a

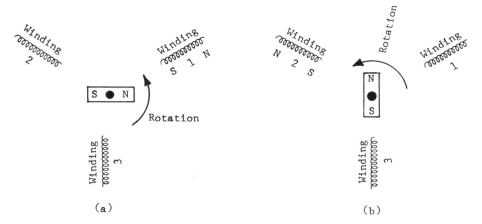

Fig. 3. Basic rotation of a brushless motor.

brushless motor. The maximum temperature limit is determined by the motor's hot armature temperature. These are quite close inasmuch as the motors develop the same approximate stall torque. Operation above this line will result in the motor's armature temperature exceeding the recommended manufacturer's maximum limit.

Further advantages of brushless motors are given in Table 1. Quantitative comparison of brushless motors with other commonly used electric motors is given in Table 2.

Control of Brushless Motors

Control of the brushless motor is accomplished by incorporating two additional output stages

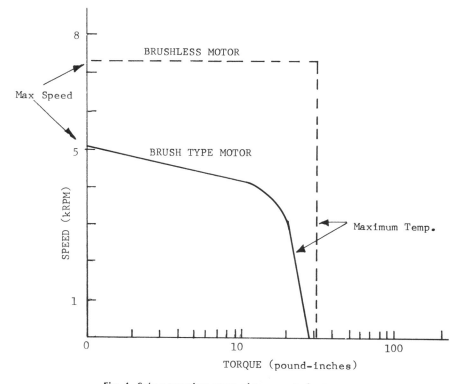

Fig. 4. Safe operating areas of two equivalent motors.

TABLE 1. Some Advantages of Brushless Motors.

Torque
Full torque over wide range of speeds. Peak torque—outperforms all other motor types. Continuous torque—much better than AC motors; comparable to PMDC motors. High torque-to-inertia ratio—high flux density of the low mass magnetics of the brushless motor provide fast acceleration capability. Low inertia—better than all motor types except low-inertia motors.

Efficiency
High efficiency—much better than other motor types but comparable to PMDC motors. AC and stepper motors have the lowest efficiency. The permanent magnet rotor of the brushless motor eliminates copper losses in the rotor assembly, resulting in high motor efficiency and thus reduced energy consumption.

Package Size
Small package size—much shorter than AC and other type motors. High flux magnetics coupled with the inside-out design results in a smaller size.

Speed and Inertia
Wide range of speed—higher speed capability than other motor types. Comparable to special-design AC motors. Stepper motors have the lowest speed range. Rare earth/ferrite magnetics reduce motor size, improve speed and torque capabilities, and reduce inertia.

Location Flexibility and Maintenance Advantages
Since electronic commutation requires no routine maintenance, the motor can be mounted in an inaccessible area. There is no electromechanical commutator and no brush parts to be serviced.

Versatility
Because of electronic commutation, the brushless motor is insensitive to line frequency changes and may be operated from either DC or 50/60 Hz. Anticog magnetics minimize the preferred rotor detent positions for smooth low-speed operation.

Life
Long life—comparable to AC and stepper motors. Outperforms the life of PMDC motors. Elimination of mechanical commutator and brush parts contributes to longer life.

Predictable Performance
System analysis and design involving brushless motors can be simplified because speed/torque and temperature characteristics can be reliably predicted.

over the conventional servo control design. Figure 5 illustrates a simplified block diagram. *Basic operation:* Once the "run" command is given, the binary decoder compares outputs from the Hall sensors (which Halls are *on*). For example, if the shaft is sitting at "zero" (see Fig. 6), input to the logic circuit (Hall sensor output) informs the logic that Hall sensors #1 and #3 are *on*. The binary decoder interprets these data and outputs a binary code. In this example the output code from the switch logic is a "5." This code will turn on appropriate OR gates, which turn on switches 3 and 2. The switches apply power from the voltage supply to motor windings (active legs T and R). As the motor rotates through 60°, the logic input will change state, as Hall #3 shuts off. The binary decoder output changes to a "1." This code will turn on appropriate OR gates, which turn on switches 2 and 5. (Note that during this transition, switch 2 has remained on, whereas switch 3 shuts off and switch 5 is turned on.) Again, the switches apply power to legs R and S. As the motor continues to rotate, the sequence continually progresses, changing current flow through the motor windings until complete rotation through 360° is attained, after which the sequence repeats. As current flow changes from winding to winding, the magnetic field also changes. In effect, the magnetic field is sequencing around the stator. The permanent magnet rotor will try to catch up to the field created but never will—because of switching of the magnetic fields as a result of the Hall-effect sensor signals.

Thus, brushless rotation depends upon the stator field rotating (similar to AC motors). The significant difference is the presence of internal shaft position feedback in the brushless design. This element gives brushless motors their linear and predictable speed-torque characteristics (similar to DC motors).

Microprocessors in Servo-Control Systems

Several types of controls have been developed over the years for DC motors, including SCR (silicon-controlled rectifier), linear, pulse-width modulation, among others. These controls have served the needs of a diverse group of applications and generally they are basic and simple. The DC approach was chosen because no economically equivalent AC package existed. The AC motor manufacturers traditionally ignored

TABLE 2. Quantitative Comparison of Characteristics of Brushless Motor with Other Motor Types of Comparable Size.

	AC Cap-Start	Steppers	PMDC	Pancake	Low Inertia	Brushless
Diameter (in)	3.9	4.2	4	4.4	4	4.25
Length (in)	6	4.74	4.8	2.4	4.5	3.7
Continuous torque (oz-in)	35	425	150	63	62	126
Peak torque (oz-in)	123	625	750	690	364	996
Inertia (oz-in-S^2)	.05	.055	.1	.003	.001	.014
Speed range (rpm)	10,000	300	4,000	4,000	8,000	10,000
Efficiency (%)	50–60	30–50	85–90	70–90	70–90	85–90
Life (hr)	20,000	20,000	5–8,000	5–8,000	5–8,000	20,000

Metric Conversions: 1 inch = 2.54 centimeters; 1 ounce-inch = 72 gram centimeters.
rpm = revolutions per minute.

Servomotor and Servosystem Design Trends

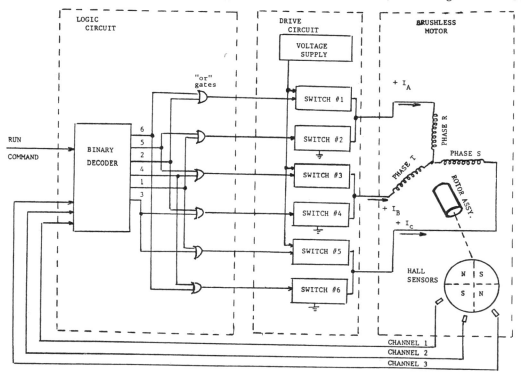

Fig. 5. Simplified block diagram of control system for brushless motor.

this market. However, after the rapid escalation of energy costs in the 1970s, the motor industry perspective changed. Controls for AC motors quickly assumed a new level of sophistication. Two basic types of control emerged and are currently available: (1) the six step, and (2) pulse width modulation. In development are more sophisticated controllers that employ SCR microprocessors and large scale integrated (LSI) dedicated electronics. In these cases the control converts 60 Hz to direct current and then synthesizes a sine wave at a frequency to produce the desired speed.

The traditional cumbersome approach with discrete components would be unacceptable when using higher technology involving brushless designs. A search for a new approach began and, in light of the microprocessor, with its power and flexibility and its increased use in numerous applications, the concept of servosystem design was revolutionized. This approach has succeeded in providing greater flexibility in design of new systems and has the potential to enhance the capabilities of existing systems. This could be termed the "intelligent" approach that utilizes microprocessor technology—an approach that impacts very favorably on design time, setup time, and implementation time.

Adjustable-Speed Brushless Control. A control of this type is self-contained, including power supply and heat sink, with the objective of driving a permanent-magnet brushless motor with Hall sensor feedback. A typical block diagram is shown in Fig. 7. The microprocessor-based control operates by energizing two of the motor's three windings and switches power from winding to winding according to the feedback from the Hall sensors, as previously described. This is coupled with pulse-width modulation drive techniques to make effective use of the output power transistors.

A user-friendly operating panel activates the control, while status lights provide easy readout. Also on the front panel is a speed adjustment pot and a digital speed readout indicator. The tight speed regulation offered by the microprocessor approach improves system accuracy. Accuracy of ± 5 RPM over speed ranges of 100 to 10,000 RPM is possible even with dynamic load variations of 50%. The digital readout also serves as a diagnostic indicator should any of the system's protection require activation to shut down the system. This provides simple, easy-to-use user interfacing.

The servosystem (servocontroller) approach is used with brushless technology in the same basic manner as it is employed with other prime mover (motor) technologies. Traditionally, these closed-loop servosystem designs will involve determination of load conditions and velocity profile, then prime mover (motor) selection, and determination of amplifier requirements. Following is the tedious task of compensation for gain and bandwidth adjustments for accuracy and response. The latter may involve a paperwork analysis prior to breadboarding a prototype. Or, if the individual components are purchased from independent suppliers, the "tweeking" of potentiometers would begin—in an effort to set up and adjust the compensation values. This approach can be difficult. As an example, in some servo amplifiers, there are ten potentiometers for a variety of adjustments. Throughout the compensation process, the main consideration is to set up the compensation for a given load condition. If the load changes, then the entire process must be repeated. To alleviate this problem, the system is designed to accommodate worst case conditions. This results in a system design that will be overdamped for all conditions (except the worst case), thus severely compromising system performance (speed and accuracy).

A much less cumbersome and effective approach is to utilize microprocessor techniques—an approach that uses digital control and digital filtering. The "intelligent" system compensates by simply inputting or programming the controller with parameters that describe the servo loop

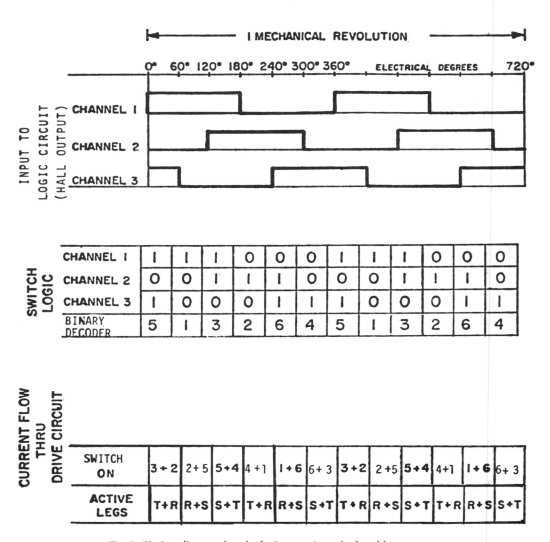

Fig. 6. Timing diagram for clockwise rotation of a brushless motor.

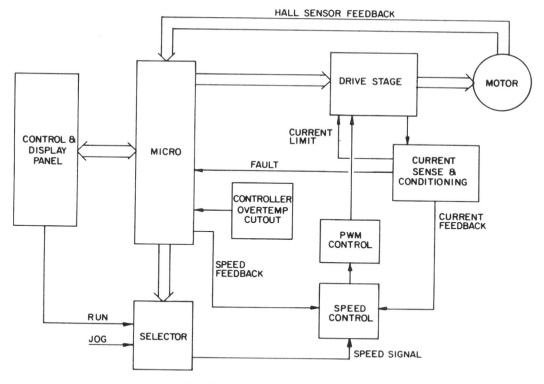

Fig. 7. An adjustable-speed brushless motor control system.

parameters. The algorithms precisely control servo loop velocity. Execution is under microprocessor control, and all servo loop compensation (motor, load, or environmental conditions) is monitored and controlled should changes occur. This simplifies design and gains flexibility for the system.

One approach is the use of a general-purpose microprocessor-based control, as shown in Fig. 8. The advantages of this approach include fewer components needed, which in turn lowers cost, reduces circuitry, and improves reliability. Components must be added to complete the system design, such as a drive amplifier. However, there are microprocessor-based controls, which include an internal drive amplifier, available today in production quantities to drive a brushless motor. These devices are termed "intelligent servo drives," as shown in the block diagram of Fig. 9.

These brushless servocontrols are software compensated, thus eliminating the need for pot adjustments. The microprocessor-based control can accept a variety of inputs—either analog or digital (serial or parallel) velocity commands. Parameters are factory-loaded in nonvolatile

Fig. 8. A general-purpose microprocessor-based control system.

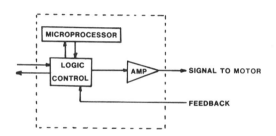

Fig. 9. Example of intelligent servo drive.

EEPROM (electrically erasable/programmable read-only memory) so the engineer receives a stable system, thus reducing design, implementation, and setup time. Although most applications do not require further adjustments, the engineer may fine tune the system by way of a hand-held pendant or any RS232C interface. The versatility of the microprocessor allows incorporation of self-protection, which makes the overall system virtually indestructible. Protection features include peak, RMS, and short-circuit protection as well as thermal protection, hardware and software protection, loss of feedback, and velocity error. A brushless resolver serves as the feedback device and replaces the Hall sensor feedback. This ensures operation at the optimum phase angle. The entire unit controls the brushless motor by way of a PWM sinusoidal driving function, thus allowing smoother operation even at very low speeds.

The "intelligent control" closes the velocity loop and directly interfaces with most readily available programmable motion controllers for position control. Basic system operation is shown in Fig. 10. The first function is to receive instruction from the motion controller and generate the speed command for the velocity loop. The algorithm looks at the *present command* of the motor, the *previous command*, and the *next instantaneous position*. This is a periodic *sample* and *compare* to a desired reference value. The difference between the speeds at two time periods serves as an indication for velocity errors and is used for velocity corrections. The feedback signals from the brushless resolver are conditioned and routed to the CPU (central processing unit) through the interrupt control. All real-time inputs have interrupt-driven software. A control manipulation is calculated and subsequently used to command the drive system. The algorithm gener-

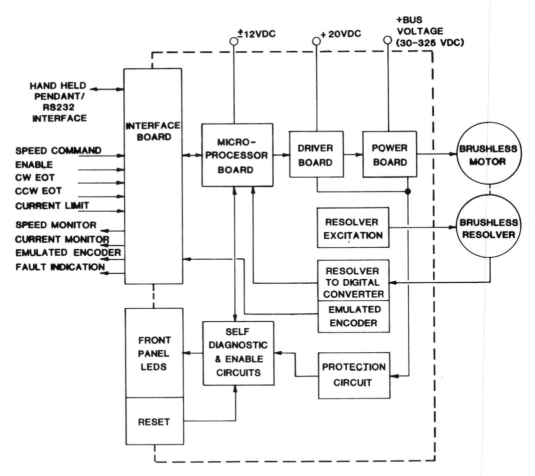

Fig. 10. Basic system operation of an intelligent brushless servo control. CW = clockwise; CCW = counterclockwise; EOT = end of travel. (*Pacific Scientific, Rockford, Illinois*)

ates a command that is the difference of a term proportional to the velocity error and a term that is proportional to the integral of the velocity error. This velocity feedforward technique, included in the microprocessor servo reference generation circuitry and applied directly to the velocity loop, allows the controller system to operate with minimal error, even during hard acceleration and deceleration. Velocity feedforwarding is valuable for maintaining wide dynamic system response. This technique stabilizes the loop and allows the system to drive the brushless motor in a smooth manner regardless of the trajectory.

The microprocessor must also determine the sign and magnitude of the current/voltage for each of the brushless motor's three phases in order to drive the motor at the appropriate torque/speed. The function of the waveform is described by a sine wave. Since the three phases are shifted by 120° and there are four poles in the motor, or two electrical cycles per revolution, the commands to the windings are:

$$\text{Phase } 1 = \text{Sin } 2wt$$
$$\text{Phase } 2 = \text{Sin } (2wt + 120°)$$
$$\text{Phase } 3 = \text{Sin } (2wt + 240°)$$

Since this calculation is accomplished only once per sample period, the appropriate weighting factor can vary considerably from the first computation time to the next. The solution is to base the weighting factors on a period base at one-half sample later. This is a velocity lead on the commutation. To optimize this, the sine functions are stored in memory. For this scheme to work, the microprocessor must "know" the relationship between the brushless motor phases and the internal drive scheme. This is accomplished by the resolver feedback. The resolver interface consists of sine and cosine reference waveforms and a phase-shifted feedback that contains absolute position data. This signal is sent directly to the microprocessor through a resolver-to-digital converter, returning a velocity per sampling period. This feedback has proven superior for improvement of dynamics and straightforwarding for controllability.

The intelligent digital microprocessor-based control system, coupled with brushless motors, constitutes an excellent solution for most applications. This combination has the ability to produce higher torque at higher speeds. The numerous specific advantages have been previously described in detail.

REFERENCES

Marurkiewicz, J.: "Brushless Motors Coming on Strong," *Electronic Products*, 61–75, September 1984.

Mazurkiewicz, J.: "Advances in Microprocessor Control for Brushless Motors," *Electronic Motion and Control Association* Conf., San Diego, California, November 1984.

Bailey, S. J.: "Servo Design Today: Software-Driven Adaptations in the Feedback Loop," *Cont. Eng.*, 67–68, February 1984.

Bailey, S. J.: "Servo Design Today: Hardware Elements Fade as Software Closes Feedback Loop," *Cont. Eng.*, 57–61, February 1985.

Note: See related articles in *Section 2* of this *Handbook*, dealing with position, displacement, motion, speed, and velocity: in Section 3, pertaining to computerized numerical control and artificial intelligence in control systems; and in *Section 4*, dealing with other types of electric motors for control systems and the actuation of robots.

Stepper Motors and Controls

Position measurement and motion control are major requirements for numerous applications in the discrete-piece manufacturing industries. Stepper motors, which are inherently digital in nature, are in tune with the digital information handling technology that typifies the modern approach to automation systems found in these industries. Fundamentally high-precision devices, stepper motors are often the method of choice for many open-loop control applications. The line of demarcation between servomotors (feedback—closed loops) and stepper motors for motion control is rapidly becoming less distinct—resulting from a change in design philosophy and improvements in stepper motors, including higher incremental resolution, more stable torques at low speeds, and a reduction of previously speed-sensitive resonances. This topic is explored in more detail later in this article. Stepper motors are also widely used in electronic equipment for disk drives, printers, plotters, medical equipment, and office automation devices.

A stepping motor is an electromechanical device that rotates a discrete step angle when energized electrically. The step angle usually is fixed for a particular motor and thus provides a means for accurately positioning in a repeatable, uniform manner. Typical step angles vary from as small as 0.72° to as large as 90°. Several means for electrically energizing stepping motors include DC pulses, square waves, and fixed logic sequence or multiple-phase square waves. Basic design types are (1) solenoid-operated (rotary or ratchet), (2) variable reluctance, (3) permanent magnet, and (4) hybrid. Variations of these types may be combined with gears or hydraulic amplifiers to provide increased-output-torque stepper motors. *Microsteppers* are also discussed in this article.

The solenoid-operated ratchet device or star wheel is not discussed in detail in this article. Its main advantages for certain applications are low cost and high nonenergized holding torque. Limitations include short life and low stepping rates. Characteristics of solenoid-operated ratchets are summarized in Table 1. Torque conversions are given in Table 2.

TABLE 1. Characteristics of Solenoid-Operated Ratchets

Normal Operation Speed at Rated Load pps	Maximum Speed (No Load) pps	Torque at Operation Speed oz•in	DC Volts	Input Current/Power Phase	Input Current/Power Watts	Steps per Revolution
To 25	—	160	12,28	1.0A	84–210	18
600	1,000	1.4–6.1	6–220	—	6	10 or 12
400	750	3.0–15.7	6–220	—	9	10 or 12
300	600	6.0–220	6–220	—	11	10 or 12

Note: For torque conversion units, see Table 2.

TABLE 2. Torque Conversions

	N•m	dyn•cm	kp•m	oz•in	lb•ft	lb•in
N•m	1	10^7	0.102	1.416×10^2	0.7376	8.851
dyn•cm	10^7	1	1.012×10^{-8}	1.416×10^{-5}	7.376×10^{-8}	8.851×10^{-7}
kp•m	9.807	9.807×10^7	1	1.389×10^3	7.233	86.8
oz•in	7.062	7.062×10^4	7.201×10^{-4}	1	5.208×10^{-3}	6.250×10^{-2}
lb•ft	1.356	1.356×10^7	0.1383	192	1	12
lb•in	0.113	1.13×10^6	1.152×10^{-2}	16	8.333×10^{-2}	1

VARIABLE RELUCTANCE STEPPER MOTORS

There are two configurations of variable reluctance (VR) stepper motors: (1) single stack, and (2) multiple stack. In the single stack version, the windings of each electrical phase share the same lamination stack. In a multiple stack motor, each electrical phase is wound on a separate stack.

The *single-stack design* is available with either three or four phases and in a number of step angles. The most common is 1.8 degrees. A single-stack VR step motor is shown in Fig. 1. Principal characteristics of the single-stack VR step motor are as follows: (1) small step angles are possible—high number of steps per revolution, (2) high slew speeds are possible, (3) no holding torque with the windings deenergized, (4) bidirectional, (5) unipolar drive, (6) low efficiency, and (7) low output power or output torque-to-size ratio.

The variable reluctance type of stepping motor offers essentially the same advantages as the permanent magnet types (described later). However, the torque capabilities have not been developed to the level of the permanent-magnet types. In the variable reluctance motor, the rotor-stator poles or teeth are aligned by the electric fields. Therefore, when the motor is deenergized, there is no residual holding torque. This is not a real disadvantage, however, in that most permanent-magnet types, as well as variable reluctance types, require that power remain on the motor between step movements. The same basic drive logic used for permanent-magnet steppers also applies to variable reluctance steppers. Representative characteristics of VR stepper motors are given in Table 3.

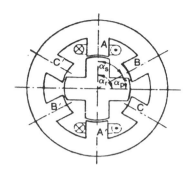

Fig. 1. Single-stack variable reluctance (VR) stepper motor. (*Stock Drive Products.*)

PERMANENT-MAGNET STEPPER MOTORS

The permanent-magnet stepper motor offers high-torque capabilities—up to 2600 oz in (19.1 N·m); for fast response, 2 ms is possible; and for a high stepping rate, up to 20,000 steps per second. Generally, the accuracy of the step may be within 3% of one step, any error being nonaccumulative. Permanent-magnet types can be operated from pulses or square waves, depending on particular motor design. The permanent magnet, which may be in the rotor or the stator—working with electrically produced flux—causes poles and teeth to align and effect rotation.

TABLE 3. Characteristics of Representative VR Stepper Motors

Normal operation speed at rated load, pps	Maximum speed, no load pps	Torque, oz·in		Rotor inertia, g·cm²	DC volts	Input current/power		Steps per revolution
		Operation speed	Detent, zero speed			Phase	Watts	
10	725	2.4	4.5	1.0	28	6.0A	15	24
18	—	12	45	8.4	28	2.7A	75	24
—	900	—	8.0	0.74	28	1.93A	100	24
400	5000	25	40	180	5.4	1.5A	—	200/72
250	3000	1.0	2.7	0.004	28	344mA	20	24
300	660	2.7	6.5	3.11	28	35mA	20	24
450	6000	0.5	0.95	0.004	28	330mA	18	24
2000	6000	640	1,920	—	28	20.0A	60	240
2000	8000	480	900	—	28	10.0A	30	240
2000	10,000	20	30	—	28	3.0A	10	160
2000	16,000	—	30	18.6	—	6.0A	36	4

See Table 2 for torque conversions.
Rotor inertia conversions:

1 gcm² = 10^{-7} kgm²
 = 1.02×10^{-8} kpm s²
 = 5.467×10^{-3} oz in²
 = 1.416×10^{-5} oz in s²

1 gcm² = 3.417×10^{-4} lb in²
 = 8.85×10^{-7} lb in s²
 = 2.373×10^{-6} lb ft²
 = 7.376×10^{-8} lb ft s² (slug ft²)

348 Stepper Motors and Controls

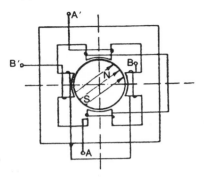

Fig. 2. Permanent-magnet rotor stepper motor (PMR). (*Stock Drive Products.*)

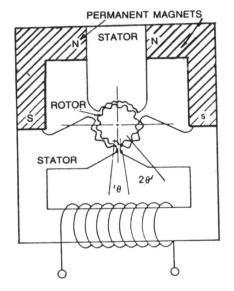

Fig. 3. Permanent-magnet stator stepper motor (PMS). (*Stock Drive Products.*)

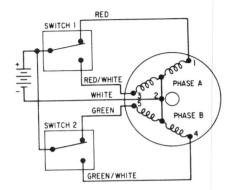

SWITCHING SEQUENCE*

STEP	SWITCH 1	SWITCH 2
1	1	5
2	1	4
3	3	4
4	3	5
1	1	5

*TO REVERSE DIRECTION, READ CHART UP FROM BOTTOM.

Fig. 4. Connection diagram and switching sequence of a phase-switched stepper motor. Diagram applies to motor having a permanent-magnet stator. (*Superior Electric.*)

The *permanent-magnet rotor* stepper motor (1) provides some holding torque with the windings deenergized, (2) possesses good efficiency, (3) is bidirectional, (4) shows high holding torque-to-size ratio, (5) has a low number of steps per revolution because of high back emf, and (6) stator flux must be reversed after each step, thus requiring bipolar drive or bifilar windings.

The *permanent-magnet-stator* stepper motor is constructed with two permanent magnets in the stator. The rotor is a toothed soft iron structure that completes the magnetic circuit. These motors are unidirectional and require drive pulses of alternate polarity for each step. Bifilar windings are often used to eliminate the need of a dual polarity power source. Principal characteristics of the permanent-magnet stator stepper motor include (1) high holding torque with windings deenergized, (2) will not free wheel, (3) has a fast response, (4) has a low slew rate, (5) has no self-damping ("rings" after each step), and (6) is unidirectional.

Permanent-magnet stepper motors are illustrated and further described in Figs. 2 through 4. Characteristics of representative PM stepper motors are given in Table 4.

HYBRID STEPPER MOTORS

The hybrid stepper motor combines the stator and rotor geometry of a VR motor with the addition of a permanent magnet in the rotor. (See Fig. 5.) To facilitate the use of a single polarity power source, the hybrid motor is usually wound in bifilar fashion. However, when using a bipolar drive, the bifilar windings can be connected in parallel to permit higher current and thus produce greater torque. Hybrid stepper motors are commonly made in the small step

TABLE 4. Characteristics of Representative PM Stepper Motors

Normal operation speed at rated load, pps	Maximum speed, no load, pps	Torque, oz·in		Rotor inertia, g·cm²	DC volts	Input current/power		Steps per revolution
		Operation speed	Detent, zero speed			Phase	Watts	
10	130	0.1	0.22	0.15	28	90mA	2.5	4
20	80	0.2	0.2	3.7	27	150mA	4	24
40	320	0.47	0.97	2.6	24	150mA	3	48
50	—	—	4.0	0.024	26	465mA	12	8
75	190	1.5	5.5	4.02	28	43mA	13	4
100	260	1.9	4.3	4.93	28	47mA	13	8
120	145	0.6	0.3	0.77	28	280mA	7.8	3
150	250	0.45	0.13	0.45	28	93mA	5.2	4
175	200	5.0	9/7/4	7.05	28	1.25mA	35	4
200	320	1.0	0.3	0.77	28	93mA	5.2	8
225	260	8.5	17/13.3	22.4	28	1.65A	46	8
250	300	2.2	3.8/3.6	1.36	28	860mA	24.1	4
300	500	0.3	0.7	0.003	26	104mA	5.4	4
360/600	—	1.3/1.8	—	34.8	20/50	—	20/1.8	12/24
400	—	12	40	180	5.4	1.5mA	—	72
500	—	200	600	7,310	—	—	—	200
600	—	100	260	1,520	4.5	3.5mA	—	200
45	—	35	90	877	3.0	4.0mA	—	500
45	—	1,200	2,500	38,010	12	3.8mA	—	200

See Table 2 for torque conversions. See Table 3 for rotor inertia conversions.

angle range of 1.8 degrees. Characteristics of a representative hybrid stepper motor include: (1) high number of steps per revolution, (2) holding torque with windings deenergized, (3) bipolar drive, (4) good efficiency, (5) bidirectional, (6) high torque, and (7) high slew speed.

Characteristics and Application Guidelines

A stepper motor provides precisely controllable speed or position. Since the motor increments a precise amount with each control pulse, it easily converts digital information to exact incremental rotation—without the need for any feedback device, such as a tachometer or encoder. Because the system is open loop, the problems of feedback loop phase shift and resultant instability common to servodrives are eliminated. If desired, a minor loop may be closed around the stepper motor with an encoder for system performance enhancement.

A stepper system requires initial "calibration" when power is applied unless strict constraints are placed on system design. Some controls have an initialization feature that always energizes the same two phases with power up. Unless the stepper system commands are limited to multiples of four steps, there can be randomly clockwise or counterclockwise movement of two steps with power up. Since the motor could have missed steps in a previously commanded sequence, even with the same-phase-power-up logic and four step-multiples command, there is a possibility of movement of the motor shaft with power up. Initialization can be accomplished automatically with a "home" position feedback device and proper logic or may be accomplished

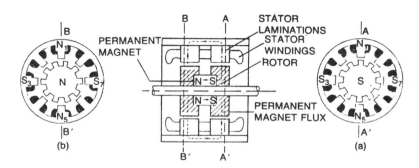

Fig. 5. Hybrid stepper motor (variable reluctance plus permanent magnet). (*Stock Drive Products.*)

manually with single step, run, and direction switches.

A stepper motor accelerates and decelerates with each control pulse, even when rotating at "full speed." This means that there is a velocity change on its output shaft rotation at all operating speeds—making a stepper unsuitable for a phonograph drive, for example.[1]

A stepper motor is not completely silent in operation. Normally, there will be a distinctly audible sound, which is a function of the driving pulse rate.

Successful application of a stepper motor requires (1) consideration of a variety of parameters, (2) calculation, and (3) experimental verification and tailoring. Parameters to be considered include: (1) distance to be traversed, (2) maximum time allowed for the traverse, (3) desired detent (static) accuracy, (4) desired dynamic accuracy (overshoot), (5) permissible time for dynamic accuracy to return to static accuracy specification (settling time), (6) required step resolution (combination of step size and gearing to the load), (7) total system friction, (8) system inertia, and (9) the speed/torque characteristics of the stepper motor.

Calculation or measurement of inertial and friction loads connected to the motor is necessary because in an open-loop stepper system the motor does not "know" if excessive inertia or friction has made the motor lose or gain one or more steps. If any steps are randomly lost or gained, the positional accuracy is lost.

Load inertia generally should be limited to more than four times motor rotor inertia for a high-performance (relatively fast) system. A low-performance system can deliver accuracy with very high inertia loads (generally up to an extreme of nine times rotor inertia). System friction may enhance performance with high-inertia loads.

Experimental verification of motor sizing is usually necessary, because of dynamic changes in system friction and inertia which cannot be calculated, as well as motor resonance effects which can change when the motor is coupled to its load. (See Fig. 6.)

Economic considerations in motor sizing include the fact that three stack motors of one size may have about the same torque as a single stack that costs less, although the former will operate at a higher pulse rate. Therefore, a design that can use a single stack motor is sometimes more economical—and it also allows increased torque by adding stacks, if required, as the design progresses.

Another economic factor in *limited-quantity applications* is the amount of engineering time and equipment needed to "fine tune" a system. In limited-quantity systems, a motor with more than adequate speed and/or torque margin may have the lowest *total* cost.

---

[1] Information in several following paragraphs furnished by *Bodine Electric Company*.

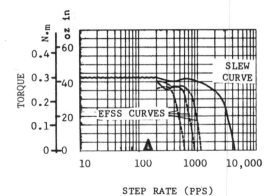

Fig. 6. Error-free start-stop (EFSS) curves. The EFSS curves (indicated by dotted lines on the speed/torque graphs) show the maximum speeds and torques at which a particular stepper motor can be operated without the need for acceleration or deceleration control. The uppermost EFSS curve includes minimal test system inertia, so an unloaded motor would run slightly faster. Curves for external load inertias of test system inertia only, and test system inertia plus one and four times rotor inertia are also included and include the range of inertia loads commonly suggested for high performance (and relatively efficient) systems. Inertial loads between 4 and approximately 9 times rotor inertia are possible in systems with somewhat limited EFSS and slew speeds. Even higher inertias up to 100 or more times rotor inertia can sometimes be handled at very slow (below low frequency resonance) speeds. Such inertias have notably adverse effects on single step response and settling time. (*Info. Source: Bodine Electric Company.*)

A stepper motor may be coupled to its load with any means that does not contribute more backlash than permitted by system accuracy specifications. Common low backlash coupling means include the timing belt or precision lead screw. While some compliance in the coupling device facilitates inertia handling capability, excessive compliance can aggravate system resonance problems. Choosing the optimum speed reduction ratio between the stepper and its load involves examination of the motor speed/torque curve. The speed/torque curve of a stepper motor depends upon the electronic control "driving" the motor. A ratio should be selected that permits operation at the optimum combination of dynamic torque and speed and that also avoids continuous operation within the resonance regions of the system.

In order to verify proper motor operation during the development of an application, some type of indexing that permits commanding a particular number of steps and subsequent (and independent) measurement of position achieved is required. It is frequently of advantage to use a control as furnished by the supplier[2].

Definitions of other performance characteristics[3] include:

1. *Step Angle*—the nominal angular displacement between adjacent step inputs.
2. *Resolution*—the number of steps a motor makes per revolution. Thus, 200 steps per revolution yields a 1.8° step angle, and the resolution is 360/400 (one 400th of a revolution).
3. *Step Accuracy*—is a noncumulative error and represents the step-to-step repeatability. This is largely dependent on manufacturing tolerances. Most step motors are accurate to ±5% of their step angle; some are as good as ±1% step angle.
4. *Position Accuracy*—this involves both the step accuracy and other system errors created by the load friction causing the motor to seek a step position offset (where the static torque equals the load friction) from the true detent position. Errors contributing to position accuracy are cumulative for the particular position in question, but not from position to position.
5. *Holding Torque*—the holding or static torque (obtained from the static torque curves) is the basic characteristic of a stepper motor. This characteristic is the restoring torque developed when the motor is forced away (displaced by a load torque) from its true detent position (energized but unloaded rest position). This torque is sometimes called *detent torque* or *stall torque*. (See Fig. 7.)
6. *Pull-out Torque*—also called dynamic torque of a stepper motor—is obtained from the torque-speed (dynamic) curves and refers to the maximum steady state load friction torque that may be applied to the motor at a given speed (RPM) without causing the rotor to lose synchronism with the input signals. (See Fig. 8.)
7. *Pull-in Torque*—represents the full-load friction torque, for a given inertia and pulsing rate, that the motor can drive. A family of curves exists for any range of inertia (one curve for a given inertia).
8. *Slew Speed Range*—slew speed is that speed over which a motor can operate in synchronism but cannot start, stop, or reverse on a single-step command. This speed is the range between pull-in and pull-out rate.
9. *Minimum Response Time*—a minimum response time curve indicates the time to travel a given displacement for a fixed inertia, while in the slew range of operation. The rotor inertia must be included with the load inertia.
10. *Single-Step Response*—this is the stepper motor's dynamic behavior (displacement, acceleration, step time, overshoot, settling time, etc.) to a single pulse input signal. Obviously, the motor characteristics, load and pulse parameters, all will affect single-step response. Knowledge of the single-step response is valuable to the design engineer in order to evaluate the damping of the system and to select proper drive circuit parameters.
11. *Multiple-step Response*—this is the dynamic behavior when a stepper motor is driven by a fixed sequence of input pulses. This is most helpful to the designer in choosing the optimum pulse characteristics to obtain the required load displacement performance.
12. *Ramping*—refers to the acceleration and deceleration of a stepper motor up to and down from slew speed. Ramping is a means for obtaining optimum acceleration/deceleration performance without exceeding the stepper motor's torque-speed characteristics. A linear ramp function can be used only if the motor's torque is constant over the speed range. However, this is usually not the case, and, therefore, the optimum ramp is usually an exponential function (reverse exponential for deceleration). System friction and the motor's retarding torque will aid deceleration, and, therefore, the stopping ramp can often be steeper than the starting ramp.

Acceleration control or ramping is normally accomplished by gating on a ramp operating circuit that provides a voltage profile for a volt-

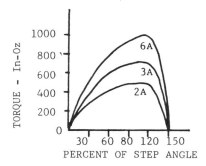

Fig. 7. Representative torque curves.

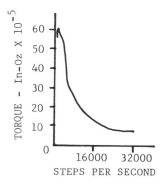

Fig. 8. Representative torque speed curve.

---
[2]For example, the Bodine *Encased Index*, *THD-1850E*.
[3]Information in several following paragraphs furnished by *Stock Drive Products*.

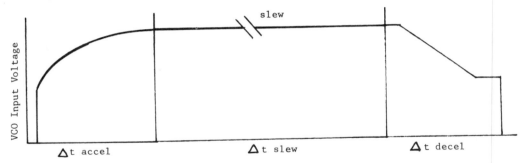

Fig. 9. Acceleration control or ramping is normally accomplished by gating on a ramp operating circuit which provides a voltage profile for a voltage controlled oscillator. Shown here is VCO input voltage from a representative acceleration control unit. The oscillator frequency varies proportionately and changes the pulse rate to the motor.

age-controlled oscillator (VCO). The input voltage (VCO) from a typical acceleration control unit is typified by the curve shown in Fig. 9. The oscillator frequency varies proportionately and changes the pulse rate to the motor. Attainable deceleration times are about 30% less than acceleration times, and the deceleration curve is linear in some commercial stepper motors. Even though some commercial models permit separate decel and accel time adjustments and adjustments of accel curve shape, initial calculations and motor selection should assume equal accel and decel times.

13. *Damping*—A stepper motor proper will oscillate (ring) and overshoot its stopping position after each step. The amount of oscillation depends upon: (1) the motor characteristics, (2) load (inertia and friction), and (3) the motor control system. The degree to which a motor can provide its own damping is a measure of its self-damping characteristics. Permanent-magnetic stepper motors have reasonably good self-damping, due to their back emf. In practice, however, most systems (motor and load) will require additional damping for satisfactory performance. Forced external damping can be accomplished in one or both of two ways: (a) add-on mechanical damping, and (b) electronic damping.

Add-on mechanical damping may be effected by the simple addition of load friction. This, however, consumes added power and perhaps increases the system cost and reduces its efficiency and performance. A *viscous coupled inertia* damping can be used, but it will also consume power (during acceleration) and may adversely affect system performance.

Electronic damping takes several forms.

External resistance damping, wherein resistors are connected to the motor windings so as to allow full current to the driven winding and a reduced current to the other windings. This will provide a damping action. (See Fig. 10.) It should be noted, however, this is accomplished at the expense of reduced motor speed and increased power loss.

Capacitive damping, wherein capacitors are placed in parallel with the motor windings. These have the effect of controlling the rates of current decay in the phase that has just been switched off. The discharging capacitors thus keep the phase energized, which creates a retarding torque. (See Fig. 11.)

One-phase-on damping. A four-phase stepper motor connected to run two-phases-on may be driven in the one-phase-on mode. Oscillation and overshoot are reduced, but at the expense of motor torque.

Two-phase-on damping. Two stator phases can be energized simultaneously, thus causing the rotor to seek a position midway between the two normal detent positions. The stator places two opposing torques on the rotor, thus providing damping.

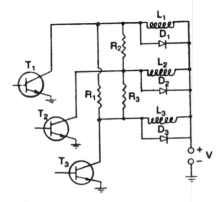

Fig. 10. External resistance damping circuit used with some stepper motors. (*Stock Drive Products.*)

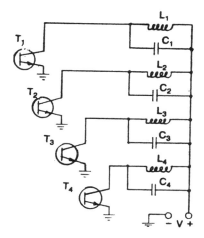

Fig. 11. Capacitive damping circuit used with some stepper motors. (*Stock Drive Products.*)

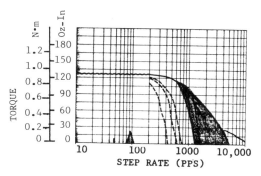

Fig. 12. Inertia effect on motor slew speed. The mid-frequency resonance phenomenon can be affected by the amount of inertia coupled to the stepper motor. Inertia is often added in an application to lend a certain stability to the low or zero torque region, and may help to reduce the effect of the mid-frequency resonance. Increasing inertia may move this resonant frequency either higher or lower, depending on the suppression circuit used. The gray shaded area in this graph shows the approximate range of variation from adding a load inertia equal to rotor inertia. Although additional inertia can be desirable, there is an accompanying reduction in the error-free-start-stop speed which it also produces. (*Info. Source: Bodine Electric Company.*)

Retrotorque damping—also called reverse-drive damping—can be accomplished by momentarily energizing the reverse phase (previously energized phase) when the next winding is energized. This method provides the fastest single step time and settling time of all damping methods. The method, also called "bang-bang" damping, is sensitive to load and drive pulse variations (current, voltage, timing, etc.) and works well for motor displacements of three or more steps.

Delayed-last-step damping (DLSD)—is an electronic damping scheme whereby the last drive pulse to the motor is delayed until the rotor has passed its desired final position. To properly function, the motor velocity must be such that upon stopping its normal overshoot is equal to one step. This method works best when the motion is three steps or a multiple thereof and with small load friction. The last and next-to-last pulses should be made adjustable for optimum results.

14. *Half-Stepping*—In a traditional 1.8° PM hybrid stepper motor, *half-stepping* can be used. With half-stepping steps of 0.9° (instead of 1.8°) are produced. The output shaft thus rotates at one-half the speed of full-step operation at a given pulse rate. With half-step drive, motor output torque may be usable in the midfrequency resonant region associated with the full-step mode and slightly decreased elsewhere. Sometimes the resultant increased motor speed capability can permit a gear reduction increase, which, in turn, increases the motor's inertia handling capability as much as two to six times through the square-law relationship. Such an additional gear reduction can also proportionately reduce overshoot measured at the load. The mid-frequency resonance phenomenon is illustrated in Fig. 12. It will be noted that the mid-frequency phenomenon can be affected by the amount of inertia coupled to the motor. Increasing inertia may move the resonant frequency either higher or lower, depending on the suppression circuit used. The gray shaded area in the graph shows the approximate range of variation from adding a load inertia equal to rotor inertia. While additional inertia may be desirable, there is an accompanying reduction in the error-free-start-stop speed that it also produces.

Whereas 1.8° steps (and half-stepping by 0.9°) were and remain the traditional approach to hybrid PM stepper motors, a concept known as *microstepping* has entered the technology. Microstepping is discussed in some detail in a separate section of this article, placed after a description of stepper motor controls.

STEPPER MOTOR CONTROLS

As shown by Fig. 13, the basic elements of stepper motor control include (1) a logic sequencer, (2) power drivers, (3) pulse current and voltage limiter, (4) motor power supply, (5) control-circuit power supply, and (6) where a feedback network for closed loop operation is required, an *encoder*.

Logic Sequencer. Sequencers may consist of only simple gating logic (unidirectional operation requires only a ring counter circuit), or more sophisticated logic may be used, depending upon the application. A typical stepper translator card will supply switching logic and power signal amplification for stepper motors. Commonly, the translator will accept direction and count pulses in half- or full-step modes to generate corresponding motor rotation. Some translators are provided with current limit protection as well as LED indicators for blown fuse, plus-and-minus limits, and home position.

Power Drivers. These devices are needed to amplify the output of the sequencer to a level of voltage and current required to energize (drive) the stepper motor. A representative driver is shown in Fig. 14. The circuit shown uses TTL logic and Darlington switch networks. The electrical load which the driver "sees" is represented by the equivalent circuit (see Fig. 15) of a typical stepper motor. Special consideration must be given in the design of power drivers to the inductance of the motor. The inductance, which varies with current and rotor positions, will limit the

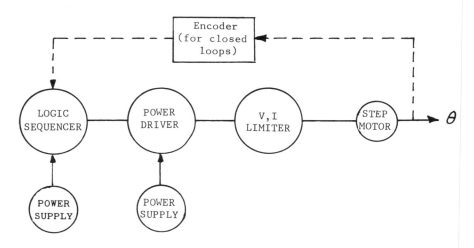

Fig. 13. Basic elements of stepper motor control system.

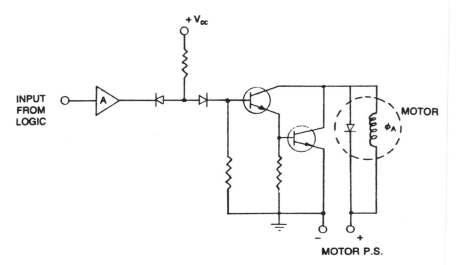

Fig. 14. Power driver stage (one used for each phase). (*Stock Drive Products.*)

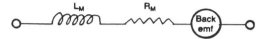

Fig. 15. Equivalent circuit of a stepper motor.

slew speed as well as cause "turn-off spikes" resulting from stored energy in the winding at the time of deenergization. This stored energy must be safely and quickly dissipated. The turn-off inductive voltage spike ($E = -L \frac{di}{dt}$) can easily exceed the maximum voltage rating of the driving transistor, thus causing breakdown.

Inductive Voltage Suppression. Diode and resistor suppression networks are often connected in parallel with the motor winding for the purpose of protecting the driver from high inductive voltage spikes.

Time Constant Reduction. The electrical time constant of a motor winding is defined as $\tau = \frac{L_M}{R_M}$. A reduction of this time constant is obviously desirable to obtain improved current and thus torque build-up time. A resistance (R_S) can be placed in series with the motor winding, thus reducing the effective time constant to $\tau = \frac{L_M}{R_S + R_M}$. Some disadvantages to the series resistance technique include: (1) increased power loss, and (2) higher drive output voltage is required for a given current.

Dual Voltage Control. This type of driver can be used in order to reduce power dissipation without the loss of motor performance. Essentially this plan uses a high-voltage output to initially energize a motor phase. This ensures maximum current rise in minimum time. When the current reaches its predetermined maximum value, the high voltage is reduced to a lower "maintenance" voltage for the remainder of the step.

Chopper Drives. This type of system limits the motor current by means of voltage modulation. It is a highly efficient means of motor control that permits an initial high voltage (10 to 20 times rated voltage) to energize the motor until the desired current is obtained. The voltage is then switched off until the current decays to a predetermined level after which the voltage is again turned on. This cycle of on-off voltage (chopping) is continued through the drive pulse time. This method is well suited for stop start operation.

Other Controls. These include special controllers for stepper motor systems with microcomputers and programmable controllers and other packaging options available. Motion control is discussed further in *Section 2* of this *Handbook*.

TRANSITION IN STEPPER MOTOR TECHNOLOGY

Continuous and incremental actuators both accept microcomputer-based control decisions that are expressed in digital commands. The requirement to generate step-by-step load positioning leads not only to motors designed to be pulsed but also to the pulsing of motors originally designed to run continuously. The sharp distinction between these two types of motors is commencing to fade. Actuation increments have shrunk in angular or linear dimension to a point where a sequentially pulsed stepper motor takes on the characteristics of a precision servomotor. Like a servomotor, the stepper motor moves smoothly in target approach yet can maintain step count to ensure command positioning, while avoiding target overshoot with a final step that can be as small as 0.432 arc minute[4].

MICROSTEPPING

By Frank Arnold[5]

Stepper motors are widely used for incremental motion control in such applications as computer peripherals, *X-Y* positioning tables, medical instruments, and specialized precision machinery requiring modest power. Some of the advantages of steppers include reduced maintenance owing to lack of brushes; simplicity of operation owing to their open-loop design rather than expensive, elaborate feedback circuitry; and adaptability to digital electronic devices.

However, the position resolution that can be obtained from a stepper motor is limited because of the mechanical design of the unit. From the earliest steppers whose rotors would assume a fixed position every 90° (four steps per revolution) to present steppers with resolutions of 3.6° (100 steps per revolution), 1.8° (200 steps per revolution), or even 0.9° (400 steps per revolu-

---

[4]Brief encapsulation of an excellent summary of stepper motor control technology (Bailey, 1985–in list of references).
[5]Vice President, *SynchroStep*[R] Technology Group, PMI Motion Technology, Division of Kollmorgen Corporation, Commack, New York.

tion), there has been steady improvement in resolution.

But for ultra-precise positioning applications, even these positional resolutions may be inadequate. For example, some equipment now being used to produce electronic integrated circuits may require mechanical motion that is precisely controlled in steps of only about 0.1 micrometer (4 microinches).

Since the minimum step that can be produced by a stepper is determined by its mechanical design—mainly the number of poles or teeth in the stator and the number of poles or teeth in the rotor—a limit is soon reached where other means must be employed to subdivide the step angle. The solution is to use electronics rather than mechanics to produce the necessary subdivisions.

The first move in the direction of higher resolution is called half-stepping, which doubles the number of controlled positions of a stepper motor. Half-stepping is accomplished easily by simply changing the current sequence in the full-step mode. However, this method is still far from what is required for ultra-precise positioning.

The solution is to use *microstepping*, in which each motor step is subdivided electronically into dozens or even hundreds of fixed positions. At present, using microstepping, it is possible to obtain up to 25,000 to 50,000 electronic steps per motor revolution. With such positional resolutions, a high-quality stepper motor with a basic 1.8° per step motion can be used to drive a lead screw to a linear resolution measured in tenths of a micrometer, and this is done simply and without the use of feedback and a closed servo loop.

Advantages and Limitations of Microstepping

The most important benefit to be gained by microstepping is increased positioning resolution. Using a 200 step/revolution motor with electronics to divide each step by 32—a common microstep divisor—the result is 6400 positions per revolution, or 3.375 minutes of arc per step.

Additional advantages include the reduction in torque ripple, especially important at lower motor speeds. Since the steps are small, they are accomplished with a minimum of overshoot and oscillation. Hence the stepper motor moves more smoothly from one position to another; it is more nearly continuous rotation than a series of discrete or jerky steps.

A third advantage is that microstepping minimizes the natural resonance problems that all stepper motors have. Typical unload steppers may resonate between 50 and 200 Hz. Load inertia adds to the motor's inertia and lowers the system resonant frequency. If the motor is operated near its natural resonance, it may oscillate, stall, jump steps erratically, or even rotate backward. By using microstepping, with its higher frequency pulse rates and smaller steps with smaller overshoot, such resonances are not triggered.

Another problem with ordinary stepper motors is the reduction that occurs in their relative torque constant (torque versus current) owing to the saturation effects of the iron rotor. If such effects are reduced, say, by the use of a noniron rotor, then these effects can be minimized.

Despite the several advantages of microstepping, one major problem that exists in the design of the typical stepper motor makes it difficult to microstep accurately. Most stepper motors, particularly the variable reluctance (VR) or hybrid types, do not have pure sinusoidal torque-versus-displacement curves. When these curves are distorted by harmonics, then it becomes difficult to employ microstepping to achieve optimum results. If compensation is applied to overcome the non-linearity of the conventional stepper, as will be shown shortly, changes in the loads on the motor may result in even greater positional errors than if no compensation were applied. Consequently, the best solution is to eliminate the cause rather than make compensation for the effect.

Microstepping Torque

It is important to understand microstepping's effect on stepper motor torque and vice versa. The typical torque specifications given on manufacturers' data sheets are for *holding torque* and *residual* (detent) *torque*.

First, microstepping controls the available torque to make one microstep, called *stepping torque*, directly with the formula:

$$T_{HMS} = T_{H2\phi} \times \sin\left(\frac{90°}{MS}\right)$$

where T_{HMS} = Differential (change in) torque from one step to the next step.
$T_{H2\phi}$ = Holding torque of the step motor with two phases on.
MS = Number of microsteps per fundamental step, i.e., 2, 4, 8, 16.......

As an example, if we have a 200-step-per-revolution stepper motor with a T_{H2} of 60 oz-in, and we want to microstep 128 steps-per-step, we find:

$$T_{HMS} = 60 \text{ SIN}\left(\frac{90°}{128}\right) = 0.74 \text{ Oz-in}$$

This result of 0.74 oz-in means that, physically, in a single step (not slew) mode, unless the load friction plus the motor's own bearing friction and residual torque loads are less than 0.74 oz-in, the motor will *not* move the microstep.

If the loads are more than 0.74 oz-in, there are two solutions available:

1. Pick a new motor with a larger holding torque, two phase on, so that T_{HMS} becomes larger, or

2. Pick a smaller number of microsteps, since the motor will be self-limiting anyway with regard to steps moved versus steps commanded to move. (In the example above, suppose the total load is 0.93 oz-in; when the command is given to move 1/128th of a full step and the phase currents are proportioned accordingly, 0.74 oz-in of torque is available—the shaft will not move. If a second command is given to move another 1/128th step, another 0.74 oz-in becomes available for a total of 1.48 oz-in. The shaft now moves driving the 0.93 oz-in load to its new position. The 1/128th commands are self-limiting to 1/64th steps.)

Second, residual torque affects microstep accuracy. In general, for microstepping, residual torque should not exceed $\frac{1}{2} T_{HMS}$ as defined above and preferably should be even less. Conceptually one can think about residual torque as an unwanted restoring torque, i.e., if the shaft is moved away from its equilibrium point, an opposing (restoring) torque is created. If we try to position the rotor in positions *between* the equilibrium points, which is precisely what we do in microstepping, the residual torque is trying to push or pull the motor away from the exact position it would otherwise assume.

Zero residual torque is the ideal, like the VR stepper. Unfortunately, the VR steppers' torque constant is a function of current squared, as mentioned below. So, where does the $\frac{1}{2}$ factor just mentioned come from?

We have already said that the sum of friction torque plus residual torque cannot be greater than one unit of differential torque, as a maximum. For safety, the residual torque must be no more than one-half of one unit, or $\frac{1}{2} T_{HMS}$. Example: For a 16 step-per-step application, we must hold residual torque to 5% of holding torque 2ϕ on; for 128 step-per-step the number is 0.6%.

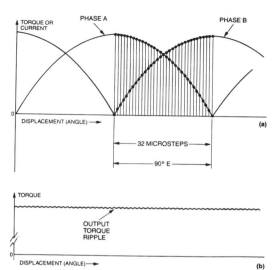

Fig. 17. Microstepper with sine/cosine proportioning. (*PMI Motion Technology, Division of Kollmorgen Corporation.*)

Correct Way to Microstep

In order to microstep a stepper motor, the currents applied to the two stator windings must be determined by the sine/cosine of the torque-displacement curves as well as the number of the microsteps. Consider the idealized torque-displacement curves shown in Fig. 16. These are shown as perfect sine waves offset by 90° (i.e., sine and cosine). The curves are also shown inverted in polarity as well in parts (c) and (d) of the illustration. The steps labeled 1 through 4 indicate the positions assumed by an unloaded rotor when the motor is operated in a one-phase-on at a time sequence. Here, one full step of mechanical (angular) displacement is equivalent to 90 electrical degrees. When these current or torque curves are rectified and combined, the theoretical result is that shown in part (e) of the diagram. The heavy line at the top of Fig. 16 (e), which is the ideal torque output of the motor as it is continuously stepped, indicates the torque ripple. The actual motion of the stepper motor in this mode is more complex than as shown in (e) and the resulting torque ripple will be larger.

In Fig. 17, we show an expansion of the Fig. 16 curves with 32 microsteps indicated. Note that with digital electronics, the number of microsteps is usually a power-of-two fraction of a whole

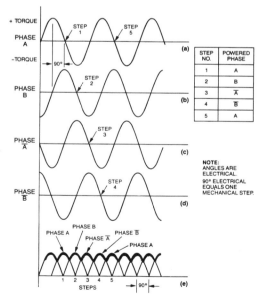

Fig. 16. Idealized torque-displacement curves for stepper motor. (*PMI Motion Technology, Division of Kollmorgen Corporation.*)

358 Stepper Motors and Controls

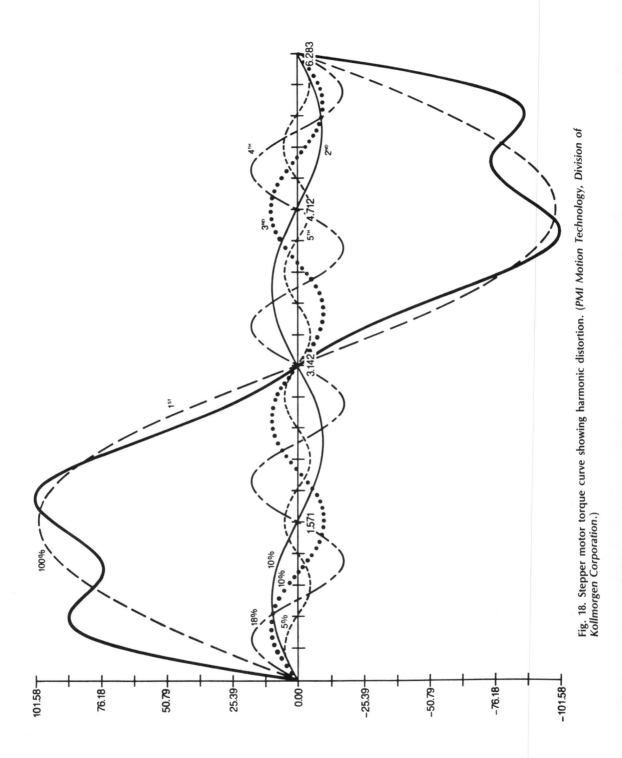

Fig. 18. Stepper motor torque curve showing harmonic distortion. (*PMI Motion Technology, Division of Kollmorgen Corporation.*)

step, such as 8, 16, 32, 64, 128, etc. In order to obtain a $\frac{1}{32}$nd step, we simply proportion the currents into the two-phase stator windings of the stepper motor as shown by the dots in the figure.

The values of these currents can reside in a PROM LSI computer memory chip, to be extracted on the command of a program or through a logical sequence. Note that as the current in phase A falls, the current in phase B rises. For example, at the sixteenth microstep, 0.707 of the rated current is applied to both phases. Since the two currents are 90° apart (in quadrature), their total is always unity throughout the entire cycle, i.e., $A^2 + B^2 = 1.0$. Note also that the output torque ripple as shown in Fig. 17 (b) is extremely small.

As previously mentioned, most variable reluctance (VR) or hybrid stepper motors do not have purely sinusoidal torque-displacement curves. Such curves are usually distorted because of the inherent design of the stepper, which may introduce second-through-fifth harmonics, as shown in Fig. 18.

When perfect sine-cosine proportioning is applied to motors with these distorted curves, the desired increased position resolution cannot be achieved. Since the motor's curves are far from sinusoidal, current level increments based on equal-angle increments of a sine wave will not produce equal-angle movements of the motor shaft. Torque ripple can also be degraded considerably.

Consider next the effects of magnetic iron saturation. This results in a variation in the torque constant depending on the level of the current in the windings. A stepper motor's torque constant (torque/ampere) is acted on by the sine-cosine proportioning currents. If, because of saturation effects, the torque constant at a fraction of the rated current is not the same as when full-rated current is applied, then even further degradation in position resolution and torque ripple occur. In fact, the torque constant of variable-reluctance steppers is a function of current squared, so that these motors are rarely used for microstepping. A typical saturation curve for

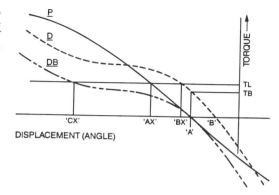

Fig. 20. Stepper motor—effect of compensation on positioning error. (*PMI Motion Technology, Division of Kollmorgen Corporation.*)

a hybrid stepper (Fig. 19) shows a 15% fall-off in torque constant as current is increased from zero to rated, demonstrating the nonlinearity of this "constant."

Compensating for Nonlinearity

To overcome the nonlinearity and magnetic-saturation effects of conventional steppers, a number of compensating techniques have been devised, some of which have been patented. The method used is to measure how much each stepper motor deviates from the ideal and then to store step-by-step correction values into computer memory.

Unfortunately, this works for only one torque-loading curve at a time, i.e., for one ratio of harmonics to fundamental and for one value of iron saturation. Also, if the compensation is realized for full-rated torque, it may not be correct for half-rated or other values of torque. Such compensation may, in fact, make the position error worse for any loading except for the one value for which the compensation has been applied. (This effect is shown in Fig. 20.)

In this figure curve P represents the ideal sinusoidal two-phase torque curve, and curve D is the torque curve distorted by third and fifth harmonics. The stable displacement position for zero torque for curve P is "A" and for curve D it is "B."

Now, we apply compensation to bring curve D into coincidence with P at position "A" by the application of an amount of torque current TB. The result is the creation of compensated curve DB.

But what happens when a different load TL is applied? In the case of the sinusoidal curve P, the stable point moves to "AX" and the offset "A-AX" will be constant at any angular position if TL remains constant. On the distorted curve D, the stable point moves to "BX" with the increased load. In this case it happens that the

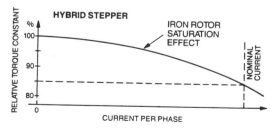

Fig. 19. Stepper motor torque constant reduction due to saturation. (*PMI Motion Technology, Division of Kollmorgen Corporation.*)

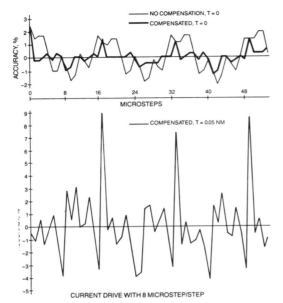

Fig. 21. Compensated and non-compensated curves of hybrid stepper. (*PMI Motion Technology, Division of Kollmorgen Corporation.*)

offset "B-BX" is even less than "A-AX." The difficulty is that because of the distortion, "B-BX" will not be constant with respect to angular position and, in addition, there is the original offset error of "A-B."

The compensated curve DB, on the other hand, which is biased for correction of the "A-B" offset now has its stable point moved a considerable distance to "CX," which is far from the correct position "AX." The unbiased curve D, in position "BX," is actually much closer to the correct position than that of the biased, compensated curve DB. Hence, it can be seen that the compensation has actually made matters worse.

Figure 21 shows the results of actual measurements of positioning accuracy of a typical hybrid stepper motor operating at 8 microsteps/step with and without such compensation. The upper curves, taken without a load, show a positioning error of around ±2% without compensation. This improves somewhat to around ±1.5% in the worst case with compensation. But as soon as a load is applied to the stepper, as shown by the lower curve, the position accuracy becomes much worse, rising to +8 to 9% and falling to about −4% position error.

In the real world, the torque loading of the stepper is rarely constant, so there is no single correct choice for the proper offset bias. As has been demonstrated, any such bias is only a compromise.

An Ironless Disc-Rotor Stepper Motor

Since the sources of the foregoing problems are distorted torque-displacement curves and iron-saturation effects, the optimum way to microstep would appear to utilize a stepper motor that is virtually free of harmonic distortion and magnetic iron saturation nonlinearity. The motor[b] shown in Figs. 22 and 23 utilizes a rare-earth permanent-magnet disc rotor with magnetically impressed poles and individual magnetic circuits for each torque-producing tooth (stator poles).

[b]PMI Motion Technology—*SynchroStep*[R].

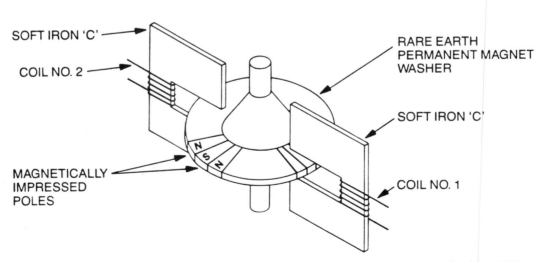

Fig. 22. Diagrammatic view of the *SynchroStep*® disc-rotor stepper motor. (*PMI Motion Technology, Division of Kollmorgen Corporation.*)

Stepper Motors and Controls 361

Fig. 23. Cut-away view of SynchroStep® stepper motor. (*PMI Motion Technology, Division of Kollmorgen Corporation.*)

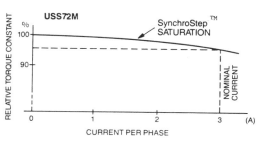

Fig. 25. Torque constant linearity of disc-rotor stepper motor. (*PMI Motion Technology, Division of Kollmorgen Corporation.*)

Through the proper design of the electromagnetic teeth or stator poles with respect to the alternating poles of the disc magnet rotor, harmonics are practically eliminated. There is no iron in the lightweight, flat-disc rotor to saturate. Also, the stator is carefully designed so that magnetic saturation effects are virtually eliminated. The motor also has extremely low inertia and high acceleration.

The actual torque-displacement curves recorded in using a *SynchroStep^R motor* (*USS-72M*) is shown in Fig. 24. Note the purely sinusoidal curves over the entire operating current range. The linearity of the torque constant of this motor is shown in Fig. 25. When this is compared with Fig. 19, the improvement over a conventional hybrid stepper can be readily observed. Figure 26 shows the measured results and benefits of the sinusoidal torque-displacement curves and low saturation of the motor as compared with a hybrid stepper. For this figure, both motors were microstepped at 32 steps/step using pure sine-cosine proportioning at rated current. Although

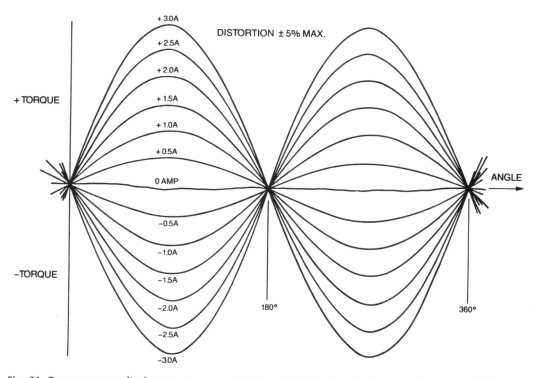

Fig. 24. Torque versus displacement curves. (*USS72M, PMI Motion Technology, Division of Kollmorgen Corporation.*)

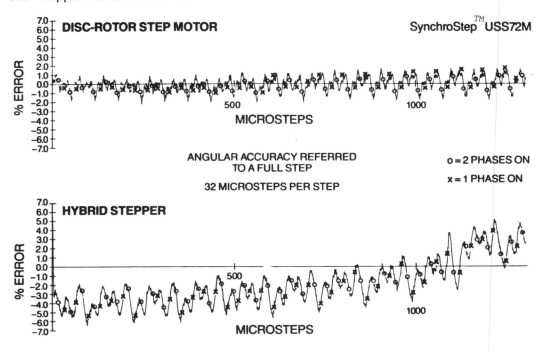

Fig. 26. Measured positional errors of disc-rotor stepper motor compared with a hybrid stepper. Both steppers operating at 32 microsteps per step. (*PMI Motion Technology, Division of Kollmorgen Corporation.*)

only a partial revolution is plotted in the figure, it can be observed that the hybrid motor error plot is riding on a sinusoidal distortion curve. The error of the hybrid stepper is more than twice that produced by the disc-rotor step motor.

In conclusion, microstepping does solve problems for the incremental motion systems designer by providing increased position resolution, reduced torque ripple, and suppressed resonances. To take advantage of these positive attributes, a stepper motor with pure sinusoidal torque curves and very small saturation effects at rated current is highly desirable. One stepper motor design resulting from consideration of these criteria has just been described.

ELECTROHYDRAULIC STEPPING MOTOR

The electric stepping motor, controlled by an external stepping motor driver, supplies the command input to the internal hydromechanical servo loop, which in turn controls the final output of the hydraulic actuator.[7] The electric stepping motor is a synchronous four-phase permanent magnet motor that converts electronic signals into angular displacement of the motor shaft—as previously described. The stepping motor driver, or translator, accepts position and velocity profile commands produced by a microprocessor controller as a variable frequency pulse stream and direction signal. These command pulses are converted by the translator into a sequence of motor winding energization patterns that cause the stepping motor to rotate one unit of angular displacement (one step) for each pulse. One step is equivalent to 1.8° in whole-step operation or, more typically, 0.9° in half-step operation.

The electric stepping motor regulates a 20, 35, or 50 GPM (~76, 132, 190 liters/minute) four-way spool type servovalve that is the flow control element of the electrohydraulic stepping motor. The servovalve is positioned by the stepping motor through a rotary-to-linear translator in a manner similar to using a nut and screw to position a machine member. The rotary-to-linear translator consists of a precisely tapped nut coupled to the stepping motor shaft and matching precision threads ground into the servovalve spool. The combination of the bi-directional step-

[7] MTS Systems Corporation, Machine Controls Division, Minneapolis, Minnesota.

Stepper Motors and Controls 363

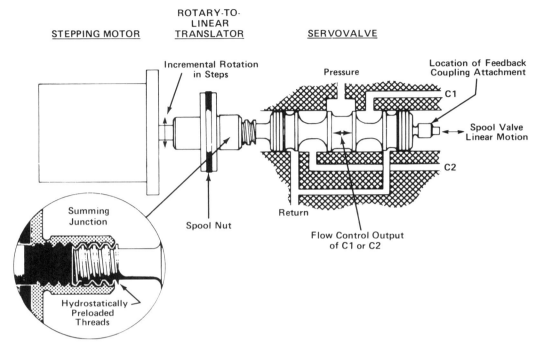

Fig. 27. Digital rotary servovalve. (*MTS Systems Corporation, Machine Controls Division.*)

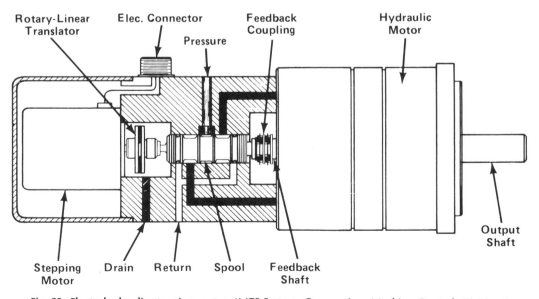

Fig. 28. Electrohydraulic stepping motor. (*MTS Systems Corporation, Machine Controls Division.*)

ping motor and the four-way servovalve provides both flow and direction control. (See Fig. 27.)

When the valve spool is displaced linearly from its null position by a stepping motor rotation, an oil path is established to the actuator element of the assembly, resulting in a rotation of the hydraulic motor. The rotor of the hydraulic motor is coupled to the servovalve spool by a torsionally rigid, axially compliant feedback coupling to create a mechanical closed loop. (See Fig. 28.) The feedback coupling is torsionally rigid to ensure accurate position feedback and axially compliant to allow the servovalve spool to stroke.

An examination of the mechanical closed loop shows that a rotation of the valve spool (feedback) in the same direction as some initial stepping motor rotation (command) will cause the spool to shift back to its null position. (See Fig. 29.) The nut/screw assembly is a mechanical

Fig. 30. External view of electrohydraulic stepping motor (electrohydraulic torque amplifier). (*MTS Systems Corporation, Machine Controls Division.*)

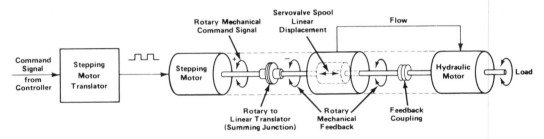

Fig. 29. Mechanical closed loop showing relationship of stepping motor translator, stepping motor, servovalve, and hydraulic motor. (*MTS Systems Corporation, Machine Controls Division.*)

TABLE 5. Principal Mechanical and Electrical Parameters of Electrohydraulic Stepping Motor*

Parameter	Specification
Hydraulic and Mechanical:	
Maximum operating pressure	3000 PSI (20.7 MPa)
Resolution (half-step operation)	0.9°
Resolution (whole-step operation)	1.8°
Repeatability	±0.2°
Backlash	0.5°
Recommended filtration	3 to 10 Microns
Relief valve setting	2250 PSI (15.5 MPa), other settings available on request
Operating temperature	40–160°F (20–142°C)
Recommended operating temperature	100–120°F (82–102°C)
Fluid type	High-grade mineral based hydraulic fluid; consult factory for use of other fluids
Fluid viscosity	75–400 SSU (17–88 Centistokes)
Recommended fluid viscosity	150 SSU at 100°F (32 Centistokes at 82°C)
Maximum allowable drain pressure	25 PSI (170 KPa)
Proof pressure (except drain port)	4500 PSI (31.0 MPa)
Electrical:	
Motor type	Superior Electric Model M091-FC-401 four-phase electrical stepping motor or equivalent
Nominal input voltage	1.7 Vdc
Current rating per winding	4.7 A
Typical response time for one step (0.9°)	3 ms
Nominal inductance per phase	1.65 mH
Connector supplied with mate	MS3106A14s-65

Specifications are subject to change.
*MTS Systems Corporation, Machine Controls Division, Minneapolis, Minnesota.

Fig. 31. Electrohydraulic stepping cylinder. (*MTS Systems Corporation, Machine Controls Division.*)

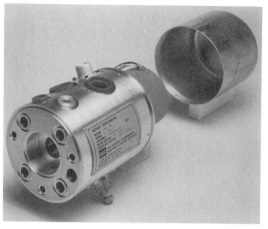

Fig. 32. Digital rotary servovalve that provides direct digital control of flow without D/A or A/D conversions. Available in 20, 35, and 50 gallons/minute flow ratings; 3000 psi (20.7 MPa) continuous, maximum operating pressure; integral dual relief valves for system protection [preset to 2250 psi (7.2 MPa) or optional user value]; directly interfaces with rear-ported hydraulic motors and cylinders; integral cross-port bleed adjustment. Valve-actuator stability enhanced by precision hydrostatically preloaded spool threads to eliminate backlash. Stepper motor is immersed in hydraulic fluid to provide effective heat dissipation. (*MTS Systems Corporation, Machine Controls Division.*)

summing junction, which causes the position of the valve spool to change linearly in response to the difference between the command position (stepping motor position) and the feedback position (hydraulic motor position).

The following sequence is typical of events necessary for motor motion: (1) As the stepping motor rotates, the rotary-to-linear translator moves the valve spool from its null position and oil is ported to the hydraulic motor. (2) As the hydraulic motor rotates, the torsionally rigid feedback coupling rotates the valve spool in the same direction. (3) When the hydraulic motor has rotated as far as the stepping motor shaft, the valve spool has again reached its null position, which stops the oil flow and consequently stops the actuator movement.

The servovalve component of the assembly is provided with integral cross-port reliefs for system over-pressure protection. The relief valves are preset to 2250 psi (15.5 MPa) with other settings optional. For applications requiring increased stability, a crossport adjustment on the servovalve allows crossport leakage to be introduced into the system, thus increasing system damping.

An external view of the electrohydraulic stepping motor is given in Fig. 30. Specifications are given in Table 5.

The electrohydraulic stepping cylinder shown in Fig. 31 provides direct digital control of a hydraulic cylinder. Standard bore sizes up to 8 inches (~203 mm) are available. Standard stroke lengths up to 36 inches (915 mm) are available. The stepping cylinder consists of a fractional-horsepower electric stepping motor, a four-way rotary servovalve (see Fig. 32) and a hydraulic cylinder (actuator). These components are linked together in a hydromechanical closed loop that allows direct digital control of the hydraulic cylinder. The unit has been designed for use in computer-controlled machine automation applications. These linear cylinders offer resolution of 0.00125 inch (0.032 mm) and repeatability of 0.0005 inch (0.013 mm), over a wide range of force and stroke capabilities.

ADDITIONAL READING

Arnold, F.: "Using Power Rate to Size an Application (Disc-Rotor Step Motor)," PMI Motion Technology, Division of Kollmorgen Corporation, Commack, New York (1985).

Bailey, S. J.: "Programmable Motion Control Paces Factory Automation Efforts," *Cont. Eng.*, **30**, 4, 107–110 (April 1983).

Bailey, S. J.: "Step Motion—Open Loop Path to Production Precision," *Cont. Eng.*, **29**, 7, 73–77 (July 1983).

Bailey, S. J.: "Step Motion Control 1985: Direct Digital Incrementing with Servo-Like Performance," *Cont. Eng.*, **32**, 8, 49–52 (August 1985).

Berris, R., and D. Hazony: "Discrete Pulses Put Induction Motors into the Stepping Mode," *Cont. Eng.*, **29**, 1, 85–86 (January 1983).

Floresta, J.: "Power Rate: A Most Important Figure of Merit in Incremental Motion," PMI Motion Technology, Division of Kollmorgen Corporation, Commack, New York (1984).

Kompass, E. J.: "Thin-Disk Rotor Improves Step Motors," *Cont. Eng.*, **29**, 10, 68–69 (October 1983).

Miller, T. J.: "Step Motor Controller Options Fit Package to Application," *Cont. Eng.*, **31**, 8, 81–82 (August 1984).

Morris, H. M.: "Microstepping a Linear Motor," *Cont. Eng.*, **32**, 2, 83–84 (February 1985).

Tal, J.: "Digital Control of DC Motors," in *Motion Products Catalog*, Stock Drive Products, New Hyde Park, New York, 1985.

Thornton, P. J.: "Introduction to Step Motors," *Data Book 757*, "Design and Application of Small Standardized Components," Educational Products, Mineola, New York, 1983.

Waldspurger, C. J., and M. A. Smith: "New Design Options (Stepper Motors) with Rare Earth Magnets," in *Motion Products Catalog*, Stock Drive Products, New Hyde Park, New York, 1985.

Linear and Planar Motors

By G. T. Volpe[1]

The fabrication of future integrated circuits will require X-Y table submicron positioning for wafer/mask alignment. This will require a servo drive with a minimum number of moving parts and the virtual absence of mechanical friction. Linear electric motors (LEMs) offer an attractive method for X-Y table positioning, since the force transducer and payload can be on the same housing. This means that high accelerations may be achieved without excessive vibrations owing to mechanical resonances.

By further utilizing compatible air slides, friction is essentially eliminated, thereby allowing extremely fine position control. This leaves the motor's equivalent magneto-mechanical spring as the major limit of performance. This spring expresses the spatial variable-magnetic reluctance between the motor's poles and stator teeth. Although the spring rate is not sufficient to maintain static holding accuracy it can be increased adequately via position feedback. In addition, position feedback significantly reduces the effects of external disturbances, such as acoustically induced noise and structural vibrations, provided they occur within the loop bandwidth.

The open loop bandwidth should then be made as large as possible, limited by either of several factors. One of the most important of these factors is disturbance and noise-induced error causing amplifier or motor saturation. Once saturation occurs, the open loop gain decreases rapidly, causing the control error to increase beyond acceptable limits. Other factors that limit loop gain are structural resonances and actuator hysteresis. Both factors present parasitic phase lag in the control loop causing the possibility of limit cycling. Lowering of the loop bandwidth offsets this effect, but does so at the cost of performance.

TYPES OF LINEAR MOTORS

There are three fundamental types of linear electric motors—alternating current (AC), direct current (DC), and linear reluctance types.

Alternating-current LEM's fall into two basic categories: (1) *linear induction motors* (LIMs), and (2) *linear synchronous motors* (LSMs). They operate from a multiphase excitation of the stator, producing a traveling magnetomotive force (MMF) wave in the air gap, as shown in Fig. 1. An MMF wave in the translator is set up either by induction (LIM) or by a fixed excitation (LSM). In either case, a slip or lag angle between the translator and the stator is required in order to develop a propelling force.

Direct-current LEMs function on the basis of a Lorentzian force (*Bil*). One type of direct current LEM, which uses its armature in much the same way as a voice coil in a loudspeaker, is shown in Fig. 2. This flux interacts with the top and bottom of an armature coil to produce an additive force in the direction of motion. Although this is a smoothly performing motor, its stator magnet is quite heavy, adding an appreciable inertia burden to the orthogonal axis, making the latter axis design cumbersome and slow-reacting. This reduces throughput.

The reluctance-type linear motor works on the principle that two magnets of opposite polarity tend to align themselves to establish a minimum reluctance magnetic path. An example of a reluctance type is the stepper motor. In a rotary stepper, as shown in Fig. 3, a three-phase stator is excited with a current pulse train in phase orders A, B, and C to rotate the moving rotor $\frac{1}{3}$ of a pitch angle θ for each pulse, where a pitch angle is defined as the angle between centerlines of adjacent rotor teeth.

A rolled-out version of the rotary stepper concept with a reversed stator-rotor becomes the basis for a single-degree-of-freedom linear reluctance motor shown in Fig. 4. In this figure only phase A is energized, causing the centerline of the phase A pole in the translator to be aligned with the centerline 1 stator tooth. The centerline spacing between teeth is the linear pitch, p.

The centerline spacing between the phases if $\frac{4}{3}p$ or $p(1 + \frac{1}{n})$, in general. (n = the number of phases employed). Consequently, when phase B alone is energized, the centerline of the phase B pole aligns with the centerline 2 of the stator, moving the translator left $p/3$, or p/n in the n-phase case. An increase in the number of phases will improve the position resolution. For typical values of $p = 0.020$ in and $n = 4$, $p/n = 0.005$ in (125 μm), which is far

[1] University of Bridgeport, Bridgeport, Connecticut.

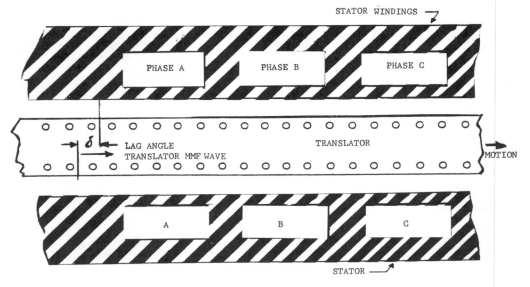

Fig. 1. Alternating-current linear electric motor (LEM).

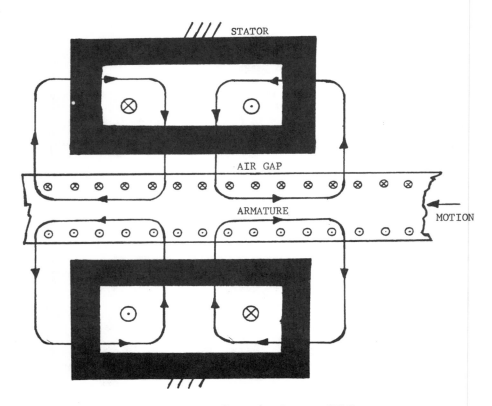

Fig. 2. Direct-current linear electric motor (LEM).

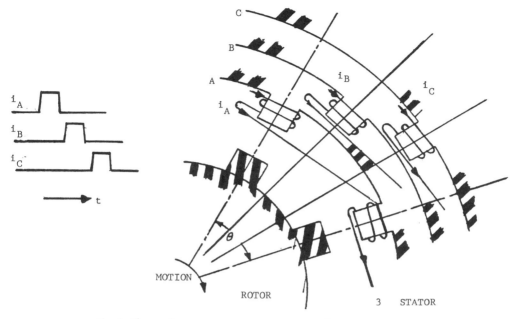

Fig. 3. Three-phase stepper motor concept with current excitation.

short of required submicron displacements. Electrical interpolation, or microstepping, can be used to improve resolution at the expense of drive circuit complexity.

Electrical interpolation for an equivalent two-phase system can be explained as follows: If an individual phase is excited with a steady DC current, a restoring force between the respective pole and its nearest stator tooth will be developed in accordance with a sine law relation. The fundamental components of developed forces in phases A and B are given by:

$$F_A = K_A \sin \lambda x \quad (1)$$
$$F_B = K_B \cos \lambda x \quad (2)$$

where

$$\lambda = \frac{2\pi}{p}$$

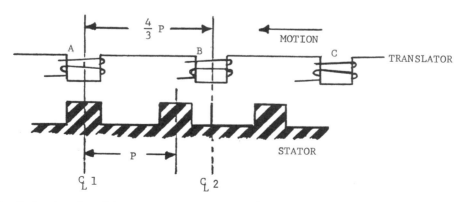

Fig. 4. A rolled-out version of the rotary stepper concept with a reversed stator-rotor becomes the basis for a single-degree-of-freedom linear reluctance motor.

370 Linear and Planar Motors

A derivation of the sinusoidal developed force is shown in the appendix and follows the work by *XYnetics* engineers, Hinder and Nocito.

If the phase force amplitudes are modulated by the rotor angle as:

$$K_A = F_M \cos \theta \\ K_B = -F_M \sin \theta \quad (3)$$

where

$F_M = K_M I$
$I =$ Stator current per phase
$\theta =$ A control variable (phase)
$K_M =$ Motor force constant

Then substituting (2) in (1) yields the net force on the translator:

$$F = F_A + F_B = F_M [\sin(\lambda x - \theta)] \quad (4)$$

which plots as shown in Fig. 5 for various values of θ. The stable reluctance point is at the negative zero crossing since this point represents a restoring spring force. The force balance equation is then:

$$F = m \cdot a + bv = m\ddot{x} + b\dot{x} = -kx; \\ k = \frac{F_M \, 2\pi}{p} \left[\frac{\text{lb}}{\text{in}}\right]$$

where

$$m = \text{Motor mass} \left[\frac{\text{lb sec}}{\text{in}}\right]$$

$$b = \text{Damping coefficient} \left[\frac{\text{lb sec}}{\text{in}}\right]$$

or alternatively

$$m\ddot{x} + b\dot{x} + kx = 0 \quad (5)$$

The solution to (5) is:

$$x(t) = Ae^{\frac{-bt}{2m}} \cdot \sin\left[\sqrt{\frac{k}{m} - \frac{b^2}{4km}} \, t + \theta\right]; \; t \geq 0 \quad (6)$$

Since from Fig. 5, the zero crossings move $\Delta x = \frac{\pi}{2\lambda}$ for $\Delta \theta = \frac{\pi}{2}$, it follows that

$$\Delta x = \frac{\Delta \theta}{\lambda} = \frac{\Delta \theta \, p}{2\pi} \quad (7)$$

Thus, for a pole pitch $p = 0.020$ in (500 μm), a one micron resolution will require 500 discrete phase steps. These steps are derived from a ROM table of sine functions for each phase of the machine.

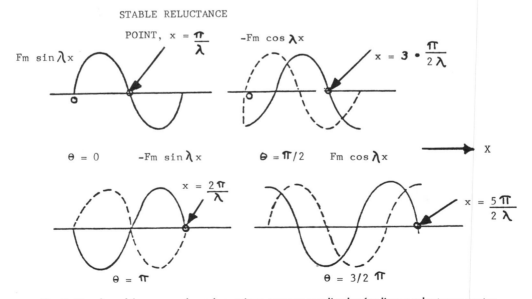

Fig. 5. Developed force waveform for various current amplitudes for linear reluctance motor.

THE PLANAR MOTOR

The concept of linear electric motors was extended to planar motion by Bruce Sawyer. It is currently manufactured by the Xynetic Corporation, Santa Clara, California, and is shown in Fig. 6. The device accomplishes *X-Y* motion in a plane without the need for gears or other mechanical couplings between the motor and stage. The stage is directly tied to two orthogonal linear reluctance motors with the assembly suspended by a thin air film over a wafflelike platen structure. Magnets on the motor are energized sequentially, so that the appropriate pole forces continually align themselves with the platen teeth, thus causing propulsion. Three-axis motion also can be accomplished as shown in Fig. 7.

Planar Motor Operation

The operation of the planar motor can be best explained for one-dimensional movement. (See Fig. 8 and caption.)

Coarse/Fine Positioning. To achieve fine submicron alignment with a coarse positioner, a limited motion fine positioner can be piggy-backed onto a stage driven by the planar motor or any conventional drive mechanism using a rotary motor and a lead screw. Because the required fine motion will be in the range of only ten micrometers, a simple flexure bearing can be utilized to mount the fine actuator, making the approach quite feasible. Of the various fine positioning actuator types, three outstanding candidates are piezoelectric, magnetostrictive, and voice coil.

A *piezo* or PZT actuator stack is a very stiff displacement actuator mount. For the required motion, close to one kilovolt may be required. Obtaining large loop bandwidth will be straightforward, owing to the PZT crystal's fast response.

Like the PZT actuator, a terballoy *magnetostrictive* actuator is very stiff and requires only low voltage (tens of volts) and a nominal current (less than one ampere) for control power. The maximum bandwidth obtainable is not so good as with the PZT because of the coil constraints, but is sufficient for most applications.

The concept of two-degree-of-freedom *voice coil* actuators having a limited motion is that of stacking two orthogonal voice coil actuator windings and allowing their armatures to move in a plane. (See Fig. 9.) The left-most (1) actuator causes *X* motion because of force F and carries the armature through the right-most actuator air gap. Similarly, the right-most actuator induces *Y* motion because of force F_y and carries the armature of the *X* actuator through the *X* actuator air gap. The common armature is flexure-supported in such a way that the *X-Y* motion is free, while *Z* motion is retarded by proper bearing design.

Having discussed the coarse/fine approach to two-dimensional position alignment, it is interesting to compare a lead screw course drive approach to the planar motor concept.

A conventional lead screw drive mechanism provides a very stiff coarse positioning and can remain inoperative during fine alignment, thus avoiding friction and wear. In addition, the problem of crossover from coarse to fine alignment is virtually eliminated, provided that the coarse alignment is well within the dynamic displacement range of the fine actuator. Heat dissipation and speed of response limit the lead screw approach. A planar motor is not nearly so stiff as a lead screw and requires power to maintain position. However, some of a planar motor's advantages are an order-of-magnitude faster response for the same motor weight, less sensitivity to external vibrations, very low friction, and, of course, no gears or other linkage coupling.

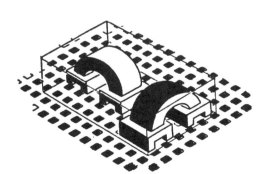

Fig. 6. The Sawyer linear motor with two-axis motion.

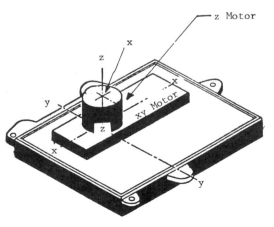

Fig. 7. Sawyer motor with three-axis motion.

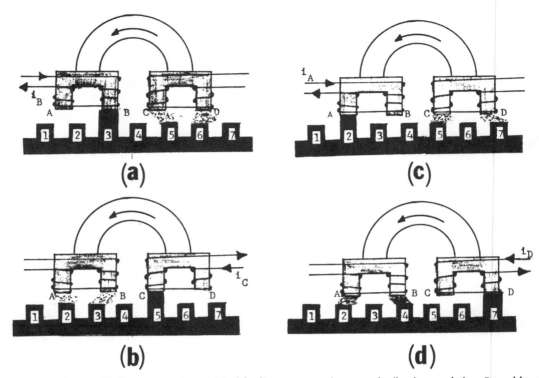

Fig. 8. Operation of a planar motor. In part (a) of the figure, current i_B causes the flux from pole face B to add to the permanent magnet bias flux and the induced flux from pole face A to cancel the bias flux. The flux path returns through the unenergized pole faces C and D. This means that pole face B aligns itself with platen tooth 3. With current i_B turned off and current i_C turned on, the same process causes pole face C to align with platen tooth 5. The net effect is one quarter pole pitch movement to the right. (For the *Xynetics* motor, the pole pitch is 0.02 in—so that a coarse step motion of 0.005 in is obtained. With current i_A off and i_C reversed to i_D, pole face D aligns with tooth 7 producing a third quarter pitch motion. Finally, when i_B is energized the cycle is completed and one pitch motion of 0.020 in will have been achieved. Fine positioning is achieved by quantizing θ into 80 steps of 6.5 micrometers each.

Closed-Loop Control

Figure 10 shows a simplified closed-loop control system that utilizes a four-phase planar motor. A position transducer provides coarse position information consistent with the motor's resolvable step size of six micrometers. A feedback counter converts the motor's movement from a reference position to an instantaneous count, which is subtracted from the digital position command word in a subtractor. The subtractor's error register is converted by a digital-to-analog (D/A) resolver to the necessary four-phase quantized sinusoids that provide the drive sequence for synchronous motor operation. To obtain fine position control to within submicron control, fine position analog error signals can be summed on each of the four quantized sinusoids. This fine position analog signal will enable the motor to modify or adjust its position to within the noise level of the sinusoidal output, which will be at least one order of magnitude less than six micrometers, or about 0.5 micrometer.

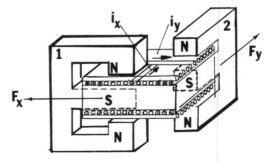

Fig. 9. Limited position translation voice-coil actuator.

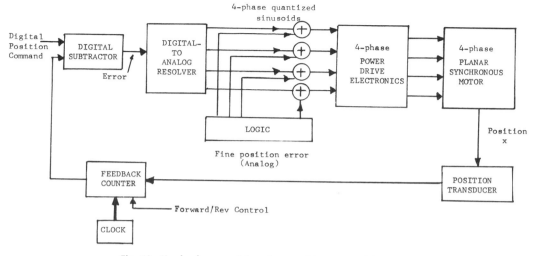

Fig. 10. Single-degree-of-freedom position control loop.

MOTOR PERFORMANCE TEST RESULTS

Measurements made on the *XYnetics* planar motor included static bearing stiffness, drift, and pressure sensitivity. Also measured were acceleration frequency spectra and time responses. Other important parameters measured included peak motor force to stall and carriage weight. These performance tests are discussed briefly in the following paragraphs.

Stiffness: Three factors were investigated:

1. *Static bearing tests*—Spring rates in three axes were calculated by applying calibrated forces against the motor forcer and measuring deflections using a Starrett mechanical gage. In the X and Y axis, applied forces of 5 and 10 pounds produced deflections of 1 and 1.5 mils respectively. The Z axis deflected 0.3 mil with an applied force of 40 pounds, yielding a spring rate of 20,000 pounds per in.
2. *Bearing drift*—An eddy current gage was used to measure Z-axis height off of the brass wafer chuck. The calibration showed a repeatable sensitivity of 0.175 V/mil = 0.175 mV/in, or 5.714 μin/mV. After the gage was turned on for several hours and stabilized, the bearing pressure was set at 80 psi and held constant for a 24-hour period. The measured voltage change was 10 mV, corresponding to a 57.14 μ-in (microinch) or a 1.452 micrometer drift in 24 hours.
3. *Pressure sensitivity*—Line pressure was varied in 5-psi increments and showed a predictable narrowing of the air-bearing gap. It was found that the gap narrowed 500 μ-in for a 50-psi decrease, or about 10 μ-in/psi, or 0.25 micrometer/psi.

Motor Parameters

The peak motor force to stall was measured in the X and Y directions at 15 and 12 pounds, respectively. This yield calculated a spring rate of:

$$K_x = \frac{2\pi \, F_{max \cdot x}}{P} = \frac{6.282 \, (15)}{0.020}$$

$$= 4{,}712.4 \text{ pounds per inch}$$

$$K_y = \frac{2\pi \, F_{max \cdot y}}{P} = \frac{6.282 \, (12)}{0.020}$$

$$= 3{,}769.9 \text{ pounds per inch}$$

which compares with measured spring rates, $K_x = 5000$ and $K_y = 6666$ pounds per inch mentioned previously. The differences are most likely attributable to magnetic retention in the platen teeth as well as to the tolerances used in the mechanical gages. The forcer weight was also measured to be eight pounds.

Dynamic Tests

Frequency Spectra. Using a three-axis accelerometer and spectrum analyzer, various frequency spectra were taken. By randomly tapping the forcer, various resonances were detected. The Z-axis dominant resonance was 1237.5 Hz with an amplitude of 81.5 μV. Using the accelerometer sensitivity of 10 mV/g, or 0.03864 in/sec^2/μV, the acceleration voltage amplitude corresponded to $81.5 \times 0.03864 = 3.149$ in/sec^2. From the velocity spectrum, the peak velocity at the same resonant frequency was 10.4 mV, which is exactly $[2\pi(1,237.5)]^{-1} \times 81.5$, that is, $\omega^{-1} \cdot a(\omega)$, as predicted by theory. This corresponds to a velocity error of 2 mil/sec. Consequently, the peak displacement error at 1,237.5 Hz is $[2\pi(1,237.5)]^{-1} \times 2 \times 10^{-3} = 0.256$ μin. This small value, of course, depends on the strength of the tapping disturbances that were on the order of a pound. Such disturbances are probably greater than those that will be encountered in practice.

The X axis and Y axis resonance peaks are 89.5 and 79 Hz, respectively. Where the Z axis resonance is probably due to a structure's compressional mode, the X and Y resonances are due to the magneto-mechanical spring plus a carriage weight of eight pounds. Using the measured spring rates reveals an average resonant frequency:

$$fo = \frac{1}{2\pi} \sqrt{\frac{k_{x,y}\, g}{w}} \qquad (8)$$

$$= \frac{1}{2\pi} \times \sqrt{\frac{(5{,}000 + 6.666)/(2 \cdot 386.4)}{8}}$$

$$= 84.4 \text{ Hz}$$

which agrees favorably with the actual measured spectra of the time response data. In acceleration-versus-time plots for various time scales, we observed 1.75 sec between turnaround accelerations. At a measured scan distance of 6.125 inches, this corresponded to the scan velocity of 3.5 in/sec.

Appendix—Sawyer Linear Reluctance Motor

Derivation of Sawyer Motor Force Equation. The following derivation follows Hinder and Nocito of the *XYnetics Corporation*. The basic mechanization of a two-phase single-axis motor is shown in Fig. 11.

Coils A and B have the function of commutating the permanent magnet flux. With full current, as in the case of coil A, all the magnetic flux circulates through 3, while poles 2 and 4 carry negligible flux. To derive the motor force equation, we start by deriving the force acting on pole 1 for a given flux circulating across the air gap. To simplify this

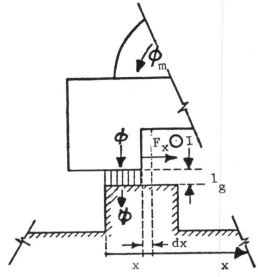

Fig. 12. Pole 1 of the Sawyer motor.

problem, the permeability of the iron will be assumed to be large enough to make the reluctance of the iron path negligible. Leakage flux will also be neglected. With these two assumptions, pole 1 can be represented as in Fig. 12.

The magnetic energy stored in the air gap is equal to:

$$E = \frac{1}{2}\frac{B^2}{\mu_0} V = \frac{B^2 h\, l_g\, x}{2\mu_0};$$

$$\mu_0 = 4\pi \times 10^{-7} \left[\frac{hy}{m}\right] \qquad \textbf{(A-1)}$$

where

E = air gap magnetic energy
B = air gap flux density
V = volume = $h l_g x$
h = electromagnetic stack length
x = pole engagement
l_g = air gap length

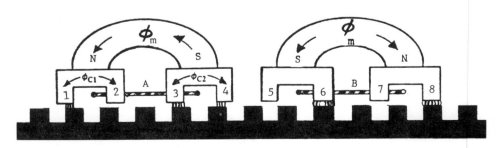

Fig. 11. Basic mechanization of a two-phase single-axis motor.

If an incremental displacement dx of the pole pieces is allowed, the tangential force F_x will perform work equal to the air gap energy change:

$$F_x \, dx = dE = \frac{B^2 \, h \, l_g}{2 \mu_0} dx$$

Then,

$$F_x = \frac{dE}{dx} = \frac{B^2 \, h \, l_g}{2 \mu_0} \quad \text{(A-2)}$$

This can be expressed as a function of the total flux ϕ crossing the air gap as follows:

$$F_x = (Bhx)^2 \frac{l_g}{2 \mu_0 h x^2} \quad \text{(A-3)}$$

The air gap flux density times the air gap threading area is the air gap flux:

$$\phi = Bhx, \quad \text{(A-4)}$$

so that

$$F_x = \phi^2 \, l_g / 2 \mu_0 h x^2 \quad \text{(A-5)}$$

Knowing that the air gap reluctance is

$$R = \frac{l_g}{h x \mu_0} \quad \text{(A-6)}$$

enables its gradient to be calculated as

$$\frac{dR}{dx} = \frac{-l_g}{h x^2 \mu_0} \quad \text{(A-7)}$$

Substituting this last expression in eqn. (A-5) gives

$$F_x = -\frac{\phi^2}{2} \frac{dR}{dx} \quad \text{(A-8)}$$

The minus sign in eqn. (A-8) comes from the fact that F_x is in the direction in which the reluctance R is decreasing.

To obtain a useful force equation, we next assume that the air gap reluctance function for a platen pitch p, neglecting higher harmonics, is equal to

$$R = R_0 \left(1 - \epsilon \cos \frac{2\pi x}{P} \right) \quad \text{(A-9)}$$

where

R_0 = average reluctance over a pitch
ϵ = fractional reluctance variation

Differentiating (A-9) and substituting the result into (A-8) yields

$$F_x = \frac{\phi^2}{2} R_0 \epsilon \frac{2\pi}{p} \sin \frac{2\pi x}{p} = K \sin \lambda \, x; \quad \text{(A-10)}$$

$$K = \frac{\phi^2 \, R_0 \, \epsilon \, \lambda}{2}$$

$$\lambda = \frac{2\pi}{p}$$

which is the describing force equation of the linear reluctance.

REFERENCES

Notico, T.K.; and T.K. Hinder: "The Sawyer Electric Motor," XYnetics Corporation, 328–331, Santa Clara, California, June 1978.

Volpe, G.T.: "Linear Motors XY Position," *Cont. Eng.*, 136–144, June 1982.

Kuo, F.: "Automatic Control Systems," 3rd Ed., 228–235, Prentice-Hall, Englewood Cliffs, New Jersey, 1975.

ADDITIONAL READING

Mawla, K.A.: "How Magnet Design Affects Linear Motor Performance," *Cont. Eng.*, 78–80, February 1982.

Morris, H.M.: "Microstepping a Linear Motor," *Cont. Eng.*, 83–84, February 1985.

Solid-State Variable Speed Drives

By Richard H. Osman[1]

The past two decades have seen rapid growth in the availability and usage of solid-state variable speed drives. Today there is a profusion of types that are suitable for virtually every type of electrical machine from the sub-fractional to the multi-thousand horsepower rating. (See Fig. 1.) Despite the diversity, there are two common properties of these drives: (1) All of them accept commonly available AC input power of fixed voltage and frequency and, through switching power conversion, create an output of suitable characteristics to operate a particular type of electric machine, i.e., they are *machine specific*. (2) All of them are based on solid-state switching devices. Even though many of the power conversion principles have been known as long as fifty years, when they were developed using mercury arc rectifiers, it was not until the invention of the thyristor in 1957 that variable speed drives became practical.

The thyristor (SCR) is a four-layer semiconductor device that has many of the properties of an ideal switch. It has low leakage current in the off-state, a small voltage drop in the on-state, and takes only a small signal to initiate conduction (power gains of over 10^6 are common). When applied properly, the thyristor will last indefinitely. After its introduction, the current and voltage ratings increased rapidly. Today it has substantially higher power capability than any other solid-state device and dominates power conversion in the medium and higher power ranges. The major drawback of the thyristor is that it cannot be turned off by a gate signal, but the anode current must be interrupted in order for it to regain the blocking state. The inconvenience of having to commutate the thyristor in its anode circuit at a rather high energy level has encouraged the development of other related devices as power switches.

DEVELOPMENT OF POWER-SWITCHING DEVICES

Transistors predate thyristors, but their use as high-power switches was relatively restricted (compared with thyristors) until the ratings reached 50 A and 1,000 V in the same device, since the early 1980s. These devices are three-layer semiconductors that exhibit linear behavior but are used only in saturation or cut off. In order to reduce the base drive requirements, most transistors used in variable speed drives are Darlington types. Even so, they have higher conduction losses and greater drive power requirements than thyristors. Nevertheless, because they can be turned on or off quickly via base signals, they are attractive candidates for drives within the scope of the transistor ratings, particularly for pulse-width-modulated inverters.

Gate-turn-off Thyristors. More recently, successful attempts to modify thyristors to permit them to be turned off by a gate signal have been made. These devices are four-layer types and are called *gate-turn-off thyristors*, or simply GTOs. Power GTOs have been around since at least 1965, but only relatively recently (1981) have devices rated more than a few tens of amps become available. Present GTOs have about the same forward drop as a Darlington transistor (twice that of a conventional thyristor). GTOs require a much more powerful gate drive, particularly for turn-off, but the lack of external commutation circuit requirements makes them desirable for inverter use. GTOs are available at higher voltage and current ratings than power transistors. Unlike transistors, once a GTO has been turned on or off with a gate pulse, it is not necessary to continue the gate signal because of the internal positive feedback mechanism inherent in four-layer devices.

Technological Base. The three aforementioned devices (thyristor, transistor, and GTO) form the technological base on which the solid-state variable speed drive industry rests today. There are other devices in various stages of development that may or may not become significant depending on their cost and availability in large current ($>$ 50 A and high voltage 1,000 V) ratings. These include: (1) the metal oxide semiconductor field effect transistor (MOSFET), (2) the insulated gate transistor, and (3) the static induction thyristor.

It has not yet been possible to construct *power MOSFETs* that have acceptably low ON resis-

[1]Engineering Manager, AC Drives, Robicon Corporation (A Barber-Colman Company), Pittsburgh, Pennsylvania.

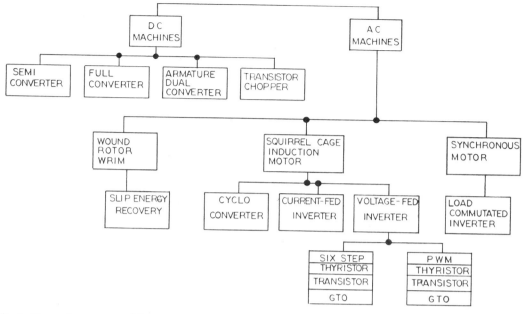

Fig. 1. General-purpose solid-state variable-speed drives. WRIM = Wound rotor induction motor; GTO = Gate-turn-off thyristors; PWM = Pulse-width modulation.

tance while still having the 1000 V rating necessary for reliable power conversion at the 500 VAC level. Therefore, their use has been limited to small drives (under 10 HP). Power MOSFETs are the fastest power switching devices (100 ns) of the lot, and they also have very high gate impedance, thus greatly reducing the cost of drive circuits.

The *insulated gate transistor* is a combination of a power bipolar transistor and a MOSFET that combines the best properties of both devices. A most attractive feature is the very high input impedance that permits them to be driven directly from lower power logic sources. Unfortunately, their ratings are not very impressive, being limited to a few tens of amperes and about 600 volts.

Static induction thyristors (SITs) are claimed to have the voltage and current ratings of GTOs, but with a much higher gate impedance so as to reduce driver requirements. The validity of these claims has not been proved in commercial use because (SITs) are just emerging from the development laboratory.

VARIABLE SPEED DRIVE HARDWARE DEVELOPMENT

Parallel to the development of power switching devices, there have been very significant advances in hardware for controlling variable speed drives. These controls are a mixture of analog and digital signal processing.

The advent of integrated circuit operational amplifiers and integrated circuit logic families made possible dramatic reductions in the size and cost of the drive control, while permitting more sophisticated and complex control algorithms without a reliability penalty. These developments occurred during the 1965–1975 period. Further consolidation of the control circuits occurred after that as large-scale integrated circuits (LSI) became available. In fact, the pulse-width modulation (PWM) control technique was not practical until the appearance of LSI circuits because of the immense amount of combinational logic required. The most significant trend in the early- and mid-1980s has been the introduction of *microprocessors* into drive control circuits. While it is doubtful whether microprocessors will significantly reduce control circuit cost, there is general agreement that they are greatly expanding the capability of drive controls. The performance enhancements include: (1) More elaborate and detailed *diagnostics* owing to the ability to store data relating to drive internal variables, such as current, speed, firing angle, and so on; (2) the ability to *communicate both ways* with user's central computers about drive status; (3) the ability to make *drive tuning adjustments* via keypads with parameters such as loop gains, ramp rates, current limit stored in memory rather than pot settings; (4) *self-tuning* drive controls; and (5) more adept techniques to overcome *power circuit nonlinearities*. The possibilities are large and are just beginning to become commercial realities.

DC DRIVES

The introduction of the thyristor had the most immediate impact in the DC drive area. Ward-Leonard (motor-generator) variable speed drives were quickly supplanted by thyristor DC drives of the type shown in Fig. 2 for reasons of lower cost, higher efficiency, and lower maintenance cost. This type of power circuit with three thyristors and three diodes in a three-phase ridge is generally referred to as a *semiconverter*. By phase control of the thyristors, it behaves as a programmable voltage source. Therefore, speed variation is obtained by adjustment of the armature voltage of the DC machine. Because the phase control is fast and precise, critical features like current limit are easily obtained. In fact, almost all thyristor DC drives today are configured as current regulators with a speed or voltage outer loop. The semiconverter is suitable for one-quadrant drives, as it produces only one direction of current and output voltage. Input power factor is dependent on speed.

Six-Thyristor Full Converter. As the cost differential between thyristors and diodes narrowed, the semiconverter was largely displaced by the six-thyristor full converter, as shown in Fig. 3. This circuit arrangement (the Graetz circuit) has become the workhorse of the electrical variable speed drive industry, as will be pointed out. The control techniques are very much the same as in the semiconverter. However, the full converter offers lower output ripple and the ability to regenerate or return energy to the AC line. The system can be made into a four-quadrant drive by the addition of a bi-directional field controller. Torque direction is determined by field current direction. Owing to the large field inductance, torque reversals are fairly slow (100-500 ms) but adequate for many applications.

Dual Armature Converter. For the best response of thyristor DC drives, the dual armature converter of Fig. 4 is preferred. This is simply two converters (as shown in Fig. 3) connected back to back. Torque direction is determined by the direction of armature current, and since this is a low inductance circuit, reversal can be accomplished in 10 ms (typically). Obviously, only one converter is conducting at one time with the other group of six thyristors not being gated. This is called "*bank selection.*"

Summary of Thyristor DC Drives. The three types of thyristor DC drives just described all share a common property in that the devices are turned off by the natural polarity reversal of the input line. This is called *natural* or *line commutation*. Thus, the inability to turn off a thyristor from the gate is no practical drawback in these circuits. Consequently, they are simple and

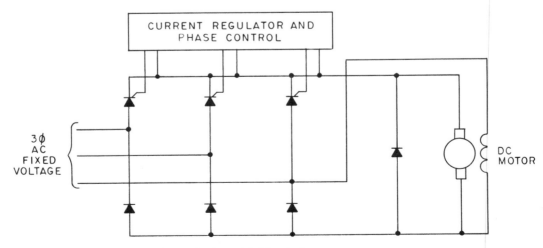

Fig. 2. Thyristor DC drive-3-phase semiconverter.

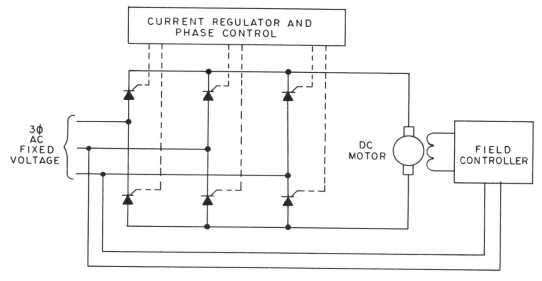

Fig. 3. Thyristor DC drive-3-phase full converter.

very efficient (typically 98.5%) because the device forward drop is small compared with the operating voltage. These drives can be manufactured to match a DC machine of any voltage (commonly 500 V) or horsepower (½ to 2500 HP, typically).

For certain types of applications, typically machine tool axis drives and tape transport drives, the response of phase-controlled thyristor drives is not fast enough. A special class of DC drives has been developed. (See Fig. 5.) A fixed DC bus is developed from the line via a rectifier and capacitor filter. This voltage source is applied to the armature through power transistors. The voltage is modulated by duty cycle (or pulse width) control. The devices usually operate in the 1–5 kHz range. These specialty drives usually operate from 120 to 240 VAC and rarely exceed 10 HP. Frequently, they are applied with permanent magnet field DC machines.

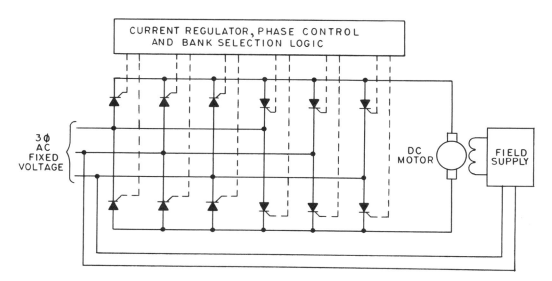

Fig. 4. Thyristor DC drive-armature dual converter.

AC DRIVES

The impact of the new solid-state switching devices was even larger on the AC variable speed drives, but it occurred somewhat later in time as compared with DC drives. AC drives are machine specific and more complex than DC drives because of the simplicity of the AC machine. Solid-state variable speed drives have been developed and marketed for wound-rotor induction motors (WRIMs), synchronous motors, and cage-type induction motors.

Historically, WRIM-based variable speed drives were commonly in use long before solid-state electronics. These drives operate on the principle of deliberately creating high-slip conditions in the machine and then disposing of the large rotor power that results. This is done by varying the resistance seen by the rotor windings.

Slip-Energy Recovery System. A more modern WRIM drive is shown in Fig. 6. This is called the *slip-energy recovery system* or *static Kramer drive*. The output of the rotor is rectified, and this DC voltage is coupled to the line via a thyristor converter. The line commutated converter is current regulated, which effectively controls torque. Efficient, stepless speed control results. Very large ($> 1,000$ HP) drives can be built, as the stator may be wound for medium voltage, while the rotor operates at 400–500 V maximum. The power conversion equipment may be downsized if a narrow range(100% to 70%) of control is adequate, for example, in fan drives. The performance drawbacks are a poor system power factor if not corrected; and no above-synchronous speed operation.

The WRIM is the most expensive AC machine. This has made WRIM-based variable speed drives noncompetitive as compared with cage induction motor (IM) drives or load commutated inverters using synchronous machines. It appears that the WRIM will become a casualty of the tremendous progress in AC variable speed drives as applied to cage induction motors.

Load Commutated Inverter. As shown in Fig. 7, the load commutated inverter is based on a synchronous machine. It uses two thyristor bridges, one on the line side and one on the machine side. All devices are naturally commutated, because the back EMF of the machine commutates the load side converter. This requires the machine to operate with a leading power factor, and, therefore, it requires substantially more field excitation and a special exciter compared with a normally applied synchronous motor. This also results in a reduction in the torque for a given current. The machine side devices are fired in exact synchronism with the rotation of the machine, so as to maintain constant torque angle and constant commutation margin. This is done either by rotor position feedback or by phase control circuits driven by the machine terminal voltage. The line side converter is regulated to control torque. A choke is used between converters to smooth the link current. Load commutated inverters (LCIs) came into commercial use about 1980 and are used mainly on very large drives (1000–10,000 HP). At these power level, multiple series devices are employed (typically 4 at 4 kV input), and conversion takes place directly at 2.4 or 4 kV or higher. The efficiency is excellent, and reliability has been very good. Although they are capable of regeneration, LCIs are rarely used in four-quadrant applications because of the difficulty in commutating at very low speeds where the machine voltage is negligible. Operation above line frequency is straightforward.

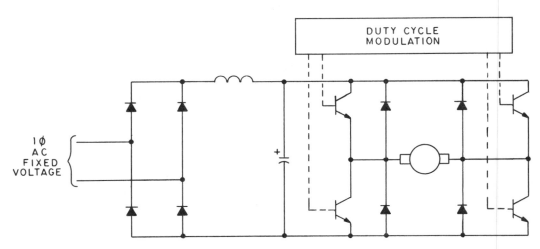

Fig. 5. Chopper DC drive-transistor bridge type.

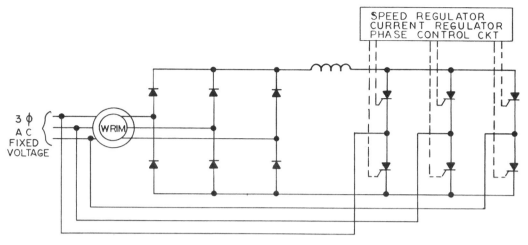

Fig. 6. Slip energy recovery system-wound rotor inductor motor (WRIM) drive. CKT = circuit.

Induction Motor Variable Speed Drives

Induction motor variable speed drives have the greatest diversity of power curcuits. (See Fig. 1.) Because the squirrel cage induction motor is the least expensive, least complex, and most rugged electric machine, great effort has gone into drive development to exploit the machine's superior qualities. Owing to its very simplicity, it is the least amenable to variable speed operation. Since it has only one electrical input port, the drive must control flux and torque simultaneously through this single input. As there is no access to the rotor, the power dissipation there raises its temperature; so very low-slip operation is essential.

Cycloconverter. One approach in an IM drive is to "synthesize" an AC voltage waveform from sections of the input voltage. This can be done with three dual converters, and the circuit is called a *cycloconverter*. (See Fig. 8.) The output voltage is rich in harmonics but of sufficient quality for IM drives as long as the output frequency does not exceed $\frac{1}{3}$ to $\frac{1}{2}$ of the input. The thyristors are line commutated, but there are 36 of them. (Sometimes half-wave circuits are used that need only 18 devices, but more harmonics result.) The cycloconverter is capable of heavy overloads and four-quadrant operation, but it has a limited output frequency and poor input power factor. For special low-speed high horsepower ($>$ 1000) applications, such as cement-kiln drives, the cycloconverter has been used.

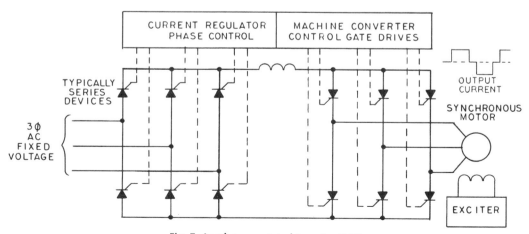

Fig. 7. Load-commutated inverter (LCI).

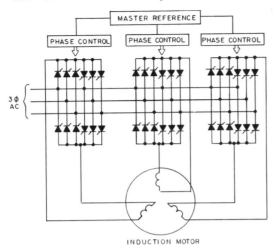

Fig. 8. Cycloconverter induction motor drive.

Autosequentially Commutated Current-Fed Inverter. Still another approach to an IM drive is to generate a smooth DC current and feed that into different parts of windings of the machine so as to create a discretely rotating magnetomotive force (MMF). This type of inverter is called the *autosequentially commutated current-fed inverter* (ASCI). (See Fig. 9.) This circuit was invented later than other inverters and is much more popular in Europe and Japan than in the United States. Once again, the input stage is a three-phase thyristor bridge, which is current regulated. A link choke smoothes the current going to the output stage. There a thyristor bridge distributes the current into the motor windings with the same switching function as the input bridge, except at variable frequency. (Notice the similarity to the LCI.) The current waveform is a quasi-square wave whose frequency is set by the output switching rate and whose amplitude is controlled by the current regulator. The capacitors and rectifiers are used to store energy to commutate the thyristors, since the induction motor cannot provide this energy and remain magnetized, in contrast to the synchronous motor. This type of drive has simplicity, good efficiency (95%), excellent reliability, and four-quadrant operation up to about 120 Hz. Harmonics in the output current are reasonably low, giving a form factor of 1.05 (same as the LCI).

Moreover, harmonic currents are not machine dependent and decrease at light load. The input power factor is load and speed dependent, but much better than that of the cycloconverter. Above 100 HP, the ASCIs are very cost effective. Because they are constructed with SCRs, they have recently (1984) become available at 2.4 and 4 kV direct conversion for very large (> 1000 HP) units. It is almost always possible to retrofit an existing motor with this type of drive. Because of the controlled current properties, this drive is virtually immune to damage from ground faults, load shorts, and commutation failures.

Since MMF (current) is directly controlled and the drive is regenerative, ASCIs can readily be equipped with field oriented controls for the most demanding four-quadrant operation.

Field-Oriented Controls. Special mention should be made of AC induction motor drives with field-oriented controls. They are the state of the art as of the mid-1980s. The control technique keeps track of the flux and MMF vectors inside the machine in order to provide a fast and precise torque response to an external reference. In addition, they are four-quadrant drives, capable of producing either direction of torque in either

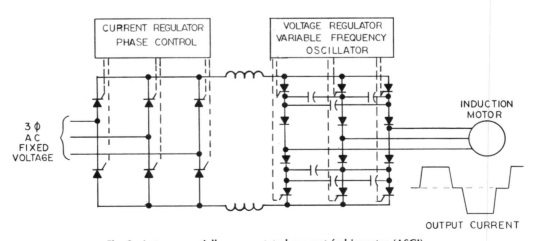

Fig. 9. Autosequentially-commutated current-fed inverter (ASCI).

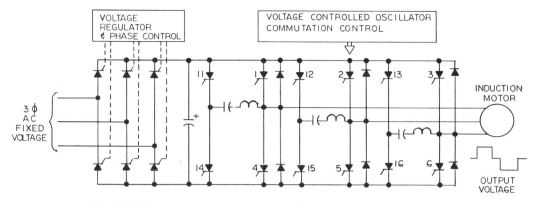

Fig. 10. Voltage-source inverter: six-step thyristor, impulse commutated.

direction of rotation. They are the AC drives of choice for the most demanding applications, such as traction drives and machine tool axis drives.

Voltage-Source Inverter. The third approach to IM drives is to generate a smooth DC voltage and apply that to different combinations of the machine windings so as to create a rotating flux. This circuit is called a *voltage-source inverter* (VSI). An implementation using thyristors is shown in Fig. 10. This circuit was the first application of thyristors to IM drives. The input is a three-phase thyristor bridge that feeds a capacitor filter bank forming a controlled low-impedance voltage source. The output stage consists of six main thyristors (1-6) in a bridge with antiparallel diodes. There are six auxiliary or commutating thyristors (11-16), which together with the L-C circuits impulse commutate the main devices. The output waveform is a quasi-square wave of voltage whose amplitude is set by the DC link voltage. Here the output frequency is determined by the output switching rate, and the output voltage is set by the voltage regulator on the input converter. In order to reduce the size of the commutating L-C, special thyristors with fast turnoff times are required—in contrast to the ordinary phase control types used in DC drives, LCIs, and ASCIs. Despite the complexity, these drives have had a reasonably good reliability record. They have good efficiency (typically 95% at full speed, full load), speed-dependent power factor, and can operate at very high frequencies 180 Hz and up). Regeneration to the line is not possible. Many of these units are in service, and they are still available from 50 to 500 HP, typically at 460 VAC. They are not available at over 600 V.

Transistors and GTOs in Voltage-Source Inverter. In order to reduce the cost and complexity of the VSI shown in Fig. 10, the thyris-

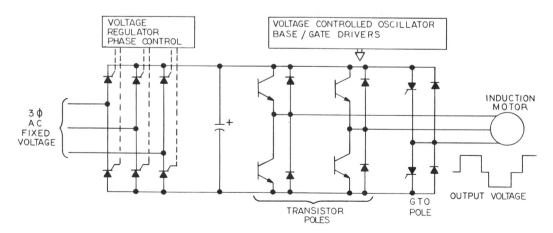

Fig. 11. Voltage-source inverter: six-step transistor or gate-turn-off thyristor (GTO).

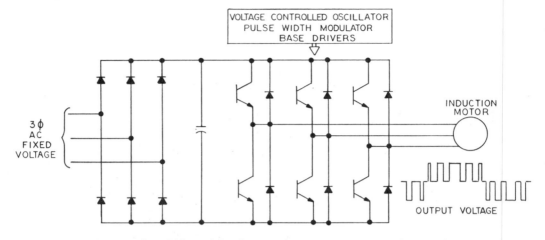

Fig. 12. Pulse-width modulated (PWM) inverter - transistor implementation.

tors and their commutation circuits have been replaced with transistors or with GTOs. The resulting circuit is shown in Fig. 11. The performance features are about the same as those of the thyristor version, but size and weight are substantially reduced. Although the conduction losses are higher because of the higher "on" voltages, commutation losses are reduced substantially—so efficiencies remain in the 95% range. The transistor version has been available since about 1982 at 460 VAC input; 230-volt units have been available since the mid-1970s. Currently, 100 HP transistor drives with single-output devices are available; up to 300 HP with parallel output devices can be obtained.

Since GTOs have somewhat higher ratings, drives from 50 to 500 HP at 460 VAC are now available. It is difficult to forecast which device will be more successful, but transistors are much more widely used up to 100 HP than GTOs. There will be a three-way competition in the future among GTO, transistor, and thyristor

Fig. 13. AC variable-frequency drives ranging from 5 to 1,000 HP. (*Robicon Corporation.*)

Solid-State Variable Speed Drives 385

Fig. 14. Representative 800 HP, 480V AC variable-frequency drive. (*Robicon Corporation.*)

Fig. 15. Representative 2,5000 HP, 2,300V AC variable-frequency drive. (*Robicon Corp.*)

drives in the 100–500 HP range to capture the bulk of the market. GTO and transistor costs will have to be substantially reduced to challenge thyristor designs at the upper end of this range.

Pulse-Width Modulated (PWM) Drives. The induction motor drives discussed thus far are all similar in that the amplitude of the output is controlled by the input converter. Another category of voltage source inverters controls both the frequency and amplitude by the output switches alone. A representative circuit based on transistors is shown in Fig. 12. Note that the input converter is replaced with a diode bridge so that the DC link operates at a fixed, unregulated voltage. The diode front end gives virtually unity power factor, independent of load and speed. This type of drive is called *pulse-width-modulated* (PWM).

An output voltage waveform is synthesized from constant amplitude, variable-width pulses at a high (1-3 kHz) frequency so that a sinusoidal output is simulated; the lower harmonics (5,7,11,13,17,19 . . .) in six-step waveforms are not present in sophisticated PWMs. One advantage is smooth torque, low harmonic currents, and no cogging. Some PWM designs merely encode the six-step square wave, but this results in having the worst features of both PWM and six-step designs. Although this approach eliminates the phase control requirements and cuts the front end losses somewhat, there are offsetting drawbacks. Since every switching causes an energy loss in the output devices and their suppressors, the total losses at high speed go up considerably over six-step (six switches per cycle) if the same devices are used. To overcome this, many versions revert to six-step at 60 Hz. The output devices are stressed much more severely than in six-step. Many PWM designs do not have a voltage regulator; at any given output frequency, they deliver a preset fraction of input voltage. If the input fluctuates, so does the output. Finally, the high-frequency switching may cause objectionable acoustic noises in the motor. There are both transistor and GTO PWM units on the market today in the range of 1–3000 HP at 460 VAC. The transistor versions have a better reliability record. As with all voltage source inverters, regeneration to the line is not inherent.

Note: Views of representative industrial solid-state variable speed drives are given in Figs. 13, 14, and 15.

REFERENCES

Bedford, B. D., and R. G. Hoft: "Principles of Inverter Circuits," Wiley, New York, 1964.
Bose, B. K.: "Adjustable Speed AC Drive Systems," Wiley, New York, 1981.
Brichant, F.: "Force-Commutated Inverters," Macmillan, New York, 1984.
Ghandi, S. K.: "Semiconductor Power Devices," Wiley, New York, 1977.
Kosow, I. L.: "Control of Electric Machines," Prentice-Hall, Englewood Cliffs, New Jersey, 1973.
Pelly, B. R.: "Thyristor Phase-Controlled Converters and Cycloconverters," Wiley, New York, 1971.
Schaefer, J.: "Rectifier Circuits: Theory and Design," Wiley, New York, 1965.
Scoles, G. J.: "Handbook of Rectifier Circuits," Wiley, New York, 1980.
Sen, P. C.: "Thyristor DC Drives," Wiley, New York, 1981.

Materials Motion/Handling Systems[1]

Aside from a few very exceptional cases, which may date back a century or more, factory automation on a rather grand scale was not taken seriously until the post–World War II era. During the 1940s and 1950s, the major steps toward automation were largely confined to various materials-handling situations. Conveying and transferring materials from one workstation to another, or in the case of the transfer machine per se, where several operations were integrated in one machine (an early version of the cell concept), were particularly suited to relay logic. Consequently, well before the appearance of the programmable controller in the very late 1960s, progress in automation tended to be measured by progress in materials handling. In fact, during those earlier days of automation, a vice-president of Ford Motor Company (D. S. Harder) in 1947 defined automation as "the automatic handling of workpieces into, between, and out of machines."

Thus, it is no surprise that the manufacturers of materials-movement systems led rather than lagged the development of other facets of automation, namely, the appearance of electronic instrumentation and control that in more recent years have introduced a high degree of automaticity to materials-handling operations. This observation applies mainly to the metalworking, mechanical assembly, and other discrete-piece product industries—because in the continuous process industries where the materials are largely in the form of bulk solids and fluids a high level of automation has been in place over many years.

While admitting of numerous engineering design refinements, most fundamental materials-conveying equipments, such as belt and roller conveyors, forklifts, and the like, have not changed dramatically—with the exception of drastically improved ways to communicate electronically with other production equipment and with centralized scheduling and programming centers. With emphasis on flexible manufacturing systems (FMS), materials-handling equipment manufacturers have learned how to approach handling problems in much more imaginative and innovative ways, but at the same time focusing on long-term cost effectiveness and the general practicality of new concepts. Conveyor manufacturers, for example, have been aware of the importance of manufacturing flexibility for many years, and hence the modular approach to conveying is not all that recent. Even the much heralded AGVS (automatic guided vehicle system) of today actually dates back well over two decades.

To employ a much overused word, today we have "smart" materials-handling systems that frequently incorporate their own microprocessors and other electronic methodologies for receiving instructions from remote control centers and for communicating real-time data back to such centers. Today, where practical and economical, materials-handling systems can be designed to operate with a minimum of manual intervention. Design objectives include not only some reduction in worker power but conservation of production time and space as well. For a comparative assessment of materials handling equipment, see Table 1.

OVERHEAD MOVEMENT OF MATERIALS

Power and Free Conveyor Systems. In this type of conveying system, the load is carried on a trolley or multiple trolleys that are conveyor propelled through part of the system and that may be gravity or manually propelled through another part. This allows for switching the free trolleys into and out of adjacent lines. The spur or subsidiary lines may or may not be powered. The multiple trolley carrier is used in heavy duty applications that call for high-weight capacities or in applications where the carrier or load bar requires a unique configuration to suit a specific material/product to be conveyed. Representative applications are shown in Figs. 1 and 2.

In one system[2], each front trolley is equipped with a retractable dog that is raised or lowered

---

[1] Robots may be considered a special form of materials-handling equipment. They are described in *Handbook*, Section 2.

[2] *Dog-Magic*R, Power and Free Conveyor System. (*Jervis B. Webb Company*.)

TABLE 1. Conveyors versus Applications.

TYPE OF CONVEYOR	Transportation	Manual Assembly	Automated Assembly	Processing	Accumulation	Sorting	Warehousing	Shipping Terminal	In-process Storage	Dispatching	Order Picking	Storage Handling	Live Storage	Horizontal Flexibility	Vertical Flexibility
Powered travel monorail	○													○	
Self-powered monotractor	●	○	●	●	●				⊕	⊕	⊕		○	○	⊕
Self-powered monotractor (Inverted)	●	○	●	●	●		○		⊕	●	⊕		⊕	○	⊕
Overhead conveyor	●	●	⊕	●		⊕			⊕	●			○	●	●
Power and free conveyor	○	●	●	●	●	○	○		●	●	⊕		●	●	●
Power and free conveyor (Inverted)	○	●	●	●	●				●	○	⊕		●	●	●
Overhead tow conveyor	●	⊕		⊕	⊕				○	●	○		⊕	●	⊕
Tow conveyor	●	○		⊕	○		●	●	○		○		⊕	●	
Roller conveyor	⊕				○	⊕	●	⊕	○	⊕	⊕		○	⊕	
Live roller conveyor	○		○	⊕	⊕	⊕	○	⊕	●	⊕	○		⊕	○	
Belt conveyor	●	○		⊕	●										
Two-strand chain conveyor	○	○	○	○											
Fixture conveyor	●	●	●	⊕											
Slat conveyor	●	●	○	⊕			○	●		●	○				
Sorter conveyor						●	●	●		●					
Storage and retrieval machine											●	●		●	●
Roller flight conveyor	⊕				●								○		⊕
Driverless vehicle	●		○	⊕			●	⊕	⊕	●	○			○	●

● Best qualified ○ Well qualified ⊕ Qualified for certain applications

Basic information source: Jervis B. Webb Company.

Fig. 1. In this view of a power and free conveyor system, axles are routed to workstations for sub-assembly. This particular system is a multiple-trolley conveyor with special load bar configuration. Trolley, chain, and channel flexibility permit the conveyor to negotiate sharp curves. A workstation control panel allows remote, automatic coding of carriers. Built-in shock absorbers aid in carrier positioning in heavy-duty application. (*Dog Magic®* power and free conveyor system, *Jervis B. Webb Company*.)

Fig. 2. In this system, custom-designed carriers move heavy coils of wire to and from storage areas. (*Dog Magic®* power and free conveyor system, *Jervis B. Webb Company*.)

390 Materials Motion/Handling Systems

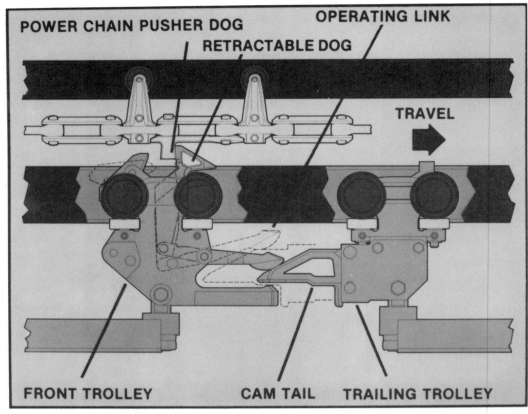

Fig. 3. Details of retractable dog and cam tail. (*Dog Magic®* power and free conveyor system, *Jervis B. Webb Company*.)

by an operating link. The trailing trolley of the preceding carrier has a cam tail behind it. (See Fig. 3.) This cam tail raises the operating link of a succeeding trolley to automatically disengage its dog from the power chain. As shown in the diagram, the front trolley has come up over the same tail of the trolley in front of it. This automatically forces the operating link up and the retractable dog down. When this happens, the power chain continues to move, but the carrier accumulates on the cam tail of the preceding carrier. In the same manner, the following carriers disengage from the power chain and accumulate behind the carriers ahead of them.

When the first carrier is released, its retractable dog rises and is contacted by the pusher dog on the power chain. As this carrier moves out, its cam tail rides clear of the operating link of the trolley behind it and allows the second carrier to be engaged by the power chain. This cycle is repeated in sequence, which allows every carrier to move up and maintain a "keep full" condition on the line. A cross section of power and free tracks and the trolley profile is given in Fig. 4.

This system is normally furnished in three standard sizes—3, 4, and 6-inch (approx. 76, 102, and 152 mm) *free tracks*. The 3-inch free-track systems are designed for medium-duty applica-

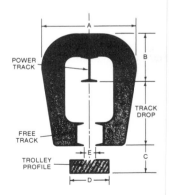

Fig. 4. Cross section of trolley assembly with important dimensions below. (*Dog Magic®* power and free conveyor system, *Jervis B. Webb Company*.)

tions. The trolley will accumulate on 15-inch (381 mm) radius turn with minimum actuating cam extension on the rear trolley. Minimum accumulation length for a single trolley is 12 inches (305 mm)—maximum with actuating cam extended is 18 inches (457 mm). The 4-inch free-track systems are designed for heavy-duty applications, and the foregoing dimensions are somewhat larger. The 6-inch free-track systems are designed for extreme service. The trolley will accumulate on a 24-inch (610 mm) radius turn with minimum actuating cam extension on the rear trolley. Minimum accumulation length for a single trolley is $18\frac{3}{8}$-inches (467 mm). Maximum with rear actuating cam extended is 22 inches (559 mm).

The range of applications and loads carried by systems of this type is wide. Among the advantages claimed for this system are: (1) *Buffer banks*—where workstations must be kept supplied with product, a patented accumulating feature allows the carrier to queue at the workstation without interrupting the flow of other carriers in the system, (2) *production rate variations*—different processes requiring different production rates can be handled because the system allows surges between processes to ensure an even throughput, (3) *on-line storage/segregation*—where there is a need to hold work-in-process or in temporary storage, carriers can be switched from the main line to recirculating loops or to dedicated spurs for batched storage, (4) *load center variations*—when different spacing between carriers is required, it is possible to switch the carriers to another power chain to maintain predetermined spacing, (5) *load speed variations*—different load speeds required at different areas of the plant can be handled in much the same manner as load center variations, where different speeds are serviced by different power chains, (6) *elevation changes*—the track cross section allows the conveyor to negotiate inclines, declines, and curves as a specific application may require, (7) *bias banking*—long loads can be banked on a bias when reduced track length is desired, and (8) *carrier indexing*—positive load control and varying chain speeds allow the intermittent indexing of carriers to accommodate different workstation requirements.

Furthermore, the system is designed to interface with programmable controllers, microcomputers, and host computers. Control interfacing is assisted by the positive mechanical control of the carriers on lines of travel, through switches, lift sections, and intra-conveyor transfers.

Underhung Cranes and Monorail Systems. Generally, as piece weights and sizes increase, there is less inclination to fully exploit the characteristics of automation. Although such systems can be adapted for nearly full automation and control, because of the size, weight, value, and inertia, there is built-in provision for rapid manual intervention and for finalizing load position in many of these systems. A single-bridge motor-driven crane that serves a variety of needs can be designed for multiple runways and interlocking applications[3]. (See Fig. 5.) Double-bridge cranes can achieve higher hook lifts in a given building height and allow greater load-carrying capacity. Their design features are similar to those of single bridge cranes. A double-bridge crane system operating at an aerospace facility is shown in Fig. 6.

Monorail conveyors are an excellent choice for a number of semi-automated plant applications. (See Fig. 7.) Basically, there are two methods of rail suspension—rigid and flexible. Each method has its merits when properly applied. Shown in Fig. 8 is a flexible suspension assembled. The adjustable beam clamp and track hanger are joined by an alloy steel intermediate hanger rod that has spherical headed nuts at both ends. The beam clamp and rail hanger each, in turn, have spherical washers as part of their assembly. When all components are assembled to the structure, the monorail or crane runways are free to move within the confinements of braces that are attached to the track and supporting structure.

For satisfactory operation of trolleys on a monorail or crane system, it is necessary to provide for freedom of certain motions, restriction of other motions, and elimination of still other damaging movements. Uneven tilt or side-to-side swaying of the load bar can cause excessive wear on trolley wheels and track. In one design[4], a "trolley rocker principle" is employed.

The cylindrical trolley rocker supports the load bar or end truck and rides in a mating rocker seat in the trolley yoke. The load bar can now swing or tilt up to 12° to compensate for off-balance loads or off-center loading. (See Fig. 9.) Trolley and rail life are increased, with savings in maintenance costs and downtime. Trolley assembly is shown in Fig. 10.

Interlocks. By allowing transfer of hoist carriers between adjacent crane runways or monorail spurs, interlocking systems can maximize floor coverage while reducing equipment expenditure. Interlocks also eliminate unnecessary duplicate handling. Cross-connected, double-locking pins assist in assuring that the safety stops will not operate until crane and connecting track are in proper alignment. Bronze end flares on conductor bars facilitate the passage of collector shoes through the gaps at interlocking points without losing electrical continuity. (See Fig. 11.) Interlocks can be provided for manual or power operation. (See also Fig. 12.)

Monorail systems are powered by motor-driven tractors that are available with a wide variety of speed and control options. A spring-loaded drive

---
[3] *Spanmaster*[R], a division of *Jervis B. Webb Company*, Farmington Hills, Michigan.
[4] *Spanmaster*[R], a division of *Jervis B. Webb Company*.

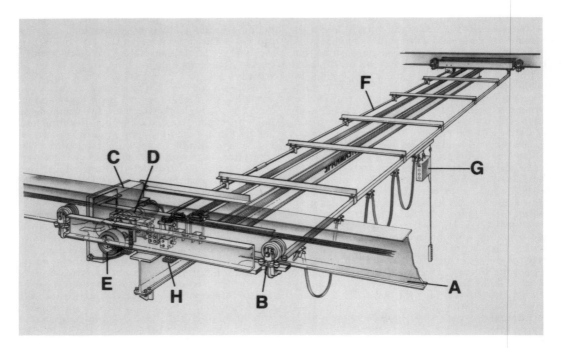

Fig. 5. Motor-driven crane: (A) Track, used for crane runways and bridge girders, is a composite section which consists of three basic pieces: (1) The lower tee is a specially rolled high carbon steel with a raised flat tread to resist peeling which, in turn, provides greater wear resistance and increased tensile strength when compared with structural I-beams. (2) the web and (3) the top flange plates are A-36 steel, computer designed to maximize the efficiency of materials used and reduce the overall weight. The three pieces are joined together continuously by an automatic welding process. (B) The end trucks are rugged structural weldments accurately squared and mounted to the crane bridge at the factory. Safety lugs are provided to limit the drop of the end truck in event of trolley failure. Special trolley features allow free movement of the crane even though runways may be slightly out of alignment. (C) The standard control package includes: (1) Fused disconnect switch, (2) magnetic mainline disconnect switch, (3) fused branch circuit protection in all three phases, (4) thermal overloads in all three phases for motor running overcurrent protection, and (5) fused control circuit on the secondary side of the transformer. (D) Drive assembly—standard single speed drive motors are rated for crane duty with a high slip rating. Gears operate in an oil bath. The triple reduction spur gears are encased in a high-strength aluminum casting to minimize internal corrosion. (E) Drive wheels are rubber or polyurethane and are mounted from each end truck providing positive traction through compression of springs to minimize skewing. (F) Drive line shaft—on single-motor cranes, the crossbridge line shaft is directly keyed to the hollow output shaft of the gear reducers and to the drive wheels. It is supported by adjustable, self-aligning intermediate bearings which are pre-lubricated and sealed. (G) Traveling pushbutton control—the control pendant terminates in a traveling junction box. A flat conductor cable joins the junction box to the main control panel. The cable and pendant are supported from steel wheel trolley assemblies with sealed bearings. The trolleys move freely in a heavy 12 gage formed steel track. (H) Runway and crossbridge electrification—conductor bars are electrogalvanized steel enclosed in a protective covering. The three-faced metal guideway insures positive tracking of collector shoes. The shoe is a composition of sintered copper and graphite which is self-lubricating and, in turn, is mounted in a collector head which is free to articulate on its support. The total assembly is spring-loaded to help insure positive electrical contact even when negotiating gaps at interlocking transfer points. (*Spanmaster®*, a division of *Jervis B. Webb Company*.)

Fig. 6. Double bridge crane system operating at an aerospace facility is equipped with radio controls to provide greater operator safety while handling heavy, bulky loads. Cranes of this type are available with many special control options. (*Spanmaster*®, a division of Jervis B. Webb Company.)

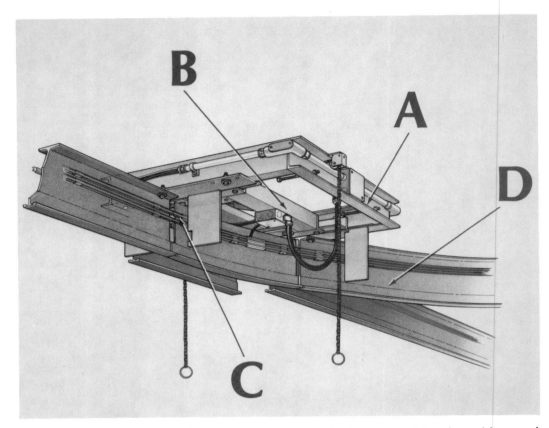

Fig. 7. Monorail conveyor: (A) *Switch outer frame*—mechanical safety baffles are an integral part of the rugged track switch. They are designed to protect against trolleys leaving the track on the switch in the event the switch is being moved from one position to another. Also, there are baffles arranged to block incoming tracks in the event the switch is set against the flow. Switches used in electrified monorail systems include a wiring harness to carry power from the incoming tracks to the conductor bars on the track switch itself. (B) *Switch inner frame*—the inner frame of the switch, which is the moving portion of the switch, is so designed that it is transported from one position to another on four rollers. When this portion of the switch is in its desired position, the inner frame drops slightly so that the weight is removed from the rollers and, at the same time, locking pins on both sides of the switch firmly hold the inner frame in place. The track switch is designed for manual, electric, or pneumatic means for shifting position. (C) *Switch*—on electrified switches, the conductor bars are formed and mounted at the factory. Bronze flared ends on each conductor bar are matched with conductor bars on incoming tracks. The incoming track conductor bars also have the same flared ends. The purpose of the flare is to allow smooth transition of the collector shoe when it passes through the gap at the switching points. (D) *Monorail curves*—these are rolled for maximum accuracy. Support points on the curve are designed for the specific load carried by the system. On electrified systems, conductor bars are roll formed and factory mounted to the curves. (*Spanmaster®*, a division of *Jervis B. Webb Company*.)

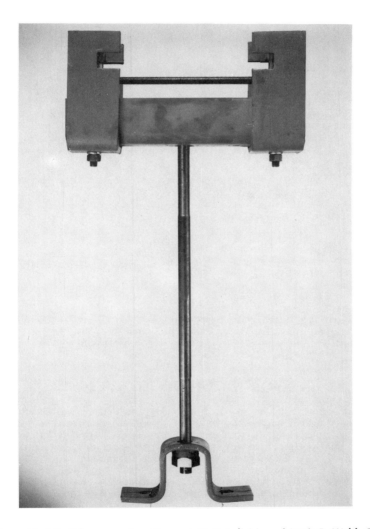

Fig. 8. Assembled flexible suspension. (*Spanmaster*®, a division of *Jervis B. Webb Company*.)

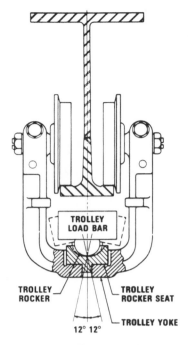

Fig. 9. Trolley design with provision for compensating off-balance loads or off-center loadings up to a swing or tilt of twelve degrees. (*Spanmaster®*, a division of Jervis B. Webb Company.)

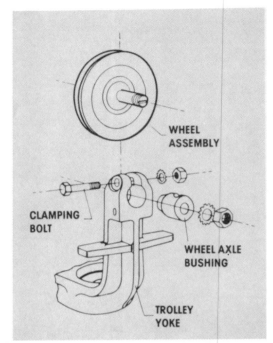

Fig. 10. Trolley assembly for monorail conveyor. (*Spanmaster®*, a division of Jervis B. Webb Company.)

wheel is keyed to the reducer output shaft, providing a positive tractive force. The gear train operates in an enclosed oil bath.

A dual-rail semi-automatic monorail system designed for transferring parts from a floor to an overhead conveyor is shown in Fig. 13.

FLOOR-MOUNTED CONVEYORS

The flexibility of the overhead power and free conveyor systems previously described has, within the last several years, been brought to the factory floor in the form of floor-mounted configurations. These conveyors have become one of the most common types used in automated and semi-automated systems. Of the track-type conveyors, they are distinguished mainly by the type of motion-generating (drive system) used. For many years, the continuous-chain drive was considered standard. More recently introduced is an entirely different form of conveyor drive, notably the "track and drive tube," which does not depend on chains[5].

Track and Drive Tube Conveyors. Decks, cars, and carriers of many configurations ride on two parallel tracks. Some of the carrier styles are shown in Fig. 14. It will be noted from these diagrams that midway between the tracks there is a spinning tube. All energy for movement of the carriers is derived from this *drive tube*. Each carrier is equipped with a *drive wheel* mounted on its underside, which is capable of moving between 0° and 45°. This drive wheel is positioned against the spinning drive tube. When the drive wheel is in the 0° position, its centerline is parallel with that of the drive tube and the wheel merely rotates in place, causing the carrier to remain stationary. As the angle between the drive wheel centerline and the drive tube centerline is increased, a thrust is developed and the carrier accelerates forward. (See Fig. 15.) Carrier speed is decreased by decreasing this angle and therefore reducing the developed thrust. Modular track sections may be straight or curved to make radius turns. Turntables are used to make turns up to 180°. Track sections can be combined to follow almost any path. Cars can be stopped at predetermined stations, diverted by turntables, and moved

[5]*Cartrac®*, *SI Handling Systems, Inc.*, Easton, Pennsylvania.

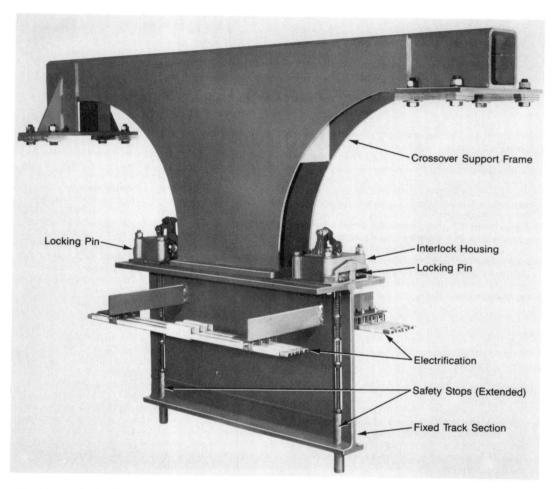

Fig. 11. Interlocks allow transfer of hoist carriers between adjacent crane runways or monorail spurs. (*Spanmaster®*, a division of *Jervis B. Webb Company*.)

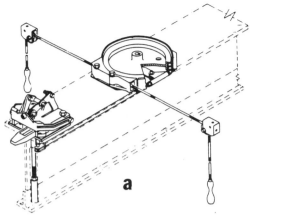 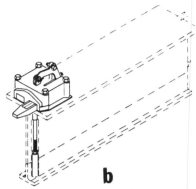

Fig. 12. (a) Interlock operator; (b) connecting interlock. As the crane approaches the interlocking point, the main locking pin is retained within the interlock body casting by a hinged keeper which covers the pin opening. As the crane comes into register with the spur track, the keeper is raised by means of a roller riding up a cam surface on the connecting interlock, clearing the locking pin opening in the interlock. When the crane has stopped, thus properly aligned, the interlock can be actuated by means of the operating handle. This causes the main locking pin to advance out of the interlock housing and into the connecting interlock housing. In so moving, the locking pin actuates the cross-connecting lever which causes the secondary locking pin in the connecting interlock to move in the opposite direction and engage a matching locking pin opening in the interlock. Simultaneously, the main locking pin acts upon levers in the connecting interlock and the secondary locking pin acts upon levers in the interlock which causes the safety stops to raise to the withdrawn or up position thus clearing the crane bridge and the spur track for trolley travel. It will be noted from the foregoing description that the safety stops are not fully withdrawn until both locking pins have completed their travel and the two units are aligned and locked together by the two heavy locking pins. (*Spanmaster®*, a division of *Jervis B. Webb Company*.) Unlatching the interlocks is the reverse of the foregoing sequence.

to different levels by vertical lifts. Control systems range from manual to highly sophisticated computer direction.

Principal characteristics of the track and drive tube system include:

1. Controlled acceleration and deceleration. Unstable loads can be moved at high speeds.
2. Speeds can be varied from 0.5 to 600 feet (0.15 to 183 meters) per minute or more. Speeds are variable. Different speeds can be used in separate sections of a system simultaneously.
3. Loads weighing in excess of 15,000 pounds (6804 kilograms) have been transported safely.
4. Vertical movement is possible for connecting multiple levels in a building. Forward and reverse movement plus right-angle and radius turns can be accomplished. A system can be mounted on or beneath the floor or hung from a wall or ceiling.
5. Cars can be automatically positioned to ± 50 thousandths of an inch (1.3 mm) for machining or processing without risk to the load. In one application, cars have been positioned to ± 5 thousandths of an inch (0.13 mm) in three planes.
6. Movement of cars is nonsynchronous, permitting independent indexing of cars.
7. Accumulation of loads is automatic—without need for support equipment.
8. Hazardous environments are tolerated (−40° to +200°F; −40° to 93°C). The systems have been installed in spray booths, drying ovens, and freezers. The flexibility of system geometry is illustrated by the diagrams of Fig. 16 and of applications in Fig. 17. The integration of several subsystems into one large, extensive system is shown in Fig. 18.

For handling loads up to 200 pounds (91 kilograms) a light-duty version of the track and drive tube system just described is available[6]. This system consists of three basic components—a track, a spinning shaft, and the carriers. The drive shaft in this case is located on one side of the track and revolves at a constant speed. Drive wheels are connected by a pivoting lever arm to the underside of each carrier. Contact between the drive wheels and the revolving drive shaft causes the carrier to move forward. The operating principles are similar to those of the heavy-

[6] *Mini-Cartrac*[R], *SI Handling Systems, Inc.*

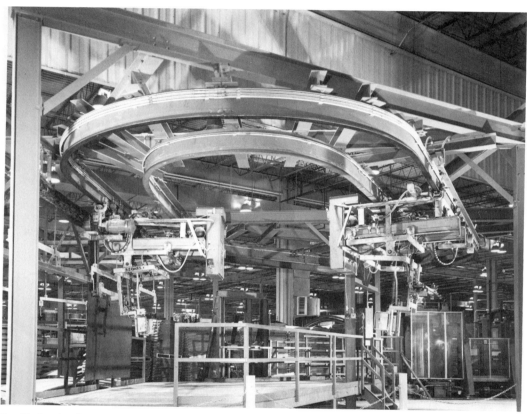

Fig. 13. Dual-rail semi-automatic monorail system designed for transferring parts from a floor to an overhead conveyor. (*Spanmaster®*, a division of *Jervis B. Webb Company*.)

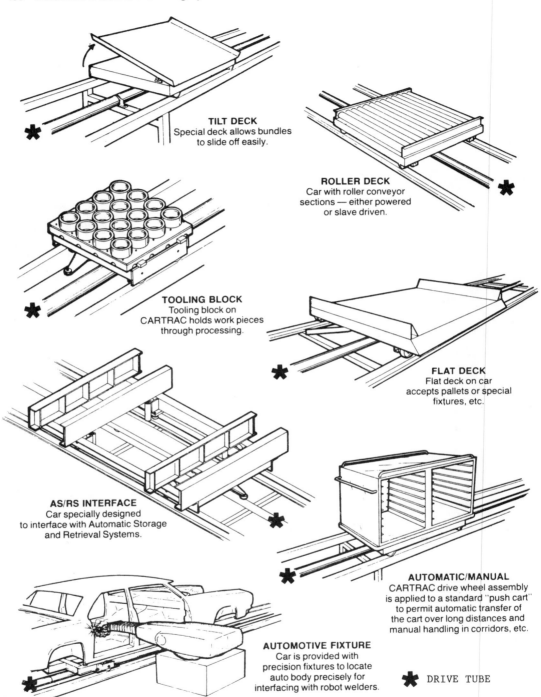

Fig. 14. Track and drive tube conveyors. Decks, cars, and carriers of many configurations ride on two parallel tracks. (*Cartrac*®, SI Handling Systems, Inc.)

Materials Motion/Handling Systems 401

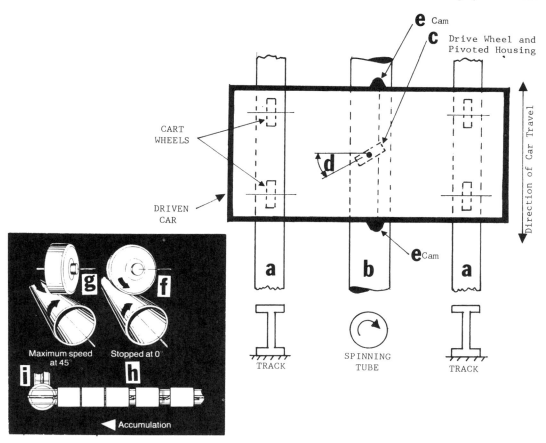

Fig. 15. Individual cars on the *Cartrac*® system travel on a simple two-rail track (a) with a spinning tube (b) mounted between the rails. The forward motion of each car is achieved through a centrally located drive wheel (c). This wheel is spring-loaded against the drive tube. The tube rotation makes the drive wheel move forward. A pivoting housing for the drive wheel permits the angle of the drive wheel against the drive tube to be varied between 0° and 45°. This angle (d) provides the ability to vary a car's speed independently of tube speed or other car speeds. When car movement is not required on a portion of the system, power to the drive tube can be stopped to conserve energy.

When the angle between the drive wheel and the drive tube is 0° there is no forward motion of the car because the drive wheel is merely spun (f) by the drive tube. By creating a small angle between the tube and the wheel, a thrust (g) develops to push the car forward at a slow speed. Increasing the angle creates more thrust and more speed until a maximum at the 45° position is reached.

Specially designed control bars along the track 'control' the pivoting housing and quickly and smoothly decelerate the cars to a precise stop by gradually reducing the angle from maximum back to 0°. The release of the control bar creates the same effect in reverse providing smooth and predictable acceleration.

A cam follower arm (e) at the front of each car, linked to the drive wheel housing and a similar cam at the rear of the car, also control the drive wheel angle and permit the accumulation of cars with controlled acceleration and deceleration. Accumulation (h) is accomplished with zero pressure between cars.

Modular track sections may be straight, or curved to make radius turns. Turntables are used to make turns up to 180°. Track sections can be combined to follow almost any path.

Cars can be stopped at predetermined stations, diverted by turntables, and moved to different levels by vertical lifts. Control systems may range from manual to highly sophisticated computer direction. (*SI Handling Systems, Inc.*)

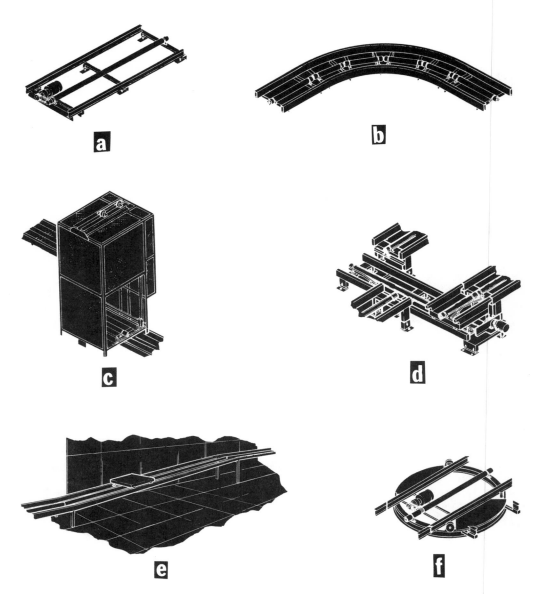

Fig. 16. Geometric flexibility of the track and drive tube conveyor system: (a) I-beams typically form the track, providing support and guidance for the car. The spinning tube moves the car, using power transmitted by the motor. Several sections of track and tube can be powered by a single drive motor. (b) Radius turn—a curved track section changes the direction of a car. (c) Vertical lift—moves car smoothly and accurately between two or more levels. (d) Transfer—moves a car horizontally between parallel or intersecting tracks. A rotary transfer rotates the car while moving it horizontally. (e) Ramp—enables cars to change elevation without a vertical lift. (f) Turntable—rotating turntable connects the intersection of two or more tracks, changing the car's direction of travel. (*Cartrac®, SI Handling System, Inc.*)

Materials Motion/Handling Systems 403

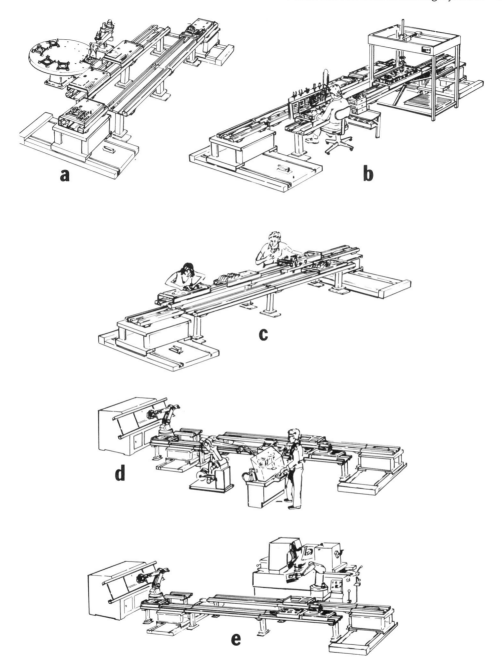

Fig. 17. Flexibility of application of the *Cartrac®* system: (a) Robotic Assembly—parts and fixture delivered to robot assembly station on same carrier and then transferred to turntable for further processing. (b) Manual/Robot Workstation—manual and robotic operations combined in same work cell. (c) Manual Assembly—progressive assembly with queuing between stations. (d) Manual Machine/Transport Interface—delivery of workpiece to human operator for machine loading. This can be arranged for manual loading initially with future upgrading to robots. (e) Machine Tool Interface—delivery of workpiece to computer numerical control (CNC) machine tool with workpiece queuing. (*Cartrac®, SI Handling Systems, Inc.*)

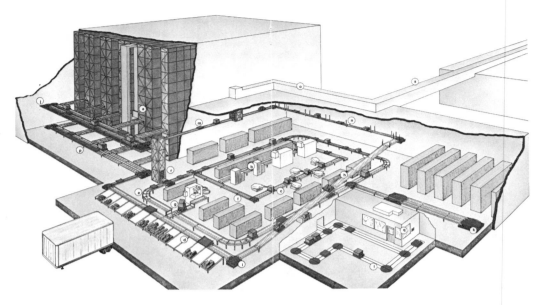

Fig. 18. Unification of multiple functions in a plant by a *Cartrac*® conveying system: (a) Automated storage and retrieval system, (b) machining line, (c) robot welding line, (d) sorting, (e) radius turn, (f) transfer, (g) rotary transfer, (h) ramp, (i) single turntable, (j) paired turntable, (k) double turntable, (l) in-floor installation, (m) wall mounted, (n) ceiling mounted, (o) roof supported, (p) accumulation area, (q) central control area, (r) lift, (s) enclosed bridge. (*Cartrac*®, *SI Handling Systems, Inc.*)

duty system. The mini system is especially useful in hard automation, machining, robotics, and manual or automatic assembly. Industries using systems of this type include appliances, electronics manufacturers, small tool manufacturers, small motor and engine assembly and test, and automotive parts producers. Components of the system are of a modular design, permitting easy installation. The system has been designed for rapid expansion or rearrangement.

Inverted Power and Free Conveyor Systems.

Reference is made to prior description of power and free conveyor systems. The inverted version is essentially an upside-down configuration of that system. The inverted systems differ quite considerably from conventional floor conveyor systems because of their greater flexibility and efficiency. Inverted power and free conveyor systems satisfy a wide range of applications, including automotive trim shops, part shops, engine assembly operations, among others. Like overhead power and free conveyors, the inverted conveyors provide positive load control over every carrier in the system. Loads can be stopped or accumulated without disrupting the flow or sequence of materials. Like conventional floor conveyors, the inverted conveyors are mounted on the floor to provide greater access to materials being transported. The load is supported on a pedestal-type carrier for complete access. This offers convenience to workers, enabling greater in-station work time.

The inverted systems eliminate the need for the automation equipment normally associated with traditional floor conveyors. Elevators and lowerators are eliminated and replaced with standard vertical curves to elevate materials for clearance or storage of the product. Ninety and 180-degree lift, transfer, and turntable devices that are normally associated with skid systems are replaced by standard horizontal turns. Side transfer conveyors, powered indexing devices, and auxiliary indexing and pusher mechanisms required by chain-on-edge systems are eliminated by standard switches and the inherent accumulation ability of inverted power and free systems. Variable chain speeds, production rate conveyors, recirculation of products, balancing of operations, and surge and stripping capabilities are all available with the inverted power and free systems.

The integration of linear induction motors offers high-speed indexing in controlled acceleration and deceleration modes. High-speed indexing is important for interfacing with production cycles of automation equipment, such as robots. To accommodate a wide variety of loads, one

Materials Motion/Handling Systems 405

Fig. 19. Inverted power and free conveyors: (a) The 6-inch (152 mm) conveyor is specifically designed for heavy-duty load capacities. Rotary floating drives are furnished as standard equipment, with linear induction motors available for high-speed indexing requirements. A retractable dog assembly is a standard feature for extreme service conditions. Typical applications are found in the automotive and foundry industries; and anywhere where heavy loads must be transported from workstation to workstation. (b) A 4-inch (102 mm) conveyor designed for medium- to heavyduty load capacities. Used in a wide variety of industries. (c) A 3-inch (76 mm) conveyor designed for light- to medium-duty load capacities. Serves a wide variety of industrial applications, including welding and assembly. (d) A 3-inch (76 mm) conveyor designed mainly for light-duty load capacities, but can be upgraded for heavier loads. The conveyor features a standard caterpillar drive, but is well suited to the integration of linear induction motors for high-speed indexing applications. Serves a wide variety of industrial applications, including assembly operations for televisions, appliances, and business machines. (*Jervis B. Webb Company*—item (d) is a *Unibilt*® Inverted P&F.)

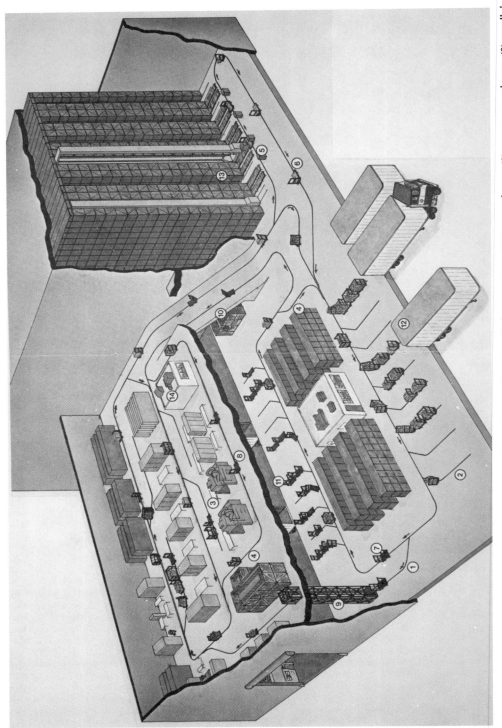

Fig. 20. Installation showing how *Low-Tow*® system facilitates the handling of multiple operations: (1) Powered spur, (2) non-powered spur, (3) parallel spur, (4) transfer, (5) automatic cart unloader, (6) roller deck cart, (7) cage cart, (8) steel deck cart, (9) lift, (10) ramp, (11) receiving dock, (12) shipping dock, (13) automated storage/retrieval system, and (14) computer control room. (*Lo-Tow*® System, *SI Handling Systems, Inc.*)

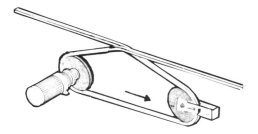

Fig. 21. Patented drive system used for towline conveyor. The vertical sprocket-drive assembly has 180° engagement of the conveyor chain and an adjustable slack take-up for positive chain control. (*Lo-Tow®*, *SI Handling Systems, Inc.*)

manufacturer[7] offers inverted power and free conveyors in 6-inch, 4-inch, and 3-inch (152, 102, and 76 mm) designs. This wide range of load accommodations is illustrated in Fig. 19.

Towlines: A towline is a simple loop of track recessed approximately three inches (76 mm) in the floor with a powered chain, which moves carts or carriers from pickup points to assigned destinations. Towlines can be controlled in numerous ways, ranging from simple manual means to highly sophisticated computer electronic techniques. Low-profile towline systems have been widely used for years as a simple, dependable, and quite versatile method of moving materials and are found in manufacturing facilities, transportation terminals, and automatic storage/retrieval systems (ASRS), among other uses. Towlines can be installed in existing or new buildings and have ranged from 135 to 18,000 feet (41 to 5486 meters) in length.

[7]*Inverted Power & Free Conveyors*, Jervis B. Webb Company.

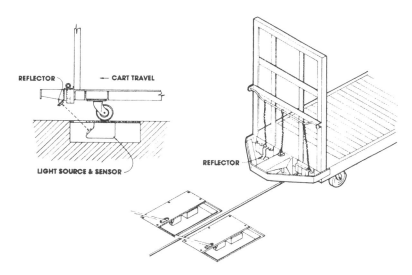

Fig. 22. Towline conveyor cart switching means that do not require probes, reed switches, or mechanical bumps. In diagram at left, the system is optical, involving above-floor optical scanners along the towline path. The scanners read coded labels which are attached to the cart, eliminating need to set codes for each cart. The labels indicate a cart's designation or ID number. When the cart passes a scanner, a signal is relayed to the control system which routes the cart and operates the civerter at the correct spur. In diagram at right, the system comprises a photoelectric reader (*Lightning Switch®*) mounted below the floor level and a code bar mounted on each cart. The reader contains an amplifier and one or more scanners. The code bar has movable reflectors for setting the codes. (*Lo-Tow®*, SI Handling Systems, Inc.)

Key factors in selecting a towline versus other conveyances, such as fork truck, tractor train, overhead trolley, automatic guided vehicles, etc., include the volume to be handled in pieces, units, and pallets and the distances over which the material must be moved. In recent years, the manner of interfacing with automatic equipment has been very important. Towlines do interface with more costly types of handling equipment, such as high-rise storage systems, automatic sorters, and automated production equipment.

A panoramic view of a towline with spurs to facilitate loading and unloading is shown in Fig. 20.

In one system[8], the track consists of a $2\frac{1}{4}$-inch (57 mm) wide by 3-inch (76 mm) high rigid channel with a $\frac{3}{16}$-inch (5 mm) thick side plate at the bottom and $\frac{3}{4}$-inch (19 mm) deep wear resistant steel top bars. The design and compactness facilitate installation, allow ample gliding surface for the automatically lubricated chain, withstand wear on towpins and top plates and permit foot and wheeled traffic to cross the conveyor.

The drive for the system is a vertical sprocket-drive assembly featuring 180-degree engagement of

[8]*Lo-Tow*[R] *System, SI Handling Systems, Inc.*

Fig. 23. An automatic guided towing vehicle with a rolling load capacity of 30,000 pounds (13,608 kilograms). Other towing vehicles in this series have capacities up to 50,000 pounds (22,680 kilograms) and down to 15,000 pounds (6,804 kilograms). Vehicles of this type are guided and controlled by low-frequency signals transmitted by in-floor wiring. Principal features include: (1) Built-in diagnostic capabilities, (2) plug-in control circuitry is of modular design, utilizing diagnostic capabilities with LED indicators. Special-function printed circuit boards can be added without altering standard board circuitry, (3) removable access panels, (4) automatic guidance with automatic gain control with either twin or single front wheel steering and low-inertia servomotor, (5) rear wheel drive capability for greater traction and load capacity, particularly on grades, (6) automatic warning devices—an amber strobe light indicates when vehicle is operating automatically. Horn sounds on all automatic starts and stops. In-motion sounding devices and turn signals are optional, (7) front-end emergency bumper—halts drive and sets emergency disk brake upon contact with obstructions, (8) dynamic braking accomplished by means of the traction motor, (9) manual controls and indicators. Vehicles are equipped with dual industrial-duty Start/Stop and Emergency Stop pushbuttons with automatic Ready and On light indicators. There are also a key-operated power switch and pushbutton horn (10). (*Prontow*®, *Control Engineering Company, an affiliate of Jervis B. Webb Company.*)

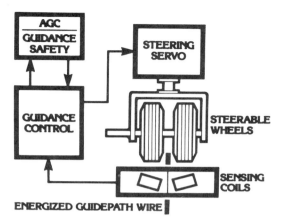

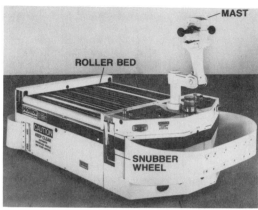

Fig. 24. Schematic diagram of proportional vehicle steering system used in automatic guided vehicle of the type shown in Fig. 23. The vehicle is guided by low-frequency signals transmitted by in-floor wiring. Right- and left-side sensing coils detect the guidewire signal, and voltages from the coils are then amplified, rectified and compared. Differences in detected signal strength cause the vehicle's low-inertia servomotor to steer right or left in proportion to the error level, thus centering the coils and steerable wheels over the guidewire. (*Prontow®, Control Engineering Company*, an affiliate of *Jervis B. Webb Company*.)

Fig. 26. Automatic guided vehicle for the automatic transfer of unit loads. Capacity of unit shown is 4,400 pounds (1,996 kilograms). Vehicles for lighter loads—1,000 and 500 pounds (454 and 227 kilograms) are also available. Designed for power-to-power friction drive or lift/lower transfer of unit loads. (*Prontow®* unit load vehicle, *Control Engineering Company*, an affiliate of *Jervis B. Webb Company*.)

Fig. 25. A custom designed automatic guided vehicle for handling rolls of products moving from production to storage. Loads up to 8,000 pounds (3,629 kilograms) can be handled. Vehicle is equipped with forward/reverse steering and is computer directed. (*Prontow®, Control Engineering Company*, an affiliate of *Jervis B. Webb Company*.)

the conveyor chain with adjustable slack take-up for positive chain control. (See Fig. 21.) Electrical controls quickly respond to system fault signals and allow fast reset after the system has been stopped for any reason. An in-line reducer increases drive efficiency by as much as 20%. Maintenance is minimal because the drive has few moving parts and all are accessible. The drive has an automatic lubricator, which oils sprockets, chain joints, and sliding surfaces to ensure lubrication of all connecting links.

For turning, an auxiliary roller chain assembly, which operates at one-half conveyor speed, is used. Forces resulting from main conveyor chain tension are transmitted through this roller chain directly to a steel backup bar. The roller turn includes a cleanout box at each end and an adjustable take-up for self-cleaning roller chain.

For switching, mechanical or electrical methods are used. In one system,[9] no electric circuitry is required. Just ahead of spurs, the carts pass over retractable "bumps" along the towline. When the properly set selector pin passes over the bumps, a diverter automatically opens, allowing the cart to enter the spur. Electrical switching is based upon a dual reed circuit actuated by a solenoid when the reed circuit is energized. With this method, computer control of the system is possible. The electrical diverter assembly is operated by a dual reed sensor. The steel diverter blade guides the towpin into the spur and is relatched by the towpin as it passes through the diverter. The sensor allows greater selectivity and offers higher reliability. By

[9] *SI Handling Systems, Inc.*

Fig. 27. An automatically-guided fork-type vehicle which is microprocessor controlled. The vehicle may operate in stand-alone service, or be fully integrated with other automated material handling and storage systems. The vehicle is designed to 'turn on a dime' and transport loads stably over bumps, around corners, and up or down inclines. No operator intervention is required to load, unload, and transport a wide variety of load sizes, weights and configurations. Various loads may be handled by modifying the load adapter, without changing the basic vehicle. The unit features a lifting mast and fork extender mechanism to pick up/deposit loads at pickup and deposit stands, storage racks, or at floor level. Self-diagnostics are incorporated for trouble-shooting and maintenance. (*Kenway GFN-25* Fork-Type Vehicle, *Eaton-Kenway, a subsidiary of Eaton Corporation.*)

Fig. 28. An automatic guided vehicle that receives directions from and reports its status to a central control and communications computer. Three types of transfer decks are available. The *GL* model features a floating lifting top which is fine-positioned by the vehicle against a P & D stand. The vehicle then raises or lowers its top to pick up or deposit the load. The *GC* model features a conveyor top that can interface with a variety of powered, static, or gravity conveyors for input or output of material for processing or storage operations. The *GS* model features a shuttle top that picks up and deposits a load alongside the vehicle without driving under the pick up and deposit stand. The shuttle automatically extends under the load, lifts and transfers it onto the vehicle. Reversing the shuttle motion deposits the load. (*GL, GC,* and *GS Vehicles, Eaton-Kenway, a subsidiary of Eaton Corporation.*)

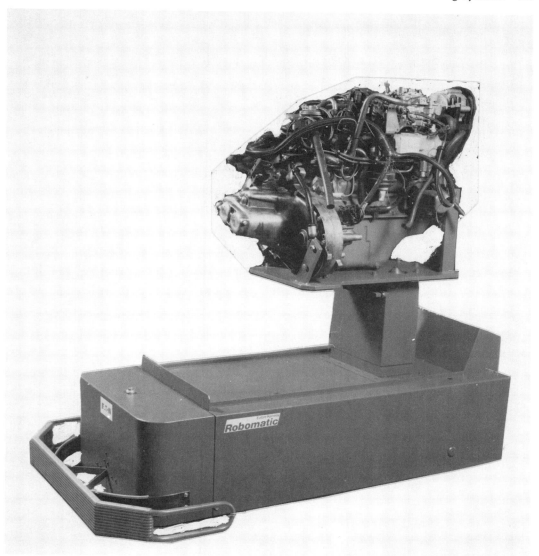

Fig. 29. A wire-guided, computer-controlled mobile workstation. Designed for easy and economical adaptation to existing assembly applications or new facilities. This vehicle takes the work to the operator, who may then elevate the workpiece to an optimum height by pushing a button on the vehicle's control panel. This is a flexible 'line balancing' vehicle for use in lieu of an assembly line. As work content varies from operator to operator, the balancing of work-in-progress assembly is not stopped because of operator tasks. The operator is not compromised by traditional synchronized production line flow. Workstation to workstation travel of the vehicle is controlled via command signals from a supervisory minicomputer through an on-board microprocessor. Command signals transmitted through the in-floor guidepath are received by the vehicle's antennae and converted into stop and destination orders for the drive unit. The vehicle defines its location through code plates in the floor and makes route decisions at those switching points. (*Kenway Robomatic Vehicle, Eaton-Kenway*, a subsidiary of *Eaton Corporation*.)

keeping the selector probes at a greater height from the floor, it minimizes the attraction of ferrous particles.

A full-width reed switch box is provided. Therefore, to change the code of any particular station, one merely removes the top plate on the box, replaces the reed board over the correct post corresponding to the correct number, and closes the top plate. This only requires a few minutes. In another system a photoelectric reader is mounted below the floor surface and a code bar mounted on each cart. The reader contains an amplifier and one or more scanners. The code bar has movable reflectors for setting of codes. (See Fig. 22.) In still another system, also shown in Fig. 22, optical switching is used. Optical scanners read coded labels that are attached to the cart, eliminating the need to set codes for each cart. The labels indicate a cart's destination or identification number. When the cart passes a scanner, a signal is relayed to the control system, which routes the cart and operates the diverter at the correct spur. There are no probes or code-address mounts, reed switches, or mechanical bumps.

AUTOMATIC GUIDED VEHICLES (AGVs)

The automatic guided vehicle is considered by some authorities as the most flexible and programmable of all materials-handling equipment, particularly the AGVs that feature automatic loading and unloading. AGVs provide unusual flexibility to a materials-handling system because they can be computer controlled, either by programmable controllers, by on-board microprocessors, or by a central computer. They can be directed and even redirected while traveling to a workstation.

In simple terms, an AGV is a driverless tractor or cart that has its own motive power aboard (usually a battery) and a steering system that is controlled by signals emanating from a wire under the floor, or optical guidance patterns sensed from reflectors affixed to the surface of the floor. Because of its lack of dependence on manual guidance and intervention, once programmed, the AGV is quite robotlike in nature, and, in fact, some people refer to it as a type of robot.

The early AGVs of twenty years or more ago were essentially powered tractors or carts and were most frequently associated with storage and warehousing situations. The AGVs initially served as more flexible replacements for trolley or floor towline conveyors and manually operated trucks and tractors, such as forklifts. Because AGVs are either computer controlled or computer directed, they have been widely used during the last several years—since the serious introduction of flexible manufacturing systems (FMS). At present, the most flexible style of AGV, and the type most often specified in FMS, is the automatic load/unload design. With the capability to self-load and -unload, these vehicles can transport a pallet or tray of product, deposit it at the appropriate station, proceed to another station, pick up a second pallet, and transport and deliver it to a third station further along in the system, and so on. AGVs have proven their adaptability to numerous handling situations well beyond the simple function of transport. This adaptability seems to be limited only by the ingenuity of the materials-handling engineer. For example, AGVs have been fitted with cartridge-type feeders designed to carry a stack of tote boxes and deliver them to a workstation by way of a magazine-style unloader. In a more recent development and an application that appears destined for very wide acceptance in FMS is the use of AGVs for carrying robotic assemblers. This concept allows the robot and its tooling to travel from station to station for work on the product, while the product per se remains fixed. The critical aspect of this scheme is high-tolerance positioning of the robot relative to the workpiece, a fundamental requirement whenever a material-handling scheme is adapted to robotic assembly.

A representative group of contemporary AGVs is shown in Figs. 23 through 29.

AUTOMATED STORAGE AND RETRIEVAL SYSTEMS (ASRS)

The modern concept of automated and semi-automated warehousing really did not get well under way until the early- and mid-1950s. The efficient storage and retrieval of parts for in-process operations or of final products had to await a number of technological advances and alterations in the basic concepts of warehousing, which had been in place and considered as standard for many years. Among these advances were: (1) high-rise, high-density storage, and (2) alterations in packaging of parts and products. These advances, of course, were markedly augmented with innovative system controls as typified by microcomputers, microprocessors, and communications from the factory floor and warehouse to host computers and vice versa in as complex and detailed hierarchies as management believed were economically viable.

Packaging of parts and products is not always given sufficient emphasis in planning raw material, in-process material, and final product storage and warehousing. Where a firm is planning to automate these operations, very early consideration should be given to packaging because obviously the nature of packaging will influence the type of ASRS system most applicable and vice versa. The traditional ways a product has been marketed and distributed, if changed drastically and rapidly, can have adverse results in the marketplace. This is an example where marketing management must have a strong voice in selecting automation methodologies because production cost savings can be overwhelmed with consequent losses in sales. Fortunately, over the last several years, a reasonable degree of flexibility has been designed into automated handling equipment so that with exception of very large installations, the amount of custom engineering (hence higher costs) required can be kept to a minimum. This, of course, is particularly important to the small- and medium-size manufacturers.

Order Selection System for Cartoned Products

In the case of cartoned merchandise, there are three basic approaches to order selection: (1) *Single-order selection*—where typically orders are selected by warehouse personnel from bins or racks and placed onto cartons or pallets. Even with the aid of mechanized vehicles, this previously traditional manual approach is highly labor intensive and limiting in volume of merchandise and number of orders that can be handled, not to mention the inaccuracies that result from human errors. Cost per order is relatively high, and the system is vulnerable to inflationary labor costs, absenteeism, and union pressures. (2) *Batch picking*—where order selectors are assisted by conveyor systems and other equipment so that one order selector can simultaneously fill several orders. Fewer order selectors are used, and these typically are obliged to work faster to achieve relatively high rates of productivity. Batch picking enables a warehouse to handle a larger volume of merchandise and more orders than with single-order selection. Accuracy still is subject to human error. (3) *Automated order selection,* which requires no selection manpower, significantly reduces total warehouse personnel and offers computer control of all warehouse functions—from receiving through shipping, resulting in better management control and greatly improving productivity and accuracy.

In one system[10], the supplier likens the system to a vending machine that is operated electronically. Instead of responding to coins, the machine is activated by electronic impulses. These impulses correspond to items on a customer's order. Each impulse releases a case from the machine. The released cases drop onto powered take-away conveyors, which deliver them to the shipping area. All product arrives at shipping in exact sequence desired, as in family groupings. The sequence can be maintained until the order is delivered. The system will handle cartons ranging in size from 6 × 6 × 2 to 24 × 30 × 20 inches (approx. 152 × 152 × 51 mm to 610 × 762 × 508 mm) weighing from 1 to 60 pounds (approx. 0.5 to 27 kilograms).

A machine of this type type consists simply of a series of tiered storage gravity flow lanes that form an expandable module. Each lane is inclined to allow cases to advance along controlled-friction runners. A release mechanism at the end of the lane allows cases to escape. Lane widths can be changed easily to fit various product sizes. (See Fig. 30.) The system works in conjunction with a variety of optional methods of backup storage, replenishment, and delivery. These support functions can vary from manual to automated. Typically, the method used relates to current output and future expansion expectations. A typical distribution center using this method is shown in Fig. 31.

The modular design of the system permits expansion as order throughput increases. This "add-on" feature minimizes initial procurement cost, while ensuring a maximum return continuously. (See Fig. 32.) A single module can be the first step toward automation. This module may select from 600 to 1200 individual items and dispense 1500 cases hourly, or more than 10,000 per shift. A dual-module system offers an hourly throughput of 3000 cases initially. A full four-module system is capable of dispensing more than 40,000 cases per shift, etc. as additional modules are provided.

Replenishment of Stock

Replenishment of supply lanes can be accomplished either manually or mechanically. With either approach, replenishment is directed by lists generated by either the system's computer or corporate computer. With manual replenishment, a forklift driver delivers pallets to a replenishment platform behind the designated lane. A warehouse worker restocks the gravity lanes. With mechanized replishment, the forklift driver brings a pallet to a mechanized depalletizing station and places the pallet on an input conveyor. The depalletizer operator sequences the pallet onto a scissors lift table. The scissors lift automatically raises to contact a vacuum head, which lifts one tier at a time. A shuttle table moves beneath the tier, and the tier is released onto the table. Cases then singulate onto a replenishment conveyor leading to the lanes. (See Fig. 33.)

In some cases, a two-worker depalletizing station is used instead of the vacuum head. The procedure is the same except that the scissors lift raises the product between the workers who remove cases onto the delivery conveyor. The work effort is still minimal inasmuch as the cases slide (no lifting required) from the pallet onto the conveyor.

Mechanized lane loaders can be used with either mechanized depalletizing method. Once depalletized, cases are counted electronically and automatically spaced apart and conveyed to a lane-loading vehicle. The spacing paces the operation. The lane loader is positioned at the correct lane, and the operator simply guides the cases into the lane. The vehicle's transfer table as well as the operator's platform raise and lower to eliminate any lifting of cases. (See Fig. 34.)

Cases exiting the ordering system may be consolidated in one of three ways: (1) Cases are placed on carts, which may be wheeled individually into trucks, or they may be formed into "trains" and moved via mechanized vehicles. (2) Cases may be mechanically palletized and removed by fork trucks. (3) Cases may exit onto an extendable conveyor leading directly into trucks.

---

[10]*SI Ordermatic*[R], *SI Handling Systems, Inc.*

414 Materials Motion/Handling Systems

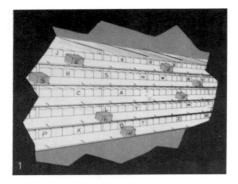
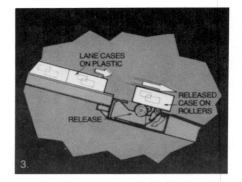
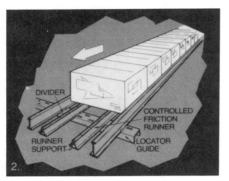
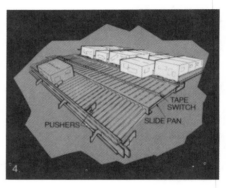

Fig. 30. General operating principles of automatic order-filling system: (1) Cartons of product which may range from groceries and frozen foods to health and beauty aids, general merchandise, electrical parts, paper good, automotive parts, toys, etc. are arranged in lanes (gravity flow) that resemble, in principle, a giant 'vending machine.' (2) general arrangement of individual lane in which cartons move by gravity over a controlled-friction runner. (3) carton release mechanism. (4) cartons exiting from supply lanes onto a collecting conveyor. Cartons reach the conveyor in exact sequence as programmed. (*SI Ordermatic®, SI Handling Systems, Inc.*)

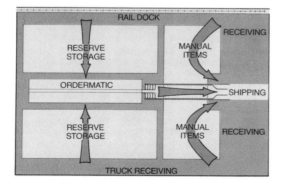

Fig. 31. Use of automatic order-filling system in a typical distribution center. (*SI Ordermatic®, SI Handling Systems, Inc.*)

System Controls

The system controls receive instructions from either a direct link to a computer (on-site or remote), paper tape, punched cards, magnetic tape cassette, or manual input from a simple keyboard. The controls consist of a shift register memory and the warehouse control system. With any of the methods of input, electronic signals release cases from the machine in precise sequence. The controls also confirm that a case has been selected and, if an item is short, relay the shortage to a printer.

The warehouse control system can be either a totally dedicated minicomputer, or it can be supplied off-line from the user's corporate computer. The warehouse control system is a major tool for inventory management and machine replenishment. It assigns slot locations in reserve storage for incoming merchandise. It also provides replenishment schedules at proper intervals to ensure the machine's lanes are always restocked as needed.

Materials Motion/Handling Systems 415

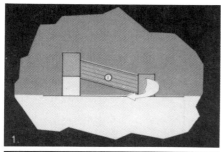

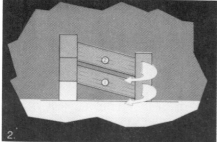

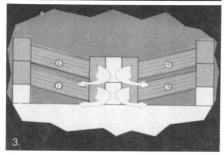

Fig. 32. Modular design of automatic order-filling system permits expansion as throughput increases. Shown are one, two, and four-module systems. (*SI Ordermatic®, SI Handling Systems, Inc.*)

Order Selection System for Non-cartoned Items

For the many storage and retrieval situations where the handling "convenience" of cartoned materials does not apply, systems have been developed to automatically handle less-than-full case quantities of parts and products of various shapes and sizes in whatever quantities that may be needed. One fully automated and self-contained storage and retrieval unit[11] provides functions that range from inventory tracking to item picking and delivery of selected items to the packing station. Each module of the system contains storage shelves divided into lanes and has the capacity to provide multi-lanes. A picking head and carriage module are positioned in front of the storage unit. Electronic impulses from the controller activate the picking head which rapidly

[11] *SI ITEMatic^R, SI Handling Systems, Inc.*

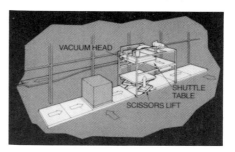

Fig. 33. With mechanized replinishment, the forklift driver brings a pallet to a mechanized depalletizing station and places the pallet on an input conveyor. The depalletizer operator sequences the pallet onto a scissors lift table, which automatically raises to contact a vacuum head which lifts one tier at a time. A shuttle table moves beneath the tier and the tier is released onto the table. Cases then singulate onto a replenishment conveyor leading to the lanes. (*SI Ordermatic®, SI Handling Systems, Inc.*)

moves to the designated product lane. A pusher plate in the lane is activated by the picking head and moves the required number of products from the lane onto a conveyor belt. The selected items then move in a rapid and even flow to the packing station until the order is completed. The picking rate is approximately fifty items per minute. An articulating conveyor carries picked items in an even flow to a second belt for delivery to the packing station. The enclosed modular design of the system contains customized lanes to accommodate a wide range of product weights and sizes. This arrangement reduces opportunities for pilferage. Replenishment is performed by a rig attached to the rear of each unit. Computer-controlled components at the packing station consist of an automatically activated diverter

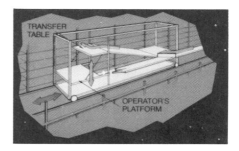

Fig. 34. Once cases are depalletized, they are counted electronically and automatically spaced apart and conveyed to a lane-loading vehicle. The spacing paces the operation. The lane loader is positioned at the correct lane and the operator simply guides the cases into the lane. The vehicle's transfer table as well as the operator's platform raises and lowers to eliminate any lifting of cases which are merely oriented properly and slid into the lane. (*SI Ordermatic®, SI Handling Systems, Inc.*)

paddle for simultaneous handling of two orders, and a split-screen CRT and printer. These components provide the packer with data on orders, pieces picked, and shortages.

The system is available in three standard lengths and capacities: (1) Unit is 10 feet (approx. 3 meters) long and contains two modules; (2) unit is 20 feet (approx. 6 meters) long and contains 4 modules; and (3) unit is 30 feet (approx. 9 meters) long and contains 6 modules. Each *shelf module* within the unit contains vertically adjustable storage shelves with lane dividers, and pusher plate assemblies. Lane widths range from $\frac{3}{4}$-inch (19 mm) to one foot (305 mm) in $\frac{3}{8}$-inch (9.5 mm) increments.

A *carriage module* provides the means for vertical and horizontal movement needed for the picking head to properly align with a designated lane. The *picking head* activates a gear at each product-filled lane for drawing the pusher plates forward and advancing the correct number of items from the lane onto the conveyor belt. Selected items are carried in a continuous flow to the *packing station*.

The *main control panel* provides the link between the selection system and the host computer, and contains the pushbuttons, selector switches, and status lights required for automated picking operations.

Controls. This is a state-of-the-art microprocessor-based system that performs a wide range of functions encompassing selection, replenishment, report generation, inventory control and diagnostics. The control system receives orders in real time from the host computer, optimizing the picking sequence and thus minimizing the order-filling time. Products are then automatically selected and delivered to the packing station.

By direct link from the host computer, the selection control system receives orders and transmits electronic signals to the picking mechanism. Order information, pieces picked, shortages, and other information are instantly displayed on the split-screen CRT at the packing station.

Continuous monitoring of all lane inventories is performed by the control system, which then issues replenishment lists at appropriate inventory levels. These lists are printed at the packing station for timely restocking of the lanes. The control system also can print a host of reports for management and maintenance personnel, providing information on product inventories, lane failures, system utilization, orders, predicted demand and restocking requirements. The control system also provides a printout or CRT display of any detectable failure, such as a drive unit, power unit, lane or timing problem.

Versatility of System

Product dimensions and weights can range from $\frac{1}{2}$-inch (12.7 mm) cube to 12-inch (305 mm) cube and between $\frac{1}{4}$-ounce (7.1 grams) and 10 pounds (4.5 kilograms). Many different packaging configurations can be handled, including glass, plastic, tubes, vials, cans, shrink or blister packs, and cartons. Among items that can be handled are assembly and automotive parts, beverages, calculators, candy, cassettes, cosmetics and drugs, electronic components, glassware, grocery products, hardware, health and beauty aids, jewelry, optical lenses, photo equipment, small appliances, tobacco products, toys and hobby items, and watches.

Large, High-Density, High-Rise AS/RS Configurations

High storage racks, which from a distance look something like a mass of scaffolding, have made high-density storage of a wide variety of materials of reality. Racks that sometimes reach the height of a nine-story building (85 to 90 feet; 26 to 27 meters) above floor level are largely replacing the multi-storied warehouses that required numerous elevators and consequently a lot of horizontal movement to store and retrieve materials on several floors. There is very little wasted vertical space in the rack structures, ranging from just a little over a foot (approx. $\frac{1}{3}$ meter) above floor level to about 15 inches (381 mm) below ceiling level. The vertical height of the rack structure is broken into several levels, depending upon the materials handled. Storage bins may be a variety of heights within the same system. These heights range from $4\frac{1}{2}$ inches (114 mm) up to 18 inches (457 mm) in six increments. The standard bin width is 2 feet (0.6 meter); the standard bin length is 4 feet (1.2 meters). However, in custom-designed systems, bins are designed to accommodate the loads with no strict limitations imposed. Storage rack structures may be 200 feet (61 meters) or more on a side. Generally, the aisles served by ASRS equipment are about 200 feet (or less) in length. The width of an aisle is usually just sufficient to allow the ASRS equipment to move along the aisle freely (a few feet).

Forklifts, which served older warehouse configurations, were limited in their vertical lift. To serve the high-density rack system, an entirely new line of materials-handling equipment was required. Commonly used today are stacker cranes or S/R machines that have very high vertical movements—sufficient to load and unload the top tier of a rack structure. The general concept of the warehouse rack structure is portrayed in Fig. 35, which illustrates the accommodation of five tiers of bins. Three unit load storage/retrieval machines are shown (at the ends of the aisles) in the view, along with AGVs which service the S/R machines.

In a recent food product warehouse[12], the ASRS system occupies 12,000 square feet (1115 square meters) and features a nine-story (85 feet;

---

[12]Frito-Lay snack-foods plant, Casa Grande, Arizona.

Fig. 35. High-density, high-rise warehouse rack structure showing three *Unit Load®* load storage/retrieval machines in the aisles and automatically guided vehicles which service the S/R machines. The setting is a modern bakery warehouse. (*Eaton-Kenway*, a subsidiary of *Eaton Corporation*.)

26 meters) high rack-supported structure with 2300 pallet bays served by seven computer-directed AGVs and three stacker cranes. It is claimed that this warehouse packs five times the capacity of a conventional warehouse into the same floor space. Except for two manual functions (palletizing and truck loading), the entire operation is computer-controlled.

High-vertical materials stackers may be captive; that is, one machine serves one aisle and must be loaded/unloaded at a transfer station at the end of the aisle by AGVs or other suitable conveyances. The other configuration is a storage/retrieval truck that can transfer from aisle to aisle under its own power without the use of a separate transfer vehicle. This arrangement offers the benefits of high-rise storage at a considerable cost savings over captive S/R machines. The storage/retrieval truck can move both horizontally and vertically with load elevated, thus increasing productivity. The S/RT thus bridges the gap between standard industrial trucks and captive, in-aisle high-rise S/R machines. Steps in the operation of a storage/retrieval truck are: (a) Truck approaches the end of the high-rise storage aisle, (b) the operator switches to DC power and drives the truck out of the aisle, (c) assisted by power steering, the operator drives the truck to the next assigned aisle, (d) the operator guides the truck into the aisle, completing the transfer in as little as one minute.

Illustrated in Fig. 36 are several S/R (aisle captive) machines that are designed for real-time computer control, semi-automatic off board control, or even manual operation with the addition of an optional operator's cab. This vehicle[13] can transport loads up to 5000 pounds (2268 kilograms) from a pickup/deposit stand or input/output conveyor, then take the load down a narrow aisle and store it in a high-rise rack. When the system computer or operator requests a load out of storage, the S/R machine automatically goes to that storage location, using an LED addressing system. Within seconds, the load is delivered to assembly or orderpicking. To store loads, the S/R machine picks them up from the end of the aisle input position and stores them in the rack. (See also Fig. 37.)

The operating instructions for storage/retrieval begin with either the computer (in automatic mode) or the operator (in semi-automatic mode). The order is received by an on-board microprocessor, which directly controls the S/R machine in storing or retrieving the load. The microprocessor communicates machine activity to the supervisory computer, which computes inventory records in real time. It will be noted from diagram that this is a single-mast design.

The S/R[14] machine, shown in Fig. 38, is aisle-captive and designed for heavy loads—up to 8000

[13]Kenway Single Mast *Unit Load*R, *Eaton-Kenway*, a subsidiary of *Eaton Corporation*.

[14]Kenway Double Mast *Unit Load*R, *Eaton-Kenway*, a subsidiary of *Eaton Corporation*.

Fig. 36. In background are several single mast *Unit Load*® storage/retrieval machines for operation in an aisle of a high-density, high-rise warehouse. (*Eaton-Kenway*, a subsidiary of *Eaton Corporation*.)

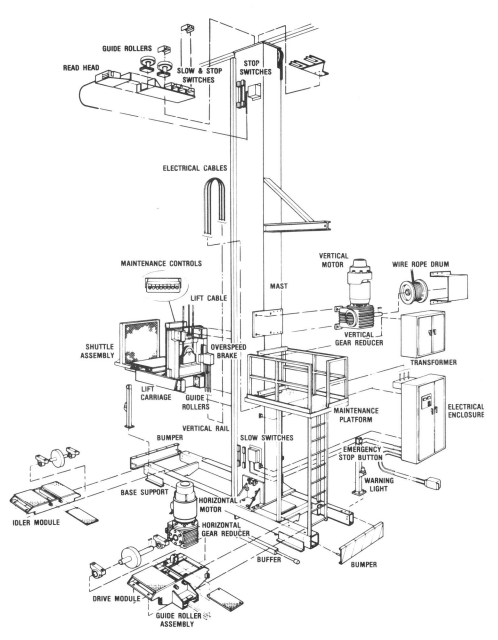

Fig. 37. Details of a single mast *Unit Load*® storage/retrieval machine with a load capacity of 5000 pounds (2268 kilograms) for operation in an aisle of a high-density, high-rise warehouse. (*Eaton-Kenway*, a subsidiary of *Eaton Corporation*.)

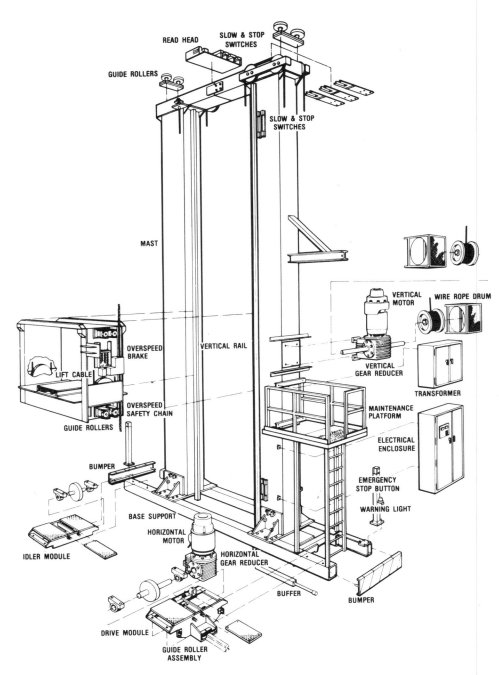

Fig. 38. Double mast *Unit Load*® storage/retrieval machine with a load capacity of 8000 pounds (3629 kilograms) for operation in an aisle of a high-density, high-rise warehouse. (*Eaton Kenway*, a subsidiary of *Eaton Corporation*.)

pounds (3629 kilograms). The double-mast construction provides the strength needed for heavy loads. General operation of the double-mast unit is similar to that of the single-mast format just described. The single rail floor track for both units is standard ASCE rail, 60 pounds/yard (24.9 kg/m). Similarly, the top guide rail and power rail are similar in both designs. The top guide rail is a 4 × 4 inch (102 × 102 mm) structural tube welded to cross-aisle ties in the rack structure. The guide is provided to stabilize the S/R machine during travel and shuttle operations. The power rail is a five-conductor bus duct, which can be mounted at the top of the aisle or at an intermediate level. This rail transmits power and communication signals to the S/R machine.

The KE-8/8 on-board microprocessor for each machine provides direct control of S/R machine motors, safety interlocks, and the logic for machine positioning and sequencing. The controller can interface with virtually any standard peripheral device to display or maintain activity and audit trail inventory status. Another control commands the S/R machine from an end-of-aisle position or remote location.

Options available include a *double reach shuttle*, which may be used to store products in rack bays 8 to 10 feet (2.4 to 3 meters) deep; a *dual shuttle* used to handle long loads, or two loads either independently or simultaneously; and a *maintenance package*, which permits maintenance personnel to troubleshoot the vehicle. The tester connects directly into the on-board microprocessor.

SECTION 5

Interfaces and Communications

Section Contents

User Interfacing to Process Computer Systems	424
Process Computing Evolution	424
Interfaces in Perspective	424
Changes in System Architecture	425
Communications and Standards	427
Graphic Display	428
User Interactions	429
Local Area Networks (LANs)	433
Needs of the Industrial Environment	436
Common Standards	436
IEEE-802.3 CSMA/CD	437
IEEE-802.4 Token Bus	437
PROWAY	438
IEEE-802.5 Token Ring	438
Communication Standards for Automated Systems	440
Need for Communication Standards	440
Early Highways	441
Developments of Late-1970s	441
Communication Systems Models	442
OSI Reference Model—Basis of MAP	443

*Persons who authored complete articles or subsections of articles or who otherwise cooperated in an outstanding manner in furnishing information and helpful counsel to the editorial staff.

CARO, R. H., Autech Data Systems, Inc., Pompano Beach, Florida. (*Development of Communication Standards*)

CONSIDINE, D. M., Editor-in-Chief.

LOYER, B. A., Semiconductor Products Sector, Motorola Inc., Phoenix, Arizona. (*Local Area Networks (LANs)*)

MCCREADY, A. K., Chevon Research Company, Richmond, California. (*User Interfacing to Process Computer Systems*)

User Interfacing to Process Computer Systems

By Arthur K. McCready[1]

This article presents a discussion of the evolution of process computing up to the current distributed architectures. The discussion asserts that the current dominant architecture, combined with recent advances in graphics technology, provides powerful tools for creating highly functional, easy-to-use, human-machine interfaces. Concepts important to the design of process computer-user interfaces and an example of a typical process user interaction, performed via an interactive graphics interface, are also presented.

INTRODUCTORY

Recent advances in computer technology and the adoption of hardware and software standards have made possible the means to provide highly functional interfaces to access the wealth of information maintained in process computing systems. This information should be available not only to process operators but also to a diverse group of users. These interfaces now can be provided to users who need no background in dealing with computers and who require only minimal training to use the system. These functional, easy-to-use interfaces can be constructed by employing *interactive graphic* techniques in a distributed computing environment.

PROCESS COMPUTING EVOLUTION

Both the hardware and software used for process computing have changed significantly in a history that dates back to the 1960s, with many of the important changes occurring since the late 1970s. However, throughout this time span there have been three overriding criteria that affect the design of these systems: (1) All process must be done in a real-time environment, (2) systems have to perform reliably, and (3) systems must be economically justifiable.

In the next few pages, a historical condensation of developments that have led to the state-of-the-art systems is presented.

Interfaces in Perspective

Until relatively recently, process interfaces have tended to avoid state-of-the-art computer technology—because it was generally perceived that the use of traditional, proven technology was required for the reliability needed. Through the 1970s, most process computer systems were centered around a modestly powered minicomputer, configured similarly to a general-purpose computing system, but with an analog input/output (I/O) subsystem added. The principal users of the systems were usually process operators, with occasional use by process and control engineers. The systems were used primarily for process monitoring, calculating variables based on field inputs, secondary alarming, historical data retention, and for advanced control of parts of the process implemented for cost reduction. The prevailing philosophy—the process could always be run without the computer and there was frequently a complete board-mounted analog backup system for the control loops.

Although there were relatively few process computer systems installed worldwide, there was considerable optimism about their potential. There was also much concern about the amount of peoplepower required to develop, install, and use them. A surprising amount of work was published concerning the improvement of the human-machine interface and other capabilities of process control computer systems (Ref. 1).

During the late 1970s, two forces revolutionized the field of process computing: (1) an enormous increase in the cost of energy, and (2) the advent of the microprocessor and inexpensive memory devices.

Energy. Increased energy costs forced older processing plants to closely monitor utilities usage and generation and implement energy-saving control loops in order to stay competitive. New plants often contained intricately coupled processing systems designed to use and

[1]Chevron Research Company, Richmond, California.

reuse any energy put into or generated by the process. These highly interactive process systems can be difficult to run stably during upsets without sophisticated control loops to provide decoupling and feedforward actions. To provide this higher level of monitoring and control capability, newer process computer systems have many more inputs and outputs than the older systems and process the points at more frequent intervals. To effectively handle the increasing amount of information and interactions, it has become imperative that operators be given higher-performance interfaces to the computer system that can direct the user to information of primary importance and to provide rapid interactions during times of upset.

Microprocessors. Introduction of the microprocessor made several actions possible:

1. The introduction of many highly functional, low-cost, computer peripherals. As examples—CRT (cathode ray tube) based video display units (VDUs), many with color and some graphics capabilities. This was a major advance because, for the first time, users could get nearly immediate feedback from interactions, at the point of interaction and at a lower cost (as opposed to board-mounted indicators).

2. Process control and data acquisition systems designed around a distributed microprocessor architecture appeared in the marketplace. The process operator was the only user of these systems. The operator's interface usually consisted of a CRT-based display device, with color and character cell graphics capability, and a function button keyboard. The available displays were often not customizable and frequently emulated the appearance of board-mounted controller and indicators. At first, these systems were used primarily for low-level processing. Examples—signal conditioning, regulatory control, and limited process alarming. Larger facilities usually interfaced these systems to a minicomputer, which often provided data historians, report generation, supervisory control, and custom graphic displays.

3. *Computer Awareness:* Probably the most significant result from introducing microprocessors was the effect on the general public—more and more people who had little knowledge of computers wanted to utilize information they contained but were frustrated because of the difficulties in using the systems. The result was a renewed awareness of the importance of the human being in the success of process system design and implementation. "User-friendly" became the most popular phrase to be used by vendors and systems designers in describing their latest products.

As awareness of the capabilities of computers increased, it became apparent that the business of the processing facility could be handled more efficiently if "other users" were provided with interfaces to the process information system. Besides the process operator, other users who could benefit from the interfaces to the systems included planners, operations management and yield accountants, process and control engineers, systems analysts, applications programmers, laboratory analysts, and instrument and electrical technicians for both the field instrumentation and digital systems. It is important to stress that different users have different needs, backgrounds, and capabilities. Each class of user should be provided with an interface to the system that is designed specifically for that user.

Changes in System Architecture. As a result of the aforementioned forces, the dominant architecture for design or process computer systems changed from the central minicomputer with analog I/O subsystem to an architecture with networks of processors running in parallel, each performing part of the total system function and communicating to the others when necessary to provide them with new information or to control their activities. When looking at one of these newer systems, the first thing an "old timer" in the field notices is that the board of indicators and controllers is gone and has been replaced by a few video display units. Because these few VDUs are the operator's only window to the process and must be used to access all the information in the system, this architecture is sometimes called *shared display architecture*.

The new architecture has its strengths and weaknesses, but with the former outweighing the latter. The *distributed nature* of the architecture allows the system to be easily expanded to very large sizes, yet it can be economically feasible for relatively small installations. Very high performance systems can be constructed by adding more, faster processors to the network and further dividing the workload between the available processors. This is the primary advantage of this architecture. It is most significant when attempting to have a large number of highly interactive user's interfaces on a single system. Indeed, a central minicomputer is simply not capable of supporting a large number of interactive users in a timely fashion while still performing the high-priority real-time tasks of monitoring and control.

Other less significant advantages of this architecture are: (1) Although these systems contain complex hardware, the hardware occupies little physical space, and much of the hardware is identical (only the software/firmware varies), (2) most hardware contains extensive built-in diagnostics, frequently isolating problems to the chip level, and (3) installations can be easily configured to handle a variety of geographical distributions.

In contrast to previous-generation systems, hardware reliability is actually enhanced by using state-of-the-art electronics technology—because parts counts are dramatically reduced and the new products tend to follow recently accepted standards for interfacing and quality assurance.

On the negative side, these new systems are very *software intensive*. Software, especially new software, tends to contain errors. Fixing these errors can be difficult in highly interactive systems, and repair time is less determinant than hardware repairs. Installation and maintenance require different skills than in previous-generation systems. Fewer electricians and hardware technicians are needed; more expertise is required in the systems and software areas. Installation of these systems tends to require more engineering to configure the hardware and the often incredibly large databases. The systems analyst is a very important user of the system. Database definition and maintenance is a significant problem of critical importance. A well-designed analyst's interface can reduce the peoplepower necessary to start up and maintain a system, improving the effectiveness of the entire system.

426 User Interfacing to Process Computer Systems

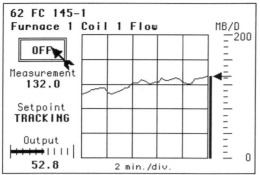

(a) The loop is off and the setpoint is tracking the measured variable. The operator moves the graphics pointer to the command state (loop status) box.

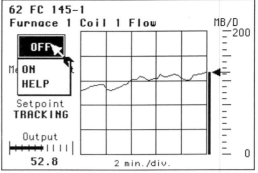

(b) The operator triggers a pick (selection) event by pressing and holding the mouse button. This causes a menu of possible changes to 'pop up' from the box. If the operator releases the button without making a selection or moves the mouse sufficiently far, the choice operation will be terminated and the display state will be as in (a).

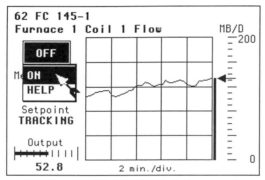

(c) The operator, still holding down the button, moves the mouse until the desired new state is highlighted.

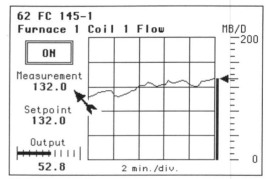

(d) The operator releases the button to put the loop into the desired state. The loop turns on and its setpoint is set to the value of the measured variable, 132 MB/D. The menu disappears and the graphics pointer returns, ready to be used for picking another object. Steps (a) through (d) have taken a little less than two seconds to complete.

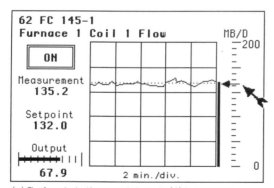

(e) The loop is in the same state as in (d) but time has passed. Note that the setpoint is trended as a dotted line. The operator now wants to set the flow rate to 110 MB/D. To initiate this, the pointer is moved to the setpoint indicator arrow by moving the mouse.

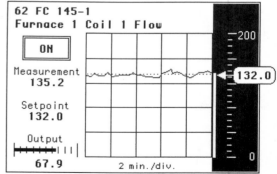

(f) The operator presses and holds the button on the mouse to pick the setpoint arrow. The system inverts the colors on the scale and replaces the pointer with a number in a box attached to the setpoint arrow. As the operator moves the setpoint arrow up and down, the number in the box changes to show the value the setpoint would have if the button were released. This is an example of GKS/PHIGS 'valuator' input. As before, if the operator moves the mouse sufficiently far to the right or left, the interaction is canceled with no change and the display will look as in (e).

Fig. 1. Example showing typical interactions that a process operator may make by using an interactive graphics interface. The example—to turn on a control loop and adjust its setpoint. The descriptions are written as if the operator were using a single-button mouse, but other types of input devices can achieve the same result. It should be pointed out that the control loop in the example is implemented with a method using 'command states' rather than the cumbersome classical method that describes a loop's state as 'auto or manual' and its

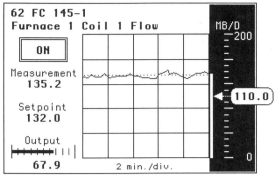
(g) The operator moves the indicator until the value shown is the desired new setpoint value.

Fig. 1. (continued)

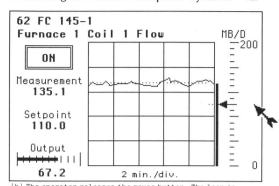
(h) The operator releases the mouse button. The loop is given the new setpoint and the graphics pointer returns, ready to be used for picking another object. Steps (e) through (h) have taken about four seconds to complete.

setpoint state as 'remote, local, or tracking.' (Ref. 11) With the command state method, the loop state is typically either 'on, off, or failed.' The setpoint state is implicit from the loop's configuration and the loop state.

Shown here is a sequence of pictures representing the contents of the display surface as the operator performs the interaction with the control loop. Details are given in the subcaptions to each of the eight diagrams.

Communications and Standards. Distributed systems are communications intensive. It is of the utmost importance that the communications aspects of process systems be designed based on industry standards. This enhances system maintainability and provides a path to utilize the higher-performance, lower-cost hardware developments that inevitably follow standardization. Within the last few years, several organizations[2] have proposed and accepted a number of standards that facilitate communications between equipment made by various manufacturers (Ref. 2). The International Standard Organization's (ISO) "Open Systems Interconnection" model defines a network architecture in terms of layers and protocols. Other organizations have adopted or proposed standards that fit within the seven ISO/OSI layers. General Motors (GM) has promoted to protocol that specifies particular standards to be used in each of the ISO/OSI layers to ensure compatibility between devices in local area networks (LANs). See Ref. 3.

Narrow Window. One of the most common complaints of shared display systems is that they give too narrow a view of the process. The user can see only what is on the VDUs and hence must know about a problem before accessing more information concerning it. No longer can the operator stand back from the panel to get a broad view of the process. Also, on several of the available systems, the user is able to access only a small part of the total system from a single screen.

USER INTERFACING CONCEPTS

The specific features employed in a user's interface should depend on the general needs of that user, the capabilities and configuration of the process computer system, and, for some classes of users, the characteristics of the process itself. In implementing an interface for a specific class of user, a number of general concepts should be evaluated to determine the extent to which they would be appropriate and cost-effective.

Ergonomics[3]

There has been considerable study directed to human-machine interfacing to process computer systems. Much of this study deals with the physical environment of the user. This environment is very important if the user is to effectively use the capabilities of the system. Every installation should give careful and professional consideration to the ergonomic aspects of the user's environment. Especially important are the physical layouts of workstations and panels, the comfort of the user while interacting with the system, and the intensity and diffusive characteristics of the lighting. Even the most functional graphic workstation will be seriously degraded by a poorly designed lighting system or a very uncomfortable chair. Other important physical aspects of the interface are the layout and labeling of the keyboard,

---

[2]ANSI—American National Standards Institute
ISO—International Standards Organization
CCITT—International Consultative Committee for Telephone and Telegraph
ECMA—European Computer Manufacturers Association
IEEE—Institute of Electrical and Electronics Engineers
NBS—National Bureau of Standards (U.S.)
EIA—Electronics Industries Association

[3]Ergonomics is a term pioneered in the United Kingdom. Until relatively recently, the term "human engineering" was most popularly used in the United States. In essence, both terms signify the practice of measuring the range of capabilities and limitations of people and designing a machine or process to accommodate the human rather than vice versa.

the feedback (audible/tactile) from keystrokes, and the readability of video displays.

There is another less understood aspect of the ergonomics of user interfacing to process computers—the *psychological aspects* of humans as processors of information. This aspect represents a significantly greater challenge to designers of user interfaces than the physical aspects. There is need for more understanding of how to improve the usability of tremendously flexible existing tools, especially video displays. More emphasis should be placed on the human skill for pattern recognition (Refs. 1, 12). Displays can be designed to signal abnormal conditions by distorting a normally regular pattern. Displays should not interfere with this skill by presenting too many shapes and patterns. Text is graphically complex. Too much text can obscure the graphical information. Text is important for relating precise information and for the labeling of entities. Graphics are superior for showing relationships between entities.

Graphic Display

Video display terminals have become the standard interface consoles for nearly all computer systems. More recently, terminals with advanced bit-mapped raster graphics capabilities have come into general usage. In systems with distributed processing architectures, one or more graphic VDUs, and their associated interaction peripherals, are physically attached to a dedicated processor whose resources are used only to interact with the users and provide a transparent interface between the user and the database of the total distributed system.

The configuration of graphic VDUs, peripherals, and a dedicated processor is usually called a *graphic workstation*. The drafting, and subsequent acceptance, of several software and hardware standards has resulted in a decrease in price and an increase in the performance, compatibility, and capability of graphic workstations. For current process systems, the use of the *Graphical Kernel System* (GKS) graphics standard to implement the graphic user interface is highly recommended (Refs. 4, 5). In the near future, another standard, the *Programmer's Hierarchical Interactive Graphics System* (PHIGS) will become more solidified and will be superior for interactive process system applications (Ref. 6). GKS allows software and two-dimensional graphic displays to be implemented in highly device-independent fashions that can be moved to different systems with different display hardware. PHIGS add three dimensionality (of limited need for process applications, but important to many other uses) and, more important, allows better utilization of graphics device capabilities, especially in the area of user interaction. PHIGS uses the same output primitives, primitive attributes, and input device model as GKS. Implementations of GKS vary widely in price, performance, and the devices supported.

For process systems, it is advisable to select an implementation that has traded off in the direction of higher performance from fewer, more capable devices.

A graphics interface to a process information system allows data to be represented and manipulated in a variety of forms. Graphic displays are formed by placing a number of graphic "objects" on the display surface where they can be viewed by the user. The appearance of the graphic objects depends on the type of object and values in the process system's database. In a GKS/PHIGS environment, all objects will be formed using the basic output primitive, such as text, lines, markers, filled areas, and cell arrays. The display software should provide additional higher-level graphics objects formed with the basic primitives.

For graphic displays on process systems, the fundamental types of objects available should include backgrounded text, bar graphs, trends, pipes, and symbols. It is desirable to allow each of the fundamental types to be freely scaled and positioned. It is also desirable for the trends to be displayable as either point plots or histograms and for both text and trend fields to be scrollable if there are more data available than can be reasonably displayed in the space allocated for the object. The system should provide libraries of symbols and a facility for the user to edit them.

Well-designed graphics displays can show large quantities of dynamic information against a static background, without overwhelming the user. The objects used to form the static portion of the display may be as simple as textual headings for a table of dynamic textual data, or a graphic representation of a process flow diagram for a section of a process, with dynamic information positioned corresponding to the location of field sensors. In a GKS system, the static portion of a display is usually stored in a device-independent form in a metafile; in PHIGS an archive file is used. From there it can be readily displayed on any graphics device in the system, including various terminals and printers.

Display software can also provide the ability to define pickable objects and a specific action to take if an object is picked. Picking an object may call up a different display, show some additional information on the same display, or initiate an interaction sequence with the user. This is discussed a bit later under "*User Interactions.*"

The graphic display software should provide the ability to update dynamic data objects at specifiable time intervals, upon significant change, or on demand. The update specification should be on a *per object* rather than *per display* basis. The updating of each object should involve the execution of arithmetic expressions with functions available to access the database of the process information system. Besides changing the values of data to be represented, the updating could result in changes to the attributes of the primitives used to form the objects. For example, an object may change color or start to blink.

For some users, especially process operators, color can add an important dimension to graphic displays, expanding the amount of information communicated. Low cost, medium resolution (approximately 600 × 400 pixels) color raster display devices are readily available. Moderately priced high resolution (1280 × 1024), color raster devices that directly execute GKS primitives are entering the marketplace. However, for a given level of performance (speed and resolution), color display devices are considerably more expensive than monochrome devices and require considerably more information to be sent to them to specify the color attributes. Also, the availability of devices and software to make hardcopies of color displays, even monochrome hard copies, in a reasonable amount of time (30–60 seconds), is currently very limited. Color should not be relied on exclusively to communicate important information, unless there is no possibility of having color-blind users.

USER INTERACTIONS

Probably the most important aspect of any computer user's interface is the means by which the user interacts with the system. It is via this mechanism that the user accesses, manipulates, and enters information to the system. It is imperative that this aspect of the interface be carefully designed and implemented to ensure responsiveness and consistency. Well-designed interaction software can eliminate the need for users to refer to printed computer manuals and can provide an interface that is pleasant to use.

Interaction systems have the greatest acceptance when users are confident there is no way of harming anything when they use the system. To achieve this, software should be provided and used that ensures that all interactions are syntactically (orderly, systematic) consistent, and all data entered by the user should be checked for validity. The system should reject invalid inputs and provide a message that the user can readily understand without external references.

User's interfaces must be responsive. The goal should be that all system responses to user actions occur faster than it is physically possible for the user to perform a subsequent action. Certain users, notably process operators, require more responsive interfaces than others and should be provided with more sophisticated hardware to obtain the additional performance.

In general, the number of interactions and complexity of each interaction should be minimized, even if it requires more sophisticated software. Most interactions on a process computer system can be performed much more easily with an interactive graphics system than with a textual dialog system. However, even with an interactive graphics system, some textual interactions will be necessary. Both concepts will be discussed.

Conversational Interactions

Systems have historically interacted with the user by carrying out a textual dialog in a dedicated portion of the display screen, often called a *dialog area*. The flow of the dialog is similar to a vocal conversation, with the system asking for information with a prompt that the user reads in the dialog area and the user providing answers by entering alphanumerical text strings or pressing dedicated function keys on a keyboard. The system provides feedback to each user keystroke by echoing the user's input to an appropriate position in the dialog area, hopefully in a different color than the system's prompt.

It is not surprising that systems were designed to interact this way. Humans are familiar with this form of interaction, and the tools to carry on the interaction are easy to construct. The main drawback to this form of interaction is that it is awkward to converse by reading and typing. Additionally, both the programmer and the user must have a similar understanding of the language used in the dialog. This can be very difficult because natural languages are imprecise and often ambiguous. The meaning of words and phrases varies not only with context but also with the geographic origin and educational background of the conversers. Users frequently misread, or do not read, prompts. Programmers have difficulty in creating meaningful prompts that are short enough to fit in the dialog area and be read by users. While voice systems have been successfully applied in some special applications, speech recognition technology must advance significantly before many users of process systems vocally interact with the systems.

In process systems the dialog area is usually at the bottom of the display surface to allow the user to view the displays while the interaction is progressing. For most users, normal display updating should continue while interactions progress. This is so the user will not mistakenly try to use old data left on the screen during an extended or interrupted interaction. If display updating is halted during interactions, the system should either obscure the data by performing the interaction in a dialog box in the center of the screen or in some way by making it obvious that the displayed data are not reliable.

Dialogs should be designed to minimize the number of questions necessary and the length of required user inputs. Prompts should be simple, pleasant, and easy to understand. They should list menus of possible alternatives when appropriate. Whenever possible, default values should be provided to facilitate rapid execution of functions. The default entries should either be the last entries made or the most common entries.

Graphical Interactions

Humans have communicated using graphics since long before any semblance of language existed. Throughout the ages, the benefits of using graphics to communicate have been recognized. The human brain evolved a tremendous capability to process images presented by the sense of sight and extract even minute details from the immense volume of information that images contain. Confucius remarked that "a picture is worth a thousand words." Unfortunately, for those persons not artistically inclined, it can take longer to draw a picture than to write a thousand words. Thus, graphics have historically been used to record or present information rather than to interactively communicate.

In the last several years, interactive graphics systems have been used for computer-aided design (CAD), computer-aided engineering (CAE), and computer-aided manufacturing (CAM), and for television and movie production, and even for video arcade games. These systems have been very expensive because of the time required to develop interaction techniques and software, and because special display processing hardware interfaced with a powerful computational processor were necessary to achieve acceptable interaction performance. Several textbooks (Refs. 7, 8) are available to teach these techniques, and microprocessors and display processors have become powerful and inexpensive enough that even most personal computers are capable of supporting very respectable interactive graphic interfaces. Indeed, through the introduction of their *Lisa*™ and *Macintosh*™ computers, which have designs based totally on interactive graphics, Apple Computer, Inc. has revolutionized the computer industry by making computers useful tools to those who do not have, or desire to have, a knowledge of computers. The user interface of these is well designed, and a novice can become productive with just a few hours of experience (Ref. 9).

Graphic Interface. A typical interactive graphic user's interface consists of (1) a bit-map graphic display device (usually with a dedicated processor), (2) a keyboard, and (3) one or more *graphics input devices*. Hence, this interface is essentially the same as the conventional user's interface except that the VDU has some additional capabilities, and a graphics input device has been added. Normally, interactive graphics capabilities are in addition to the conventional capabilities, and the user may use either.

Graphics input devices are a combination of hardware and software that allows the positioning of a graphic pointer or cursor to a desired location on the screen and the triggering of an "event" to perform some action based on the selected pointer position. The event is normally triggered by pressing and/or releasing a button. Several types of hardware can be used, and each has been found to be superior in some interactions and inferior for others. Examples include: (1) light pen, (2) joystick, (3) trackball, (4) mouse, (5) graphics tablet, (6) touchpad, (7) touchscreen, and (8) keyboards. Much research has gone into studying the problems of handling input devices, particularly from the viewpoints of both systems and users (Refs. 4, 5, 6, 7, 8, 13, 14, 15, 16). Once a device is chosen, it is important to consider how to use it most effectively. (Refs. 14, 17) The way the user interacts with the input device often determines the effectiveness of the user interface. Appropriate devices on process computer systems are a keyboard, a mouse, and a touchscreen.

Keyboard—This should be available with an alphanumeric section for entering text and an array of function buttons for certain specialized functions. In contrast to nongraphical interactions, the alphanumeric section will be used infrequently. On process systems, the function button section of the keyboard is important even in a graphic environment because the buttons can provide immediate generation of commands or random access to important information. There is often a need for keyboards with excess of 100 function buttons on a single workstation. There is also a need for lamps associated with the keys that can be used to prompt the user. Keys should provide some kind of feedback when actuated. Tactile feedback is desirable, but not necessary. Visual feedback from every keystroke is essential.

Touchscreen—This is probably the simplest graphic input device because it requires no associated button for triggering events. This can be done by sensing a change in the presence of a finger or some other object. A touchscreen occupies no desk or console space and appears, to the user, to be an integral part of the screen. It is especially useful to naive users for selecting one of several widely spaced items displayed on a screen. Because of calibration problems, CRT screen curvature, and the coarse resolution of touchscreens and of human fingers, implementations using touchscreens must be carefully planned, and applications are limited. It is extremely important that systems be designed to trigger the event upon removal of the finger from the screen rather than approach of the finger. This allows the user to correct the initial (often inaccurate) position of the cursor by moving the finger until the graphics pointer is properly positioned over the desired object, at which point the finger can be removed, triggering the selection of the object. Touchscreens should not be used for applications requiring accurate positioning or interactions over an extended period of time.

Mouse—This is a considerably more useful and accurate device than the touchscreen and is less expensive. The mouse consists of a box, about the size of a deck of cards, with one or more buttons for triggering events. The mouse requires a small, flat, empty surface on which to move it. The graphics pointer moves on the display surface relative to the motion of the mouse on the desk surface. It is very easy to achieve pixel resolution for position and selection with a mouse. As previously shown in an example, the mouse is exceptionally well suited to building complex interactive mechanisms that are extremely responsive and easy to use.

Graphics Software—Graphics packages should provide device software to allow various graphics input devices to be treated similarly in a device-independent fashion (Ref. 10). GKS and PHIGS do this by defining six classes of logical input devices. The classes and input values they provide (somewhat simplified) are: (1) LOCATOR—a position on the display surface, (2) STROKE—a sequence of positions on the display surface, (3) VALUATOR—a real number, (4) CHOICE—an integer number representing a selection from a number of available choices (as from a menu or pad or function buttons), (5) PICK—two integer numbers representing an object and part of an objected selected (this is a little different with PHIGS), and (6) STRING—a string of characters (usually entered via a keyboard).

Each logical input device can be implemented using any of the previously described physical devices, and multiple logical devices can share the same physical device. In fact, an applications program using one of the logical input devices will run without modification regardless of the physical device actually used. This allows a manager using a touchscreen to select information from a menu using exactly the same program as an engineer using a mouse. The actual images seen by the two users may vary slightly because of differences in implementing the logical device with the different physical devices, but the syntax of the interactions and the result accomplished is the same.

Using carefully designed logical input devices rather than allowing applications programmers to directly access the physical input devices is extremely important. Not only does this allow device independence and enhance portability of software, but it also makes the software much easier to write and maintain. Even more important, it enforces consistency in the interactions throughout the system. This consistency allows users to quickly gain familiarity with the basic interactive concepts of a system so that they can feel comfortable in using new features, even for the first time, with little or no instruction.

Additional Considerations

An extremely wide variety of interactions is possible via interactive graphics. It is important that designers avoid attempting to emulate the interaction mechanisms of previous-generation systems—just because people are used to them. Of course, previous work on user interfacing should not be ignored, but it is now established that interactive graphics and distributed processing provide a tool that is powerful enough to implement many of the techniques that had been suggested in the literature over the past twenty-five years but that were too difficult to implement on process systems.

Aside from the functional characteristics of the actual user's interface, there are other design and implementation factors that can affect the acceptance and usefulness of a process computer system. These include adequate training in the proper use of the system. Users should not be trained before the system is available for use. Otherwise, they will not have an opportunity to practice their new skills, and they will be quickly forgotten. Also, the system should work properly from the time that users first have access to it, or the users will lose confidence in the system. It is difficult to regain lost confidence in a process computer system.

Field data should enter the system at a sufficiently fast rate to ensure that operators are not uncertain about the values they see displayed. Historically, this has been a problem with temperature multiplexers and crudely designed links to other computers. The field data must be reasonably accurate. This requires a dedicated effort to maintain the field instrumentation. The computer should include software to aid in the calibration of field instruments. Additionally, systems software should be designed to indicate failed and questionable values and should propagate the failure or questionable status to results of calculations that use these values. It is also desirable for systems to provide a means of inferring an approximation for selected failed values. The substituted values must clearly be indicated as questionable.

The use of event-driven software architectures, instead of brute force, time interval scanning architectures, can significantly increase the performance and throughput of the entire system, including the interfaces. Event-driven architectures are especially effective in networked, distributed systems.

REFERENCES

1. Edwards, E., and E. P. Lees: "Man and Computer in Process Control," The Institute of Chemical Engineers, London, 1972.
2. Staff: "Networking for Control Series," *Instruments & Control Systems*, Chilton Publishing Co., Philadelphia, Pennsylvania (February-August 1985).
3. GM MAP Task Force: "General Motor's Manufacturing Automation Protocol Version 1.2," Warren, Michigan, March 31, 1985.
4. ANSI Tech. Committee X3H3: American National Standard Graphical Kernel System (GKS) (ANS X3.124-185), American National Standards Institute, New York, 1985.

5. ISO Working Group SC21/WG5-2: ISO Graphical Kernel System (ISO DP 7942), International Standard Organization, 1984.
6. ANSI-Tech. Committee X3H3: American National Standard for the Functional Specification of the Programmer's Hierarchical Interactive Graphics Standard (PHIGS) (draft), ANSI document X3H3/85-21, American National Standards Institute, New York, February 18, 1985.
7. Foley, J. D., and A. Van Dam: "Fundamentals of Interactive Computer Graphics," Addison-Wesley, Reading, Pennsylvania, 1982.
8. Enderle, G., Kansy, K., and G. Pfaff: "Computer Graphics Programming: GKS—The Graphics Standard," Springer-Verlag, Berlin, 1984.
9. Apple: "Inside Macintosh," Apple Documentation Group, Apple Computer, Inc., 1984.
10. Rosenthal, D. S. H., et al.: "The Detailed Semantics of Graphics Input Devices," *Computer Graphics*, **16** (3) 33–38, July 1982.
11. McAmis, J. H., and R. C. Sorenson: "Chevron's Approach to Advanced Process Control," *Pcdgs of Control Expo—85*, 257–269 (1985).
12. Dallimonti, R.: "Challenge for the 80s: Making Man-Machine Interfaces More Effective," *Cont. Eng.*, 22–60 (January 1982).
13. Foley, J. D., Wallace, V. L., and P. Chan: "The Human Factors of Computer Graphics Interaction Techniques," *IEEE Computer Graphics and Applications 4.11*, 13–48 (November 1984).
14. Buxton, W.: "Lexical and Pragmatic Considerations of Input Structures," *Computer Graphics*, **17**, (1) (presented at the SIGGRAPH Workshop on Graphical Input Techniques, Seattle, Washington, June 1982.
15. Olsen, D. R., Jr., Dempsey, E. P., and R. Rogge: "Input/Output Linkage in a User Interface Management System," *SIGGRAPH '85 Pcdgs.*, 191–197 (1985).
16. Buxton, W., et al: "Towards a Comprehensive User Interface Management System," Computer Graphics, **17** (3) 35–42 (July 1983).
17. Buxton, W., Hill, R., and P. Rowley: "Issues and Techniques in Touch Sensitive Tablet Input," SIGGRAPH '85 Pcdgs, 215–224 (1985).

Local Area Networks (LANs)

By Bruce A. Loyer[1]

As automation in the factory increases, the need for communication between computers, controllers, and other "intelligent" machines has become critical. In the past, when factory automation was limited to the use of programmable controllers (PCs), numerical control (NC) of machines, and similar traditional approaches, communication was not a major limiting factor. Each tool was essentially self-contained, and the communication requirement was primarily a user interface for controlling and updating machine operation. With the accelerated growth of automated tools and processes, however, communication between these entities is required to control not only their operation but their interrelationships as well. To this is added the desire to overlay environmental control, energy management, and materials requirements planning (MRP) to the factory operation. The end result is that intercomputer/controller communication has become the largest single problem to be addressed for factory automation. These interrelationships are shown in Fig. 1[2].

Early communication needs were served with point-to-point data links (Fig. 2). Standards, such as TTY (teletypewriter) current loops and RS232 appeared and were accepted, which allowed different equipment to interface with one another. From that, the *star topology* (Fig. 3) was developed so that multiple computers could communicate. The central or "master" node uses a communications port with multiple drops, as shown in Fig. 4. The master is required to handle traffic from all the nodes attached, poll the other nodes for status, and, if necessary, accept data from one node to be routed to another. This heavy software burden on the master is also shared to a lesser degree among all the attached nodes. In addition, star topology requires routing a separate wire for every piece of equipment attached. This makes it difficult to wire and even more difficult to change. Also, the star topologies are inflexible on the number of nodes that can be attached. Either one pays for unused connections (for future expansion), or a system results that cannot grow with a business.

To overcome some of these shortfalls, multi-drop protocols were established and standardized. Data loop, such as SDLC (synchronous data link control), were developed as well as other topologies, including buses and rings. Some of the early standards are shown in Fig. 5. The topology of these standards makes it easy to add (or subtract) nodes on the network. The wiring is also easier because a single wire is routed to all nodes. In the case of the ring and loop, the wire is also returned to the master. Inasmuch as wiring and maintenance are major costs of data communications, these topologies have virtually replaced star networks. These systems, however, have a common weakness; i.e., one node is the "master," with the task of determining which station may transmit at any given time. As the number of nodes increases, throughput becomes a problem because: (1) a great deal of "overhead" activity may be required to determine who may transmit, and (2) entire messages may have to be repeated because some protocols allow only master-slave communications; i.e., a slave-to-slave message must be sent first to the master and then repeated by the master to the intended slave receiver. Reliability is another problem. If the master dies, communications come to a halt. The need for multinode networks without these kinds of problems and restraints led to development of *Local Area Networks* (LAN) using peer-to-peer communication. Here no one node is in charge, and all nodes have an equal opportunity to transmit.

A LAN is a distributed communications network with the following characteristics: (1) peer-to-peer oriented (no master), (2) two to 200 data devices (nodes) may be incorporated, (3) distance is limited to less than two kilometers (1.2 mile), and (4) 1 to 20 M bits/second data rates.

In a local area network, each node is an independent computer system. Since there is no master to control traffic, each node must deter-

---

[1] Systems Engineer, Semiconductor Products Sector, Motorola Inc., Phoenix, Arizona.
[2] Similarly, in the instrumentation and control of the fluid/bulk processing industries, the electrical and pneumatic transmission systems that tied field sensors and field-mounted instruments to central control panelboards, characterized by their simplified interfaces, no longer suffice for all but the simplest installations requiring a few open or closed loops. In the automated plant, whether in the discrete-piece manufacturing field or in the continuous processing category, controls are but one portion of a vast information transferring network where layers of management information requirements have added to the control functionality; or vice versa! Information transfer is synonymous with communication.

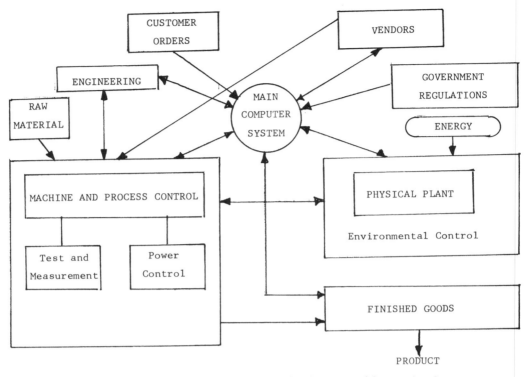

Fig. 1. Intercommunications in a generalized automated factory situation.

mine when it has the right to transmit. In a typical system, the host computer of the node is left free to perform its job while the LAN protocol unit is moving data on and off the network. (See Fig. 6.)

As previously shown in Fig. 1, the need for local area networks is being driven by the proliferation of computer functions. The real motivating factor is that computers are now so cost-effective and relatively inexpensive that they are used throughout the factory floor, processing plant, and office. These computers must talk not only to the giant mainframes but also to each other. The demands of MRP, such as scheduling, inventory, management, et al., require constant monitoring and data acquisition. Further, factory floor management requires coordination of machine operation; environmental control requires constant monitoring, among other factors that go well beyond the traditional tasks of

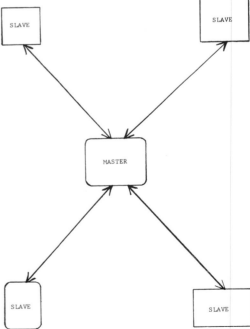

Fig. 2. Point-to-point communication.

Fig. 3. Star topology.

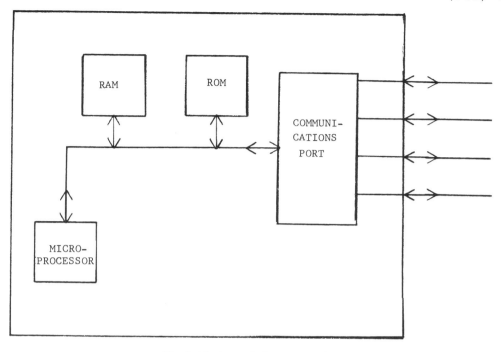

Fig. 4. Master node for star topology.

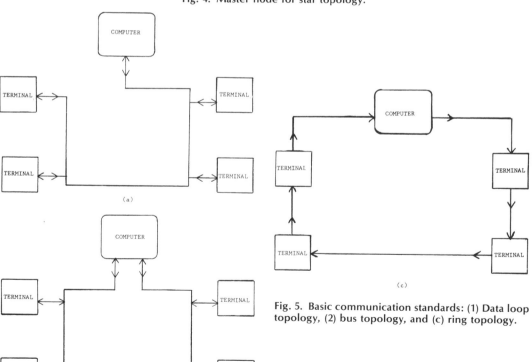

Fig. 5. Basic communication standards: (1) Data loop topology, (2) bus topology, and (c) ring topology.

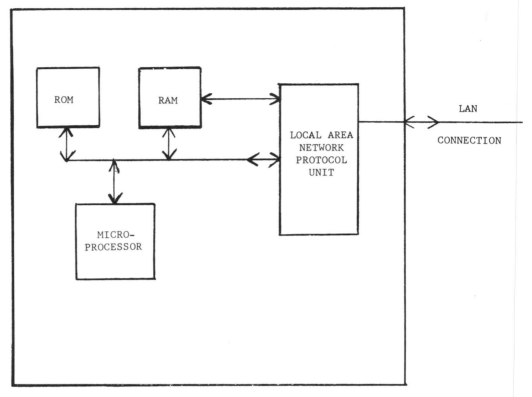

Fig. 6. Local area network (LAN) node.

regulatory or sequencing control. The need to communicate status between devices (for a total integrated environment) and the sharing of expensive resources (large-capacity disk storage, line printers, etc.) have driven the requirement for LAN communication networks.

NEEDS OF THE INDUSTRIAL ENVIRONMENT

Although the need for LANs exists both in the office area and on the factory floor, the latter environment requires some special characteristics. The needs of the harsh industrial environments are demanding and include:

1. *Noise*. Since a local area network will have long cables running throughout the factory, the amount of noise pickup can be large. Therefore, an industrial LAN should be capable of working well in an electrically noisy area. The physical interface should be defined to provide a significant degree of noise rejection and the protocol must be "robust" to allow easy recovery from data errors.

2. *Response*. An LAN in an industrial situation should have a guaranteed maximum response time; that is, the network must be able to transmit an urgent message within a specified time limit. The real-time aspect of industrial control demands communication within a known timeframe that has a maximum value.

3. *Priority Message*. Another aspect of real-time control is the ability to send out a priority message. On the factory floor, both control and status data will be carried over the same network. A control message should have a higher priority and be transmitted before other messages.

Common Standards

A LAN standard that serves the harsh factory environment can also be used in the less demanding office and administrative areas. Unless the requirements for the factory add too much cost, the factory floor standard can be the choice for the entire network. A common standard is advantageous inasmuch as system and information-handling elements located in both office and factory environments must communicate.

THE NEED FOR STANDARDS

To meet current communication needs, different types of local area networks are possible. Many of them already have been developed. All of these LANs provide the same basic service, which is to allow computers to pass data. However, since the major function of a network is to connect many different computers, *standards* are needed. The standard should describe not only how nodes are connected but the protocol followed in transferring data as well. Ideally, the standard should be sufficiently comprehensive to permit any computer following it to pass data to any other computer that follows the same standard.

Currently, there are no widely accepted standards that fully meet this requirement, and even though they are evolving, some years ahead will be required to satisfy these requirements. Currently, most of the work is being done on the physical problem of getting data from one computer to another. How the receiving computer can use that data has not been fully developed. The ISO (*International Standards Organization*) has broken the problem into seven layers, of which only the standards developed for moving data within the lowest two layers are described in this article. See other articles on communications in this *Handbook* section.

STANDARDS CURRENTLY AVAILABLE

The major focal point of LAN standards as of this writing is the IEEE (*Institute of Electrical and Electronics Engineers*), which is supporting the *IEEE-802* "Local Area Network Standards Committee." This Committee has concluded that one standard LAN will not fit the needs of all applications and therefore has proposed three standards: (1) *802.3* (CSMA/CD), (2) *802.4* (Token Bus), and (3) *802.5* (Token Ring).

IEEE-802.3 CSMA/CD

This standard is patterned after *Ethernet*, a popular LAN developed by *Xerox Corporation*. It is the first LAN to be standardized and has the longest history of use. This standard uses CSMA/CD (*Carrier Sense Multiple Access with Collision*), which is a baseband system with a bus architecture. (See Fig. 5(b).) Baseband is a term used to describe a system where the information being sent over the wire is not modulated; that is, the "ones" are respresented by one voltage level and "zeros" by another voltage level. Normally, only one station transmits at any one time. All other stations hear and record the message. The receiving stations then compare the destination address of the message with their address. The one station with a "match" will pass the message to its upper layers, while the others will throw it away. Obviously, if the message is affected by noise (detected by the frame check sequence), all stations will throw the message away.

The CSMA/CD protocol requires that a station listen before it can transmit data. If the station hears another station already transmitting (carrier sense), the station wanting to transmit must wait. When a station does not hear anyone else transmitting (no carrier sense), it can start transmitting. Since more than one station can be waiting, it is possible for multiple stations to start transmitting at the same time. This causes the messages from both stations to become garbled (called a "collision"). A collision is not a freak accident but is a normal way of operation for networks using CSMA/CD. The chances of collision are increased by the fact that signals take a finite amount of time to travel from one end of the cable to the other. If a station on one end of the cable starts transmitting, a station on the other end will "think" that no other station is transmitting during this travel time interval and that transmission can be resumed. After a station has started transmitting, it must detect when another station is also transmitting. If this happens (collision detection), the station must stop transmitting. Before quitting transmitting, however, the station must make sure that every other station is aware that the last frame is in error and must be ignored. To do this, the station sends out a "jam," which is simply an invalid signal. This jam guarantees that the other colliding station also detects the collision and quits transmitting. Each station that was transmitting must then wait before trying again. To make sure that the two (or more) stations that just collided do not collide again, each station picks a random time to wait. The first station to time out will look for silence on the cable and retransmit its message again.

IEEE-802.4 Token bus

This standard was developed with the joint cooperation of many companies on the *802 Committee*. Since becoming a standard, it has been selected by *General Motors* (GM) to be used in its *Manufacturing Automation Protocol* (MAP) as the local area network to interconnect GM's factories. It is also a bus topology but differs in two major ways from CSMA/CD. First, the right to talk is controlled by passing a "token," and, second, the data on the bus are always carrier modulated. In the token bus system, one station is said to have an imaginary "token." This station is the only one on the network that is allowed to transmit data. When this station has no more data to transmit (or it has held the token beyond the specific maximum limit), it "passes" the token to another station. This token pass is accomplished by sending a special message to the next station. After this second station has used the token, it passes it to the next station. After all the other stations have used the token, the original station is passed the token again. A station (for example, station A) will normally receive the token from one station (B) and pass the token to the same third station (C). The token ends up being passed around in a logical token ring (A to C to B to A to C to B . . . .). The exception to this is when a station wakes up or dies. For example, if a fourth station (D) gets in the logical token ring between A and C, station A would then pass the token to D so that the token would go: A to D to C to B to A to D . . . . Only the station with the token can transmit so that every station gets its turn to talk without interfering with anyone else. Obviously, the protocol also has provisions that allow stations to enter and to leave the logical token ring.

The second difference between token bus and CSMA/CD is that with the token bus, data are always modulated before being sent out. The data are not sent out as a level, but as a *frequency*. There are three different modulation schemes allowed—two single-channel and one broadband. Single channel modulation permits only the token bus data on the cable. The broadband method is similar to CATV (community antenna television) and allows many different signals to exist on the same cable including video and voice in addition to the token bus data. The single channel techniques are simpler, less costly, and easier to implement and install than broadband. Broadband is a higher-performance network, permitting much longer distances and, most important, satisfying the present and future communications needs by allowing as many channels as needed (within the bandwidth of the cable).

As of this writing, token bus has only recently become a standard and is still evolving. The major thrust of this evolution has been to incorporate the ideas of another standards group, the *PROWAY Committee*, into *802.4*. PROWAY has been working on LANs specifically for industrial applications. Since the PROWAY proposal is also a token bus protocol, it would be better to have one token bus standard for all markets. To this end, both committees have been working to clear up the differences between the two protocols. The *802.4* standard is being modified to fit the requirements outlined by PROWAY. This will be incorporated into a revised standard.

Additions from PROWAY: These include:

1. Provide for the immediate acknowledgment of a frame. If station A has the token and uses it to send a message to station B, station B can be requested to send an acknowledge message back to station A. Station B does not wait until it gets the token, but instead, station B "uses" station A's token for this one message. Station B, in essence, is telling station A that it received the message without any errors. This idea is allowed for in the current *802.4*, but all details are not yet spelled out.

2. The capability must exist to remotely initialize and control a station by sending a message over the network. Every token bus protocol handler would have an output to control the rest of the node.

3. Every station would have some predefined data that are available upon demand. Any other station can request the data, and this station would send the information out immediately. Again, the station sending the response is using the other stations' token.

4. The access time of the network must be deterministic, and an upper bound must be predictable for any given system. This means that if a station on a network wants to send a message, one should be able to predict the maximum possible delay before it is sent out. *802.4* has a maximum access time per node, but no upper bound on the number of nodes.

5. PROWAY also specifies some slightly different signalling power levels and other physical considerations.

6. Provide a method of monitoring membership in the logical token ring. If a station dies, for example, every other station should be aware of it. Every station in the logical ring would have a list of all the stations that are in the ring.

7. Provide for the accumulation of network performance statistics. It should be possible to find out how the network is performing over time.

IEEE-802.5 Token Ring

At this writing, *Token Ring* is the last proposed standard from the *802 Committee*. Originally, token ring and token bus used the same protocol with different topologies. IBM then submitted a draft proposal for a different token ring protocol to the committee. This proposal was accepted and made the basis for the current Token Ring protocol (thereby forming two token protocols). IBM has been working on the needed additional clarification of the specification. In general, the token ring protocol is very similar to token bus. As the name implies, the topology is that of a ring. As shown in Fig. 5(c), this means that any one node receives data only from the "upstream node" and sends data only to the "downstream" node. All communication is done on a baseband point-to-point basis. This seems to imply that one node can talk only to its downstream node. However, this is not the case inasmuch as each station repeats what it hears from the upstream station to the downstream station. Since the last station is connected to the first (forming the ring), any station can send data to any other station. Of course, some precautions must be taken to prevent a small message from being retransmitted around the ring forever. This is done by having the transmitting station remove its messages from the ring once they have gone around one time.

The "right to talk" for this network is also an imaginary token. Token ring has simplicity in that the station with the token simply sends it to the next downstream station. This station either uses the token or lets it go on to the next station.

Currently, only two of the three methods proposed by the *802 Committee* have become standards: (1) CSMA/CD and (2) Token Bus. The third, Token Ring, still is in committee but may be approved soon. Of the three methods, 802.3 (CSMA/CD) is the most widely known. CSMA/CD is very simple and has been used for a long time in other local area networks. Because of the driving force behind *Ethernet*, there are several LSI silicon solutions available today. This makes CSMA/CD the most cost-effective standard currently available. But, as previously mentioned, CSMA/CD does pose some problems in the industrial setting.

The CSMA/CD protocol is designed for a lot of short messages (bursty traffic) and works very well as long as there are not a lot of collisions. However, if too much traffic is being sent, there is a lot more overhead because of the increased number of collisions. Therefore, the efficiency of CSMA/CD decreases as the load increases. In contrast, with token passing protocols, there are no collisions and thus the overhead (passing the token) as a percentage of total time goes down as the load increases.

More important, by its very nature, CSMA/CD is probabilistic, that is, every station has a finite chance of hitting some other station every time it tries to send. This means that there is no way to guarantee that a message might not be held up forever. In the factory a condition of not being able to send a message across (such as "somebody's leg is in the robot") can be catastrophic. In a token passing system, on the other hand, there is a "guaranteed access time," which is the maximum waiting time possible before an important message can be sent. This access time is the time it takes a token to be passed to everyone (the token rotation time). This is equal to the number of stations on the network multiplied by the time each station is allowed to hold the token. Another advantage of a token passing scheme is that it permits multiple priorities on messages. Therefore, if there are several important messages that must be transmitted, they will go out before routine messages. Finally, the Token Bus protocol is very robust and able to recover from errors easily. For example, if a token message is destroyed by noise, the node passing the token will resend it.

The other major difference between CSMA/CD and Token Bus is that Token Bus is carrier modulated. CSMA/CD, as a baseband system, means that the digital data to be sent out are represented as discrete levels on the cable. Several advantages of modulating the data include: (1) The effect of noise is reduced because the receivers (and transmitters) can use filters to reject frequencies outside of the expected range. (2) The maximum length of the cable is increased. CSMA/CD has length limits because of the contention problem. In Token Bus, the cable length is limited only by the attenuation of the signal in the cable. (3) Token Bus allows three different modulation schemes. These schemes use the same protocol but permit the network designer to make a decision based on cost versus performance. The highest-performance method is broadband, which can have many different functions on the same cable. It can carry multiple voice and video channels at the same time as the data. This makes it the highest-performance LAN standard available at the present time.

Token Ring has the advantages of a token passing protocol, namely a deterministic access and priorities on messages. Token Ring is a baseband system like CSMA/CD, but it has the advantage that if the cost is justified, Token Ring can use fiber optic cables. In addition, since every station retransmits the data, the maximum physical length of a network can be very large. To prevent an off station from bringing down the network, every station must have an automatic bypass that activates when power to a station is turned off. Finally, a ring structure requires more complex wiring than a bus because every station must have separate input and output cables, the last station must be connected to the first station to form the ring, and an amount of redundant cabling is recommended for fault tolerance because a single break or down station can stop all data transfer.

ADDITIONAL READING

Capel, A. C., and G. F. Lynch: "Proway: The Evolving Standard for Process Control Data Highways," *Instrumentation Technology*, 91–94 (September 1983).

Kozlik, T. J.: "Local Area Networks in Control," *Cont. Eng.*, 76–78 (July 1984).

Laduzinsky, A. J.: "Local Area Networks Expand the Horizons of Control and Information Flow," *Cont. Eng.*, 53–56 (July 1983).

Laduzinsky, A. J.: "Local Area Networks; Is MAP It?" *Cont. Eng.*, 73–75 (July 1984).

Morris, H. M.: "New Controller Distributes Intelligent Control," *Cont. Eng.*, 96–97 (June 1984).

Rosenberg, R., and T. E. Feldt: "Local Nets Arrive in Force," *ElectronicsWeek*, 67–73 (August 6, 1984).

Spenser, K., and F. Raines: "High Level Industrial Command Language Simplifies Control Communications," *Cont. Eng.*, 150–152 (April 1984).

Various: "Special Report: Interfacing Plant Controls and Management's Computers," *Cont. Eng.*, 83–107 (June 1985).

Various: "Special Edition: Manufacturing Automation Protocol," *Cont. Eng.*, 19–72, (October 1985, 2nd Edition).

Communication Standards For Automated Systems

There was a time not many years ago when the *transmission of information* to and from controlling devices and the logging of manufacturing and processing data were simply a routine part of the control engineering task. Without question, today, the tail is wagging the dog, so to speak—with communication systems, including the quest for communication standards, now demanding much of the creative thinking of the practitioners of automation technology. And this situation, of course, has arisen for many excellent reasons that have been repeated many times over in the professional automation literature.

In terms of communication systems, automation has played a dual role of cause and effect. Electronic data processing, including computers, now offers a tremendous ability for plant managers to collect, analyze, and base their decisions on vast amounts of information—information that not only is of high quality and reliability but also costs less on a $/unit basis than was true less than a decade ago. The type of systems that were pioneered successfully for office automation initiated a thirst for information that has not been satiated and that has spilled over into manufacturing and processing automation. Thus, the needs for data exchange between automated production equipment no longer are confined to controlling and coordinating the equipment per se. Rather, all manner of information must be generated on the factory floor—to become parts of databases housed in remote hierarchical computers. A large percentage of this information is not directly related to machine or process control and thus represents a major superimposition of information requirements and consequently much greater loads on the communication systems.

Currently, industry may be in a stage where the needs for information have been somewhat overstated. One of the weakest links in the information system is that involving the interpretation and use of information, including that of readying information for easy comprehension and digestion by the minds of management. Selecting the right information at the right time for management decisions goes far beyond simply formating information in a variety of graphs and colors. Considerable attention has been given to the machine/process/operator interface, involving such specialties as human factors engineering, ergonomics, et al. Much less attention has been paid to the manage/information interface. Help along these lines may come from the application of expert systems and artificial intelligence (AI) techniques. There would seem to be limitless opportunities in this area.

Not mentioned very frequently concerning management's needs for greater and greater information is what might be termed the *gluttony phenomenon*, that is, the tendency to amass information that really is not needed simply because data can be generated so easily as the result of the great advancements in electronic data processing. As considerably more experience is gained over the next few years, perhaps there will be a shift of emphasis to information simplicity rather than complexity as the target to pursue. Perhaps the needs for information will be qualified more carefully—indispensable information, essential information, helpful information, marginal information... valueless information. In retrospect, the tendency to collect meaningless information has always been present in the industrial instrumentation field, long before the entry of solid-state physics and cheap electronics.

Putting philosophy aside, the modern, sophisticated information systems are in a surprisingly overloaded state. Massive information handling is the mode of the day and will continue to require much engineering attention.

The Need for Communication Standards

Inadequate standards have plagued emerging technologies since time immemorial. Engineers living today did not have to cope with establishing the standards for the common products of industry and commerce—screw and pipe threads, rivets, gears, piping, tubing and fittings, metal castings, plate, sheet, strip, rail, and wire; glass and masonry; pumps and valves; electric motors... among many other products whose strength, dimensions, geometry, and other physical parameters are taken for granted today even though many hundreds of different suppliers may vie for the same market. One can easily envision the chaos that prevailed prior to the establishment and enforcement of standards for these products. A number of engineering institutions and societies rose to greatness partially based upon their leadership in establishing standards.

Some authorities point to the present effort to establish communication standards for use in automated systems (à la CIM, CAD/CAM, etc.) as perhaps an order of magnitude more complex than the past work required to create standards for other technologies. A review of the constantly changing literature on communication standards tends to strengthen this observation. Much progress has been made, particularly in the 1980s, to formulate a communication standard that satisfies most of the parties involved, but numerous changes and refinements can be expected within a relatively short period.

Early Highways[1]. As long ago as 1972, the very first *serial data* communications highways were introduced. At that time, the only purpose of the data highway was to allow host computers to adjust setpoints or, in some cases, perform direct digital control (DDC), while providing measurement data to the host computer. (Reference to article on "Control System Architecture" in *Handbook Section 3* is suggested.) Later, in 1975, the use of these same data highways was extended to provide for graphic CRT-based operator consoles apart from host computer operation. Also, at the same time, the first distributed control system was announced. It featured a digital data highway.

The early data highways were relatively simple and of moderate speed. One used a simple multidrop technique with the computer end being the only master. Another system used a more novel multimaster technique that depended on a centralized single point of control. At the speed of operation, 50-250 Kbps, neither data highway required elegant line driving technology.

Developments of the Late-1970s. In 1979, the next evolutionary step in process data highways was announced. This system also had a central point of control and a multimaster protocol, but it used CATV cable at 1 Mbps. At these data rates, even in baseband, the transceiver design was based on RF technology. The topology of this network used a local star cluster with the CATV interconnecting all clusters. The data communication within the cluster was bit serial, byte parallel.

These later networks offered dual redundant mechanisms so that if one data highway ever failed, a second data highway would take over. The second highway was unused except for integrity diagnostics. To be sure that these data highways were in good order, elaborate mechanisms were implemented apart from the data communications to ensure cable and station integrity.

First Call for Standardization. With several instrument companies producing or about to produce data highways and all with similar desires for integrity, high speed, and multimastership, a standardization effort was called for. A committee under the IEC (International Electrotechnical Committee) Technical Committee 65A was formed as Working Group 6 (IEC/TC 65A/WG6). This committee, formed in 1976, was composed of representative members from many countries. The American, European and Japanese data communications committees (TC5) of the International Purdue Workshop on Industrial Computer Systems (IPWICS) provided the nucleus of this group and also met in local countries as technical advisory committees. The TC5 group of the American Region became ISA (Instrument Society of America) SP72. The work of the international group became known as PROWAY, for *Pro*cess Data High*way*. From this international effort first came a set of requirements for the desired PROWAY network characteristics. The network was to operate with the undetected bit error rate better than one in 10^{15} bits. Mastership was to be distributed without a central point of control (a peer level network).

As the committee began to develop its standard documents, it solved the multiple mastership problem with a method called "baton passing," where each station took turns in rotation being the network master for a short period of time. This method was selected over Ethernet since it was deterministic rather than statistical.

Enter the IEEE. At about this time a similar activity also began in IEEE. A combination of three vendors had proposed that the Ethernet specification be made an IEEE standard. The committee that formed, IEEE 802, recognized that Ethernet, while satisfactory for most office communications, did not satisfy the needs for many commercial and industrial local area networks. Since the chairmanship and several members of IEEE 802 were also members of the US PROWAY Committee, a separate subcommittee was formed to address the preparation of a baton or token passing based standards document to counter Ethernet. It eventually become obvious that there was a sufficient number of different applications and thus there was a need for both the CSCM/CD (Carrier Sense, Multiple Access with Collision Detection) methods of Ethernet and for token passing. The CSMA/CD method became IEEE 802.3 and was approved in 1981 with minor differences from Ethernet. At that time, several semiconductor vendors committed silicon designs for support of IEEE 802.3 in its most complete form. The significant fact was that this standard was a proven reality, both in its collision detection protocol and its speed of implementation (10 megabits per second) on a high-grade of 50 ohm coaxial cable.

The most significant fact about either the IEC draft specification for PROWAY (1 megabit per second on CATV cable), or the various proposed implementations of IEEE 802.4 token passing,

---
[1]The next several paragraphs were furnished by Richard H. Caro, Autech Data Systems, Inc., Pompano Beach, Florida.

was that they had never before been implemented. Of all the combinations, only the 1 megabit baseband 75 ohm coaxial cable media had even been used, but without token passing. Also, the media options seemed like a cafeteria of choices; baseband 75 ohm coaxial cable at 1 megabit; baseband 75 ohm CATV cable at 5 and 10 megabits; and broadband 75 ohm CATV cable at 5 megabits. At least all shared the same protocol, token passing.

Meanwhile, the U.S. PROWAY committee, recognizing that the 802.4 effort was very close to its needs, began a joint effort to fuse PROWAY requirements into 802.4. At the completion of this effort, the IEEE 802.4 Token Bus standard was approved with the PROWAY extensions included as an Appendix B. The extensions that were deemed to be necessary for industrial control included: a positive acknowledgment of all point-to-point messages and an ability to request data from nontoken holder (slave) devices. Also at about this time, one vendor was able to supply a proven token passing broadband modem.

With the agreement of both the commercial and industrial sectors on a common standard that shared many features with the rest of 802, the semiconductor vendors began to commit silicon chip designs for implementation. While token bus was the protocol, the station management functions, the maximum speed requirements, and the cyclic redundancy codes were all common with 802.3. This was the breakthrough needed to provide an economic process data highway using this new technology.

Enter General Motors. When General Motors began to analyze its in-plant data communications needs, it had to add several more factors. First, many GM plants had, for some years, been using a network of CATV cable for closed-circuit television. To this had been added several channels of low speed serial data communications, which involved RF modems to operate in the television broadband spectrum. To get a plant-wide high-speed information network standard established for GM, the firm formed an internal committee called MAP (Manufacturing Automation Protocol). This well-publicized committee invited proposals from many vendors and circulated several papers on their requirements.

The primary purpose of the plant-wide network (the "backbone" network) was not process control, but to allow the two-way flow of high-level production data. Because of the approval of IEEE 802.4, the current installation of CATV cable and the availability of at least one modem supplier, MAP selected the CATV broadband, 5 megabit version of IEEE 802.4 for the backbone network. While GM indicated to all vendors that every GM manufacturing automation system must provide an interface to this backbone network, GM was not requiring its use within those systems. While there was a preference for these local systems to use a compatible protocol, it was recognized that the automation systems should not use the backbone network for control information, but only for communications to upper-level computers or other systems. Otherwise, the whole bandwidth of the backbone network would easily be consumed with local traffic. GM had defined a true hierarchical network environment, and had clearly endorsed token bus and CATV, but not for process control.

Communication Systems Models

Backtracking a bit, it is interesting to note that the concept of control centers in the process industries is several decades old, dating back to the 1930s. In contrast, the general practice in the discrete-piece manufacturing industries was essentially to confine instrumentation and control to the factory floor. As mentioned earlier, the dedication to sophisticated automation (largely started in the automotive and aircraft industries) essentially commenced in the late 1970s. Partially to meet intense competition, improve product quality, and cut costs, an intense desire for more information developed among factory managers. Thus, the incorporation of intensive, modern communications into automation systems is now taken for granted in both the discrete-piece manufacturing industries and in the fluid/bulk processing industries.

Over the years several idealized concepts or models of communication systems have been proposed—these ranging from three to seven or more layers. One example is a three-level system:

Level 1—Lowest-level. Linking groups of people and/or machines. The link involves the flow of information among different workstations at a department level. These local networks must have a fast payback because it is in this area where most manufacturing and processing alterations are made.

Level 2—Mid-Plateau. Facility-wide networks that permit all departments within a facility to share data. Examples may include: (a) obtaining employee vacation data at a moment's notice so that shift assignments can be made without delay; (b) tracing the history of a quality control problem immediately when a failure is noted; and (c) permitting a service supervisor to check the current readiness of production equipment.

Level 3—Highest Level. Corporate-wide communications, where all multi-factories and departments can exchange information and report to a central information-processing site. Manufacturing automation, such as CAD/Cam, materials management, and automated machine tools, can be linked to management information processing, including such functions as financial data, sales reporting and planning, among many others. This arrangement enables managers to check the flow of production and product lead times—as only one of many ex-

TABLE 1. OSI Reference Model (Basis of MAP)

No.	Layer Name	Uses/Applications	Protocols
1	Physical	Electrical, mechanical, and packaging specifications of circuits. Functional control of data circuits	ISO/IEEE 802.4 Broadband 10Mbs data rate, 2-channel midsplit format preferred (MAP 2.1). IEEE 802.4 Phase Coherent Carrier Band, 5Mbs data rate (modified by IEEE 802.4B Committee, July 1985) based on media specs (ISA S71.Q2, 1985)[1]
2	Link	Transmission of data in local network—message framing. Establish, maintain, and release data links, error and flow control.	*Media Access:* IEEE 802.4 Token Bus, 48-bit addresses. *Link Control:* IEEE 802.2 Type 1 Connectionless service.
3	Network	Routing, switching, segmenting, blocking, error recovery, flow control. System addressing and wide-area routing and relaying.	ISO DIS 8473 Connectionless Network service (CLNS), ISO DAD 1/8028.[2]
4	Transport	Transparent data transfer, end-to-end control, multiplexing, mapping. Provides functions for actual movement of data among network elements.	ISO Transport, Class 4[3]. ISO 8073 (IS)
5	Session[4]	Communication and transaction management. Dialog coordination and synchronization. Administration. Administration and control of sessions between two entities.	ISO Session Kernel[5]. ISO 8237 (IS)
6	Presentation[6]	Transformation of various types of information, e.g., file transfers. Data interpretation, format, and code transformation.	null/MAP transfer. ISO 8823 (DP)
7	Application[7]	Common application service elements (CASE)	ISO 8650/2 (DP)
		Manufacturing message service (MMS)	RS-511
		File transfer and management (FTAM)	ISO 8571 (DP)
		Network management	IEEE 802.1
		Directory service	GM working draft

NOTES: Bottom layer (#1) defines mechanical connections that tie personal computers together in a network. This includes electrical requirements for establishing and terminating transfers of information between nodes. The second layer (#2), the data-link layer, defines procedures and control protocols for operating the communication lines. It offers a means of detecting and correcting message errors, which enables the model's higher layers to direct the circuits within the bottom layer. The link layer also incorporates the network topology, which defines how the nodes are tied together. For example, a network can have a star, ring, or bus topology. The network layer (#3) determines how data are transferred between personal computers. To date, network vendors conform to the OSI model and IEEE 802 standards through the first three levels. The remaining four levels still vary among vendors and are a source of confusion in some instances for the potential user of a local network.

The MAP Users Group has defined two basic architectures for MAP: (1) MAP Backbone Architecture, which is also specified by GM's MAP 2.1; and (2) MAP Cell Architecture. The backbone architecture supports MAP end systems. These, in turn, implement the MAP suite of OSI protocols (layers 6 and 7). This enables communication with all other MAP and systems, which may be located within the same plant or at remote sites by way of the international protocol.

A basic intent of GM, questioned by some users and suppliers, is that all but the simplest factory devices will support the protocol suite selected for MAP end systems—as an effort to simplify network management.

Cell architecture includes nodes that support the MAP end system protocol suite, plus a reduced protocol suite, which is based upon the confirmed Data Link service provided by PROWAY. An advantage of the PROWAY suite is that it incorporates necessary real-time responsiveness needed by control applications, in what are known as MAP/PROWAY nodes. An advantage of the cell architecture is that it accommodates situations where the reduced protocol suite may provide an efficient connection of relatively low-cost units (microprogrammable controllers and smart sensors, as examples) to the MAP network. MAP provides for two systems of interconnects: (1) bridges and routers, which are used to interconnect parts of a plant-wide MAP subnetwork, and (2) gateway for connecting MAP systems to non-OSI networks. MAP bridges use an algorithm operating over a common Data Link protocol to join LAN segments.

SPECIFIC NOTES:

[1] Specification developed from testing by Eastman Kodak. Adoption by IEEE 802.4 and ISA SP72 committees expected.

[2] Within a single subnetwork, inactive (null) implementation. To nodes on a different subnetwork, subset (specified by MAP 2.1) is used.

[3] ISO IS 8072/8073 Class 4 Connection Oriented Transport Service (COTS), but with restrictions per MAP 2.1, Chapter 3. Class 4 is needed for operation over the CLNS (Connectionless Network Service).

[4] The session layer in open systems architecture provides functions and services that may be used to establish and maintain connections among elements of the session, to maintain a dialog of requests and responses between the elements of a session, and to terminate the session. (Weik reference)

[5] ISO IS 8326/27 Kernel subset is restricted by provisions of MAP 2.1—Chapter 3. Included are TWS (Two Way Simultaneous) and TWA (Two Way Alternate) dialog structures.

(NOTES continued next page)

amples. This level of networking is more complex than the local area networks (LANs) or the facility-wide nets (WANs), in part because of the heavy information-exchange load.

OSI Reference Model

The OSI (Open System Interconnections)[2] Model was developed as a joint effort of several groups in the 1970s, including the International Standards Organization (ISO), the American National Standards Institute (ANSI), the Computer and Business Equipment Manufacturers Association (CBEMA), and the U. S. National Bureau of Standards (NBS). It has been estimated that this work represents many thousands of man years of effort by leading communications experts worldwide. OSI standards were later targeted as the basis for MAP previously mentioned.

The OSI reference model was developed with an orientation to telephony systems and features seven separate functional specifications or layers as depicted in a condensed manner in Table 1.

REFERENCES[3]

Babb, M.: "New Industrial Control Networks: Can They Survive the Factory Floor?" *Cont. Eng.*, **33** (3) 51-55 (1986).
Blickley, G. J.: "Process Industries Take One Step at a Time," *Cont. Eng.*, **32** (6) 83-84 (1985).
Capel, A. C., and G. F. Lynch: "Proway: The Evolving Standard for Process Control Data Highways," *Instrn. Techy.*, **30** (9) 91-94 (1983).
Griem, P. D.: "Security Functions in Distributed Control: Vendor and User Responsibilities," *Instrn. Techy.*, **30** (3) 57-59 (1983).
Harrington, J.: "Computer Integrated Manufacturing," Krieger, Marabar, Florida, 1979. (Historical reference)
Hoagland, J. C.: "Privacy and Authentication: New Priorities for Industrial and Commercial Communication," *Instrn. Techy.*, **29** (5) 47-51 (1982).
Hurst, Mark: "Evaluating Network Solutions," *Electronics Week*, **57** (18) 60-64 (1984).
Knipp, R. S.: "Closed Circuit Television: Cost-Effective Communications for Industrial Plants," *Instrn. Techy.*, **30** (8) 45-47 (1983).
Kozlik, T. J.: "Local Area Networks in Control," *Cont. Eng.*, **31** (7) 76-78 (1984).
Laduzinsky, A. J.: "Microprocessor-Based Terminals Provide Masterless Distributed Control," *Cont. Eng.*, **29** (5) 102-104 (1982).

[2] As pointed out by Weik (See Reference List), open systems architecture enables system design, development, and operation to be described as a hierarchical structure or a layering of functions. Each layer provides a set of accessible functions that can be used by the functions in the layer above it. Layers can be implemented without affecting the implementation of other layers. The concept is useful inasmuch as it permits the alteration of system performance without disturbing the investment in existing equipment or procedures because alterations are allowed at the higher levels or at the lower levels.

[3] For those readers who may wish to trace the development and background of present communication standards in more detail than is permitted here, a number of references of 1982/1983 vintage are included in this list.

Laduzinsky, A. J.: "Local Area Networks Open New Avenues for Control and Management Information," *Cont. Eng.*, **29** (8) 53-56 (1982).
Laduzinsky, A. J.: "Local Area Networks Expand the Horizons of Control and Information Flow," *Cont. Eng.*, **30** (7) 53-56 (1983).
Laduzinsky, A. J.: "Local Area Networks: Is MAP it?" *Cont. Eng.*, **31** (7) 73-75 (1984).
Laduzinsky, A. J.: "Factory Automation/USA: Just the Beginning," *Cont. Eng.*, **32** (6) 88-89 (1985).
Laduzinsky, A. J.: "Factory Automation/USA and the Land of Oz," *Cont. Eng.*, **32** (6) 98-103 (1985).
Miller, T. J.: "Distributed Acquisition System Gathers Data in Hazardous Environments," *Cont. Eng.*, **30** (8) 79 (1983).
Miller, T. J.: "Supervisory Unit Links Controllers to Form Distributed Control Systems," *Cont. Eng.*, **30** (11) 93 (1984).
Morris, H. M.: "New Controller Distributes Intelligent Control to Farthest Ends of Telemetry Systems," *Cont. Eng.*, **31** (6) 96-97 (1984).
Morris, H. M.: "Is Anyone Interfacing Process Controls and Management's Computers?" *Cont. Eng.*, **32** (6) 92-95 (1985).
News: "Vendors Ready Multichip Packages that Shrink T1 Interfaces," *Electronics Week*, **57** (15) 48 (1984).
News: "French Firm Opts for Token Passing," *Electronics Week*, **57** (17) 24, 29-30 (1984).
Rosenberg, R., and T. E. Feldt: "Local Nets Arrive in Force," *Electronics Week*, **57** (17) 67-73 (1984).
Spenser, K., and F. Raines: "High Level Language Simplifies Control Communications," *Cont. Eng.*, **31** (4) 150-152 (1984).
Staff: "Special Report: Manufacturing Technology," *Cont. Eng.*, **32** (4) 173- (A through PP) (1986).
Staff: "Special Report: Interfacing Plant Controls and Management's Computers," *Con. Eng.*, **32** (6) 81-108 (1985).
Thurber, K. J.: "The LOCALNetter Designers Handbook," Architecture Technology Corp., Minneapolis, Minnesota, 1983.
Weik, M. H.: "Communications Standard Dictionary," Van Nostrand Reinhold, New York, 1983.
Zollo, S.: "Server Links Ethernets with IBM Mainframes," *Electronics Week*, **57** (19) 77-78 (1984).

[6] The presentation layer in open systems architecture provides the functions, procedures, services, and protocol that are selected by the application layer. Functions may include data definition and control of data entry, data exchange, and data display (Weik reference). This layer comprises: CASE (Common Application Services), SASE (Specific Application Services), and management protocols required to coordinate the management of OSI networks in conjunction with management capabilities that are embedded within each of the OSI layer protocols.

[7] The application layer in open systems architecture is directly accessible to, visible to, and usually explicitly defined by users. This layer provides all of the functions and services needed to execute user programs, processes, and data exchanges (Weik reference). For the most part, the user interacts with the application layer, which comprises the languages, tools (such as program development aids, file managers, and personal productivity tools), database management systems, and concurrent multiuser applications. These functions rely on the lower layers to perform the details of communications and network management (Hurst reference). Until recently, network vendors provided a proprietary operating system for handling functions in the upper layers of the OSI model. This resulted in systems with unique features and interconnection difficulties.

Index

A

Acceleration, stepper motor, 350, 351
    servomotor, 336
Accident prevention, detectors for, 73
Accuracy, stepper motor, 350, 351
Active monitoring switch, programmable controller, 167
Adaptive control, 216–217, 220, 229, 232, 318
Adaptive hardware concept, machine vision, xv, 89
Adaptive robots, xvii
Adapt state, controller, 219
Adhesive application, robotic, 299 (*See also Sealing*)
Adjustable-speed brushless motor control, 341
Advisory control, 139
AGV (*See Automatic guided vehicle*)
AI (*See Artificial Intelligence*)
Aisle-captive stacker/receiver, 418
Alarm, control system, 137, 139
Algebraic robot dynamic modeling program (ARM), 328
Algorithm, brushless servomotor control, 344–345
    control, 141, 149
    machine vision, xv, 110
    robotic control, 321–325
    self-tuning controller, 223
Allen-Bradley programmable controller, 169, 171, 176–177, 180–181
Alternating current linear motor, xvii, 367–370
Alternating current servomotors, xvii, 336
Alternating current variable-speed drives, xviii, 380–385
Alternative branching, 202–203
American National Standards Institute (ANSI), xix, 427, 443
AMRF (*See Automated Manufacturing Research Facility*)
Analog input/output, programmable controller, 162
Androids, 260
ANSI (*See American National Standards Institute*)
Application layer, communication system, 443
Architecture, control system, xv, 136, 425
Area algorithm, machine vision, 111
Area-type scanner, machine vision, 89
ARM (See Algebraic Robot Dynamic Modeling Program)
Articulation, robot, 262–264
Artificial intelligence (AI), 7, 88, 125, 139, 149, 189
ASCI (*See Autosequentially commutated current-fed inverter*)
ASCII interface, control system, 141, 164
ASRS (*See Automated Storage and Retrieval System*)
Assembler language, 190
Assembly operations, conveyors for, 387, 388, 393, 403
Assembly, robots for, xvii, 299, 301, 307, 316–317
Automated Manufacturing Research Facility, 123
Automated order selection, 413
Automated storage and retrieval system (ASRS), xviii, 412, 421
Automatic gaging, 64–68
Automatic guided vehicle (AGV), xviii, 387, 408–412
Automaticity, scale of, 10
Automatic tuning, controller (*See Controller, self-tuning*)
Automation, advantages/limitations of, 3, 4
    fixed, 10, 260
    hard, 8, 260
    incentives for, 9
    manufacturing, 5
    partial, 3
    productivity from, 3, 4
    profitability from, 3, 4
Automobile springs, gaging of, 54
Autosequentially commutated current-fed inverter (ASCI), 382
Axes drives, numerical control, 231
Axes motion, robot, 262

B

Backbone architecture, MAP, 443
Back-drivable coupling, robot, 287
Backlash, stepper motor, 350
"Bang-Bang" damping, stepper motor, 353
"Bang-Bang" robot, 265, 266
Bank selection logic, DC drive, 379
Baseband, communication system, 437
Baseband signals, programmable controller
  use of, 165
BASIC language, 148
Batch-end report, 138
Batch picking, 413
Batch processes, 136–137, 145, 149, 260
Battery backup, 154, 169, 172
Bearing drift, XYnetics planar motor, 373
Belt conveyor, 387
Beta-radiation thickness gage, 69
Bifilar winding, hybrid stepper motor, 348
  permanent-magnet-stator stepper motor, 348
Binary machine vision system, 90
Bin-of-parts problem, robot, xv, 119, 319
Block diagram, programmable controller, 153
Boolean language programming, 158
Boring, numerical control, 237
Broadband network, 438
Broadband signals, programmable controller
  use of, 165
Brushless servomotors, xvii, 337–340
Buffer bank, conveyor, 397
Building costs, automation effects on, 4
Bulk materials industries, 5, 9
Bus configurations, programmable controller, 160–161, 165
Bus sharing, programmable controller, 167
Bus topology, 435
Butterworth filter, 225

C

CAD (*See Computer-Aided Design*)
Cadmium sulfide photocell, 77
CAE (*See Computer-Aided Engineering*)
Calculating ability, control system, 146
CAM (*See Computer-Aided Manufacturing*)
CAM-1 (*See Computer Aided Manufacturing-International, Inc.*)
Camshaft gage, 64
Capability ratio, 58

Capacitive damping, stepper motor, 352
Capacitive proximity sensor, xiv, 73
Capacity, control system, 146
Carrier sense multiple access with collision
  detection network protocol
  (CSMA/CD), xix, 437–439, 441
Cartesian coordinates, robot, 262–264
CASE (*See Common Application Services*)
Casting, numerical control machining of, 236
CATV (*See Community Antenna Television*)
CBEMA (*See Computer and Business
  Equipment Manufacturers
  Association*)
CCITT (*See International Consultative
  Committee for Telephone and
  Telegraph*)
Ceiling mount, robot, 263
Cell architecture, MAP, 443
Cell controller, 186–188
Cell, machining, 234
Central processing unit, programmable
  controller, 154
Cermet potentiometer, 56
Chain conveyor, 387
Charge coupled and injection devices, 89
Chester, G. L., 169, 170
Chopper drive, DC motor, 380
  stepper motor, 355
CIM (*See Computer-Integrated
  Manufacturing*)
Circuit boards, machine vision inspection of, 95
C language, 148
Classification, robots, 262
Closed-loop control, numerical system, 231
  planar motor, 372
  robot, 266
CNC (*See Computer Numerical Control*)
Color differentiation detector, 77
Color raster, 429
Color, sorting by, 73
Command interpreter, VAST, 323
Common Application Services (CASE),
  communication, 443
Communication gap, human/machine, 192
Communication hierarchy, robot, 298
Communication port/multidrop network
  topology, xix, 427–444
Communication protocol, xix, 427–444
Communication system, automated, 427–444
  levels of, 442–443
  models of, 442
  standards for, xix, 427, 436–444

Communication theory, 6
Community antenna television (CATV), 438, 441
Commutator, DC drive, 378
  DC servomotor, 337
Compare functions, programmable controller, 196
Compensation (nonlinearity), stepper motor, 359
Compliance, robot, 287
Compound torque, robot, 327
Computer-aided design (CAD), xiii, 7, 9-10, 430
Computer-aided design and manufacturing (CAD/CAM), xiii, 7, 9-11, 430
Computer-aided engineering (CAE), xiii, 7-9, 10, 430
Computer-aided manufacturing (CAM), xiii, 7, 9-10, 430
Computer Aided Manufacturing-International, Inc. (CAM-I), 9-10
Computer and Business Equipment Manufacturers Association (CBEMA), xix, 444
Computer awareness, 425
Computer evolution, 424
Computer-integrated manufacturing (CIM), xiii, xvi, 8, 11-12
Computer kinetics, 10
Computer numerical control (CNC), xvi, 10, 12, 232-255
  Dynapath, 244-252
  Numeripath, 253-255
  state-of-the-art systems, 235-255
  three/four axis system for milling, 249-253
  two-axis system for turning, 244-249
  workstation, 68
Conaway, J., 11
Concentricity, checking of, 55
Conductive-plastic potentiometer, 56
Connectionless service, communication, 443
Connection oriented transport service (COTS), 443
Console, control system, 142
Console software, 144
Contact-type object detectors, 73
Containers, inspection of, 73
Continuous-path numerical control, 231
Continuous-path robots, xvii, 267, 327
Continuous processing industries, 433
Control, advisory, 139
  automated warehousing, 417-421
  automatic guided vehicle, 408-412
  brushless motor, 339-344
  direct-digital, 141
  distributed, 141
  hybrid, 141
  order selection, 413-416
  photoelectric, 74
  robot, 265-268
  single-loop, 141
  supervisory, 141
Controller, multiloop, 141
  programmable, 136-188, 189-200
  self-tuning, 216-229
  sequence, 201-215
Control system, architecture of, xv, 136, 425
  CRT stations for, 143
  database configuration of, 140
  data highways for, 139
  enhancement of, 145
  flow chart programming of, xvi, 191;
  hierarchical, 136
  higher level, 144
  input/output systems, 136
  isolation of, 140
  microprocessor-based, 141
  protection of, 140
  reliability of, 140
  security for, 140
  trends in, 147
Conversational interactions, display system, 429
Conversion table, stepper motor rotor inertia, 347
  torque, 346
Converter, DC drive, 378-379
Conveyor jam detection, 73
Conveyor selection chart, 387
Conveyor systems, 386-422
Convolution, machine vision use of, 90
Cooling, programmable controller, 163
Coordinate systems, robot, xvii, 262-264
Coriolis forces, 321
Cost, robotization, 4, 265
COTS (*See Connection Oriented Transport Service*)
Counter functions, programmable controller, 195-196, 213
Counting, photocells for, 79
Crane, 391-396
CRT displays, control system, 137, 143
  design of, 147
  numerical control systems, 244, 246-247, 249-251

448 Index

CRT displays (*Continued*)
    operator workstation, 143
CSMA/CD (*See Carrier Sense Multiple Access with Collision Detection*)
Cursor, 430
Cutler-Hammer programmable controller, 172–173
Cycloconverter, AC motor drive, 381
Cylindrical coordinates, robot, 264

D

Dahlin control, 223
D'Alembert principle, 331
Damping-overshoot relationship, controller, 218–220
Damping, stepper motor, xvii, 352–353
Dark-operated (DO) detector, 74
Darlington switch network, stepper motor, 354
Darlington transistor, 376
Data acquisition, control system, 136
Data compression, 148
Data entry, numerical control, 232
Data highways, 139, 144
    early, 441
Data historian, 138, 144, 425
Data loop topology, 435
Data processing, machine vision, 90
Data quantity, machine vision, 88
Data storage, control system, 148
Data transmission (*See Communication*)
Database, CAD and CAM, 11
Database maintenance, 148
Database modules, VAST, 323
Dead storage, 11
Dead time, 136
Dead-time compensation, controller, 216
Deburring, robotic, 308
Deceleration, stepper motor, 350–351
Degradation, graceful, 167
Degrees of freedom, robot, 262
Delayed last-step damping (DLSD), stepper motor, xvii, 353
Depalletizing station, 413
Design, computer-assisted (*See Computer-aided design*)
Design costs, automation effects on, 4
Design modeling, 10
Desk calculator, 6
Destination, robot, 263
Detent, stepper motor, 350
Detent torque, stepper motor, 351, 356

Deutsche Industrie Normenausschluess (DIN), 191
Deweighting, control model use of, 228
Diagnostics, control system, 146
    conveyor system, 408, 416–421
    numerical control, 254
    programmable controller, 156
    variable-speed drive, 377
Dialog area, display, 429
Die casting, robotic, 297, 310, 311
Diffuse scanning, photoelectric, 75
Digital motion and position control, 346–366
Digital rotary servovalve, 363
Dimension, measurement and control of, 51–72
    automatic gaging systems, 64
    eddy-current inspection system, 66
    impedance-type gaging transducer, 55
    interferometer, 51
    linear potentiometer, 55
    linear transformer, 55
    linear variable differential transformer, 52, 72
    linear variable reluctance transducer, 53
    measuring machine, 65
    optical gages, 57
    positioning tables, 68
    statistical methods for, 57
    strain gage load cell, 56
    surface finish inspection, 72
    thickness measurement, 69
    ultrasonic thickness gage, 70
    x-ray thickness gage, 71
Dimensional tolerance determination, machine vision for, 95
Direct current linear motor, xviii, 367–370
Direct current noise suppression, programmable controller use of, 166
Direct current servomotors, xvii, 337
Direct current variable-speed drives, xviii, 378–379
Direct digital control, 141
Direct model, self-tuning controller, 225
Direct process control, 136
Discrete inputs, programmable controller, 161
Discrete-piece manufacturing industries, 5, 260, 346, 433
Disk drives, stepper motors for, 346
Disk-rotor stepper motor, 360
Dispatching, conveyors for, 387
Displacement, measurement and control of, xiv, 51–57
    impedance-type gaging transducer, 55

Index 449

linear potentiometer, 55
linear transformer, 55
linear variable differential transformer, 52, 72
linear variable reluctance transducer, 53
optoelectronic system, 57
strain gage load cell, 56
Display, control system, 143, 425–431
dimensional gaging system, 61–64
Distance limitation, stepper motor, 350
Distributed communication network, 433
Distributed control, 141, 145, 147
graceful degradation of, 167
history of, 7
Documentation, 200
Double-reach shuttle, 421
Downtime, automated system, 4
Drilling, numerical control, 237
Driverless vehicle, 387 (*See also Automatic Guided Vehicle*)
Drive, variable-speed, xiii, xvii, 376–385
Drive tube, conveyor, 396, 400–403
Drive wheel, conveyor, 396, 400–403
Dry reed switch, xiv, 81
Dual shuttle, 421
Duty cycle, machine, 3
Dynamic function blocks, 141
Dynamic gaging, 60
Dynamic matrix control, 136, 223
Dynamics, robot, 283–287

E

EAROM (Electrically alterable PROM), 176, 177
Eastman Kodak Company, 443
ECMA (*See European Computer Manufacturers Association*)
Eddy-current dimension gage, 66
Edge guidance and control, 73
EEPROM (Electrically erasable programmable read-only memory), 154, 344
EIA (*See Electronics Industries Association*)
Einstein photoelectric law, 74
Elbow analogy, robot, 262
Electric drives, robot, 281–282
Electrohydraulic stepper motor, 362–365
Electronic assembly, robotic, 283
Electronic damping, stepper motor, 352
Electronics Industries Association (EIA), 427
Encapsulated reed switch, 82

Encoder, xiii, 28–35, 231, 354
End-effector, robot, xvii, 288–293
Energy cost, monitoring of, 424
Engineer's workstation, 145
Equilibrium, multilegged robot, 331
Ergonomics, xviii, 148, 427
Error checking, programmable controller, 164
Error detection and correction (in fault tolerance), 167
Error-driven control, 223
Error-free start-stop curve, stepper motor, 350
Ethernet, 147, 439
European Computer Manufacturers Association (ECMA), 427
Evaluation, robotic system, 318
Excursion range, linear transformer, 55
Excursive phenomena, measurement of, 54
Expert systems, xvi, 149, 216–220
Explicity model, control system, 225
Extraction, machine vision data, 87

F

Factory management, 11
Fail-safe electronics, control system, 145
Fault tolerance, 167
Feedback, brushless motor control, 341
gate-turn-off thyristor, 376
history of, 6, 9
keyboard, 430
linear electric motor, 367
Feed cut-off controllers, 73
Feedforward control, 216
Filter, noise, 166
Fine tuning, stepper motor, 350
Finishing, robotic, 299, 310, 311, 313
Finite element analysis, 10
Fixed automation, 10, 260
Fixed-wire programmable controller, 163
Fixture conveyor, 387, 400
Flame cutting, robotic, 299, 313
Flatness, checking of, 55
Flaw detection, machine vision for, 92
Flexibility, automation, 4, 260, 265
Flexible manufacturing system (FMS), xiii, xvi, 4, 7, 8, 12, 186, 233, 386
Floor-mounted conveyors, xviii, 396–411
Flowsheet graphics, 136
Fluid processing industries, 5, 9
Flyball governor, 6
Ford Motor Company, 3, 386
Forearm analogy, robot, 262

FORTRAN, 144, 148
Freedom, robot, 262
Frequency shift keyed signals, programmable controller, 165
Frequency spectra, *XYnetics* planar motor, 373
Frictional load, stepper motor, 350
Function block, 198

G

Gaging, dimension, 51–72
Gain scheduling, controller, 216
Garment industry, 6
Gate-turn-off (GTO) thyristors, xviii, 376–378, 383
Gateway, control system, 144
    MAP, 443
General Conference on Weights and Measures, 51
General Motors Corporation, xix, 87, 149, 437, 442
Geometric arithmetic parallel processor (GAPP), 124
Geometric location, robot, 262
Geometric modeling, 10
Geometric variables, 15–85
Geometry, image, 87
GKS (*See Graphical Kernel System*)
Global control, 140
Global database, 149
Gould programmable controller, 171–173, 177–180, 182–183, 185–187
Governor, engine, 6
Gracefet programming language, 158
Graceful degradation, 167
Graetz circuit, DC drive, 378
GRAPH-5 language, 201
Graphic displays, xviii, 428
    control system, 137, 143
        numerical control systems, 244, 246–247, 249–251
Graphic Kernel System (GKS), 428
Graphics tablet, 430
Graphics workstation, 428
Gray-scale system, machine vision, 110
Gripper, robot (*See End-effector, robot*)
Grooving, numerically-controlled, 245
Grounding, programmable controller, 165
Group technology, 4

H

Half-bridge transducers, 64
Half-stepping, stepper motor, 353
Hall-effect object detectors and switches, xiv, 73, 78
Hall sensor feedback, brushless motor control, 343
Hamming codes, 167
Hand analogy, robot, 262
Hand-held program loader, 159
Handling, workpiece, 3 (*See also Robots*)
Hands, robot, xvii, 315
Hard automation, 8, 260
Harder, D. S., 3, 386
Harmonic distortion, microstepper, 358
Harrington, J., 11
Heat dissipation, electronic equipment, 163
Heat exchanger control, 216
Heat sinking, programmable controller, 163
Hexapod vehicle, 330
Hierarchy, control system, 136
    robot communications, 298
High-density, high-rise automatic storage/retrieval (AS/RS) systems, xviii, 416–421
High-level process control, 144
High-level programming languages, 158
Highway, data, 139, 144, 427–444
Histogram, 58, 63
Historian, data, 138, 144
Holding torque, stepper motor, xvii, 351, 356
Hollerith card, 6
Hopper level, detection of, 73
Horizontal handling, conveyors for, 387
Host computer control system, 147
Hughes Aircraft Company, 142
Human engineering, 189, 192, 427
Human labor, inadequacies of, 4, 7
Hybrid brushless servomotors, 338
Hybrid distributed control, 141, 145
Hybrid stepper motor, xvii, 348–352, 362
Hydraulic drives, robot, 282

I

IBM card, 6
Identification models, self-tuning controller, 225
Identifier, control system model, 227

IEC (*See International Electrotechnical Committee*)
IEEE (See Institute of Electrical and Electronics Engineers)
IEEE 488, in machine vision systems, 90
IEEE 802.1, network management, 443
IEEE 802.2, 443
IEEE 802.3, communication protocol, xix, 443
IEEE 802.4, xix, 437–438, 442, 443
IEEE 802.5, token ring, xix, 438–439
Image, machine vision, 87
Impedance-type gaging transducer, 55
Implicit model, self-tuning controller, 225
Incremental motion control (*See Stepper Motors*)
Inductance, stepper motor, 354
Induction alternating current servomotor, 336
Induction motor variable-speed drives, 381–384
Inductive proximity sensor, 81
Inductive voltage suppression, stepper motor, 355
Inductosyn plates, 65–66, 230
Inertia, alternating current servomotor, 340
  permanent-magnet direct current motor, 340
  stepper motor, 340, 350
Information theory, 7
Information collection, control system, 136
Inherent capability of the process (Cp), 57
In-process storage, conveyors for, 387
Input noise filtering, 166
Input/output bus sharing, 167
Inputs, control system, 136
  graphic display, 430
  programmable controller, 160, 167–188
Inspection, automatic gaging systems, 64–68
  machine vision, 92
  numerically-controlled probe, 232
  robotic, xvii, 299, 308, 310–311, 313
Institute of Electrical and Electronics Engineers (IEEE), xix, 427, 437, 441, 438–439
Instrumentation and automatic control, 5
Instrumentation, distributed, 141, 149
Instrument Society of America (ISA), xix, 441, 443
Integrated distributed control, 149
"Intelligent" control, brushless servomotor, 344
Interactions, user, 426, 429
Interactive graphic techniques, xviii, 424, 428–431

Intercommunication, automated systems, 434
Interface, automated systems, xviii, 136, 144, 423–443
  numerical control, 232
Interferometer, 51
Interlocks, conveyor, xviii, 391, 397–398
Internal mode controller, 136, 224
International Consultative Committee for Telephone and Telegraph (CCITT), 427
International Electrotechnical Committee (IEC), xix, 191, 441
International Organization for Standardization (ISO), 147
International Purdue Workshop on Industrial Computer Systems, xix
International Standards Organization, xix, 427, 437, 443
Inventory control, automated, 4, 11, 387, 413–416
Inverted power-and-free conveyors, 387, 404–405
Inverter, direct current motor drive, 383
Ironless disc-rotor stepper motor, 360
ISA (*See Instrument Society of America*)
ISA $^S71.Q2$, 443
Island, automation, 260
ISO (*See International Standards Organization*)
ISO DIS 8473 connectionless network service, 443
Isolation, programmable controller, 165
ISO Session Kernel, 443

J

Jacquard loom, 6
Jamming detection, conveyor, 73
JIT (*See Just-in-Time System*)
Jointed-arm coordinates, robot, 262
Joystick, 148, 430
Jump and call functions, 197, 202
Just-in-Time System (JIT), xiii, 3, 4, 11

K

Kalman filter, 228
Keyboard, xviii, 147, 173, 430
Kinematics, multilegged robot, 330–331
Kramer static drive, AC motor, 380
Kurtosis, 58

L

Label inspection, 73, 111
Laboratory data collection, 138
Labor, attitudes toward automation, 4
    Egyptian concept of, 3
    Florentine concept of, 3
    Greek concept of, 3
Labor-intensive industries, 6
Ladder logic, xvi, 152, 158, 171, 182, 192, 204–208
Lagrange-Euler robot dynamics, 322
LAN (*See Local Area Network*)
Language, assembler, 190
    GRAPH-5, 201
    machine-oriented, 190
    natural conversation, 189
    programmable controller, 157
    problem-oriented, 190
    STEP, 190–193, 195, 200
Laser inspection system, 68
Laser interferometer, 52
Lathe, numerically-controlled, 235–238, 244–245
Layering, database, 11
    communication system, 442–443
    control system, 140
Lead/lag, 136
Lead screw drive, 371
Least weighted squares model identifier, 227
Legged robot, xvii, 321–329
Length, definition of, 51
Leptokurtic curve, 58
Library, algorithm, 141
Light curtain photoelectric detector, 75
Light-emitting diodes, photoelectric system use of, 74
Light pen, 430
Light sources, photoelectric system, 77
Limited sequence robot, 265
Limit switches, xiv, 83
Linear algebraic equation (LAE), 323
Linear electric motor (LEM), xviii, 367–370
Linear induction motor (LIM), 367–370
    conveyor use of, 404
Linear positioning table, 69
Linear potentiometer, 55
Linear reluctance motor, 367–370
Linear sequence, 202
Linear synchronous motor (LSM), 367–370
Linear transformer, 55
Linear variable differential transformer, 52, 72
Linear variable-reluctance transducer, 53
Line scanner, machine vision, 89
Link control, communication, 443
Link layer, communication system, 443
Liquid crystal display (LED), 159
Lisa computer, 430
Lithium battery backup, 169, 172
Live roller conveyor, 387
Live storage, conveyor, 387
Load, electrohydraulic stepper motor, 364
    frictional (stepper motor), 350
    inertial (stepper motor), 350
    permanent-magnet stepper motor, 349
    solenoid-operated ratchet, 346
    variable-reluctance stepper motor, 347
Load capacity, robot, 281
Load cell, 56
Load commutated inverter, AC drive, 380–381
Loader, programmable controller program, 158
Loading, robot, xvii, 281
Local area network (LAN), xviii–xix, 7, 433–439
Local feedback loop, controller, 220
Location determination, machine vision for, 94
Logic processing, 141
Logic sequencer, stepper motor, 354
Loop gain, linear electric motor, 367
Lorentzian force, DC linear motor, 367

M

Machine automation, 10
Machine code (MC), 189
Machine loading, robotic, xvii, 299, 303, 308, 312, 313
Machining center, numerically-controlled, 238–239, 241–243
Machine-oriented programming language, 189–190
Machine tool interface, conveyor, 403
Machine vision, xiv–xv, 84, 86–132
Macintosh computer, 430
Macro instructions/codes, numerical control, 232
Magnetic proximity switch, xiv, 78
Magnetic saturation, stepper motor, 359
Magneto-mechanical spring, linear motor, 367
Magnetostrictive actuator, 371

Magnetostrictive limit switch, 83
Maintenance cost, automation effects on, 4
Management, process, 138
Management expertise, automation demands for, 4
Manipulator, robotic (*See Robots*)
Manual assembly, conveyors for, 387
Manufacturing automation, 5
Manufacturing Automation Protocol (MAP), xix, 7, 147, 149, 187, 437-438, 442-443
    backbone, 188
    gateway, 187-188
    MAP/PROWAY nodes, 443
MAP (*See Manufacturing Automation Protocol*)
Master nodes, communication system, 433, 435
Master/slave network, programmable controller, 164-165
Matching, parts, 64
Materials handling, robotic, xvii, 299-300, 303, 308, 310, 312-313, 316
Materials motion/handling systems, xviii, 3, 386-421
    floor movement, 396-414
        automatic guided vehicle (AGV), 412
        inverted power-and-free conveyor, 404-407
        towline, 407-412
        track and drive tube conveyor, 396-403
    high-density, high-rise systems, 416-421
    order selection systems, 413-416
    overhead systems, 386-395
    storage and retrieval systems, 412-421
Materials requirements planning (MRP), xiii, 4, 433, 434
Mean time between failures (MTBF), 166
Mean time to repair, 166
Measurement systems, history of, 6
Measuring machine, 65
Mechanical damping, stepper motor, 352
Mechanization, history of, 6, 9
Medical equipment, stepper motors for, 346
Memory capacity, robot-system, 266-280
Memory programmable controller (MPC), 189
Metal-oxide varistor, noise suppression with, 166
Meter, definition of, 51
Microprocessor applications, communication system, 435-436
    controls, 141
    conveyors, 416, 421
    electrohydraulic stepper motor, 362
    instruments, 64
    microstepper, 359
    numerical control, 233
    process/operator interface, 425
    servomotors, 340-345
    stepper motors, 353, 355-362
    variable-speed drives, 377
Milling, numerically-controlled, 237, 243
Minimum variance control, 224
Minimum variance with detuning, 224
Mitsubishi programmable controller, 169, 172
Mobile workstation, 403, 411
Modbus, 183-185
Model algorithmic control, 223
Model, multilegged robot, 331
Model, robot, 321-325
Model-based self-tuning controllers, xvi, 216-217, 220-229
Modeling, control system, 138
Monitoring, process, 136
Monorail conveyors, xviii, 387, 391-396
Motion control, stepper motors for, 346-366
Motion control systems (*See Materials motion/handling systems*)
Motion system, simulation of, 321-327
Motor control circuit, 153
Motor controls, history of, 6
Motor sizing, stepping system, 350
Motorola MC 68008 microprocessor, 175-176
Mouse, 426, 430
Movement, image, 87
MRP (*See Materials Requirements Planning*)
Multilegged robot, xvii, 321-329
Multiloop controller, 141
Multiple channels, programmable controller, 165
Multiple drop, communication system, 433
Multiple-instruction multiple-data, 126
Multiple-stack variable-reluctance stepper motor, 347
Multitasking, programmable controller, 156
Mumford, Lewis, 3

N

National Bureau of Standards (NBS), xix, 9, 123, 427, 444
National Electric Manufacturers' Association (NEMA), 152

454 Index

Natural commutation, DC drive, 378
Natural conversation language, 189
NC (*See Numerical Control*)
NCR Corporation, 124
Neighborhood processing, machine vision, 90
Network layer, communication system, 443
Network, local area (*See Local area network*)
Network, wide area, 433-444
Newton-Euler formulation, 322
Node, communication system, 433
Noise, communication system, 436
Noise suppression, 166
Noncontact object detectors, 73
Nonlinearity, control system, 216
    stepper motor, 359
    variable-speed drive power circuit, 378
Nonlinear spring, quality control of, 53
Nonpresence, detection of, xiv, 73
Nonservo controlled robots, 265
Normal distribution curve, 58
Nuclear accident, robots for, 330
Nuclear fluorescence thickness gage, 70
Nuclear radiation dimension gage, 69
Number processing capability, control system, 146
Numerical control (NC), xvi, 10, 12, 230-255, 437
NumeriProbe inspection system, 66

O

Object detection, xiv, 73
Office automation, 5 stepper motors for, 346
Ohio State University *Hexapod* vehicle, 330
One-phase-on damping, stepper motor, 352
Open-loop control, numerical system, 231
    robot, 265-266
Open System Interconnection (OSI) communication model, xix, 443
Operator station, control system, 142, 143, 148, 232
Optimization, steady state, 138
Optoelectronic dimension gage, 57
Order-filling system, xviii, 387, 413-416
Order picking, conveyors for, 387, 413-416
Origin, robot, 263
Origin shift, numerical control system, 254
Oscillation, robot, 283-285
Overdamped response, controller, 216
Overhanging roof detector, 73
Overhead conveyors, xviii, 387-396

Overshoot-damping relationship, controller, 218-220
Overshoot, stepper motor, 350, 353
Overview display, control system, 143

P

Painting, robotic, 299, 310-311, 313-314
Pallet shuttle, 236
Parallel input/output system, 160
Parallelism, checking of, 55
Pareto chart, 59, 63
Partial automation, 3
Part matching, 64
Part program, numerical control, 230
Parts handling and transfer, robotic, xvii, 301, 308, 310-312
PASCAL, 148
Path modification, robot, 269
Pattern Processing Technologies, Inc., 91
Pattern processor, machine vision, 86
Pattern recognition, in expert system control, 217-218
    in machine vision, 86
Peer-to-peer communication, 164, 433
Permanent-magnet DC servomotor, 338
Permanent-magnet stepper motor, xvii, 347-348
Personal computer, 149, 177, 186
Phase coherent carrier band, 443
Phase controlled AC outputs, 166
Phase control logic, DC drive, 379
PHIGS (*See Programmer's Hierarchical Interactive Graphics System*)
Photoconductivity, 74
Photodiodes, 77
Photoelectric detectors, xiv, 73, 74
Phototransistors, 77
Photovoltaic effect, 74
Physical layer, communication system, 443
Pick-and-place robot, 265
PID control mode, 136, 162, 171, 216-229
Piezo (PZT) displacement actuator, 371
Piezo stack positioning device, xviii, 371
Pilot duty outputs, programmable controller, 161
Pipelined image processing engine (PIPE), 123
Pitch, robot, 262
Planar motor, xviii, 371-375
Plantwide control, 149

Platykurtic curve, 59
Playback concept, robot programming, xvii, 267
Plotter station, 10
Plotter, stepper motors for, 346
Plotting, process variable trends, 137
Pneumatic drive, robot, 282
Pneumatic gaging, 65
Point-to-point communication, 164, 434
Point-to-point robot, xvii, 267, 327
Polar coordinates, robot, 264
Pole-cancellation control, 223
Position accuracy, stepper motor, 351
Position measurement, numerical control, 230-231
Positioning table, motors for, 68, 367-375
Post-objective scanning, 121
Potentiometer, linear, 55
Power-and-free conveyors, xviii, 386-391
Power drive, stepper motor, 354
Power rating, solid-state switching devices, 376-377
Power steering, 6
Power supply programmable controller, 153
Power switching devices, xviii, 376-377
Presence/absence detection, machine vision for, 73, 95
Presentation layer, communication system, 443
Press loading, robotic, 310-311
Pressure sensitivity, XYnetics planar motor, 373
Printed circuit board, machine vision inspection of, 100-108
Printed circuit potentiometer, 55
Printer, stepper motors for, 346
Priority message, 436
Problem-oriented programming language, 190
Process capability coefficient, 59
Process data, reconciliation of, 139
Process dynamics, 216
Processing, conveyors for, 387
Processing industries, 433, 519
Process management, 138
Process, monitoring of, 136
Processor redundancy, programmable controller, 167
Processor software, programmable controller, 155
Processor station, programmable controller, 153
Process planning, 11
Process variables, trend plotting of, 137

Productivity from automation, 3, 4
Product quality, automation effects on, 4
Program loader, programmable controller, 158
Programmable controllers, xv-xvi, 152-166, 260, 433
  large-size, 182-188
  micro, 169-174
  mid-size, 177-182
  small-size, 174-177
  state-of-the-art, 167-188
Programmable read only memory (PROM), 173, 180, 182
Programmer's Hierarchical Interactive Graphics System (PHIGS), xviii, 428
Programming, distributed control system, 146
  levels of, 267-280
  microprocessor-based, 271
  numerical control system, 230
  robot, 267-280
  sequence controller, 201-215
Proportional-integral-derivative control (*See PID Control Mode*)
Protocol, communication system, xix, 188, 437-439, 441-443
PROWAY, xix, 438, 441, 443
Proximity switches, xiv, 73, 78
Psychology, user-interaction, 428
Pull-in and pull-out torque, stepper motor, xvii, 351
Pulse-echo type ultrasonic gage, 71
Pulse frequency modulation (PFM), 337
Pulse width modulation, variable-speed drive, 377, 384-385

R

Radiation environment, robots for, 330
Radio control, conveyor, 393
Ramping, stepper motor, xvii, 351
Random access memory (RAM), 169, 171-173, 176-178, 180, 183-186, 188
Random walk, 228
Rare-earth magnet, stepper motor use of, 360
Ratchet, solenoid-operated, 346
Reach, robot, 262-264
Read-only memory (ROM), communication system, 436
  control system, 142
Recognition process, machine vision, 90
Reconciliation, process data, 139

Recursive least weighted squares model
    identifier, 227
Reduced instruction computer (RISC), 129
Reduncandy, programmable controller, 167
Reference location, robot, 262
Reference model, communication system, 443
Reflective scanning, photoelectric, xiv, 75
Registration control, photoelectric, 77
Regulator control, 136
Relay ladder logic (*See Ladder logic*)
Reliability, control system, 140, 145, 147
    programmable controller, 165
Reliance Electric programmable controller,
    174–176
Reluctance-type linear motor, xviii, 367–370
Remote door opener, 73
Remote information collection, control
    system, 136
Repeatability, robot, 265, 286–287
    stepper motor, 351
Report generation, control system, 137, 144
Research and development, robotic systems,
    319
Resistance damping, stepper motor, 352
Resolution, graphic display, 429
    robot, 286
    stepper motor, 350–351
Resolver, xiii, 22–27, 230
    brushless, 344
Resonance, microstepper, 356
    stepper motor, 350
Resonance-type ultrasonic gage, 71
Response time, communication system, 436
    programmable controller, 164
    stepper motor, 351
Restricted complexity control, 224
Restricted complexity delay identification,
    228
Retooling, 260
Retrieval and storage system, automated,
    413–421
Retrofitting, numerical control, 230
Retroflective photoelectric sensor, xiv, 75
Retrotorque damping, stepper motor, 353
Revolute coordinates, robot, 262
Ring topology, communication, 435
RISC (*See Reduced Instruction Computer*)
RMS protection, brushless servomotor, 344
R-Net, 175
Robot, xvi–xvii, 10–11, 260–320
    actuator dynamics, 322
    advanced, 318
    axes, 262

Cartesian coordinates, 262
classification, 262
compliance, 287
continuous path, 267
control systems, 265–267
coordinates, 262–264
cylindrical coordinates, 262
degrees of freedom, 262
dynamic properties, 283
electrically driven, 281
electromechanically driven, 283
end effectors, 288–293
evaluation of, 318
grasp, 290
history, 260
jointed-arm coordinates, 262
load capacity, 281
magnetic clamping, 292
multifinger, 317
multilegged, xvii, 330–335
non-servo controlled, 265
oscillation, 284
overhead, 297
playback concept, 267
pneumatically driven, 282
point-to-point, 267
polar coordinates, 264
programming, 267
repeatability, 286
resolution, 286
revolute coordinates, 252
rotary joints, 286
servo controlled, 265
simulation of, 321–324
sliding joints, 286
"Smart", 262
spherical coordinates, 264
stability, 283
tool coordinates, 264
tool fastening wrists, 293
vacuum clamping, 290
VAST system, 324
work cells, 298
work envelope, 265
workplace configuration, 293
world coordinates, 263
Roller conveyor, 387
Roller deck conveyor, 400
Roll, robot, 262
Rossum's Universal Robots, 260
Rotary axes, robot, 263–265
Rotary joints, robot, 262, 286
Rotary stepper, rolled-out version, 369

Index 457

Rotating parts, gaging of, 54
Rotation principles, brushless motor, 339
Rotor geometry, hybrid stepper motor, 348
Roughing and finishing, numerically-
    controlled, 245
Roundness, checking of, 55
RS-232 serial transmission, 177, 433
RS-511, 443
Rubber actuators, robot, 317
Run chart, 60
Run-out, measurement of, 54

S

Safety features, automatic guided vehicle
    (AGV), 408
    object detectors for, 73
Sample-and-compare system, brushless
    servomotor, 344
SASE (*See Specific Application Services*)
Sawyer linear reluctance motor, 371, 374–376
Scanner, photoelectric system, 75
SCR (*See Thyristor*)
SDLC (*See Synchronous Data Link Control*)
Sealing, robotic, xvii, 299, 308, 312, 313, 316
Second-generation distributed control, 150
Security, control system, 147
Self-adaptive control, 216–217, 220–229
Self diagnostics, 147, 169, 170, 172
Self-powered monotractor, 387
Self-tuning controller, xvi, 216–229
    variable-speed drive, 378
Semiconvertor, DC drive, 378
Sensing distance, inductive proximity
    detector, 83
Sensitivity, photocell, 77
Sensors, history of, 6
Sequence controllers, xvi, 201–215
Sequence-of-events recording, 136
Sequencer-controlled robot, 266
Sequencing blocks, 141
Sequencing control, 146
Serial data communication, 441
Serial input/output, programmable
    controller, 161
Serviceability, programmable controller, 166
Servocontrolled robots, 265–267
Servomotors, xvii, 336–345
    AC motors, 336
    brushless motors, 337–340
    DC motors, 337
    microprocessors for, 340–345

Servopower, history of, 6
Session layer, communication system, 443
Settling time, stepper motor, 350
Shannon, C. E., 6
Shape verification, machine vision, 94
Shared-display architecture, 425
Sheet thickness, control of, 69
Shipping terminal, conveyors for, 387
Shoulder analogy, robot, 262
Silicon phototransistor, 78
SIMATIC programmable controller, 191
Simulator, robot, xvii, 321–324
Simultaneous branching, 202–203
Sine/cosine proportioning, microstepper use
    of, 357
Single instruction multiple data (SIMD), 1,
    126
Single-loop controller, 141
Single order selection, 413
Single-stack variable-reluctance stepper
    motor, 347
Size control, automatic, 64–68
Sizing, stepper motor, 350
Skewness, 60
Slat conveyor, 387
Slave-to-slave message, 433, 434
Slew rate, stepper motor, 348, 350–353
Sliding joints, robot, 286
Slip energy recovery system, AC drive, 380
"Smart" materials handling systems, xviii, 386
"Smart" robots, xiii, 260, 262
Smith predictor, 136
Software, costs of, 4
    display system, 425, 428–429, 431
    motion control system, 274
    numerical control, 232–233
    programmable controller, 155–156,
        196–200
    sequence controller, 208–212
Solenoid-operated ratchet, 346
Solid-state camera, 89
Solid-state switching, 376–377
Solid-state variable-speed drive, xviii, 376–385
Sorter conveyor, 387
Sorting, detectors for, 73
Specific Application Services (SASE),
    communication, 443
Specifications, computer numerical
    controllers (CNC), 248, 252, 255
    conveyors, 386–415
    electrohydraulic steppers, 364
    high-rise, high-density storage/retrieval
        systems, 416–421

Specifications (*Continued*)
    microsteppers, 355–359
    permanent-magnet steppers, 349
    programmable controllers, 167–188
    robots, 272–273, 304, 306–308, 310–311, 313–314
    solenoid-operated ratchet, 346
    variable-reluctance stepper, 347
Specular scanning, photoelectric, 75
Speed, control system, 147
    electrohydraulic stepper, 365
    ironless disc-rotor stepper motor, 361–362
    microstepper, 356, 359
    permanent-magnet stepper, 349–351
    robot, 265
    solenoid-operated ratchet, 346
    variable-reluctance stepper, 347–348
Speed of response, photocell, 77
Speed-reduction ratio, stepper motor, 350
Speed/torque characteristic, stepper motor, 350
Spherical coordinates, robot, 264
Spidergraph programmable controller selection chart, 169–170
Springs, automatic gaging of, 53–54
Squareness, checking of, 55
Stability, robot, 283–285
Stacker/retrieval systems, 416–421
Stand-alone controller, 141
Standard deviation, 60
Standards, communication system, xix, 435–444
Star topology, communication network, xix, 433–434
Star wheel, 346
Statement list (SL) programming method, xvi, 192–193
Statically indeterminate structure, 331
Static bearing test, *XYnetics* planar motor, 373
Static gaging, 60
Static induction thyristor, 376
Static Kramer drive, AC motor, 380
Statistical quality control, 57
Stator field, DC servomotor, 337
Steady state gain, controller, 222
Steady state optimization, controller, 138
Steam engine, control of, 6
Stepper motors, xvii, 346–366
    application guidelines, 349–353
    calibration of, 349
    characteristics of, 340
    controls for, 353–355
    electrohydraulic, 362–365
    equivalent circuit of, 355
    hybrid motors, 348–349, 362
    ironless disc rotor type, 360–362
    microstepping, 355–362
    permanent-magnet type, 347–348
    robot use of, 276
    solenoid-operated ratchet, 346
    stall torque, 351
    step accuracy, 351
    step angle, 346, 351
    stepping torque, 356
    technological trends, 355
    three-phase, 340
    torque, conversion table, 346
    variable-reluctance type, 347
STEP programming language, 190–193, 195, 200
Stiffness, *XYnetics* planar motor, 373
Storage and retrieval systems, xviii, 387, 412–421
Storage, control system data, 148
    conveyors for, 387
    material in-process, 387, 389
Stored-program control, 213–215
Strain gage load cell, 56
String encoding, machine vision, 90
Struthers-Dunn programmable controller, 176–178
Supervisory control, 141
Suppliers' role, automated system, 150
Surface finish, measurement of, 72
Surface flaw detection, machine vision, 92
Swing, robot, 262
Switching, conveyor system, 392, 394
Switching sequence, phase-switched stepper motor, 348
Switching, towline, 407
Swivel, robot, 262
Synchronous AC servomotor, 336
Synchronous Data Link Control (SDLC), 433
Systemization, history of, 6
System overview display, 143
Systolic array, 126

T

Tachometer, xiv, 35–40
Tactile feedback, 430
Tactile sensor, robot, 287

Index   459

Tank-turret analogy, robot, 264
Taper, checking of, 55
Tapping, numerically controlled, 237
TC5 Communication Committee, 441
Teach mode, robot programming, 267–269
Telenet, 188
Telescoping tubes, robot use of, 263
Teletypewriter, 433
Texas Instruments programmable controller, 177, 180–184
Thickness, measurement of, 69–72
    linear-variable differential transformer, 72
    nuclear radiation gage, 69
    surface-finish gage, 72
    ultrasonic gage, 70
    x-ray gage, 71
Thin-film potentiometer, 55, 56
Thru-scanning photoelectric detector, xiv, 74
Thyristor (SCR), 376–379, 383
Tilt deck, conveyor, 400
Time constant reduction, stepper motor, 355
Time-driven multitasking system, 157
Timing functions, programmable controller, 194–195
TIR (*See Total indicated runout*)
Token bus communication protocol, 437–438, 442
Token ring communication protocol, xix, 438–439
Tool changing, numerically-controlled, 233, 238
Tool coordinate system, robot, 264
Tool mounting, robot, 293
Tool radius compensation, 244
Topology, communication system, xix, 433–436
Torque conversion table, 346
Torque, servomotor, 336, 340
    stepper motor (*See Stepper motors*)
Torque ripple, microstepping reduction of, 356
Total indicated runout (TIR), 54
Touchscreen, xviii, 147, 430
Towline conveyor, 387, 406–408
Track-and-drive-tube conveyor, xviii, 396, 398, 400–404
Trackball, 430
Track configurations, conveyor, 402, 405
Traffic, communication, 433
Training, robotics, 318
Trajectory, robot, 325–326
Transfer line, 237, 239–243

Transformer, linear, 55
Transistor-bridge chopper DC drive, 380
Transistor power-switching devices, 376–377
Transmitted beam system, photoelectric, 75
Transport layer, communication system, 443
Tranzorbs, noise suppression with, 166
Traveling magnetomotive force, 367
Traverse distance, stepper motor, 350
Trending, process variable, 137
Trends, control system, 147
Trolley, crane conveyor, 395–396
    power-and-free conveyor, 386, 390–391
Trolley mount, robot, 263
Tuning rules, expert system, 216
Turning, numerically controlled, 235–238, 244–245
Twinaxial cable, 165
Twisted wire pairs, 165
Two-out-of-three voting strategy, control system, 145
Two-phase-on damping, stepper motor, 352
Two-way alternate (TWA) communication, 443
Two-way simultaneous (TWS) communication, 443

U

Ultrasonic proximity switch, 83
Underhung crane, xviii, 391–196
Unified data base structure, control system, 146
Unimation, Inc., 86
Universal Common Language (UCL), 183
Unstable response, controller, 216
Upgrading, human labor, 4

V

VAL-II robot programming, 269
Variable-frequency drive, 385
Variable-reluctance proximity switch, 78
Variable-reluctance stepper motor, xvii, 347
Variable-speed, solid-state drives, xviii, 376–385
Variation, concept of, 60
Varistor, noise suppression with, 166
VAST (*See Versatile Robot Arm Dynamic Simulation Tool*)
VAX 11/750 minicomputer, 323
Velocity loop, servomotor, xvii, 336–345